ASPECTS OF
ENERGY CONVERSION

Proceedings of a Summer School held at Lincoln College, Oxford, 14-25 July 1975

Other Pergamon Titles of Interest

BRATT	Have You Got the Energy?
DIAMANT	Total Energy
DUNN	Energy Conservation
HUNT	Fission, Fusion and the Energy Crisis
JONES	Energy and Housing
KOVACH	Technology of Efficient Energy Utilization
MESSEL & BUTLER	Solar Energy
MURRAY	Nuclear Energy
SIMON	Energy Resources
SMITH	The Technology of Efficient Electricity Use
SPORN	Energy in an Age of Limited Availability and Delimited Applicability

*Cover picture reproduced by kind permission of Harlequin
(Leisure Magazine UKAEA).*

ASPECTS OF ENERGY CONVERSION

Summer School, Lincoln College, Oxford. 14-25 July 1975

Organised jointly by Science Research Council, Energy Technology Support Unit and Harwell Education Centre.

ASPECTS OF ENERGY CONVERSION

Proceedings of a Summer School held at Lincoln College, Oxford, 14-25 July, 1975

edited by

I.M.BLAIR
Energy Technology Support Unit, Harwell, England

B.D.JONES
Rutherford Laboratory, Chilton, Didcot, Oxon, England

A.J.VAN HORN
Physics Dept., Harvard University, Cambridge, MA 02138, U.S.A
(formerly of Rutherford Laboratory)

ORGANISED JOINTLY BY
SCIENCE RESEARCH COUNCIL,
ENERGY TECHNOLOGY SUPPORT UNIT AND
HARWELL EDUCATION CENTRE

PERGAMON PRESS

OXFORD · NEW YORK · TORONTO · SYDNEY · PARIS · FRANKFURT

U.K.	Pergamon Press Ltd., Headington Hill Hall, Oxford OX3 0BW, England
U.S.A.	Pergamon Press Inc., Maxwell House, Fairview Park, Elmsford, New York 10523, U.S.A.
CANADA	Pergamon of Canada Ltd., P.O. Box 9600, Don Mills M3C 2T9, Ontario, Canada
AUSTRALIA	Pergamon Press (Aust.) Pty. Ltd., 19a Boundary Street, Rushcutters Bay, N.S.W. 2011, Australia
FRANCE	Pergamon Press SARL, 24 rue des Ecoles, 75240 Paris, Cedex 05, France
WEST GERMANY	Pergamon Press GmbH, 6242 Kronberg-Taunus, Pferdstrasse 1, Frankfurt-am-Main, West Germany

Copyright © 1976 Pergamon Press Ltd

First edition 1976

Library of Congress Cataloging in Publication Data

Main entry under title :

Aspects of energy conversion

1. Power (Mechanics)-Congresses. 2. Power resources-Congresses. 3. Energy policy-Congresses. I. Blair, Ian M. II. Jones, B. D. III. Van Horn, Andrew James, 1945- IV. Great Britain. Science Research Council. V. Harwell, Eng. Atomic Energy Research Establishment. Energy Technology Support Unit. VI. Harwell, Eng. Atomic Energy Research Establishment. Education and Training Centre.
TJ153.A748 1976 333.7 75-45417
ISBN 0-08-019975-5

In order to make this volume available as economically and rapidly as possible the author's typescript has been reproduced in its original form. This method unfortunately has its typographical limitations but it is hoped that they in no way distract the reader.

Printed in Great Britain by A. Wheaton & Co. Exeter

CONTENTS

v

PREFACE

This publication contains texts of lectures presented at the Summer School "Aspects of Energy Conversion" together with edited transcripts of the discussions which followed each lecture and of the panel discussions which were held at intervals in the proceedings (we are grateful to the students for providing notes of many of the discussions). We must stress that the lecturers are not responsible for the accuracy of information given in these transcripts as, in the interests of early publication, we have not confirmed the validity of statements made therein; we do feel that the main points made in the discussions have been preserved, however. Some of the students felt stimulated to produce their own papers at the School and these are reproduced in the Appendix.

Our fellow-members of the School Organizing Committee are listed separately and we are indebted to them for helping to make the School and this publication possible. We also thank the staff of the Education and Training Department at AERE Harwell, in particular Ms. Julie Carpenter, whose painstaking secretarial work contributed greatly to the smooth running of the School, and Mr. Fred Major for his audio-visual aid, and the staff of Lincoln College and the Examination Schools, Oxford. We are grateful to Dr. George Kalmus for his interest and it is a pleasure to thank Ms. Judy Williams for her invaluable assistance with the massive effort needed to prepare the material for publication, Ms. Elaine Knowlton for her enthusiastic typing and, of course, Carol, Jill and Lynda.

Finally, we wish to express our gratitude to the lecturers for their cooperation in producing texts of their lectures for publication and to both lecturers and students for uniting to make the School a stimulating and informative experience. We hope that readers will derive similar benefits from these Proceedings.

I. M. Blair
B. D. Jones
A. J. Van Horn

INTRODUCTION

The recent massive increases in the price of oil have focused attention on
the importance of energy in society. They have reminded us of the fact that
our traditional fuel reserves are of finite extent, and that the continuous
growth in their use over recent decades, due mainly to their relative cheapness, cannot continue definitely. There is, therefore, at the present time
much concern over the wise husbandry of these traditional fuels, and over possible alternative energy sources to replace them. In particular the most imminent alternative source, nuclear power, is being subjected to a most detailed
scrutiny.

It is most appropriate that the Science Research Council and the Energy
Technology Support Unit at Harwell should collaborate at this time to hold a
Summer School on "Aspects of Energy Conversion". The energy problem is complex, involving scientific, technological, economic, environmental, sociological
and political issues. It is also international. We are pleased to note that
all these facts have been borne in mind in the preparation and presentation
of the School Programme, and that the material presented at the School can
now reach a wider audience through these Proceedings.

Sir Sam Edwards
Chairman, Science Research Council

Dr Walter Marshall
Chief Scientist
Department of Energy

ORGANIZING COMMITTEE

A. P. Banford, Science Research Council, Swindon, Wilts.

Dr. I. M. Blair (Chairman), Energy Technology Support Unit, Harwell, Oxon.

Dr. R. T. Hall, Esso Research Centre, Abingdon, Oxon.

Dr. B. D. Jones, Rutherford Laboratory, Chilton, Didcot, Oxon.

W. G. Knowles, Education and Training Department, AERE, Harwell, Oxon.

Professor N. Kurti, Clarendon Laboratory, Oxford, Oxon.

C. J. Preuveneers, Education and Training Department, AERE, Harwell, Oxon.

Dr. D. T. Swift-Hook, Marchwood Engineering Laboratories, Southampton, Hants.

Dr. A. J. Van Horn, Rutherford Laboratory, Chilton, Didcot, Oxon.

School sponsored by the Science Research Council and by the Energy Technology
Support Unit, Harwell.

ORGANIZING COMMITTEE

LECTURERS AND STUDENTS

LECTURERS

Atkinson, F.J., Dept. of Energy, London.

Bainbridge, Prof. G.R., The Energy Centre, University of Newcastle Upon Tyne.
Balogh, The Lord, Dept. of Energy, London.
Banford, A.P., Science Research Council, Swindon.
Blair, Dr. I.M., Energy Technology Support Unit, Harwell.
Brinkworth, Dr. B.J., Dept. Mechanical Engineering, University College, Cardiff.
Brookes, L., United Kingdom Atomic Energy Authority, London.
Buss, B., Electrical Research Association, Leatherhead.

Chapman, Dr. P.F., Energy Research Group, Open University, Milton Keynes.

Dale, W.B., Fast Reactor Systems Directorate, U.K.A.E.A., Risley.
Dawson, Dr. J.K., Energy Technology Support Unit, Harwell.
Davoll, Dr. J., Conservation Society, London.
Ducret, Dr. C -G., Geneva, Switzerland.
Dunn, Prof. P.D., Engineering and Cybernetics Dept., University of Reading.

Eden, Dr. R. J., Energy Research Group, Cavendish Lab., Cambridge University.

Feld, Prof. B.T., Physics Dept., Massachusetts Institute of Technology, U.S.A.

Garner, Prof. P.J., Chemical Engineering Dept., University of Birmingham.
Gaskin, Prof. M., Political Economy Dept., University of Aberdeen.
Gray, Dr. J. A., R. & D. Division, British Gas Corporation, London.

Hadlow, M.E., Electrical Research Association, Leatherhead.
Harris, Dr. P.S., Energy Technology Support Unit, Harwell.
Harrison, J.S., R. & D. Dept., National Coal Board, Harrow.
Holder, Prof. D.W., Engineering Science Dept., University of Oxford.

James, C.G., R. & D. Division, British Gas Corporation, London.
Jones, Dr. B.D., Rutherford Laboratory, Chilton, Didcot, Oxon.

Kenning, Dr. D.B.R., Lincoln College, Oxford.
Kurti, Prof. N., Clarendon Laboratory, University of Oxford.

Leach, Dr. S.J., Building Services & Energy Divn., Building Research Sta., Watford.
Leslie, Prof. D.C., Nuclear Engineering Dept., Queen Mary College, London.
Littler, Dr. J.G.F., Architecture Dept., University of Cambridge.
Long, Dr. G., Energy Technology Support Unit, Harwell.

Martin, Prof. F.M., Social Administration & Social Work Dept., Glasgow University.
Mellanby, Prof. K.M., Monks Wood Experimental Sta., Abbots Ripton.
Merrick, D., R. & D. Dept., National Coal Board, Harrow.
Munby, D.L., Nuffield College, Oxford.

Newton, G.E.H., Reed Engineering & Development Services Ltd., Maidstone.
Norris, Dr. W.T., Central Electricity Research Laboratories, Leatherhead.

Oxburgh, Dr. E.R., Geology & Mineralogy Dept., University of Oxford.

Peters, C.M.D., Total Energy Co. Ltd., Oxford.

Roberts, F., Energy Technology Support Unit, Harwell.
Roberts, Dr. K.V., Culham Lab., U.K. Atomic Energy Research Group, Abingdon.
Roberts, Dr. P.C., Systems Analysis Research Unit, Dept. of Environment, London.
Rothberg, Prof. J.E., Physics Dept., University of Washington, Seattle, U.S.A.

Sabel, W., Science Department, Oxford Polytechnic.
Smith, Dr. I.E., School of Mechanical Engineering, Cranfield Institute of Technology.
Swift-Hook, Dr. D.T., CEGB, Marchwood Engineering Lab., Southampton.

Van Horn, Dr. A.J., Rutherford Laboratory, Chilton, Didcot, Oxon.

White, Dr. D.J., Ministry of Agriculture, Fisheries and Food, London.
Williams, E.C., Nuclear Inspectorate, Health and Safety Executive, London.
Wilson, Prof. R., Physics Dept., Harvard University, U.S.A.
Wilson, S.S., Engineering Science Dept., University of Oxford.
Wright, Dr. J.K., Research Dept., Central Electricity Generating Board, London.

STUDENTS

Abbott, B.F.M., Cavendish Laboratory, Cambridge.

Barnard, Dr. R.D., University of Salford.
Barnes, R., Leicester Polytechnic.
Barnett, R.J.T., Kingston Polytechnic.
Binns, Dr. D.F., University of Salford.
Black, Dr. M., University of Salford.
Blega, Miss S., Technical University of Denmark.
Bodansky, Prof. D., University of Washington, U.S.A.
Brainch, G.S., University of Aston.
Brearly, I., University of Birmingham.
Buckingham, J.F., University of Reading.
Britton, L.G., University of Leeds.

Chester, Miss M., University of Sheffield.
Clayton, Dr. B.R., University College, London.
Common, M.S., Programmes Analysis Unit, Harwell.
Critchley, B.L., University of Surrey.
Critoph, R., University of Southampton.

Dorling, J.A., Imperial College, London.

Evans, N.L., University of Birmingham.

Faiman, Dr. D., Weizmann Institute, Israel.

Gibbons, Dr. T.G. (and others), National Coal Board, London.
Gray, M.F., University of Aston.
Gudmundson, J.S., Atomic Energy Research Establishment, Harwell.

Hale, Dr. D.K., National Physical Laboratory, Teddington.
Hall, Dr. R.T., Esso Research Centre, Abingdon.
Hammond, G.P., Cranfield Institute of Technology.
Haskew, D.W., University of Reading.
Hodgkinson, M.G., University of Sheffield.
Honstvet, I.A., University of Birmingham.
Hough, D.W., Programmes Analysis Unit, Harwell.

Janikowski, H.E., Cranfield Institute of Technology.
Johnes, Dr. G.L. (and others), Esso Research Centre, Abingdon.
Jones, J.B., Department of the Environment, London.

Kato, K., London School of Economics.
Knight, F.I., University of Cambridge.

Labat, P., TAAS, University of Toulouse, France.
Lewis, Dr. J.A., Department of Energy, London.
Liddell, H.L., Hull School of Architecture.
Lipman, Dr. N.H., Rutherford Laboratory, Chilton, Didcot, Oxon.
Lister, Mrs. J.R., Westfield College, London.

Madeley, G.D., British Gas Corporation, London.
Mann, Mrs. J., National Coal Board, London.
Manning, H.J., Cavendish Laboratory, Cambridge.
McMahon, M., University of Leeds.
Melvin, A., British Gas Corporation, London.
Metcalfe, C.I., University of Aston.
Morgan, Dr. J., University of Edinburgh.
Munroe, M.M., University of Bradford.

North, P.H., King Wilkinson (International) BV, Holland.

Parry, G.E., University College, Cardiff.
Patel, R., Imperial College, London.
Patterson, W., Friends of the Earth, London.
Perkins, L.J., University of Birmingham.
Priestley, M., Esso Research Centre, Abingdon.
Pullin, D.J., Cavendish Laboratory, Cambridge.

Quarini, G., Queen Mary College, London.

Ramsay, Dr. W.J., Lawrence Livermore Laboratory, California, U.S.A.
Randle, J.M., Architect.
Rouse, L., Brunel University.

Salooja, Dr. K.C., Esso Research Centre, Abingdon.
Saluja, C.L., University of Wales, Institute of Science and Technology.
Simon, Prof. A., University of Rochester, U.S.A.
Smith, K.G., Sunderland Polytechnic.
Souster, C.G., University of Leicester.
Stewart, W.R., University of Keele.

Tajuddin, A., Queen Mary College, London.

Walker, P.R., University College, London.
Waller, R.F., The Open University,
Williams, P., Pugwash Conferences, London.

Yamba, F.D., University of Leeds.

SECTION 1

FUNDAMENTALS OF ENERGY RESOURCES
AND CONSUMPTION

THE ENERGY PROBLEM AND THE EARTH'S FUEL SITUATION

G. R. Bainbridge

The Energy Centre, University of Newcastle upon Tyne

Introduction

The sources to alleviate the energy situation in the world and in Britain are sufficient to supply all foreseeable needs. Conservation of energy and rationing in some form will however have to be practised by most countries, to reduce oil imports and redress balance of payments positions. Meanwhile development and application of nuclear power and some of the traditional solar, wind and water energy alternatives must be set in hand to supplement what remains of the fossil fuels.

Oil Dependence

In the industrial countries windmills and waterwheels have given way to cheap and convenient coal. Britain has had coal in abundance, Fig. 1. Over 200 million tons p.a. were produced in the 1950s,[1] compared with about 120 million tons p.a. now. Deceptively cheaper oil has proved a temptation too great to resist despite increasing import bills. Compensating revenues have come from shareholdings in the international oil companies. In a few countries a proportion of hydro-power has been introduced. In others some natural gas has been found.

Then in the 1960s nuclear power programmes began to be realised, Fig. 2. Britain however now relies for almost 50% of her energy supplies on imported oil,[2] Europe over 60% and Japan 70%. Even the USA which is rich in indigenous energy resources imports about 15% of energy use, including 35% of its oil.

In the early 1970s changes in the oil supply scene increased the need for oil importing countries to protect their energy and economic futures. The Organisation of Petroleum Exporting Countries brought pressures to bear on the international oil companies. These OPEC countries are responsible for 90% of

3

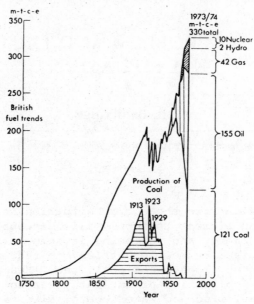

Fig. 1   BRITISH COAL PRODUCTION AND EXPORTS

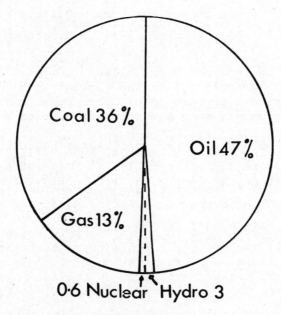

Fig. 2   BRITISH ENERGY %

the international oil trade,[3]   Table 1. They control 55%
of proven world oil reserves, equating with 10 times the USA
reserves and 20 times those in Europe.

TABLE 1. WORLD OIL PRODUCTION & RESERVES [t]

1973	Annual Production (m-t-c-e (m-t-o))		Reserves x 10^3 proven	
Middle E.	2100	(1140)	90	(50)
Africa +	540	(300)	17	(9.5)
Asia	270	(150)	4	(2.2)
Europe	36	(20)	4.5	(2.5)
USA	860	(480)	9	(5)
Canada et al	340	(188)	6.8	(3.8)
+ Venezuela	32	(18)	3.6	(2.0)
Russia et al	880	(490)	27	(15)
Total World	5000	(2800)	160	(90)

[t] million tons coal (oil) energy equivalent.
International oil companies were obliged [4]   to accept abrupt
upward revisions of prices, Fig. 3, to improve the exporting
countries' revenues. They have been driven to accept the
countries where the oil wells are located increasingly into
partnership. In the Autumn of 1973 the Arab nations applied
a reduction of up to 25% of all oil output. Holland and the
USA, classed as friendly to Israel, had to suffer an embargo
on some oil supplies from the Middle East.

Fuel and Living Standards

There has been increasing concern that high priced fuels may
restrict improving living standards. The world has a
voracious appetite for energy, Fig. 4. Without some surprising
change of attitude the demand for more energy seems likely to
continue.

All countries have sought economic growth [5]   in terms of Gross
National Product per head, (the annual value of cars, buildings,
refrigerators and other goods and services produced, averaged
over the population). For the advanced countries, such as the
USA and Sweden, the GNP/head is large, Fig. 5, and it is still
rising. The poor countries have a low, perhaps decreasing,
GNP/head. But even there a substantial part of the GNP is for
military equipment and some people are very wealthy with cars
and television sets, contrasting with the undernourished many.

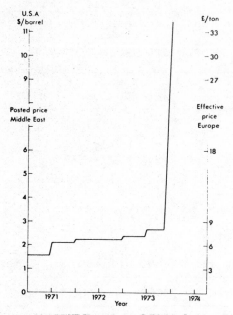

Fig. 3   TREND OF MIDDLE EAST OIL PRICES

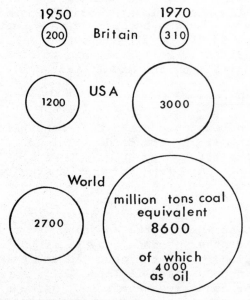

Fig. 4   GROWTH IN ENERGY USE

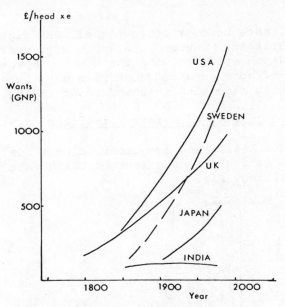

Fig. 5   GROWTH OF NATIONAL PRODUCT

To **effect** economic growth generally means [6]   to apply more energy, Fig. 6.

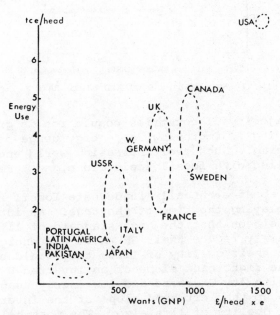

Fig. 6   NATIONAL PRODUCT AND ENERGY USE

Germany and France however achieve their GNP figure with less
energy than Britain. Sweden is a low energy user compared with
Canada, and Japan compared with the USSR. Improvement in the
energy usage of some advanced countries may therefore be
possible without reduction in standard of living.

Birth Rate and Quality of Life Implications

Education, advertising and amendment of political and religious
dogmas have relatively less effect on birth rate [5]   than has
standard of living, Fig. 7.

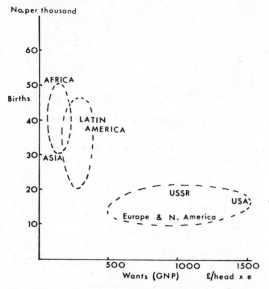

Fig. 7 NATIONAL PRODUCT AND BIRTH RATE

In the great majority of countries population is growing and
standard of living is improving, so energy usage per head and
in total is rising. Such ever increasing world energy use
cannot be supplied with fossil fuels alone, [7]   Fig. 8.

People worried about eventually inadequate fossil fuel supplies
seek ways of delaying the time of shortage. An interesting
proposal [6] has been that the high standard of living countries
might progressively reduce their GNP/head and energy usage but
thereby improve their quality of life; they would be represented
as moving up the right hand slope of an inverted U curve, Fig.
9. Countries with low GNP/head and low energy usage might then
be able to move up the left hand slope of the inverted U,
improving their quality of life also. The possibility of this

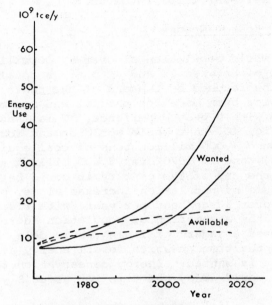

Fig. 8 FOSSIL FUEL SUPPLIES

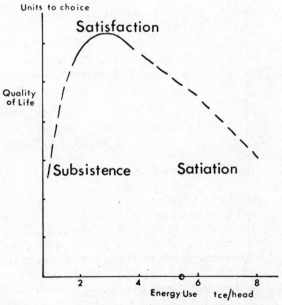

Fig. 9 QUALITY OF LIFE AND ENERGY USE

happening is of course remote due to the great organisational
problems and self-sacrifice involved.

Realism about Energy Needs

With present world population of over 3,000 million people and
a growth rate of between 2% and 3% p.a., a total of over 6,000
million will be reached by the end of the century,[8]   due
allowance having been made for natural and man-made catastrophes
of magnitudes within past experience. This leads to the
conclusion, Fig. 10, that while world demand for energy in the
1950s was around 3,000 million tons of coal equivalent (1 tce/
head per year), and in 1970 over 8,000 million tce (2 tce/head
per year) by the end of the century it could be 16,000 million
tce on the basis of population increase alone, or 25,000
million tce (over 3 tce/head per year) with rising standard of
living. Thus the assumption of population increase alone
without improved standard of living would delay by only a
decade or two the time at which fossil energy available is
less than what is wanted. Energy conservation to the extent
of 10% of use could delay the time for about 3 years.

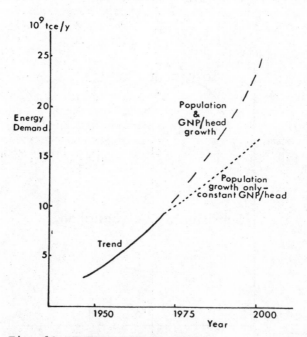

Fig. 10 ENERGY DEMAND GROWTH

Increasing Demand for Limited World Fossil Fuels

Fig. 11 suggests [9] that supply difficulties with oil would
occur on the basis of present use before the middle of next
century.

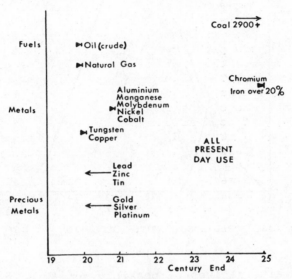

Fig. 11   MATERIALS SCARCITY PROSPECTS

Oil and coal use seem more likely to increase in the face of
demand and then decrease [7]   as sources get depleted, Fig. 12.
Many factors, e.g. alternative energy resources, scarcity prices,
geographical distribution of resources and limitations on
extraction and transport will in fact cause adjustments. With
increasing usage, oil may be scarce within a few decades, and
coal in a few centuries. World proven oil reserves at 160,000
million tce equate with little over 30 years of present
extraction at around 5,000 million tce. Discovery of new
resources of the magnitude of the North Sea or Alaska fields
will be required almost annually in the years ahead to prolong
the period of plentiful availability of oil. The recent fall
off in world oil production due to a transient reduction in
world demand will delay only slightly the time in the next few
decades when a real problem in oil supply will occur.

There is no indication that the fossil fuels can be prevented
from running out. Allowance has to be made for a greater
proportion of future energy being supplied as electricity and
the slow moves to reduce wastage of heat rejected from the
thermal cycle, amounting to two thirds of the total heat

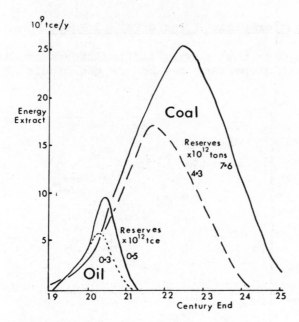

Fig. 12   WORLD FOSSIL FUELS DEPLETION

supplied.   The alternatives of possibly supplying process heat
and urban heat as well as electricity from large power stations,
or of using smaller total heat units tailored to the energy
needs of individual factories, will because of the cost and
difficulty of implementation, modify the trends of growing
energy usage, Fig. 13, only marginally during the next 25 years.
There is the assumption that a substantial capacity of nuclear
powered electricity will fill the gap.   But the nuclear
component for Britain, Fig. 14, and a few others among the
leading industrial countries is providing only a few percent of
total energy now, Table 2, and the competition between the
energy supply industries for scarce development resources seems
likely to intensify.

So in looking for sources of energy for the world and for Britain
it is for increasing amounts above today's figures that one must
be concerned.   One cannot yet see the possibility of any leading
country going alone and continuously on the path to constant or
decreasing energy use while still maintaining or increasing
production of those things necessary for the home and export
markets.   One can however see more clearly the wisdom of
diversifying energy usage to reduce the risk and cost of
dependence on imported fuels.

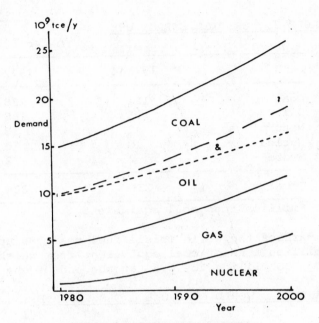

Fig. 13   FUTURE WORLD ENERGY DEMAND

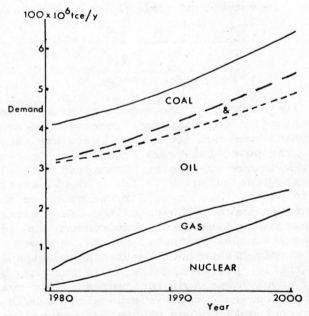

Fig. 14   FUTURE BRITISH ENERGY DEMAND

TABLE 2.   BRITISH ENERGY USE

m-t-c-e*

	1973/4	1972/3
Coal	121	128
Oil	155	158
Natural gas	42	38
Nuclear	10	10
Hydro	2	2
Total	330	336

*million tons coal equivalent

Table 3 summarises the world fossil fuel reserves upwards of 5,000,000 million tons of coal equivalent from which to supply demand of 25,000 million tce p.a. at the end of the century.

TABLE 3.   WORLD COAL, OIL & GAS RESERVES

	prospective 10^{12} tce
Coal	4.3 - 7.6
Oil*	0.32 - 0.56
Gas	0.2 - 0.45
Total	4.8 - 8.6

*Ex. sands & shales

Dividing the first number by the second gives an upper limit for the time to exhaustion only if extra reserves are not found and the rate of demand does not increase. There are many interposing difficulties. The potential users of fuels do not coincide geographically with the suppliers. Transport for oil by sea is established world-wide but, much as for coal, oil reserves are not always close to ports; this is so for the vast and much discussed tar sands of Canada and oil shales of the USA. Extraction from them can be very disruptive to the environment and so may be limited in practice. Gas transport by sea is growing but liquefied gas supply plants and the transport ships are comparatively costly, depressing the price that can be paid at the source for supply into a distant market. For coal large scale transportation would require expansion of bulk carrier fleets not yet contemplated, or plants for conversion of coal to oil or gas in the mining areas to raise the value and ease transport problems for the product.

British Fuels Summary

Britain has over 10,000 million tons of coal equivalent for an
end of century demand of 600 million tce p.a., Table 4.

TABLE 4.   BRITISH COAL OIL & GAS RESERVES

$$10^9 tce$$

Coal*	3.5	Classified working
Oil[+]	3.1	Licensed probable
Gas	2	
Total	8.8	

*Ex.sea plus 2.5 unclassed 20 accessible

[+]4.6 Licensed possible

The definition of fuel reserves is a science in itself [10]   as
Table 5 illustrates.

TABLE 5.   BRITISH NORTH SEA OIL RESERVES

$$10^6 \text{ Tons Oil (Equivalent m-t-c-e)}$$

	Proved	Probable	Probable Total	Possible	Possible Total
Commercial (10 fields)	895	165	1060	100	1160
Appraisals		230	230	160	390
Total found	895	395	1290 (2000)	260	1550 (2400)
Future possible (Licensed)		700	700	700	1400
Total (Licensed)	895	1095	1990 (3100)	960	2950 (4600)

One hears view that there may be 2 to 3 times the amount of oil
shown as possible in this official summary [11]   and other views
that less than half of that shown will be extracted.
Intensified exploration, particularly out under the North Sea,
has already substantially increased the amount of workable coal
above that given in Fig. 15.

The reserves of the fossil fuels extractable depend on many
factors,[12]   including product price, taxation and technology.

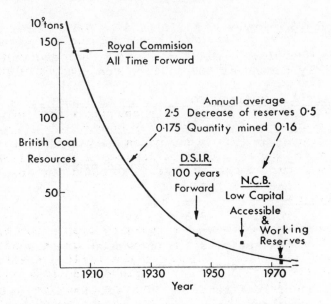

Fig. 15 VIEWS ON BRITISH COAL RESERVES

British oil discovered so far, if applied only to the extent
required by Britain to avoid net oil imports could be used up
in a few decades, Fig. 16.

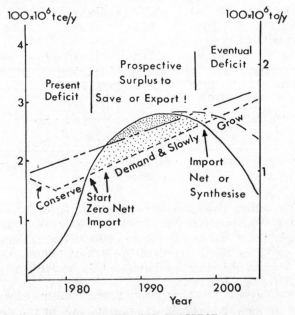

Fig. 16 NORTH SEA OIL PROSPECT

There are nevertheless interesting strategy options for it provided the other fuels, gas, coal and nuclear can be marshalled in support. British oil can be conserved or exported during a possible 10 to 20 year period of production in excess of needs.

British coal can also supply this country with a proportion of between a quarter and a half of her energy for the next 50 years. If the undersea reserves are successfully exploited there would be a very much longer period of assured coal supplies. There could be coal for export as well as oil and perhaps natural gas.

Nuclear Fission

Nuclear power can provide an increasing proportion of the energy needed in the years ahead [13]; without it quite modest expectations of economic growth in the world are almost bound to be frustrated. This energy supplement depends on the extent of uranium and thorium supplies and the reactors in which they are used. Little has been reported on the amounts of these materials indigenous to Britain. Supplies can however be purchased and stockpiled well ahead of need, so uranium and thorium, and in due course plutonium, can be treated as indigenous even if imported. The world quantities of fissile fuels at low cost are illustrated in Table 6.

TABLE 6. WORLD URANIUM & THORIUM RESERVES

10^6 tons up to $/lb	15	30
Uranium	0.7 - 1.5	1.3 - 3.3

(In sea water 4×10^9 tons)

1 million tons = 26×10^9 tce thermal

in reactors   (2.6×10^{12} tce fast)

| Thorium | 1.4×10^6 tons | |

Three million tons of uranium are thought to be available [13] at a price below $30 USA/lb. Used in the world's thermal reactors that would last 30 to 40 years for electricity generation. The energy equates with about a fiftieth of the coal reserves. Thorium would add half as much energy again.

Breeder reactors can further supplement energy resources, making more efficient use of the U238 in natural uranium by converting it into P_u239; also Th 232 to convert into U233. The energy

worth of low cost uranium reserves alone can so be brought up
to the same order as the coal reserves.

$$U238 + n \longrightarrow U239 \longrightarrow Np239 \longrightarrow Pu239$$
$$Th232 + n \longrightarrow Th233 \longrightarrow Pa233 \longrightarrow U233$$

In these discussions account has not been taken of the many
millions of tons of uranium and thorium salts in sea water.
Though extractable the cost would be high due to the large
amounts of water which would have to be pumped in the process.

Any remaining problems or doubts about nuclear power stations
and their supporting chemical plants must be resolved speedily
if they are to form a basis for growth of energy production.
Safe and economic commercial designs of breeder reactors
suitable for industrial siting will be necessary in the 1980s
in order to make the best use of fissile fuel reserves.

A Note on Nuclear Fusion

With fusion, nuclear fuel resources are expanded effectively
beyond limit.

The first fusion power plants are expected to use deuterium -
tritium reactions.   The tritium would have to be obtained
from neutron reactions in the lithium breeder blanket.

$$Li\ 6 + n \longrightarrow Li\ 7 \longrightarrow He4 + T3$$
$$D\ 2 + T3 \longrightarrow He4 + n$$

Range of Renewable Energy Resources

The fossil, fissile and fusile resources (coal, uranium and
thorium, deuterium and lithium) together appear adequate for
man's needs far ahead in time. Allocation of some effort to
other alternatives from the renewable sources nevertheless
seems worthwhile, [14]   both as a speculation and to provide a
measure of insurance. We in Britain are experiencing delays of
a year or two in the planned landing of North Sea oil. The coal
industry has for years been run down and is not proving easy to
run up again. Slowness in decision making and hesitant
organisational and construction planning has delayed nuclear
projects. Prevalent objections from environmental protection
organisations, irrespective of their validity, could force
delays or adoption of more costly energy provision arrangements
at any time.

Solar, geothermal, hydro, tidal and wind energy have all been

harnessed previously, though usually in small capacity units.
It is thought that it may also be possible to extract energy
from waves and ocean temperature gradients. The large plants
needed could not however be built without some loss of amenity.

In each of the following specific renewable energy summaries
the relevance if any for Britain of such energy sources will be
introduced.

Solar Radiation

It is possible with relatively simple flat plate solar
collectors, Fig. 17, to provide warmed water and enable some
space heating for homes and offices which is particularly
useful when the buildings are well insulated and thermal
capacity sufficient for the carry over of energy from day to
night is arranged.

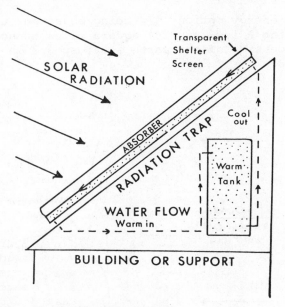

Fig. 17   SOLAR WATER WARMER

Although solar radiation incident on Britain alone is of the
order of twice the total world energy usage it is only a third
of what is received on the same area of South Mediterranean land.
In this country large scale collection could not be economic but
a few tens of square yards of collector surface can suffice to
give basic home heat input.

Only about half of the solar radiation incident on the earth's
atmosphere reaches the earth's surface to give about 22 watts per

square foot available - more near the equator and less at high
lattitudes or in unfavourable weather or terrain. However, the
whole surface receives about 20,000 times present world energy
usage so the effect of man's incremental input to the environment
is quite small. Some of the solar radiation could be collected
and converted directly to electricity rather than heat with
panels of silicon or cadmium cells. Regrettably at present the
cost would be very high and the efficiency of conversion very
low.

In good conditions over a surface area of 4 to 5 miles square of
earth one might generate 1000 MW with a convertor plant. That
is about half as much as a large modern power station and could
save 2,500,000 tons of coal each year. The problems, even in
favourable areas, are essentially those of maintaining the
collectors clean and working while integrating the electricity
yield before distribution.

A large scale energy option [15] being assessed in the USA is
that for putting a few (say 25) square miles of solar cells some
20,000 miles outside of the earth's atmosphere in stationary
orbit, Fig. 18.

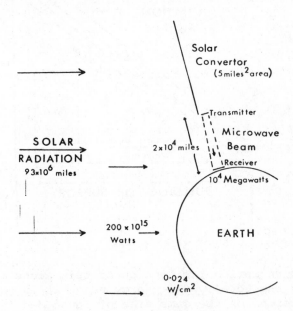

Fig. 18 SOLAR ENERGY PROSPECT

Beaming down the energy to earth with micro-waves this sort of
plant might supply 10,000 MW or a tenth of the electricity needs
of Britain. The problems of establishing such a system and

maintaining it are space-programme in scale, but with the
demonstrated capability in this technology prospects of the
kind described cannot be ignored.

Geothermal Energy

Geothermal steam [2]   has been used in volcanic regions in many
countries to generate electricity. The operating world capacity
in many plants to date however only adds up to the amount
produced by the Wylfa nuclear power station - a bit over 1000
MW. There is world scope for about 50 times that amount to be
developed over the next half century in select geographical
locations, but none in Britain.

Deep Earth Heat

Deep drilling of the earth[16]   and fissuring of the
subterranean rock there in order to get increased steam raising
surface and reduce pumping power is thought to have prospects.
Heat, thought to come from radioactive decay at greater depths
might be usefully extracted at about 6 miles down, as
temperature rise with depth is of the order of $40^{\circ}C$ per mile.
Within that distance under the continents there is available
about 5000 times the heat content of the coal reserves, or 6
million times the present world annual usage. Selective use of
rocks, using only the best, could possibly provide a large
fraction of energy needs for many years to come.

Tidal Energy

Tidal power prospects[17]   are regularly investigated for usable
shallow seas and estuaries which amplify the regular rise and
fall of sea levels due to moon/earth interactions. The La Rance
Scheme in France is unique as a development from the tidal energy
available world-wide. Tides of over 3 yards rise are desirable
compared with below 1 yard on average. Plans are being made to
build a 2000 MW tidal plant, 5 times as big as La Rance, in the
Bay of Fundy, Canada. Reassessment of the Bristol Channel, Fig.
19, has indicated a potential capacity almost twice that of
Fundy and it could incorporate useful pumped storage capacity,
freshwater storage and 2 cross channel motorways. It is quite
attractive if it can be made environmentally acceptable.

Wave Energy

Wave energy [18]   has also been in the news recently. There is
about 140 Megawatts per mile available round British coasts. It
could make a useful contribution to our needs; about twice that
of the UK generating system is available provided we are

prepared to forego the pleasure of waves on our shores. Had it
been easy or low cost of course it would have been done before.
The building of large ships and oil rigs in the North Sea may now
give engineering data to establish if wave energy extraction
equipment can be made to withstand the elements.

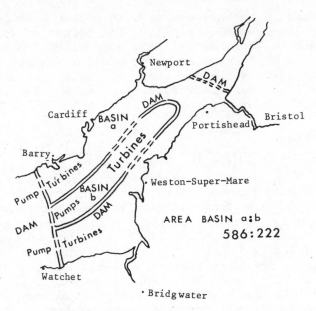

Fig. 19 SEVERN TIDAL - PUMP STORAGE SCHEME

Wind Energy

For local industry wind[19]   has been a motive force for
machinery, particularly for cornmills and irrigation for many
hundreds of years. In remote locations it is used now, for
example to trickle charge electric battery storage and pump
water from below ground to the surface.

Windmills to drive 1 MW or 2 MW electric generators are readily
feasible. The problems are in maintenance and control of 200ft
diameter blades at the tops of tall towers and having them
adjust to both violent gales and dead calms. As for several of
the renewable energy sources alternative supplies have to be
available for those periods.

Location of the many thousands of units that would be needed for
national distribution would not be easy, though perhaps a few
tens of locations could be found for say 100 to 200 MW total at
each. Windmills to supply the horticultural industry where many
individual fossil fuel boilers provide heating energy seem a
logical line for investigation.

Windvanes for local domestic energy supply of a few kilowatts
must be produced at a low enough cost for wider use than has so
far happened - rather like the growth in use of washing machines
and motor lawnmowers it is not impossible to visualise. Some
80% of household energy is used as heat so the associated energy
systems need not be fully electricity generating.

Energy Policy

Energy policy as it is emerging, particularly in Britain, appears
to be to produce more indigenous fossil fuels and near indigenous
nuclear energy in the future, but more immediately to make some
effort to economise in energy usage.

In Britain the coal industry is trying to adjust to its new task
of producing about 15% more coal than previously,[20]   reversing
the trend of the last decade. With coal at a low price relative
to oil for the first time in many years the opportunity for
providing more of this country's energy needs is open to the coal
industry.

To achieve the long term growth in energy use of 2 to 3% p.a.,
which even the most conservative forecasters predict to be
necessary for national well-being, (world growth in energy usage
exceeds 5% p.a.) nuclear power has to play a major part. With
just under 10% of Britain's electricity coming from 9 (Magnox)
nuclear power stations with a capacity of under 5000 MW, and
another 6 (AGR) stations of 6000 MW being built, interest in
taking the next step has been regrettably slow. A Government
declaration of interest was made in 1974 to build up to 4000 MW[21]
using a new (SGHWR) design but the work and progress towards
an actual order for 4 or even the first 2 of the 600 MW units
seems too slow for the resolution of an urgent energy and
financial crisis.

The oil trading role of British and international companies with
North Sea interests is going to increase quickly now.[22]   The
first 2 million tons of oil is said to be due to come ashore this
year rising to 20 million tons within a few years. The
technology of North Sea fuel exploitation is however difficult
and the rate of development to a peak over the next 20 years or
so could be slower than expected, as well as more costly.
Uncertainties about Government financial support and taxation
rules just do not yet give international companies that easy
confidence they need to operate positively in production.
Uneasiness about financial returns can lead to less oil being
produced and more left unrecovered. The same can be said for
natural gas production even where the success in recovery of
North Sea resources has been good and continuous over several

years past. Given a fair return for their entrepreneurial
activities the producer companies will maximise the return on
capital investments they have made which, hopefully, will
coincide with maximising total extraction rather than rate of
extraction to the benefit of both the companies and the country.

Until coal, nuclear energy, oil and gas are available at the
planned time and in the amounts required for energy self
sufficiency Britain's dependence on imported oil will continue.
Reducing energy use by taxes (on petrol) and regulations (on
vehicle speed) will always be disliked by those who have to pay
or obey, particularly if the burden falls unevenly on some
people. Encouraging energy saving by subsidies on efficiency
improvement projects will be more effective and better policy
provided there is a net saving of resources, energy being just
one among labour, other materials and capital.

Greater endeavour to produce more indigenous energy sooner but
firmly for the longer term seems a better preferred line for
priority however than that aimed at many and small, some of
them transient savings. Plans made now to achieve indigenous
energy self sufficiency for Britain in the 1980s, would be a
worthwhile contribution to the solution of world energy problems.

References

1. National Coal Board Annual Reports & Statistics.

2. Energy Prospects to 1985, Organisation for European
    Co-operation & Development, 1974.

3. Exploring Energy Choices, Energy Policy Project of
    the Ford Foundation, 1974.

4. Beyond the Energy Crisis, Maddox, J. Hutchinson 1975.

5. The Carrying Capacity of our Global Environment,
    Randers, J. and Meadows, D. H. Toward a Steady State
    Environment, Daly, Herman E. Freeman 1973.

6. Energy and Humanity, Thring, M. W. Mankind and the
    Engineer Vol. 2., Peter Peregrinus.

7. Energy Resources, Hubbert, M.K. Resources & Man,
    Freeman 1969.

8. The Limits to Growth, Meadows, D. H. et al, New York
    Universe Book, 1972.

9. Mineral Resources in Fact and Fancy, Cloud, P.,
    Environment, Resources, Pollution and Society, Sinauer
    Associates Inc., 1971.

10. Department of Energy, Production and Reserves of Oil
    and Gas in the United Kingdom, 1974.

11. The North Sea Oil Province - A Simulation Model of
    Development, O'Dell, P. R. and Rosing, K. E. Energy
    Policy, Vol. 2. No. 4. pp.316, December 1974.

12. World Coal Resources and their Future Potential,
    Armstrong, G. Phil.Trans.R.Soc.Lond. A276 439-452
    (1974).

13. Nuclear Power - The Future, Marsham, T. N. and Pease,
    R. S. Atom 196 February 1973.

14. Survey of Energy Resources World Energy Conference 1974.

15. Space Solar Power, Glaser, P.E. Paris International
    Congress 1973.

16. Geothermal Energy, Muffler, L. J. P. and White, D.E.
    The Science Teacher (USA) 1972 Vol. 39 No. 3 p.40.

17. Tidal Power, Hubbert, M. K. Resources and Man, Freeman
    1969.

18. Wave Power, Salter, S. H. Nature Vol. 249, p.720 June
    1974.

19. Wind Power, A New Term Partial Solution to the Energy
    Crisis, Heronemus, W. E. EASCON of Institute of
    Electrical & Electronic Engineers, 1973.

20. Coal Industry Examination Final Report, 1974.

21. Report to Parliament by the Secretary of State for
    Energy - Nuclear Reactor Systems for Electricity
    Generation. Atom July 1974, p.198.

22. Department of Energy, Development of the Oil and Gas
    Resources of the United Kingdom, 1975.

DISCUSSION

Question: Should the need for fossil fuels for chemical feedstocks and food
            production be a stimulus for conservation?
Answer: Yes, but we must persuade the policy makers that this is so. In
            any case, the proportion of fossil fuel used for feedstocks etc.
            is small, about 5%, and will only grow as fuel substitutes such as
            nuclear power develop.

Question: Are there substitutes for fossil feedstocks?
Answer: Yes, hydrocarbons produced using nuclear power and hydrogen from
            sea water, for example.

Question: Will there come a time when one could not afford to burn fossil
            fuel?
Answer: Yes. At the moment the chemical industry could afford to pay much
            more for natural gas.

Question: What are the costs and outputs for the 3 tidal power systems that
            you mentioned?
Answer: For the Bristol Channel project the cost per kw is greater than
            for a conventional fossil fuel station (£150/kw) but less than
            that of a nuclear power station (£250/kw). The output would be
            3500 MW (i.e. about 5% of the current CEGB installed capacity).

Question: Is more energy put into large projects such as the Bristol Channel
            project or nuclear power programmes, than is output? Nuclear
            power stations may last longer (50-60 years) than the 25 years
            that Dr. Chapman used in his initial work, while industry may
            learn how to build power stations more cheaply having built the
            earlier ones.
Answer from I never stated that there is an energy deficit in building one
Dr. Chapman:nuclear station; indeed, the energy output over the station's
            lifetime is 10 times greater than the energy input. An energy
            deficit will arise in the earlier years of a sufficiently rapidly
            growing nuclear programme. It is important to arrange that the
            deficit occurs when it can be afforded and when institutional
            problems will allow the rate of growth.
Comment: According to recent analysis at Cambridge a programme with a
            doubling time of 4 years will be in energy deficit only until 3
            years after the commissioning of the first power station (assuming
            a 6-year building and commissioning period for each station).

Question: To what extent is it justified to use fossil fuel for food produc-
            tion, since fossil fuel is "bottled solar power" with a large time
            investment while food can be produced in a few months using direct
            solar power?
Answer: The use of fossil fuels is determined by economics and the problem
            is how to get food cheaper from the field than from fossil fuels.
Comment: In the USA 3-10 times more energy is used in food production than
            the calorific value of the food.

Question: What is the status in the UK of the conversion of coal to gas?
Answer: Work was done in the last war. The UK used to make gas from coal
            before the advent of natural gas. The National Coal Board and
            British Gas have contracts for the development of gasification
            processes for use in the USA. ICI is doing work on oil gasifica-
            tion but there is no gas production method in the UK which is
            cheaper than natural gas.

Question: How are your energy equivalents calculated, how reliable are they,
            and which bodies produce them?
Answer: One can never get a perfect number, because coal has many forms.
            I have used 1.8 tonne of coal equivalent to one tonne of oil.
            Some people use 1.4 because this is for power station coal. The
            United Nations uses 1.3 in its work.

Question: You envisage the necessity to site breeder reactors in industrial
            areas to make use of their waste heat. Is this a personal opinion,
            and what sort of use do you envisage?
Answer: This must happen if the reactors are to be used most efficiently
            and economically. The waste heat presently discharged into the
            sea or rivers should be taken out as steam, hydrogen, or fresh
            water which could be transported up to 15 miles into an 'indus-
            trial hinterland'.

Question: What is the efficiency gain if use were to be made of waste heat
            from power stations?
Answer: The overall efficiency would rise from 30% to 80% or 90%.

Question: Should solar plants be built in OPEC countries?
Answer: Yes. Iran will develop indigenous alternative energy sources,
            viz. nuclear power plants, coal and solar, to enable more oil to
            be exported. There is great scope for agricultural development
            possibly linked with solar power.

Question: Why did you not comment on the environmental hazards of nuclear
            power?
Answer: Solar power has environmental hazards as well. I believe you will
            be discussing many of these aspects later in the School.

ENERGY DEMAND AND SUPPLY
IN THE UNITED KINGDOM

R. J. Eden

Energy Research Group, University of Cambridge

1. Introduction

In this lecture I shall begin by outlining the historical data on the pattern
of energy use in the United Kingdom. Then I will discuss the data on
different sectors, namely the domestic energy demand, commercial and public
buildings,transport and industry. The demand for primary energy as opposed
to energy supplied to the consumer depends strongly on the mix of energy
supply so I will discuss the energy industries and the amount of energy
required for conversion from primary fuels to energy supplied for the
customer. Finally I will look briefly at some possibilities for future energy
demand and supply and some of the difficulties that might have to be
considered in the next 25 years.

Energy Units

The standard international unit for energy is the joule denoted by the symbol
J. This unit is too small for convenience in discussions of commercial
energy so I shall use the giga joule denoted by GJ which is a thousand
million joules. The international unit for power is the watt denoted by W
which is one joule per second. I shall use this unit or multiples of it as
the unit of power. Approximate conversion factors and thermal equivalents
are shown in Table 1.

TABLE 1.   UNITS AND APPROXIMATE CONVERSION FACTORS

---

aapg denotes 'annual average percentage growth'
Energy unit: giga joule, denoted GJ, equal to a thousand million (10^9)joules;
mega joule, denoted MJ, equal to one thousandth of a GJ.
Power: watt, kilowatt, etc, denoted W, kW, etc
Mass: metric tonne (written tonne), 1000 kg
Approximate conversion factors
1 GJ   = 0.95 million BtU = 9.5 therms
              = 278 kWh   = 239 million calories
1GJ/tonne = 1 MJ/kg
1000kWh   = 3.6 GJ[a]
1 TWh   = 3.6 million GJ
Approximate thermal equivalents
1 tonne coal[b]   = 26.6 GJ(range 22 GJ to 32 GJ)
1 tonne crude oil   = 44 GJ = 7.3 barrels = 255 Imperial gallons
1000 cubic metres natural gas = 38GJ
1GJ   = 0.035 tce (tonne coal equivalent) = 35kg coal
                                   = 0.023 toe (tonne oil equivalent) = 23kg oil

---

[a] 1 kWh = 1000 watts for 1 hour
          = 3,600,000 joules
          = 0.0036 GJ
[b] 26.6 GJ per tonne of coal is the UK average, but the common international
     convention is to use 28.8 GJ per tonne of coal

2.   The Pattern of Energy Use in the United Kingdom

The inland consumption of primary fuels in 1973 was equivalent to 346
million tons of coal. The long run growth rate has been approximately 1.8%
per annum but it has been increasing slowly over the past 20 years both in
an absolute sense and relative to the rate of increase of GDP in the U.K.
The ten year average of the energy coefficient from 1953 to 1963 was 0.6 but
from 1963 to 1973 it was 0.8. The energy coefficient is defined as the ratio:
average rate of increase in primary energy consumption to the average rate
of increase of gross domestic product at factor cost and constant prices.
Some illustrative figures for primary energy supplied in the U.K. are shown
in Table 2.

TABLE 2.   PRIMARY ENERGY SUPPLIED IN THE UK, 1955-1973

Year	1955	1960	1965	1973.	1960- 73 aapg
Primary energy,10^6tce$(^a)$	255	269	302	346	2.0
Primary energy,10^9GJ	6.78	7.15	8.02	9.20	2.0
Energy supplied to final consumer,10^9GJ	5.32	5.33	5.66	6.43	1.5
Energy overheads	1.46	1.82	2.36	2.77	3.3

Source:*UK energy statistics*
(^{a}) 1 tce = 26.6GJ.

Fuel substitution can have a major effect on greater efficiency or lower costs or better amenity values, and some changes are illustrated in Table 2 which shows a large shift from the dominant position of coal in 1955 to its position today where it supplies only about one third of the UK market. It should however be noted that an improvement in efficiency resulting from fuel substitution may be due primarily to the new equipment that is installed rather than any basic advantage of the new fuel. Some illustrative figures for primary fuels and electricity are shown in Table 3.

TABLE 3. PRIMARY FUELS AND ELECTRICITY AS PERCENTAGES OF TOTAL UK ENERGY[a]

Year	% using thermal content of fuels suppied			
	1955	1960	1965	1973
Coal	86	74	62	38
Petroleum	14	25	35	46
Natural gas	0	0.1	0.4	12
Nuclear	0	0.3	2.0	3
Hydro	0.4	0.6	0.8	1
% of total primary energy used for public electricity generation(b)	14	18	23	30

Source:*UK energy statistics*
(a) Percentage using thermal content of fuels supplied
(b) The average percentage efficiencies for **steam** station   conversion from primary fuel to electricity were 24.2(1955) , 26.7(1960), 27.4 (1965), and 30.0 (1973).

The energy requirements for the conversion of primary fuels to the form of
energy required by the customer form a major part of United Kingdom energy
use. The largest energy overheads come from conversion of primary fuels to
electricity but these overheads are also quite significant for other
conversion processes. Some examples of the energy overheads are shown in
Table 4.

TABLE 4.   ENERGY SUPPLIED AND OVERHEADS FOR UK ENERGY INDUSTRIES, 1973

Energy Industry	Gross energy input 10^9GJ	Net energy output 10^9GJ	Energy Overheads 10^9GJ	Gross/net
Petroleum	4.32	3.98	0.34	1.087
Coal	3.57	3.46	0.10	1.029
Natural Gas	1.17	1.08	0.09	1.084
Manufactured fuels	1.12	0.78	0.34	1.43
Electricity	3.02	0.82	2.20	3.70

If the energy overheads are taken into account the pattern of energy demand
in the different sectors of the United Kingdom economy is significantly
different from the pattern of energy supplied as can be seen from the
figures shown in Table 5. The total energy requirements are often referred
to as "gross energy" demand and the energy delivered to the customer on a
heat supplied basis is often called the "net energy".

TABLE 5.   NET ENERGY USED AND GROSS ENERGY REQUIRED FOR MAJOR SECTORS IN THE
UK, 1973

Sector	Net energy used 10^9GJ	% of total Primary energy	Energy Overheads 10^9GJ	Gross energy required 10^9GJ	% of total
Industry	2.70	29	1.11	3.81	41
Household	1.58	17	1.07	2.65	29
Transport	1.36	15	0.14	1.50	16
Other users	0.78	8	0.52	1.30	14
Energy Industries	2.85	31	Included under supply	Included above	Included above
Total primary energy	9.26	100		9.26	100

3. Domestic Energy Demand

The energy supplied to households is illustrated in Table 6. The net energy
supplied has been almost constant for the past fifteen years but there has
been an increase in the gross energy requirements due to the increasing use
of electricity in the domestic sector.

TABLE 6.   ENERGY SUPPLIED TO THE DOMESTIC SECTOR

	1960 10^6 GJ	1965 10^6 GJ	1973 10^6 GJ	1960/73 aapg
Solid fuel	1192	1025	562	-5.6
Town gas	137	197	168	1.6
Natural gas	–	–	340	(c)
Petroleum	71	102	176	7.2
Electricity	121	206	329	8.0
Net energy total	1520	1530	1574	0.3
Electricity overheads (a)	311	530	846	8.0
Gross energy total (b)	1831	2060	2420	2.2

Source: *UK energy statistics*
(a) Electricity overheads on the basis of a constant 28 percent conversion
efficiency, thus they are 2.57 times the net electrical energy.
(b) Overheads for primary fuels are omitted from this total.
(c) The total gas supply increased at an average annual rate of 13 per cent
from 1965 to 1973.

It is remarkable that the net energy used in households has remained almost
constant despite the fact that the number of households have increased by
about 20% since 1960 and there has been a substantial increase in the number
of houses with central heating. These figures illustrate that improvements
in the efficiency of the use of net energy more than compensated for the
energy required for additional appliances and for the improved standards of
heating. The main reasons for improved efficiency are: the improvement in the
efficiency of heating appliances which have been changed at a rapid rate due
in part to the fuel substitution that has taken place. Improved insulation
and draft proofing in new dwellings compared with those that they replace.
Many new dwellings have lower ceiling heights than those they replace.
Central heating systems are often more efficient than the open fires that
they replace. There may also be other changes not related to efficiency that
have tended to reduce energy demand per household, for example, the relative
shift of population towards the South of the United Kingdom.

The improvements in efficiency of the use of energy of households has taken
place in part because of the extensive fuel substitution over the past
fifteen years. The benefits from this substitution will be less in the future
than in the past and one may therefore expect there to be an increase in the
energy supplied to the domestic sector over the next fifteen years. After
that time the saturation of appliance use in households may begin to have
some effect and it is possible that the energy demand would begin to level
off again.

This potential increase in the total energy demand by households could be
reduced by an extensive program of energy conservation involving, for example,
improved insulation of existing houses. It is, of course, important that new
houses should have higher standards of thermal insulation but these higher
standards for new houses alone would not have a very substantial effect on
the rate of growth of energy demand for households in total. There would be a
substantial financial cost for the nation in improving the insulation of
existing houses and this cost should be compared with the costs of providing
the energy that would otherwise be required. The financial costs of thermal
insulation should   be compared with the possible benefits that one might
get by making similar investments in manufacturing industry. On these
questions I take the view that we are concerned with the overall optimum use
of resources rather than the optimisation of a single resource such as energy.

In estimating the costs and benefits of improved efficiencies in heating houses
one should also take into account social benefits from warmer houses as well
as financial benefits from the fuel savings that are obtained. It is however
important to note that there is a slow change over in the housing stock in
this country and insulation standards for new houses now should be related to
expected costs of energy for those houses in the future perhaps ten or more
years ahead.

4.   Commercial and Public Buildings

This sector includes schools, hospitals, office buildings, shops, hotel etc.
Their energy demand has seen a major shift from coal to other fuels and
electricity during the past fifteen years. The net energy supplied is shown
in Table 7.

TABLE 7.   NET ENERGY AND FUEL SUBSTITUTION IN PUBLIC SERVICES AND
                 MISCELLANEOUS GROUPS

	1960	1965	1973
Solid fuel,%	56	44	14
Gas, %	9	10	17
Petroleum, %	25	32	47
Electricity, %	10	15	22 (a)
Total, %	100	100	100
Net energy, 10^6GJ	563	609	690

Source: *UK energy statistics*
(a) The gross energy (including overheads) required for electricity comes to
    nearly half the total gross energy demand for the sector.

It is interesting to note that the substitution towards electricity in this
sector has taken place in spite of the high cost differential. On a heat
supplied basis coal and oil prices in 1973 were each about 3p per therm to
the industrial and commercial users whereas electricity was about 20p per
therm. In some commercial and public buildings lighting takes over half the
total of electricity whereas in households it takes only a few percent.
This is partly due to the fact that many office buildings have been over-lit
and there is undoubtedly some scope for energy conservation in this area.

There is also evidence to suggest that substantial energy savings could be
achieved by having better control systems and possibly modified heating
systems in office buildings. A major lead in this area has been taken by the
Property Services Agency of the Department of the Environment which manages
a large number of government buildings. By installing advanced heating
control systems in some of their larger properties they have achieved fuel
savings of 30 to 50 % compared with previous systems.

5. Transport

The net energy demand for transport sectors in the U.K. is illustrated in
Table 8.

TABLE 8.   NET ENERGY DEMAND FOR TRANSPORT SECTORS IN THE UK

	Net energy required, 10^6GJ			Percentages of transport energy		
	1960	1965	1973	1960	1965	1973
Rail	301	135	59	32	14	5
Road	483	689	1052	52	69	77
Water	60	57	46	7	6	3
Air	84	113	201	9	11	15
Net total	928	994	1358	100	100	100

Source : *UK energy statistics*

The energy for road transport is provided almost entirely by oil based
products and demand has increased at an average annual of 6% since 1960 so
that it now represents nearly 80% of the total energy for transport. Energy
for air transport has increased at 7% per annum and is now 15% of the total
transport energy, and it should be noted that this energy depends to a large
part on international air traffic which is dependent on world economic
factors as well as on the special situation in the United Kingdom. The
dominant role of petroleum products in providing energy for transport will be
a cause for concern in the medium to long term future when supplies of
petroleum become scarce on a world wide scale. The onset of a period of
scarcity is unlikely to be more than fifteen to twenty years from now and
may be sooner. It would therefore be prudent to start now to examine options
for changes in methods of transport that give a more efficient use of energy.

The lead times for significant savings in the use of energy for transport
range from a few years to many decades. For example, a more economical petrol
engine introduced into 20% of new cars this year, and 40% next year and in
the following years is an unlikely speed of market penetration, but it would
not affect as much as 1 car in 5 until about 1981. Major changes in transport
patterns are dependent on the patterns of residence and industry, and
changes in this area take many decades rather than many years. In looking to
new transport modes that will save energy we should note that our concern is
with overall savings of all resources rather than concentrated on a single
resource alone. For example, there is little value in theoretical studies
that show that railways can transport goods with a smaller use of energy per
ton-mile than for road transport if in fact the use of rail transport implies
transport by road to the railway terminal, rail between terminals and road
at the far end of the journey. Under such conditions it is important to take
into account the costs and benefits relating to all resources that are
involved in a change from one system to another.

6. Industry

Fuel substitution in the United Kingdom industry is illustrated in Table 9
and the percentage energy used by the main industrial groups is shown in
Table 10.

TABLE 9.   FUEL SUBSTITUTION IN UK INDUSTRY(excluding iron and steel)

	% of total industrial energy including iron and steel		
	1960	1965	1973
Iron and steel total %	30	29	23
Other industries: Solid fuel	35	25	10
Gas	3	3	13
Petroleum	13	21	28
Electricity (including overheads)(a)	19	22	26
Other industries total %	70	71	77

Source : *UK energy statistics*

(a) Electricity overheads are included on the basis of a constant 28 per cent
    conversion efficiency.

TABLE 10.   ENERGY USED BY MAIN INDUSTRIAL GROUPS [a]

	% of total industrial energy		
	1960	1965	1973
Engineering and other metal trades	17	17	19
Food, drink and tobacco	7	8	8
Chemicals and allied trades	13	12	17
Textiles, leather and clothing	8	7	6
Paper, printing and stationery	5	6	5
Bricks, tiles etc	4	4	2
China, earthenware and glass	3	3	3
Cement	4	4	4
Other trades	9	10	13
Total 'other industry'	70	71	77
Iron & Steel	30	29	23

Source : *UK energy statistics*

(a) Electricity overheads for 28 per cent efficiency are included, other
    energy overheads are neglected.

Petrochemical feedstocks are not included in the above figures

In process industries it is possible to identify a steady improvement in the
efficiency with which energy is used to produce the end product. For example,
in steel-making there has been a steady improvement of about 1.6% per year
in the specific energy required per ton of crude steel. Such improvements are
however strongly dependent on the rate of new investment in the industry, and
this would be of benefit during a period of rapid expansion but would not
lead to such high savings if an industry became static in its output. If the
present investment program of the British Steel Corporation is completed over
the next 10 years then it is probable that there would be an annual improvement in specific energy per ton of steel of about 2% per annum during this
period.

The chemicals industry is very complicated with multiple intermediate products
and multiple final products. The industry is also integrated to some extent
on a European scale and this makes it extremely difficult to analyse the flow
of materials, let alone the energy requirements at different stages of
chemical processes. The output from the industry has been growing very
quickly over ten to fifteen years and in some sectors the rate of growth has
been as high as 9% per annum. One of the fastest growing sectors that is

important for the future is the petrochemical industry in which feedstock
requirements increased at a rate of about 10% per annum between 1965 and 1972.
The high value added obtained by the petrochemical industry make it likely
and desirable that an increasing proportion of petroleum should go into this
industry and that major economies should be made in other aspects of the use
of petroleum. This would indicate that over the next 20 years there will be
substantial changes in refinery design and patterns of output. We might
expect petrochemicals and transport to dominate the use of petroleum within
the next 20 to 30 years.

In order to make a detailed analysis of the future demand for energy in the
industrial sector, it is desirable to look at the energy required for final
products and relate the demand for these final products to the growth of the
economy. In this manner one can feed into the analysis the effects of
improvements in technical efficiency for industrial production and changes
in product demand as the economy develops. This underlines the importance of
energy analysis. However, one should also emphasise that in order to carry
out energy analysis, considerably more information is required of the actual
way that energy is used in industry. It is going to be a major task to
evaluate the use of energy in such detail as one would require for the
effective use of energy analysis.

Energy conservation in industry will be considered in other lectures but I
would like to note one or two points that arise in these connections. Ideally
one would like to have an analysis of potential energy savings as a function
of the cost of investment to achieve these energy savings. For some industrial establishments such dependence has been evaluated and it has been found
that 5% of the energy used can be saved at very little capital cost but it
requires some changes in management and operation techniques. The next 5%
can be saved with investments that might amount to the cost of fuel that would
be saved within one year so there is nearly a 100% return on the investment.
Another 5% might be saved at a cost which would be economically worthwhile
provided we were not in a time of economic difficulty for industry.
Engineers in the chemical and petrochemical industry have given examples of
energy savings where the benefits and costs are of the order of magnitude
of the percentages that I have just mentioned. However it would be wrong to
calculate energy savings across industry by supposing that similar percentage
savings can be achieved in all sectors. It is possible that even in those

sectors where the savings could be achieved, not all factories will adopt the necessary changes. It should also be borne in mind that improvements in the efficient use of energy have been taking place in the past, and therefore one cannot count all such improvements as gains compared with the trend curve that can be found from historic data. If the energy savings are dependent on changes in management and work practice then it is not clear that such changes could be maintained in a period when there did not seem to be an energy crisis. Taking all these factors into account one must be very cautious in estimating the savings that might be achieved in industry by energy conservation over a long term future. However an energy saving of 1% per annum compared with the trend curve would lead to a saving of 10% in 1985 and this would be of significant national benefit. If over the next ten years the economic performance of the country improves so that manufacturing industry produces a higher output, then without better energy efficiencies one would expect industrial energy demand to increase at a faster rate than in the past. Industrial demand for energy has been increasing at about 2% per annum over the past 12 years or so and given a significant improvement in economic performance one might expect the trend to amount to an increase of 3% per annum, which might then be reduced to a growth rate of about 2% per annum due to energy conservation. If the economy improves so that there is an increased rate of investment in industry this would permit the use of more efficient equipment. This could lead to energy savings that were of greater significance in relation to output than we would achieve if the economy was improving at a slower rate, thereby inhibiting investment in manufacturing industry.

7.   The Energy Industries

The energy industries play a most important role in influencing energy demand and in affecting the investment pattern in the United Kingdom. One of the most difficult tasks in energy policy questions must be the long term eval-uation of investment and expansion or contraction of the different energy industries. It appears to have been fully recognised only relatively recently that it is equally important to take into account the scope for energy conservation as an investment criterion to be compared with the investment required for increasing energy supplies. There does not appear to be adequate information available at the present time for an overall assessment of the proper balance between different energy industries and energy conservation. One of the difficulties of course is the uncertain future in that one does

not know at what stage petroleum will become scarce and one is not sure of
the rate of substitution that would be possible from other energy sources
such as coal or nuclear power. Considerable difficulties arise from the long
lead times that are required for change in the energy industries. These lead
times are increasing due to the higher costs of alternatives that are now
being considered as replacements for crude oil from the Middle East. The
greater investment required per barrel of daily output from the North Sea is
such that there is a much greater lead time in developing North Sea oil fields
than there was in developing equivalent sized fields in the Middle East. For
example, the capital investment required in the North Sea for one barrel of
daily output of oil is about 5000 dollars, whereas in the Middle East the
similar capital requirement would be about 200 dollars.

It is probable that by the turn of the century, the need to develop energy
industries based on coal and nuclear power will be apparent on a world-wide
scale. It is possible also that major developments in other energy sources
such as solar power will be underway before the end of the century. The
advantage of solar power for heating is that it can be developed for heating
purposes on a small scale so that one does not need a massive industrial
build-up for central power supply in order to have the rapid development of
this energy source. Electricity from solar power looks like being a different
problem in that it could well require major industrial developments at high
costs before it can be achieved. We should not under-estimate the importance
of intermediate technology energy developments such as solar heating but at
the same time we should not under-estimate the need for major changes in
central energy supply systems. In particular developments in nuclear power
supply require the development of industrial skills and industrial equipment
that sets limits on the rate of expansion. Even if one assumes a rapid rate
of expansion for nuclear power supply in advanced countries, it is unlikely
that this expansion would be fast enough to eliminate the necessity of
another alternative to oil and gas supplies. Even with moderately optimistic
estimates about North Sea oil and gas, this situation holds in the United
Kingdom. On a world-wide scale it is likely that there will be a need for
coal production to increase so that it becomes a major part of world energy
supply early in the next century. It is unlikely that adequate expansion of
the world coal industry could take place if this was to be based on deep
mined coal, and we shall therefore be dependent on the development of surface

mining. This will be constrained by environmental considerations, and there
is doubt whether total energy supply will be adequate towards the end of the
century without a significant rise in energy prices. Planning for the energy
industries in the United Kingdom has to be made in the light of all the
uncertainties associated with the pattern of future energy supplies and the
pattern of economic growth. If economic growth is slow, then there will be
less demand for energy but there will be less investment available for new
energy sources. If economic growth is high, the demand for energy will
increase but the economic growth would permit a greater investment in alternative energy sources. In considering alternative futures it is of course
desirable to balance the costs that would be required for alternative energy
sources against the costs that would be involved in a programme for energy
conservation.

8. Future energy demand and supply

Energy policy should not be based on the assumption that we are able to forecast the future, neither should it be based on a plan for energy that sets
out what the country should do over the next 25 years. Instead of energy
forecasting, one seeks to make a variety of projections of future energy
demand based on various assumptions about economic growth, improvements in
energy efficiency, fuel substitution and changes in energy sources. Against
each of these projections one should have a variety of plans or alternatives
by which the energy requirements in these projections could be met by a
variety of types of supply. The choice of type of supply to meet a given
projection should depend on future patterns of costs as they are perceived
at the time when decisions have to be made. Under these conditions a policy
on energy supply or energy conservation should be continuously evolving and
under continual revision. At any given time it would not be possible to
assess a most probable future demand projection. However, one should bear in
mind future demand can to some extent be guided by control of the prices for
different forms of energy, and it could also be controlled by rationing
systems though these are more likely for industry than for the private
consumer. In general terms, energy policy should be designed so as to avoid
undue risk from the future turning out differently from what was implied by
the decisions on energy supply that had been made perhaps many years previously. The avoidance of risk implies flexibility but it also implies the use of
game theory in which one tries to minimise losses from the unexpected. One

thing that seems sure about the future is that the unexpected will occur.

Some illustrative figures for energy demand in the United Kingdom that were produced by the Department of Energy for the Parliamentary Select Committee on Science and Technology are shown in Table 11. The figures in this Table may be compared, for example, with the idealised profiles for possible North Sea oil production shown in figure 1. The profiles in this figure are based on an assumption of total reserves in the UK sector of 4×10^9 tons of oil. It is also assumed that production at the plateau is held until the reserves to production ratio is 15 years and thereafter this ratio falls slowly to a value of 10. For illustration a median estimate of demand for oil is also shown in figure 1 (this estimate differs from the one given in Table 11). If one makes similar estimates for North Sea gas supplies and if one assumes also that UK coal production will not be significantly above the present level by the year 2000 then it appears unlikely that the United Kingdom will still be self-sufficient in fossil-fuel by that date. One cannot say how unlikely this would be, nor can one obtain figures for import requirements at that time for fossil fuels. However, the possibility of scarcity of fossil fuels arising in the year 2000 certainly needs to be taken into consideration in energy policy decisions, particularly since some of these decisions have to be made now in relation to the expansion of alternative sources of energy or enhanced methods of energy conservation. The long lead times for developing alternative energy sources and for developing the industries and the skilled man power required are such that it is necessary now to take decisions that may not produce major industrial and commercial impact for some decades ahead.

References

UK Energy Statistics (issued annually by the Department of Energy) HMSO, London.
Energy Conservation in the United Kingdom, (Report by the National Economic Development Office) HMSO London 1974. (This report contains a list of further references).

TABLE 11. EXAMPLES OF THE POSSIBLE RANGE OF ENERGY DEMAND IN THE UK[*]

* For source see next page

The limits shown by the model runs in 1974　　　　　　　　　　　mtce

	1972/73	1975	1980	1985	1990
Energy uses					
Coal	128	119–133	107–154	84–199	77–196
Natural Gas	34	46–48	54–87	62–109	56–68
Nuclear	10	14	25	33–44	50–106
Hydro	2	2	2	2	2
Oil	158	148–163	122–200	126–258	146–317
TOTAL ENERGY USES	332	344–349	368–391	407–441	464–505
Non-energy uses					
Oil(inc.Bunkers)	25	27–32	39–44	48–56	60–68
Natural Gas	4	4	6	7	6
TOTAL ALL USES	361	375–385	413–441	462–504	530–579
of which					
North Sea Oil	–	8	170–238	170–255	170–255
All Other Fuels	361	367–377	175–271	207–334	275–409
North Sea Oil as PERCENTAGE OF TOTAL FUEL REQUIREMENTS	–	2	39–58	34–55	29–48

Notes:

    (i) The ranges set out above cover 32 scenarios of the 1974 Energy Review.
    (ii)The "states of the world" covered were:
       Economic Growth 2.7 per cent and 3.3 per cent per annum. (*Note:*These
       are not predictions but working assumptions chosen to bring out the
       effect on forecasts of a half percentage point shift in average growth
       rates).
       Oil price greater than coal price *and* oil price less than coal price.
    (iii)The "strategies" covered were:-
       High coal investment, mimimum coal investment.
       High gas absorption, strict gas conservation
       High nuclear development rate, low nuclear development rate.

    (iv)The high coal investment strategy examined the economic *demand* for
coal in certain conditions and was not therefore constrained to likely levels
of domestic production.

    (v) The high and low assumptions in each case were not precise reflections
of available "real" strategies but generalised approximations designed to
test the interrelationships of broad policy options.

    (vi)The columns do not necessarily add up to the totals shown- because
for example low figures for one fuel will not necessarily be associated with

TABLE 11 continued

low totals for other fuels.

6. From this very wide range of figures a single set has been chosen
for illustrative purposes and to act as a sample projection against which
variants can be assessed. It is not necessarily optimal in economic or
energy policy terms, and the figures for 1990 have to be regarded as
particularly speculative.

*Sample Projection**   mtce

	1973	1980	1985	1990
Coal	131	130	135	150
Oil	159	145	170-180	175-225
Natural Gas	40	80	80	65
Nuclear	10 }	25	35-45	50-100
Hydro	2			
TOTAL ENERGY	342	380	430	490
NON-ENERGY	30	45	55	65
TOTAL ALL USES	372	425	485	555
NORTH SEA OIL	-	170-240	170-255	170-255

Total Energy Demand

7. The main single determinant of the growth of energy demand is the
rate of economic growth. The second major determinant is the relationship
between the level of energy prices as a whole and the general level of
prices in society. This factor merges in its effects with the direct effects
of energy conservation measures. The sample projection associates a 2 per
cent per annum growth of energy demand with a 3.3. per cent average (though
not constant) growth of GDP.

  * The sample projections have been rounded to the nearest 5 mtce

Source: The Select Committee on Science and Technology (Energy Resources Sub-
        Committee) Minutes of Evidence, Wednesday 12th March 1975, (from
        Annex C of the Memorandum submitted by the Department of Energy),
        HMSO, London.

FIGURE 1.   UK NORTH SEA OIL: IDEALISED PRODUCTION PROFILES

Assumptions: Total Reserves 4×10^9 tons

Production held at plateau until R/P is 15 years;

thereafter R/P falls to 10 according to

$(R/P) = 5\exp[-0.16(T-T_o)]+10$

where T is date and T_o is date when R/P is 15.

A median projection of UK oil demand is shown as a broken line (note that this line is purely illustrative and is not a forecast, for example, if relative prices were suitable, a substantial part of the oil demand could be met by extra demand for coal if supplies were available).

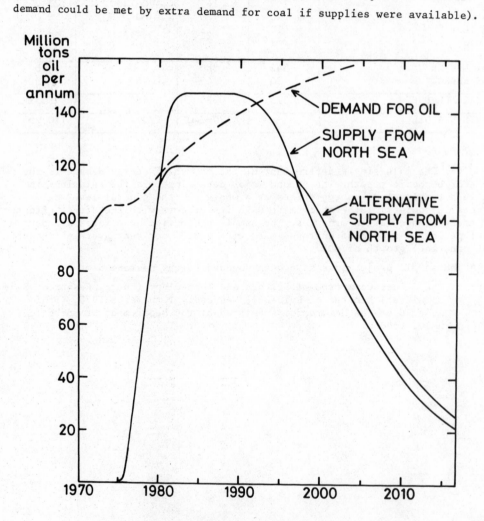

DISCUSSION

Comment: In relation to what you said about useful energy from electricity,
            I think the efficiency of use may be as low as 60%, not close to
            100%, because of the use of storage heaters.

Response: This is not true. I quoted data for 1963 and 1975. In 1963
            storage heaters were not widely used. In 1975 at least half of
            the domestic electricity use is for cooking and water heating with
            very high efficiencies, so the effect of storage heaters on the
            overall efficiency is much less than you suggest. I agree however
            that storage heaters are not a very satisfactory means of space
            heating.

Comment: I.H.V.E. quotes figures of 70% for the efficiency of storage heaters.
            Their main disadvantage is the difficulty of controlling them.

Response: I take the point that electricity is not always efficiently used
            for space heating. What I did try to show was that, even for
            space heating, electricity can be more efficient than other methods
            in special circumstances. For some industrial processes, e.g.
            induction heating, electrical methods can be extremely efficient.
            It is unsound to make broad statements to the effect that electri-
            cal heating is always a bad thing. Further, if we are moving
            towards an electrical economy, we should start to move in that
            direction now.

Question: Can you give a figure for the percentage of oil consumption used
            by the private motorist?

Answer: Energy used for road transport is about 10% of national energy use.
            It is not clear just how much of this is used by private motorists
            but I would guess about half. As far as conservation is concerned
            a combination of more careful driving, speed restrictions and other
            government measures might save 20% of this or 1% of national use.
            Though this may sound a small amount, to save the same amount in
            other ways might require substantial capital investment and/or effort.

Question: I have two questions. (1) You want to fill the 'energy gap' with
            nuclear power. What about solar energy, based possibly on solar
            parks in OPEC countries? (2) You predict an increase in domestic
            energy consumption as people try to improve their thermal comfort.
            Could not the same effect be achieved by changes in clothing?

Answer: My estimates of nuclear growth assume the introduction of breeder
            reactors. This is not a value judgement, merely a possible way
            to fill the gap. In view of the uncertainty about how fast solar
            energy can be introduced this is a sensible course. Personally,
            I would prefer rather more solar energy use. I am aware of the
            subjective effects of comfort standards. It appears to be the
            case that people invariably adjust heating levels rather than vary
            their clothing and this seems to be a fact that one must accept.

Question: Two questions. What capital costs do you assume for breeder
            reactors? Why is your North Sea production profile flat-topped?

Answer: The current cost of nuclear plant is around £300 per kW , but I
            haven't made any careful estimates of future costs. I mention
            breeder reactors because uranium supply problems would make these
            necessary in a nuclear programme of the size envisaged. This pro-
            gramme would require quite a large proportion of the fixed capital
            investment of the UK at the time. As to the second question, the

flat top maintains self-sufficiency as long as possible. I agree
that this may not be the optimum procedure.

Comment: I suggest that breeder reactors may cost around £1,000/kW.

Response: This may be true. It seems likely that breeders will be used as
much as a fuel factory as a power source, thus offsetting the
high cost.

Question: What assumptions do you make about the capital investment required
to achieve this North Sea production?

Answer: I've taken government figures.

Question: You've shown national trends and not the results of imposed policy
except for North Sea oil. I was worried about what you said about
the efficiency of electricity use. Why can't there be some policy
to cut down on electric heating? All electric appliances except
heaters have recently had their VAT increased.

Answer: I didn't put in very strong government measures, but I did assume
a plausible amount of conservation. Of course if there were a
mismatch between supply and demand the government would take
crisis measures. People usually respond to a crisis better than
in anticipation of one. I would hope analyses of the sort I have
given might enable us to anticipate and possibly avoid difficulties
common to a number of trends. This may be politically difficult
though. One should also remember that we currently have considerable
spare electricity generating capacity. We also need to get a
nuclear programme underway rapidly. Given these two points we
don't want to encourage too strong a swing away from electricity.

Question: Can you state how you arrive at your figure for North Sea oil
reserves?

Answer: I've used the Department of Energy's "Production and Reserves of
Oil and Gas in the UK". I don't include the continental shelf,
the Rockall trench or the deep oceans. My figure corresponds to
the oil likely to be found, and to be exploitable, to 1990. There
will undoubtedly be more North Sea oil but it will be more expensive. In making these predicitons it is best not to be too
optimistic.

ENERGY CONSUMPTION AND CONSERVATION IN THE UNITED STATES

J. E. Rothberg

Physics Department, University of Washington

1. INTRODUCTION

"United States energy use unexpectedly fell 2.2 percent in 1974 after rising steadily for more than two decades." Thus began a Department of the Interior news release[1] on April 3, 1975. Although the Secretary of the Interior hoped that this was "the start of a new trend" it is more likely to be a short-lived shoulder on a rising curve. It probably does mean that earlier predictions of a continued annual 4% increase in energy use are too pessimistic. This is a dangerous time to try to predict future trends in energy use but it is clearer than ever that wide-ranging government efforts to limit consumption, decrease waste, and improve energy conversion efficiency together with continued technical development can have a significant impact on energy use. Conservation efforts alone are not likely to obviate the need for new energy sources and for massive new energy supply programs but they can give us time to develop new facilities at a reasonable rate and to ensure that environmental damage is held to a minimum.

In this review we will examine energy consumption patterns in the United States. We will focus on those end-uses which are the largest consumers of energy and for those, try to assess the impact of some conservation options. The emphasis will be on conservation measures which can have some effect during the next 10 or 15 years. This is a critical time period since new energy sources will not yet have made an important contribution to filling the gap between domestic supply and demand. See Fig. 1.1.

Any detailed consideration of energy sources or supplies is outside the scope of this review; we are strictly dealing with energy demand. Management of solar energy influx is sometimes a conservation technique and sometimes an energy source but we will not say much about it. Using solar energy for building heating and cooling is an exciting and viable possibility but it will not have much impact before 1990 at best. Similarly the use of fuel cells to supply both heat and electricity and "total energy systems" fall into both categories but are not near-term solutions.

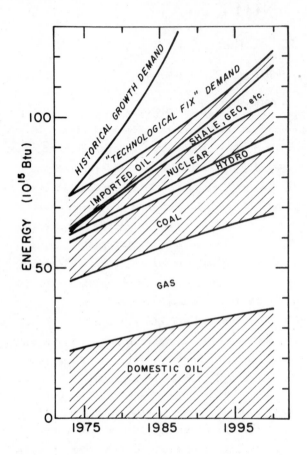

Fig. 1.1. Current and projected energy demands and
sources. The projections are
those made by the Ford
Foundation Energy Policy
(see Ref. 2) for the
"Technical Fix" scenario
with reduced imports. The
demand in the "Historical
Growth" scenario is also
shown; it would reach 187
Quads by 2000 corresponding
to an annual growth rate of
3.4%. 1 Quad - 10^{15} Btu.
See Appendix for conversion
factors.

A persistent source of confusion in energy consumption tabulations concerns
how to convert electrical energy usage so that it can be compared with
energy derived directly from fossil fuels. How to account for hydroelectric
power and nuclear reactor generated electricity are particular problems.
For point-of-use types of comparisons it may sometimes be useful to
convert electrical energy (kWh_e) to thermal energy (Btu_e) at the rate of
3414 Btu_e per kWh_e (simple conversion of units). This conversion ignores
the efficiency of electrical generating plants (~30%) and is quite misleading if one is interested in overall energy consumption. Hydroelectric
power is sometimes converted at the 100% efficiency rate, a procedure which
ignores the fact that saving hydroelectricity ultimately saves fuel. The
Bureau of Mines (Department of the Interior) procedure is to assign to

hydroelectric power the average conversion efficiency of fossil fuel plants. This varies somewhat from year to year but is sufficiently precise for our purposes to follow the Bureau of Mines procedure and to convert at the rate of 10,400 Btu_t per kWh_e. In general we have tried to provide reasonably accurate numbers rather than attempting to resolve all conflicts and providing the highest precision.

In general when we discuss "efficiency" we are referring to ordinary energy conversion efficiency. A useful concept for evaluating conservation strategies and new technologies is the "second law efficiency". This is described in detail in a recent report,[3] "Efficient Use of Energy", sponsored by the American Physical Society. The second law efficiency is the ratio of the thermodynamically minimum amount of energy needed to perform a certain task to the amount of energy actually used for that task. For tasks involving heating one can estimate the second law efficiency by imagining the task to be carried out with a perfect fuel cell and a perfect heat pump. As an approximation we can take the energy content of the fuel to be the same as the electrical energy input to the heat pump. In these terms conventional heating systems have efficiencies below 5%. As heat pump and fuel cell technologies improve it becomes more relevant to use such an efficiency criterion.

In compiling energy consumption data one can categorize usage according to a number of different schemes:
1. Traditional sector - industrial, transportation, etc.
2. End-use - space heating, process steam, etc.
3. Final demand - total energy consumption related to automobiles, to food, etc.
4. Energy source - oil, coal, etc.
5. Energy form at point of use - electric drive, low temperature heat, etc.

Unfortunately recent statistical information on all of these subcategories is not available. The Department of the Interior publishes energy consumption figures by sector and source and other government agencies publish statistics relevant to their own purview. The most detailed and comprehensive report is the well-known "Patterns of Energy Consumption" published by the Stanford Research Institute[4] (SRI) in 1972. The data which they

used is for 1968 and earlier and so the report is badly outdated. Furthermore some categories of energy use were incorrectly assigned to sectors.
Recently a massive report has been published by the Federal Energy
Administration,[5] the Project Independence Report. They have compiled
much data on both energy supply and consumption and have made new extrapolations. Assignment of total energy consumption to a final product or
demand is a relatively new and interesting way to look at energy use. Some
recent estimates for automobiles, food, and appliances have been made either
by a detailed study of the processes or using input-output analysis. Not
enough data are available yet and there are still some accounting problems
so we have not tried to summarize the results of this approach.

A breakdown of energy consumption by sector is given in Table 1.1.

TABLE 1.1 Energy Consumption in the U.S. - 1974

| Sector | Energy Consumed 10^{15} Btu | | | |
	Fossil Fuel	Electrical	Total	%
Residential and Commercial	13.80	11.24	25.0	34
Transportation	18.27	0.05	18.3	25
Industrial	21.20	8.12	29.3	40
Total	53.27	19.4	73.	100.

Source: Ref. (1).

Electricity usage has been assigned to the consuming sector at the rate of
10,400 Btu$_t$ per kWh. Energy consumption summaries are also given in
Figures 1.2 and 1.3.

The per capita energy consumption in 1974 was:
$$3.5 \ 10^8 \text{ Btu}$$
$$\text{or } 0.37 \ 10^{12} \text{ J.}$$
The rate of energy consumption per capita was:
$$11.7 \text{ kW}$$
$$\text{or } 242,000 \text{ kCalories per day.}$$
This rate of consumption is about 100 times that due to food intake. The

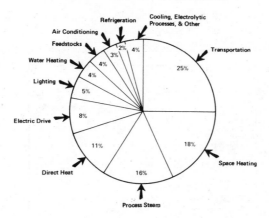

Fig. 1.2.   U.S. energy consumption in 1973 subdivided according
            to end use.   Energy consumed in generating electricity
            has been included.   Source: Ref. (6).

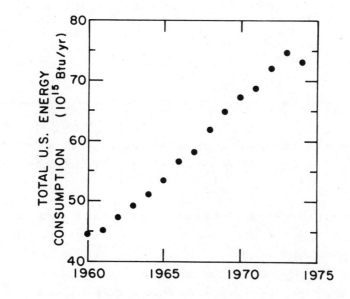

Fig. 1.3.   Total U.S. energy consumption since 1960. The last
            point shown (1974) indicates a decline of 2.2% from
            the previous year. Source: Ref. (7,8)

rate of consumption of energy for the past few years is shown in Fig. 1.4.

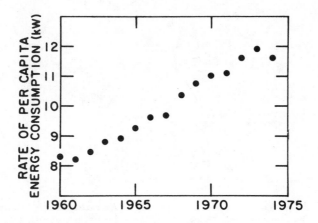

Fig. 1.4.   The per capita rate of energy consumption in kW.
            The present rate is just under 12 kW per person.

The major components of demand and of the growth rate will be examined for
each sector and some comments and evaluations of conservation options will
be presented.

Although any strategy that seems to save energy is probably worth implementing there are a number of consequences that one has to be very
cautious about. Reducing energy consumption in a certain domain may have as
a direct result that employment is reduced; furthermore the money not spent
on energy may be spent on other goods and services which may or may not be
equally energy intensive (in the sense of energy consumed per dollar of
product). The results of reduction in demand for certain goods or services
have been investigated and the conclusions are sometimes paradoxical[27].

When one type of fuel becomes scarcer or more expensive there is usually
a high rate of conversions to another type. A striking example has been
the extraordinarily high rate of conversions from oil heat to electrical
resistance heating in the Pacific Northwest (where electricity is cheap).
With the present mix of fuels for electrical generation such a conversion
is not energy conserving and may not even be oil conserving. In the
longer term, of course, one may want to exercise demand management
strategies and convert various fossil fuel uses to electrical power.

Conversion factors, definitions, and energy content of fuels are listed in
the Appendix.

2.   RESIDENTIAL AND COMMERCIAL SECTORS

A.   Patterns of Consumption

The total energy demand of the residential and commercial sectors in 1974
was 25.0 10^{15} Btu or about 34% of the total U.S. consumption. Fuel energy
used for electrical generation is included in this figure.

Energy consumption for 1970 in this sector, broken down according to end use,
is displayed in Table 2.1 and Fig. 2.1.

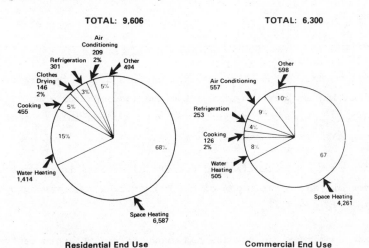

Fig. 2.1. Residential and commercial energy consumption in 1970 sub-
divided according to end-use. The electrical energy component has been
included taking into account only the electrical energy expended at
point-of-use. The energy units are 10^{12} Btu. Source: Ref. (6).

Table 2.1. Residential and Commercial Energy Use - 1970

Residential Sector	Energy 10^{15} Btu	% of Sector Energy	% of U.S. Energy
Space Heating	8.5	58.7	12.6
Water Heating	2.4	16.5	3.6
Cooking	0.65	4.5	1.0
Air Conditioning	0.48	3.3	0.7
Other	2.5	17.0	3.7
Sub-total	14.5	100.	21.5
Commercial Sector			
Space Heating	2.3	41.2	3.4
Lighting	1.3	23.1	1.9
Air Conditioning	.45	8.1	0.7
Refrigeration	.39	6.9	0.6
Other	1.2	20.7	1.8
Sub-total	5.6	100.	8.3
TOTAL	20.1		29.8

Source: Ref. (9).

It is important to notice that the two applications of low temperature heat, namely space heating and water heating, together account for about 22% of total U.S. energy consumption, an amount of energy comparable to that used by the entire transportation sector.

Fossil fuel use in the residential and commercial sectors is about equally divided between oil and natural gas with electricity a major and growing component. See Table 2.2.

Residential and commercial energy use has grown in recent years faster than either population or number of households as can be seen in Table 2.3. It

TABLE 2.2.   Residential and Commercial Energy Consumption by Source - 1974.

	Energy 10^{15} Btu	% of Sector Energy	% of U.S. Consumption of this source
Coal	0.29	1.2	$\sim$0
Oil	6.39	25.5	19.1
Natural Gas	7.12	28.4	32.1
Electricity	11.24	44.9	57.3
Total	25.04	100.	

Source: Ref. (1).

TABLE 2.3.   Growth in Residential and Commercial Energy Consumption.

	Pop. 10^6	No. of households 10^6	Persons per household	Res. & Comm. Energy Consumption 10^{15} Btu	Energy per household 10^8 Btu	Energy per person 10^8 Btu
1950	151.3	42.9	3.5	9.25	2.16	0.61
1960	180.0	53.0	3.4	14.01	2.64	0.78
1970	203.8	63.4	3.2	23.13	3.65	1.13
1971	206.2	64.8	3.2	24.00	3.70	1.16
1972	208.2	66.7	3.1	25.30	3.79	1.22
1973	209.8	68.3	3.1	25.62	3.75	1.22
1974	210.4			25.04		1.19

Source: Ref. (7,10).

is useful to identify the components of growth for extrapolation purposes.
As is the case for the transportation sector, changing lifestyles and increasing family income have directly resulted in higher energy consumption.
The direct residential and commercial energy consumption per person has
doubled since 1950 although the growth is now leveling off.

A striking feature of the energy growth in these sectors has been the increasing reliance on natural gas (for space heating) and on electricity for
heating, appliances, and air conditioning. The rapid growth in electricity
consumption by households which all tend to heat, cool, and cook at the same

time has caused serious peak load problems for electric utilities. The
growth in energy usage according to source is displayed in Fig. 2.2 and
Table 2.4. Electricity consumption has increased by nearly a factor of
7 since 1950, corresponding to an average rate of about 8% per year. In
the same interval population has grown by only a factor of 1.4.

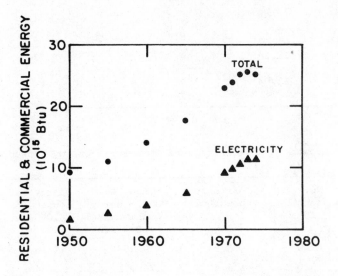

Fig. 2.2. Total residential and commercial energy use since 1950.
           Fuel consumed in generating electricity has been in-
           cluded and hydroelectric energy has been converted at
           the average efficiency of fossil fuel plants according
           to the Bureau of Mines procedure. The fraction of total
           energy consumed as electricity has grown rapidly.

The amount of energy used for space heating per residential unit has grown
by a factor of 1.4 from 1950 to 1970. The biggest components of this
growth are due to conversion to central heating and increased residence
size and exposure; these trends are not expected to continue over the next
15-year interval.[9]

TABLE 2.4.   Growth in Energy Consumption by Source Type. 10^{15} Btu/yr.

	Coal	Oil	Natural Gas	Electricity	Total
1950	2.91	3.04	1.64	1.66	9.25
1960	0.98	4.92	4.27	3.84	14.01
1970	0.43	6.45	7.11	9.14	23.13
1971	0.41	6.44	7.37	9.78	24.00
1972	0.39	6.67	7.64	10.60	25.30
1973	0.30	6.69	7.32	11.31	25.62
1974	0.29	6.39	7.12	11.24	25.04
Ratio of 1974 to 1950.	0.1	2.1	4.3	6.8	2.7

Source: Ref.(7).

Several of the most important components of energy use in this sector will be examined in some detail so as to identify conservation possibilities and project future use.   Although space heating is by far the largest consumer of energy, improvements in efficiency may be difficult to achieve very quickly because of the large number of individual dwelling units, their very diverse design and construction, their durability, and the initial expense involved in making major improvements.   However, it is such a large consumer of energy and is often carried out so inefficiently that it is important to examine in detail.

B.   Space Heating

The energy demand for space heating depends on the size, type, and construction of the residence, the average outdoor temperature, the indoor temperature that is considered comfortable, and the efficiency of the heating system.   This dependence can be written approximately as

$$E = \sum_{j} \frac{(T-T_j)}{e} \sum_{i} h_i$$

where E is the total annual energy demand per residence, T is the indoor temperature, T_j is the average outdoor temperature on a particular day (j), e is the overall efficiency of the heating system (if electricity is used

waste heat at the generating plant should be accounted for), Σh_i represents
a sum over heat losses through various mechanisms: conduction through
walls, windows, air infiltration, etc. The dimensions of h_i are Btu per
degree per day. This expression can be simplified if we define a "degree-
day" (DD)*; in the U.S. the standard indoor temperature used in the
definition of the degree-day is 65°F (18°C) corresponding to an actual
temperature of 70°F (21°C). The temperature increment of 5°F which does not
have to be supplied by the heating system is provided inadvertently by
other appliances in the house and by the occupants. The expression then
simplifies to $E = \frac{DD}{e}\Sigma h_i$. Omitted from this approximate expression are heat
loss mechanisms which are not linear with temperature difference, solar
energy influx, and the effect of the time dependence of outdoor temperature.
Heat load calculations are usually based on such simple expressions with
appropriate heat conductance factors.[11]

We will specify the average of some of these parameters for the United
States and indicate which have been responsible for the past growth in
space heating energy consumption. Conservation options and their expected impact can then be assessed.

Climate. The number of degree-days per year in the United States varies
from just a few hundred in Florida to over 10,000 in Minnesota. The
population-weighted average is 4,540. Because of the wide range in heating
demand the same sorts of heating systems and dwellings are not optimum in
different parts of the country. The population-averaged number of degree-
days has been calculated for four major regions in the U.S. to assist in
making energy demand models.[9] Some climatic data for these four major
heating regions are given in Table 2.5. Summer climate is specified in
terms of the number of hours per year during which cooling is required.
A rough average for the U.S. is 1,000 hours per year of cooling. Cooling
degree-days are sometimes tabulated but are not so useful since cooling
load is mainly due to solar influx and electrical load and not to heat
conduction.

* The number of degree-days per year is the sum of (65°F minus the average
daily temperature) over the days for which the average is below 65°F.

TABLE 2.5. Climate in U.S. by Region

Region	Pop. 10^6	No. of households 10^6	Degree-days per year	Cooling hours per year
Northeast	49.0	15.3	5470	300
North Central	56.6	17.4	6345	500
South	62.8	19.1	2795	1,600
West	34.8	11.1	3515	1,600
Total	203.	63.	4540	~1,000

Source: Ref. (7).

Type of Residence. The majority of people live in single family houses.
Despite movement from cities to suburbs there has been a net decline in the
number of single family houses over the past 10 or 15 years which has been
attributed to the increasing proportion of younger, smaller families and
single person households. New construction of mobile homes and low density
multiple dwellings has increased substantially. Table 2.6 lists the U.S.
distribution of housing types and of new construction.

TABLE 2.6. Housing Types

Type of Residence	Inventory (1970) per cent	New Construction (1970-80) per cent
Mobile home	3	18
Single Family	66	48
Low Density-multiple	16	18
Low Rise Apartments	10	10
High Rise Apartments	5	5
	100	100

Source: Ref. (7).

The average size of new single family houses has increased from 1,000 ft^2
of floor space to about 1,600 ft^2 in the interval from 1950 to 1968, but
since then the average size has remained steady. Heating energy demand is
approximately proportional to floor space for a given type of residence and

is, of course, lower for multiple dwellings. Table 2.7 summarizes the
heating energy demand for various types of fuel and residences in the
Northeast region. In that region 48% of the dwellings are single family
type and 52% are apartments of one sort or another. The winter climate in
that region is close to the U.S. average with about 30 degree-days per day
during the winter.

TABLE 2.7. Heating Demand in Northeast - 1970.

| Type of Residence | Energy Demand · 10^5 Btu per unit per year. | | | |
	gas	oil	electricity[a]	electricity[b]
Single family - detached	1,800	2,100	660	2,000
Single family - attached	1,100	1,300	400	1,200
Multi family - low rise	740	860	240	730
Multi family - high rise	680	800	210	640

a) at point of use   b) including generating efficiency

Source: Ref. (7).

A rough overall average energy demand is about 1.3×10^8 Btu per unit per
year (1,300 therms) or about 26,000 Btu per degree-day. For single family
detached houses the demand is about twice that for an apartment house and
corresponds to approximately 20 Btu per ft^2 per degree-day or about 10^6 Btu
per day.

Efficiency. Table 2.7 also reveals that the energy demand for electric
heat (when calculated at point of use) is dramatically lower than for fossil
fuels. This comes about because of improved construction and insulation in
dwellings using electric heat, efficient furnaces, and local room control
of temperature. There is a clear message here about how much the efficiency
of central heating systems can be improved.

Very rough furnace efficiency figures for several heating systems are given
in Table 2.8, although these values depend on the age and upkeep of the
furnace.

TABLE 2.8.   Inventory of Furnaces and Efficiency

System	% of Residences (1970)	Efficiency (%)
Gas	56	60
Oil	27	50
Electricity	8	33 (at point of use: 100%)
Heat Pump	∿0	54 (at point of use: 167%)
Other	9	
Total	100.	∿55%

Source: Ref. (7).

While it is clear from the point of view of overall energy conservation that electric heat is a very poor choice,in 1974 it was installed in 50% of new construction.   However, the use of electric heat and its cost in most of the U.S. has encouraged better construction, insulation, the use of double-layered windows, etc. Particularly in regions where cooling is desirable and with relatively mild winters the use of heat pumps is much to be preferred. Conversion from fossil fuel systems to electrical resistance heating should be discouraged.

Conservation. The energy savings possible in a typical residence are usually quite large. It has been estimated[7] that 25% savings in space heating energy are possible in existing single family residences and that up to 50% savings are possible in new construction using existing technology and at reasonable initial cost. If 10% of existing dwellings were retrofitted to save 20% of the space heating energy the total savings would be about $0.3 \ 10^{15}$ Btu/year or 0.4% of total U.S. energy consumption. It is clearly important to try to reach a larger fraction of existing dwellings by applying pressure with tax incentives, publicity, improved and less costly retrofit materials, and realistic energy prices.

In order to gain some insight into the sorts of energy conservation methods that are likely to be effective it is useful to estimate the sizes of the most important heat loss mechanisms. Figure 2.3 and Table 2.9 show the heat balance for a typical house on a mild, clear day (indoor-outdoor

temperature difference (ΔT) is 30°F (17°C). Infiltration plays a remarkably important role especially if the outdoor humidity is low.

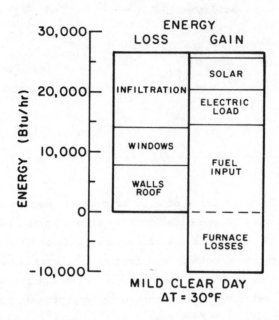

Fig. 2.3. The energy loss mechanisms for a typical house are shown
for a day in which the outdoor temperature is 30°F below the indoor
temperature (ΔT = 30°F) and the outdoor humidity is 40%. Also shown
are the energy inputs to the house. The total fuel requirement is shown
as well as that fraction which is lost due to furnace inefficiency.
In an improved house losses could be reduced by 30% and greater
advantage taken of solar influx reducing fuel requirement by at least
50%. Source: Ref. (3).

For windows facing south, solar influx far exceeds heat loss through the
window during daytime and may nearly equal the 24-hour average loss (at
least at latitudes of the Northeast United States). The energy gain from
electric load includes that from the refrigerator, water heater, etc.
Unfortunately much of this heat may not be provided to the part of the
house where it is needed.

Heat losses in houses are divided nearly equally among walls and roof,
windows, and infiltration; the furnace inefficiency is an additional major
energy cost. In apartment buildings heat loss through windows plays a

TABLE 2.9.   Heat Balance for a House on a mild, clear day.($\Delta T=30\,^{\circ}F$)

Energy losses	Btu/hr	%
Walls and Roof	7,200	27
Floor	1,100	4
Windows	6,600	25
Incoming Air (heating)	8,000	30
Incoming Air (humidification)	3,800	14
Total	26,700	100.
Energy Gains		
Solar through windows	5,200	
Three people	1,000	
Electric load	5,700	
Fuel Input (gas)	24,660	
Furnace efficiency (60%)	-9,860	
Total	26,700	

Source: Ref. (3).

proportionately much larger role ($\sim$50%) while in office buildings forced air ventilation may be responsible for half of the heat loss.

The important role played by windows is not surprising if one remembers that the effective heat transfer coefficient (U) for single pane glass is about[11] 1.1 Btu/ft^2-hr-$^{\circ}$F while that for walls is only about 1/10 that; windows comprise 10 to 12% of wall area. Double glazing could reduce window losses by 50% (U = 0.65 Btu/ft^2-hr-$^{\circ}$F) providing a 12% reduction in heat load. Similarly reducing air infiltration by sealing cracks, etc., could have at least as big an effect in houses. In office buildings the fresh air ventilation rate could be adjusted to provide adequate but not excessive amounts of outside air.

Nearly 25% of single family houses have no insulation at all and nearly 50% have no storm windows.[2] The fraction of houses lacking these energy saving improvements is much larger among poorer families with the result that the heating cost per household is almost the same for the poor as for the upper middle income group in spite of the larger average dwelling size of the higher income families. Subsidized energy conserving house improvements[12] and widely available measuring equipment and information could have a significant impact in upgrading existing buildings. Energy conserving building codes are needed to ensure that new construction is not designed solely to minimize first costs for the builder.

Reduction of indoor temperature could save a significant amount of energy. Most heat loss mechanisms are proportional to ΔT, the indoor-outdoor temperature difference, so one might expect to save a fraction $\Delta t/\Delta T$ of the heat load where Δt is the indoor temperature reduction. One can also save some energy by reducing thermostat settings at night or while away. Here the savings are smaller since they are due to the reduced heat transfer rate while ΔT is lower and depend on the thermal time constant for the structure. Detailed calculations[13] have been carried out to estimate the potential savings resulting from thermostat setback.

Table 2.10 shows the effect of several set-back options for a house in a 5,000 degree-day climate and also the national impact of such options if widely adopted. It has been assumed here that the normal thermostat setting is $72^{\circ}F$ ($22^{\circ}C$).

TABLE 2.10. Effect of Thermostat Setback - Normal setting is $72^{\circ}F$.

Option	% Savings for 5,000 DD house	% Reduction in total U.S. energy use.	Total energy saved 10^{15} Btu
$70^{\circ}F$	10%	1.2%	0.9
$68^{\circ}F$	20%	2.3%	1.7
$68^{\circ}F$ - day, $60^{\circ}F$ - night	31%	3.6%	2.7

Source: Ref. (13).

Although these are very optimistic estimates, significant savings are easily achieved with relatively little discomfort. Greater availability of time-control thermostats would be a help in getting widespread acceptance of this option. Readily available information about determining thermal time constants would encourage diagnosis and improvement of "leaky" houses and make thermostat setback less uncomfortable.

Although savings of 25% per residential unit are possible, mainly with
technical improvements such as insulation and double glazing, the number of
units for which we can expect these improvements to be implemented is
quite small. The extent of implementation depends on the cost of fuel, the
initial cost of the retrofit, the "payback" period, and the consumer's
response to these variables. Some estimates[7] have been made of costs
and benefits of conservation measures and of the percentage of market
acceptance as a function of payback period. With reasonable assumptions
about the cost of heating fuels the conclusion is that about 3% of the
heating energy consumption could be saved annually; in the case of new
houses a 5% increase in efficiency is expected by 1990. The problem is
that even conservation measures which have a payback period as short as
one year would be implemented by only 15% of the residential consumers. On
the other hand, commercial consumers are more perceptive about economic
benefits and 30% of that group would implement a one-year payback improvement.

   C.   Other Energy Consumption
Water Heating. This is a significant contribution to energy use but the
potential for reduction is not very large. Some possible improvements
include better insulation on tank and pipes, reduction of tank temperature,
local water heaters in dishwashers, etc., and electrical ignition of gas
heaters. It has been estimated that perhaps 25% of present consumption per
unit could be saved but only with government pressure.

Cooling. Air conditioning is a small consumer of energy at present but is
growing rapidly. In 1970 27% of residential dwellings had at least a room
air conditioner and this is expected to increase to about 90% in 1990 with
the South being completely saturated by then. Air conditioning consumes
11% of residential electricity but with a very poor duty factor.

The heat load on air conditioners in residences is mainly due to solar influx, especially through windows. In the commercial sector the heat load
due to lighting is over 40% of the total with an additional contribution
from ventilation. Reducing the excessive light levels by, say, 50% in
office buildings would result in significant savings and would help
alleviate the summer peak power crisis in many cities.

Labelling of efficiency or "life-cycle" costs will provide incentives for
manufacturing better units. The use of heat-pump-air conditioners will be
an important and economical development particularly in the mild winter-
hot summer parts of the country. In the longer term solar assisted air-
conditioning using absorption cycles will be an exciting development.

Appliances. Of the major appliances: ranges, T.V., refrigerators, freezers,
washers, dryers, dishwashers, only freezers, clothes dryers, and dishwashers have not, as yet, penetrated the market completely. Table 2.11
shows the percentage saturation and average energy consumption for these
appliances. By 1990 it is expected that those appliances which are not yet
saturated will be found in about 80% of households. In some cases consumption per unit will increase due to increased size or new features

TABLE 2.11. Appliances

Appliance	% Saturation 1970	Avg. operating energy per unit per year (10^5 Btu)	% of Avg. space heating demand (1,300 therms) per unit
Range	100.	∿130	10.
Refrigerator	100	∿130	10.
Freezer	28	∿200	15
T.V. (black & white)	99	48	4.
T.V. (color)	51	66	5.
Washer	71	14	1.
Dryer	42	∿120	9.
Dishwasher	19	50	4.

Source: Ref. (14).

(frost-free refrigerators, color T.V., etc.). The operating energy needed
for small electrical appliances is quite negligible; for those the manufacturing energy cost is often comparable to the operating cost.

Although energy labelling and improved efficiency will reduce consumption per
unit in some cases, it is unlikely that the overall energy savings will be
significant.

3.   TRANSPORTATION

A.   Patterns of Energy Consumption

Direct energy consumption for transportation of people and freight accounted for 25% of the total U.S. energy consumption in 1974.   More than half of the transportation energy is used by automobiles and about 70% is consumed in the form of gasoline.   See Fig. 3.1.

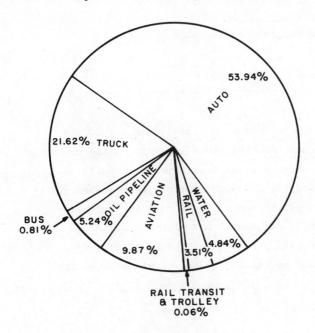

Fig. 3.1. Energy utilization in the transportation sector according to mode of transportation. The dominance of motor vehicles is evident despite the high energy intensiveness of air travel.

In Table 3.2 we have summarized data on energy use, traffic, and energy intensiveness for various transportation modes and end-purposes.   Data for automobiles has also been tabulated separately (See Table 3.1).   Urban and inter-city travel place very different demands on the automobile and it is rather unfortunate that the same device is used for both purposes.

J. E. ROTHBERG

TABLE 3.1. Automobile Usage - 1970

	Traffic 10^9 PM	Mb/d	Energy 10^{15} Btu	Btu/PM	PM/VM	Btu/VM	Fuel Economy mpg
Inter-city	970	1.6	3.3	3400	2.4	8,160	16.7
Urban	690	2.7	5.6	8100	1.4	11,340	12.0
Total	1660	4.3	8.9				

PM - Passenger Miles

Mb/d - Million barrels of oil per day equivalent

VM - Vehicle-miles

mpg - miles per gallon

Source: Ref. (15).

Energy use for transportation has been growing rapidly since 1950 and a major component of this growth has been the great increase in the number of automobiles per capita. The number of miles driven per automobile has not increased very much (10%) and the fuel consumption per mile has gotten only slightly worse (9%). This growth can be attributed to increasing income, movement to suburbs, and other social and economic factors. At this point there is nearly one automobile per person of legal driving age. In Table 3.3 we have summarized data pertaining to the growth of automobile energy consumption.

TABLE 3.3. Growth in Automobile Use.

Year	Pop. (10^6)	No. of autos (10^6)	auto/cap	miles/auto	fuel economy mpg	auto. energy	total transport. 10^{15} Btu
1950	151.3	40.3	.27	9020	14.9	3.3	8.64
1970	203.8	89.3	.44	9978	13.6	8.9	16.36
1971	206.2	92.8	.45	10,121	13.6	9.4	17.07
1972	208.2	96.9	.46	10,184	13.5	9.9	18.06
1973	209.8	101.2	.48			10.55	18.91
1974	210.4					10.16	18.27

mpg - miles per gallon

Source: Ref. (16).

TABLE 3.2.   U.S. Transportation Modes – 1970.

	PM(10⁹) TM	% of Traffic	% of Energy for end use	Energy Mb/d	Energy 10¹⁵ Btu	% of Total Transport. Energy	EI Btu/PM or Btu/TM
Inter-city passenger Traffic							
Air	110	10	22		0.92	5.6	8,400
Rail	11	1	1		0.03	0.2	2,900
Bus	25	2	2		0.04	0.25	1,600
Auto	970	87	77	1.6	3.3	20.	3,400
Total	1,116	100	100	2.1	4.3	26.05	3,800
Urban Passenger Traffic							
Mass Transit	20	3	1	0.04	0.08	0.5	3,800
Auto	690	97	99	2.7	5.6	34.2	8,100
Total	710	100	100	2.7	5.7	34.7	8,000
Inter-city Freight Traffic							
Air	3.4	0.15	6		0.1	0.8	42,000
Rail	770	35	21		0.5	3.1	700
Truck	410	19	48		1.2	6.9	2,800
Waterways	600	27	17		0.4	2.5	570
Pipeline	430	19	8		0.2	1.2	450
Total	2,213.4	100	100	1.15	2.4	14.5	1,100
Urban Freight Truck			100	1.1	2.3	14.2	
Military Air				0.3	0.6	3.8	
Other				.6	1.2	6.8	
Total Transportation Energy				7.9	16.5	100.	

Source: Ref. (15).

Notes to Table 3.2.

TM - Ton-miles   PM - Passenger-Miles

EI - Energy intensiveness

Mb/d - Million barrels of oil per day equivalent.

Load factors: 49% was used for passenger air traffic.

               28% for air freight.

               For automobiles in urban use 1.4 persons per car, in inter-
               city driving 2.4 persons per car.

               37% for railroads.

B.   Conservation Options

The energy consumption of automobiles is so large and cars are used so inefficiently that the major impetus for conservation should be in this area.
Automobiles are individually owned and therefore difficult to regulate; airlines are easier to regulate if one wanted to. On the plus side is the
fact that the lifetime of an automobile is short compared to that of a house,
office building, or factory so that one might hope to replace existing automobiles with "better" ones in not too long a time. However a very long
production lead time gives the existing fleet of automobiles great
longevity and slows down the implementation of technical improvements. The
development of energy-efficient engines has been slow because of the constraints of low cost, light weight and, most important, acceptable emission
levels for hydrocarbons, CO and nitrogen oxides. We will attempt a more
systematic review of the conservation options available in the transportation
sector with the emphasis being placed on passenger transportation.

The expression for the energy used directly in transportation in terms of
conventional variables is

$$E = \sum_i PM_i/u_i e_i$$

where the sum over i is over modes of transportation and the variable, PM,
refers to the number of passenger-miles, u is the average load factor in
units of passengers per vehicle, e is the fuel efficiency in vehicle-miles
per gallon(or per Btu).

We can categorize conservation strategies as they apply to the elements of
this expression. Each strategy must then be evaluated as to its ultimate

impact, time scale, social and economic costs, environmental side-effects, level of technology required, etc.

Some conservation options are:

1.   Shifts in transportation modes. Likely possibilities to investigate might include (a) converting short air trips to bus or train; (b) converting urban car use to mass transit; (c) truck freight to rail freight; (d) inter-city auto trips to rail or air; (e) multi-mode transportation terminals.

2.   Reduction of demand (passenger-miles). (a) Economic disincentives to travel; (b) driving bans; (c) decreasing dwelling-work distances; (d) fuel rationing.

3.   Increase vehicle occupancy. (a) Improve load factors in flights; (b) encourage car-pools; (c) revise tolls, parking fees, etc.

4.   Improving fuel efficiency. (a) Speed reduction; (b) reduction of rolling friction and air drag; (c) increased efficiency of engine and transmission; (d) engine size reduction; (e) increased proportion of small cars.

The amount of government coercion required to achieve conservation objectives varies widely, especially in the transportation area so it is interesting to classify conservation strategies according to the level of government pressure required. A few examples in order of increasing government intervention:

1.   Informed publicity.

2.   Specific suggestions for conservation measures - car-pools, airline speed reduction.

3.   Positive incentives - Tax credit on efficient new cars, subsidies to mass transit.

4.   Indirect economic pressures - gasoline tax, toll schedules.

5.   Business practices laws - labeling laws, price controls.

6.   Fuel rationing.

7.   Prohibitions on individual activity - speed limits, Sunday closings, driving bans.

8.   Production or sales restrictions - fuel economy standards.

   C.   Strategies for Energy Conservation

We will discuss and present evaluations for several energy conservation

schemes. These may involve technical innovations, changes in habits, and
changes in costs of transportation. We will finally conclude that the
most effective strategies entail technical developments for automobiles and
mandatory fuel economy standards. Nevertheless it is useful to examine some
of the other conservation schemes that have been proposed to see what
roles they can play in reducing consumption in the near future. A useful
conservation scheme should reduce energy usage by an amount which is significant compared with the energy consumption of the transportation sector,
16.5 10^{15} Btu/year (or 16.5 Quads).

1. Shifts in transportation modes

When we consider mode changes as a strategy for conserving energy it is not
really sufficient to consider the differences in direct consumption of fuel
by the respective vehicles. Each transportation mode has associated with
it indirect energy costs; in the case of automobiles these include manufacture, road construction and maintenance, etc., and similarly for air or
rail travel. Some of the indirect energy cost may be difficult to assign
and furthermore in some applications one may want to consider only the
marginal indirect cost. Table 3.4 lists some estimates for the total energy
cost (direct plus indirect) of automobile and air travel. Because of insufficient data and some ambiguities we will in general only consider direct
energy usage.

TABLE 3.4. Total Energy Intensiveness

	- Btu/PM -	
	Direct	Total
Air (average)	8,400	11,200
Auto(IC)	3,400	5,800
Auto(IC) + driver only	8,160	13,920
IC - Inter-city	PM - Passenger-miles	

Source: Ref. (17).

(a) Short air trips. The most heavily traveled air route in the U.S. is
the New York to Boston run, only 191 miles long (See Table 3.5).

TABLE 3.5.   The three most heavily used short (<400 miles) air trips.

	dist(m)	pass. (10^6)	PM(10^9)	% of Total PM	Energy (10^15 Btu)
NY - Boston	191	2	0.395	0.4	0.0047
NY - Washington	215	1.7	0.370	0.3	0.0044
LA - San Francisco	355	0.89	0.304	0.3	0.0026

Source: Ref. (18).

The total direct energy used in the New York to Boston run is only 0.005 Quad. or about 2,000 barrels of oil per day, compared to the total transportation energy of 16.5 Quad. Even if most of this could be saved by encouraging bus or rail travel the amount of energy involved is completely negligible.

That the savings are small from converting short air trips to bus remains true even if several of the shorter routes are considered. The cumulative energy expenditure for the shortest trips is shown in Fig. 3.2. All trips of less than 400 miles, although they serve nearly one third of all passengers consume only 10% of the air travel energy budget.

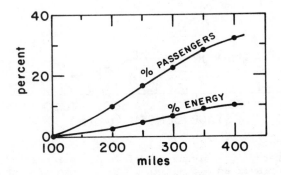

Fig. 3.2. The percentage of air passengers making a trip shorter than a specified number of miles. Also the fraction of airline energy consumption directly used for trips shorter than specified. Source: Ref. (18).

On the other hand, if short stages of longer flights were also curtailed the savings might be worthwhile. At present short flights are subsidized and moves are afoot to raise their cost to reflect true operating cost. Pilati[19] estimates that if half the short flight stages under 200 miles

were shifted to buses then a net energy saving of about 6% of the airplane
sector fuel usage could be saved. Such a cutback, however, might have the
effect of reducing airplane load factors, since short hops are scheduled
for traffic reasons. The impact of multi-mode transportation terminals
needs to be investigated.

Eliminating or raising the cost of short air trips carries with it the
danger that the mode switch will be to automobiles. The automobile occupancy
for persons who might have flown is probably 1 per car so that the automobile energy intensiveness is about 8,160 Btu/PM. Since the energy intensiveness for air travel increases for short flights (see Fig. 3.3) we
are probably slightly better off driving but if total energy cost is considered the car is worse and we have increased road traffic in addition.

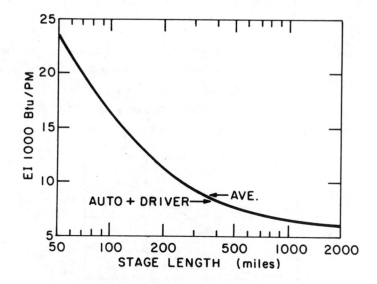

Fig. 3.3. The energy intensiveness (EI) in Btu per passenger-
mile(PM) for air travel as a function of the length of a stage of
a flight (the distance between take-off and landing). This represents an average for various aircraft in common use. The average
value for all trips is 8,400 Btu/PM. Source: Ref. (19).

     (b) Converting urban automobile use to mass transit. Only about 3%
of urban travel currently uses mass transit; of this about a third are
passengers on subway systems. If the number of bus riders could be doubled
and if this increment came entirely from auto users then the maximum possible

savings would be 0.04 Mb/d or about 0.08 Quad. Since about 34% of vehicle-miles are used to commute to work, it is attractive to consider a massive shift to public transit. Since load factors are usually very high during rush hours this would entail major expenditures for new equipment. Such an expanded transit system would involve many social and institutional changes and is unlikely to be developed in the near future. Although shifting to mass transit may become more attractive when the price of fuel goes up, the cost of gasoline and even the variable costs of driving are not large compared to the cost of using mass transit at present. See Table 3.6.

TABLE 3.6.   Cost of operating a 1973 Compact Automobile.

	cents/mile	% of cost
Fixed costs	6.22	48
Gasoline	3.26	25
Other variable costs	3.37	26
Total	12.85	100.
Cost of Mass Transit	8.3	

Source: Ref. (10,15)

Past experience has shown that modest improvements in rapid transit facilities do not reduce automobile use very substantially. New transit riders are former walkers, car-poolers, etc. In fact, expanded mass transit may sometimes make outlying regions of a city more attractive as places to live and may result in increased automobile use. Improved mass transit must be accompanied by disincentives for driving: restricted parking, high fuel costs, auto-free regions, etc. Of course mass transit may have many beneficial side effects such as reduced pollution, fewer traffic jams, more attractive cities, etc.

(c) Rail shipment of automobiles. This is being done across Canada and also between Washington, D.C. and Florida. The energy savings per car appear to be substantial since the average energy intensiveness for rail is 700 Btu/ton-mile or about 1,000 Btu per automobile mile as compared with 8,160 Btu per vehicle-mile when driving. The passengers are presumably carried on the same train at a very low incremental energy cost. Such a

scheme could make a significant dent in the inter-city auto energy usage if
convenient terminal facilities were available and if car carriers could be
made compatible with railroad tunnels.

2. Reduction of Demand

(a) One day per week driving ban. It has been estimated that this
would reduce automobile travel by about 5% at most, accomplishing a savings[20]
of about 0.2 Mb/d or about 0.4 Quad. If such a ban were instituted, its
effectiveness might be short-lived as people learn to shift travel needs to
other days. The administrative machinery needed to enforce the ban and to
allow for exceptions would be cumbersome. Closings of service stations on
Sundays had a significant effect when first instituted but is expected to
produce a negligible reduction in driving if it were reestablished. Vacation
travel represents only 4.7% of the Sunday vehicle-miles of travel; other
Sunday trips add up to a total of about 33 miles of travel which, of course,
is easily accomplished with one tankful of gasoline. Assuming that all
vacation travel is affected but is not shifted to other days and also that
the trips exceed the range of the automobile the maximum expected savings
would be 0.03 Mb/d or about 0.06 Quad.

(b) Gasoline rationing. It could be arranged that rationing would
reduce the consumption by 1 Mb/d or about 2 Quads. Such a plan[21] would
allocate about 36 gallons of gasoline per month to each registered driver
(432 gallons per year), whereas the present usage is about 50 gallons per
month. The implementation of rationing could cost nearly 2 billion dollars
and require 17,000 employees. This system could result in serious inequities to the disadvantage of low-income groups since ration coupons would
be traded on the open market. Procedures for distribution of coupons could
be found to reduce the inequities. A rationing system with a 1 Mb/d reduction in gasoline consumption would enable the U.S. to reduce the import of
crude by 1.5 Mb/d, but products(fuel oil) might have to be increased by about
1/2 Mb/d. On the whole this may be one of the surest solutions in the short
term, especially when combined with fuel economy standards and augmentation
of the public transportation sector.

(c) Manipulation of price. The plan[12] of the present administration,
the Energy Independence Act, involves a $2 per barrel import fee on petroleum,
a 37 cent per thousand cubic feet tax on domestic natural gas, decontrol

of domestic petroleum and gas prices, a profits tax on domestic petroleum
production, and a rebate to consumers of the fees and taxes that are
collected. The effect of these proposals would be to increase the price of
gasoline from about 52 cents per gallon to 62 cents (19% increase) and the
price of heating oil from 37 cents per gallon to 47 cents (27% increase). A
demand model has predicted that the short term effects of the tax program
would be to reduce oil imports by 1.6 Mb/d (3.2 Quads) in 1977; the projected
imports would be about 8 Mb/d if no new action were taken. Imports may be
controlled to reach this goal.

The plans include rebates to offset the disadvantage to low-income families.
Poor families with an income below $3,000 spend 5% of their income on gasoline
while the well-off (income greater than $15,000) spend 2.2%, corresponding
to a household consumption more than four times as large. See Table 3.7.

TABLE 3.7 Gasoline Consumptions for Income Levels (1972)

Income Group	No. of hslds 10^6	Avg. no. of cars per hsld.	Annual Expend. on gas & oil $	% of income spent on gas & oil	% of total gas energy used.
$3,000 & less	8.6	0.4	$96.	5.14	3.8
3,000 - 6,000	10.4	0.8	198.	4.39	9.5
6,000 - 10,000	13.4	1.3	312.	3.89	19.4
10,000 - 15,000	13.9	1.6	410.	3.32	26.4
over 15,000	13.4	1.9	499.	2.24	31.0
income not reported	11.6		182.		9.8
Total	71.2	1.17	302.	4.24	100.

Source: Ref. (20,22).

The rebates would be adjusted so that the poor and lower middle income groups
benefit slightly and the highest group suffers a slight disadvantage from the
increased fuel prices. The lower income groups are very minor consumers of
gasoline compared with the others although the price elasticity in their case
might be greater. A significant decrease in consumption will have to come

from the high income groups since they are the major consumers. In addition
to the increased direct cost of fuel many goods are expected to rise in cost
as well.

Rationing or allocation will have unequal impacts on different parts of the
country since gasoline consumption varies widely between large urban centers
and more sparsely populated regions. See Table 3.8.

TABLE 3.8. Gasoline Consumption in several regions of U.S.

	gal/person/year	% of U.S. average
New England	302.9	82
Middle Atlantic	282.2	76
West Coast	385.4	104
Mountain States	427.1	115
National Average	370.	

Source: Ref. (22).

e.   Increasing Vehicle Occupancy

(a) Improving load factors in flights. Passenger load factors decreased from nearly 70% in 1950 to about 50% in 1970. The load factor rose
again to about 59% in mid-1974 following curtailments in flight schedules
by many airlines. Since the incremental fuel usage per passenger is quite
small the fuel consumption is inversely proportional to the load factor. It
has been estimated that a load factor of 70% would result in essentially no
further inconvenience to passengers. A change of this size could save about
0.25 Quad. of fuel annually in direct consumption. It would also reduce indirect consumption of energy but might have an adverse effect on employment.

(b) Increasing automobile occupancy. An increase in urban automobile
occupancy for trips to work (which comprise 34% of automobile mileage) might
result in a significant overall increase of the urban load factor. See
Table 3.9.

TABLE 3.9.   Automobile Usage Patterns

Purpose	Avg. Trip Length (miles)	Avg. Occupancy	% of Vehicle-miles	% energy consumed
Work (commuting)	9.4	1.4	34	36
Work-related	16.	1.6	8	7
Shopping	4.4	2.0	7.6	9
Other family business	6.5	1.9	17	19
Vacations	165.	3.3	2.5	2
Social, recreation	11.6	2.5	31	28
			100.	100.

Source: Ref. (23).

The car occupancy rate for trips to work is below 1.4 and perhaps could be raised to 2.0 with massive efforts to promote car-pools. A change of this magnitude could result in a saving of nearly 0.5 Mb/d or about 1 Quad. Although it might be possible to improve the load factor for a short time, the improvement is unlikely to persist. Even during the gasoline shortage of 1973–74 the car occupancy rate changed by less than 5% for a short time. Nevertheless encouraging car-pools with favorable tolls, lanes, parking rates, etc. would have beneficial side effects. Legislation has been proposed to give car-pool drivers tax benefits.

4.   Improving Fuel Efficiency

(a)   Speed reduction. The energy utilization in automobiles can be attributed to various loss mechanisms. Fig. 3.4 assigns such losses for a car operating in the Environmental Protection Agency (EPA) composite Urban and Highway test driving cycles. The loss due to air drag is important at speeds above about 35 mph while rolling friction is the dominant frictional loss mechanism below this speed for a 2,500 pound compact car. For heavier cars in the 5,000 pound class the speed at which the two loss mechanisms become equal is about 41 mph. The fuel economy as a function of speed is shown in Fig. 3.5.

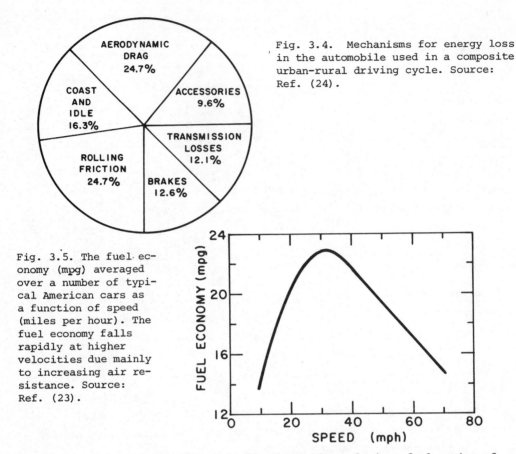

Fig. 3.4. Mechanisms for energy loss in the automobile used in a composite urban-rural driving cycle. Source: Ref. (24).

Fig. 3.5. The fuel economy (mpg) averaged over a number of typical American cars as a function of speed (miles per hour). The fuel economy falls rapidly at higher velocities due mainly to increasing air resistance. Source: Ref. (23).

A reduction in speed from 70 mph to 55 mph would result in a fuel saving of about 18%. To take advantage of the savings possible at reduced driving speeds a nationwide 55 mph speed limit was put into effect early in 1974 superseding state-controlled speed limits ranging from 60 to 75 mph. Prior to the imposition of the lower speed limit 56% of vehicle-miles were traveled at speeds above 60 mph on major roads. Immediately afterwards the corresponding fraction was 40% but the average speed has slowly risen since then because of lack of enforcement and decreasing energy-consciousness on the part of the public.

The total number of vehicle-miles traveled on major rural roads is about 400 10^9 per year leading to an expected gasoline saving of about 2 10^9 gallons per year based on realistic speed distributions[23]. With a total motor vehicle consumption of about 10^{11} gallons per year (1973) we have then effected

a 2% savings in motor vehicle energy use or about 0.4% (0.3 Quads) of total
U.S. consumption of energy. The actual savings that could be attributed to
the lower speed limit was indeed of the order of 2 to 3% immediately
following the enactment of the 55 mph limit but by the second quarter of 1974
the savings (attributable to speed reduction) were eroded. The total
gasoline consumption dropped by a larger amount but this was mainly due to a
reduced amount of driving (presumably because of difficulty in buying
gasoline). The reduced speed limit seems to have been associated with an
entirely different beneficial result, namely a 19% decline in traffic fatalities per mile of travel. A substantial decrease in fatalities has persisted
despite the gradual increase in vehicle speeds. This large decline could,
in part, be attributed to the narrowing of the vehicle speed distribution and
to a non-linear dependence of accident rate on traffic volume. The overall
conclusion reached by a detailed study[23] is that while the reduced speed
limit was a "contributing factor" in the reduction of fatalities it was "not
a major factor" in the reduction of gasoline consumption.

(b) Technical improvements in cars and engines.

The inefficiencies in the internal components of a typical automobile have
been described by Kummer[25]. A fuel economy of 200 miles per gallon could
be achieved if all of the available work[3] of the fuel could be used to
drive the rear wheels of the car. Table 3.10 lists the factors contributing
to the inefficiency of the automobile. Only 14% of the available work is
delivered to the rear wheels to be used in overcoming rolling friction and air
resistance. In fact the 14% efficiency has not taken into account energy
required to run accessories (air-conditioners are a particularly heavy load).

The energy delivered to the rear wheels is all dissipated in one of three
ways: rolling resistance, air drag, brakes. The proportion lost in each of
these is very dependent on the type of driving cycle. In urban driving, for
example, about 32% of the energy is dissipated in braking and almost none in
air resistance.

From this table we see that there are several general areas where improvement could result in significant energy saving: engine, transmission,
rolling resistance, air drag. Regenerative braking is an attractive idea but
still requires much development.

TABLE 3.10.   Sources of inefficiency in the automobile engine and
related components.   Urban driving cycle; 2,500
pound automobile with 2 liter, 4-cylinder engine.

Source of Inefficiency	Efficiency Factor	Percent of Energy Available
Fuel Available Work		100.
Heat of Combustion	0.96	96.
Ideal gas, air, Otto cycle engine (comp. ratio= .8)	0.56	54.
Fuel, air Otto cycle (stoichiometric)	0.75	40.
Burning, cylinder wall, exhaust losses	0.8	32.
Frictional losses, pumping of air, exhaust	0.8	26
Engine running at partial load	0.75	19.
Automatic Transmission	0.75	14.
Output to Rear Wheels		14.

Source: Ref. (3,25).

A report[24] issued by the Department of Transportation and the Environmental
Protection Agency has considered realistic improvements that could be implemented by 1980 or 1985 and still meet emissions standards.

Table 3.11 and Fig. 3.6 summarize what could be expected for 1980 vehicles
with an "improved engine" and with four-speed automatic transmission, radial
tires, weight reduction, air drag reduction, and improved accessories.

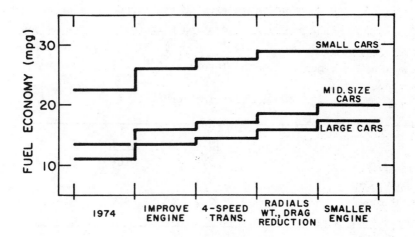

Fig. 3.6. The effect on fuel economy (miles per gallon) of various
improvements to automobiles that could be implemented by
1985. Typical values for 1974 cars in three size classes
are also shown. Source: Ref. (24).

TABLE 3.11. Fuel economy expected from several technical improvements by
1980. Miles per gallon figures for a composite EPA driving
cycle.

System	Fuel Economy (miles per gallon)		
	Large Size	Mid-Size	Small Size
Baseline(1974)	10.7	13.1	22.3
1. Improved Engine	13.4	15.7	25.6
2. System 1 + 4-speed auto. transmission	14.3	16.9	27.6
3. System 2 + radial tires + wt.red. + air drag red. + acces. improvement.	15.7	18.5	28.8
4. System 3 with engine resized.	17.3	19.8	28.8 (same size)

Source: Ref. (24).

The largest gains are obtained by improving the Otto cycle engine with
modulated fuel injection, programmed exhaust recirculation and a double
catalyst system.   Reducing the engine size on the larger cars saves 10 to 15%
of fuel at the expense of poorer acceleration performance.   The weight
distribution of cars produced in 1974 is shown in Table 3.12 along with a more
desirable weight distribution for the 1980 model.

TABLE 3.12.   Weight Distribution of Cars.

	1974	1980
large size	27%	10%
medium size	45%	50%
small size	28%	40%

Source: Ref. (24).

Using the 1974 production distribution the weighted fuel economy should be
20.3 mpg by 1980 as compared with 14.0 in 1974.   If the sales size distribution can be changed to the more desirable mix shown above the fuel economy
would be 22.2 mpg by 1980, corresponding to a fuel consumption only 63% that
of 1974.

By 1985 one can expect to see the impact of some newer technologies: the
introduction of lightweight improved Diesel engines and stratified charge
engines.   Table 3.13 specifies some automobile improvement scenarios for the
1980's and Figs. 3.7 and 3.8 show the reduction in automobile fuel consumption expected with each of these scenarios. The gains described here could
be achieved with the aid of direct government specification of fuel economy
standards.   These standards could be applied to each size category or else
the average fuel economy of the entire fleet of cars produced by a manufacturer could be specified.   An alternate scheme would be to tax new car
sales according to fuel economy, alternatively an annual tax could be imposed
on all cars encouraging replacement of older, less efficient cars with new
ones.   To realize the various scenarios listed above would require different
amounts of government intervention and different types of incentives.

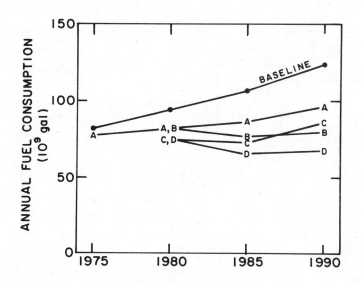

Fig. 3.7. Annual fuel consumption for automobiles (in units of
10^9 gallons of gasoline) for various improvement scenarios.
See text and Table 3.13. The "baseline" case represents
no improvements beyond 1975. Source: Ref. (24).

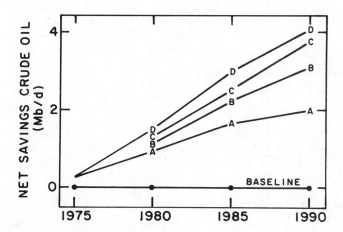

Fig. 3.8. Net savings of crude oil over the "baseline" case (in
units of million barrels per day) for various improvement
scenarios. See text and Table 3.13. Source: Ref. (24).

TABLE 3.13. Scenario Summaries

Scenario	Percent Gain in Fuel Economy (mpg) 1980	1985	Fuel Economy Improvements
Baseline	0	0	No improvements in fuel economy relative to 1974 vehicles. Minimum changes to meet statutory emission standards.
A. Modest Improvement	28%	27%	Optimized conventional engines, radial tires, slight weight and aerodynamic drag reductions (in line with announced industry goals). No improvements after 1978.
B. Gradual improvement thru 1980's.	33%	52%	Steady technological improvement through the 1980's: Weight reduction through materials substitution and minor redesign during the 1970's; further changes (unitized body) in the 1980's. Some aerodynamic drag reduction and substantial transmission improvements full accomplished by 1984. Diesel engines phased in for larger cars from 1981 to 1989 plus some stratified charge engines for smaller cars. No performance degradation.
C. Maximum improvement by 1980.	43%	44%	Maximum rate of improvement through 1980 with little further gain during the 1980's. Rapid weight reduction, aerodynamic drag reduction, and transmission improvements. Displacement reduction of optimized conventional engines, but no diesel or stratified charge engines.
D. Scenario B plus shift to smaller cars.	63%	84%	Same as B with 1980 sales mix assumed at 10% large cars, 25% intermediates, 25% compact, and 40% subcompact.

Source: Ref. (24).

4.   INDUSTRY

A.   Patterns of Consumption

The energy consumption by the industrial sector in 1974 was 29.3×10^{15} Btu out of a total U.S. consumption of 73.1×10^{15} Btu or 40% of national energy consumption. The 1974 preliminary figures shown a 0.9% decline in this sector compared with 1973.

The industrial sector in the U.S. has been divided into 20 major industrial groups (these are called "Standard Industrial Classification" groups or SIC's). Six of these SIC groups account for about 2/3 of industrial consumption of energy. Table 4.1 and Fig. 4.1 show the energy consumption of these large users. As usual, purchased electrical energy has been converted to consumed energy at the rate of 10,400 Btu per kWh to account for generating efficiency.

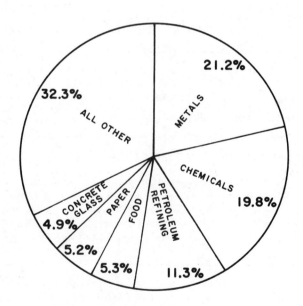

Fig. 4.1.   Energy consumption in the industrial sector for
            1968 subdivided according to industry.
            Source: Ref. (4).

TABLE 4.1. Energy Consumption of Major Industrial Groups - 1968

SIC Group	Industry	Energy 10^{15} Btu	% of sector energy	% of U.S. energy consumption
33	Primary metals	5.30	21.2	8.7
28	Chemicals & allied products	4.94	19.8	8.2
29	Petroleum refining	2.83	11.3	4.7
20	Food & kindred products	1.33	5.3	2.2
26	Paper & allied products	1.30	5.2	2.1
32	Stone, clay, glass, concrete.	1.22	4.9	2.0
	All Other	8.05	32.3	13.3
	TOTAL	24.96	100.	41.2
	Total U.S. Consumption	60.58		

Source: Ref. (4).

Industrial energy use can also be broken down according to the mode of energy use rather than product type. Both types of subdivision are needed to assess areas in which conservation efforts are worthwhile. Table 4.2 summarizes industrial energy use according to type of process.

TABLE 4.2. Industrial energy consumption by process - 1968.

Process	Energy 10^{15} Btu	% of Sector Energy	% of U.S. Energy Consumption
Process Steam	10.13	40.6	16.7
Direct Heat	6.93	27.8	11.5
Electric Drive	4.79	19.2	7.9
Feed Stock	2.20	8.8	3.6
Electrolytic processes	0.71	2.8	1.2
Other	0.20	0.8	0.3
TOTAL	24.96	100.	41.2
Total U.S. Consumption	60.58		

Source: Ref. (4).

Since they form substantial fractions of U.S. energy consumption one should
question whether the heating and steam handling processes in industry are
being carried out with high efficiency. It has been the recent experience
of companies that have looked into this that savings of 20% can be achieved
with relatively minor improvements in procedures and monitoring. In the past
energy conservation was not one of the major concerns of industrial
managers; capital costs and especially labor costs were more significant. A
tabulation of the fuel types employed by industry shows that there is a
heavy dependence on petroleum and natural gas and hence reason to worry about
future energy supplies and costs. See Table 4.3.

TABLE 4.3.   Energy Consumption by Fuel Type - 1974.

fuel	Energy consumption 10^{15} Btu	% of total sector energy
Anthracite	0.03	0.1
Bituminous coal & lignite	4.18	14.3
Natural Gas	11.13	38.0
Petroleum	5.83	20.0
Hydropower	0.03	0.1
Electricity(purchased)	8.12	27.7
TOTAL	29.3	100.

Source: Ref. (1).

The rate of growth of energy use in the industrial sector has declined, at
least temporarily. Of course, it is very dependent on the general state of
the economy which makes extrapolations unreliable. Table 4.4 shows the
gross annual energy consumption in the industrial sector over the past 15
years. Fig. 4.2 shows the per capita consumption of steel and cement in
both the U.S. and Western Europe for this century. In recent years the
growth rate in Europe has exceeded that in the U.S. for these two key
products.

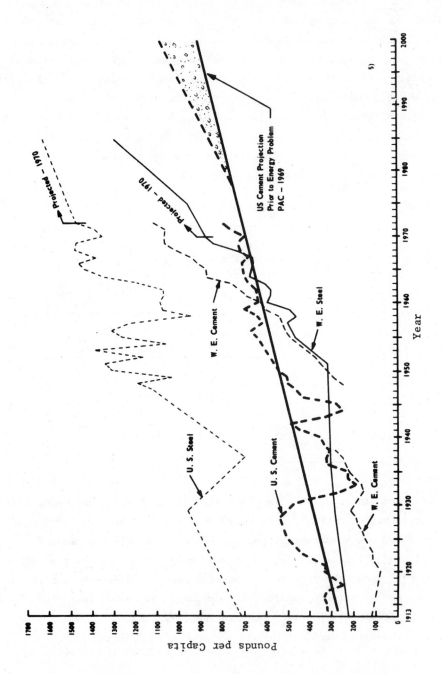

Fig. 4.2.   The per capita consumption of steel and cement (in pounds per person) in the
            U.S. and Western Europe. Recent growth rates in Europe exceed those in the
            U.S.. Source: Ref. (26).

TABLE 4.4.   Energy Consumption in the Industrial Sector since 1960.
GNP is in constant (1958) dollars.

	GNP 10^{12}	Energy Consumption 10^{15} Btu	Energy per GNP$ 10^3 Btu/$
1960	0.488	18.6	38.1
1970	0.723	27.0	37.3
1971	0.745	27.0	36.2
1972	0.791	28.1	35.5
1973	0.837	29.5	35.2
1974		29.3	

Source: Ref.(7,10)

B.   Energy Intensive Industries

The energy and labor intensiveness of an industry can be displayed by plotting
the energy consumed per dollar of product against the employment per dollar.
This is plotted for a few industries in Fig. 4.3.

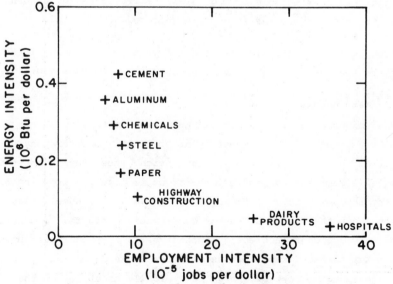

Fig. 4.3. The Energy Intensity(10^6 Btu per dollar) for several industries plotted against the Labor Intensity (10^{-5} jobs
per dollar of product).   Source: Ref. (27).

Although one might have supposed that the U.S. cement industry, being one of
the most energy intensive, might have made great efforts to reduce its energy
consumption by implementing new processes, in fact, here as in many other
industries the main effort has gone into reducing labor costs. Since 1950
labor costs have increased by about a factor of three and at the same time
the labor intensiveness (man-hours per ton) has dropped by the same factor.
Industrial processes have simply not been optimized for low fuel consumption
during the era of rapidly rising wages. Energy costs are only about 14% of
the selling price of cement. On the other hand some industries, for example,
the aluminum and the steel industries, have been somewhat more aggressive in
trying to cut energy costs by developing new processes and procedures.

In the past year or two a major chemical corporation[28] claims to have reduced its energy consumption per unit of output by 20% by careful monitoring
of energy flow in chemical processes, by improved day-to-day operations, by
reusing energy rich waste materials, and by general careful housekeeping. This
program has taken a massive effort on the part of management and some fairly
new technology; infra-red cameras, heat recuperation devices, etc. It may be
that the big rewards will come from generally improved energy husbandry
rather than from new processes in certain selected industries.

We will look a little more closely at a few of the most energy intensive
industries to see where energy is used and how it could be saved.

   1. Primary Metals

Steel and aluminum production dominate the primary metals subsector in terms
of both energy consumption and quantity of product. Together they account
for over 7% of the total U.S. energy consumption, steel taking about 6% and
aluminum, 1%. The aluminum production process is highly energy intensive
and furthermore uses energy in a very "high grade" form, namely electricity,
and has thus been the target of energy conservation efforts. Similarly
steel production, though less energy intensive, is often carried out with
inefficient furnaces and repeated wasteful heating and cooling cycles. For
steel and aluminum the cost of energy is typically about 10% of the selling
price. We will discuss energy use in steel and aluminum production
separately in order to assess the impact of improved processes and manufacturing procedures.

Steel. In Table 4.5 we list the major energy consuming steps in steel
production.

TABLE 4.5. Energy Consumption in Steelmaking (1970).
The energy required to make one ton of raw steel is shown.

Production step	Energy per ton 10^6 Btu/ton	% of energy
Mining, beneficiation of ore	1.34	7.0
Coke production	3.51	18.3
Pig Iron production (blast furnace)	5.14	26.7
Steel Production	1.95	10.1
General Utilities for mill	5.10	26.5
Other steps	2.18	11.3
TOTAL	19.22	100.

Source: Ref. (29).

In addition to the energy requirement shown in Table 4.5 approximately
17 10^6 Btu per ton of finished metal are required for mill processing. The
total production of raw steel in 1973 was 150 10^6 tons with mill shipments[10]
amounting to about 110 10^6 tons. Consequently the total energy consumed in
steel making through the milling stage is then about 4.8 10^{15} Btu or about 6%
of U.S. energy consumption with an energy intensiveness of about 40 10^6 Btu
per finished ton.

Major improvements have taken place in the steel production process as the
old open hearth method has been replaced by the basic oxygen process. The
energy requirements in these processes are:

open hearth process   3.4 10^6 Btu/ton
electric furnace   6.6 10^6 Btu/ton
basic oxygen process   0.7 10^6 Btu/ton

Fig. 4.4 shows the quantity of steel produced with each of these processes.

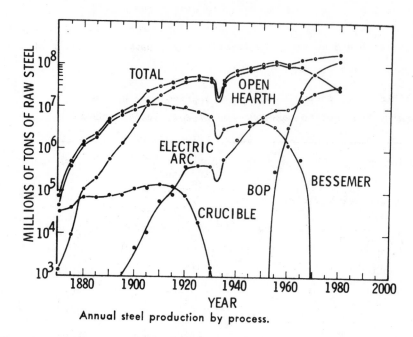

Annual steel production by process.

Fig. 4.4.   The production of steel in the U.S. using various
types of furnaces. The basic oxygen process (BOP) is
replacing the more energy intensive open hearth process.
Source: Ref. (30).

The basic oxygen process has begun to dominate the industry. In 1973 it
accounted for 55% of steel production while the open hearth was employed in
27% of production and the electric furnace in 18%. The basic oxygen process
has a serious disadvantage in that it can tolerate much less scrap than the
old open hearth process. The processing of scrap steel requires one quarter
to one half as much energy as is needed for primary metal production. In
current practice the electric furnace can be used for scrap processing while
the basic oxygen process is used for primary metal. It has been estimated [31]
that $27 \cdot 10^6$ tons of steel could be recycled at a saving of $13 \cdot 10^6$ Btu/ton
or a total saving of $0.35 \cdot 10^{15}$ Btu/yr. (0.5% of U.S. consumption) Recycling
on such a large scale would require government intervention and economic
incentives to overcome the technological problems involving separation of
materials and transportation of scrap.

There are savings that could be achieved with waste heat recovery from furnaces, etc. and while large plants take advantage of exhaust fuels and gases the heat content may be low, making recuperation difficult or un-economic for smaller operations.

An important improvement in steel making procedures has been the introduc-tion of continuous casting now used in production of about 23×10^6 tons (in 1974). Cooling and reheating of steel prior to rolling is eliminated in this process saving nearly half of the total processing energy.

Aluminum. Most of the energy utilized in aluminum production is for electrolytic reduction of aluminum oxide. The electrical energy demand is 16,500 kWh_e per ton ($\sim$ 8 kWh per pound) of which about one-third is supplied by hydro-electric power. If we convert this electrical demand to total energy consumption we require 133 10^6 Btu/ton for electrolysis.

The total energy required for aluminum production and processing amounts to about 160 10^6 Btu per raw ton. The production in 1973 was 5.5 10^6 tons for a total production energy of 0.9 10^{15} Btu or 1.2% of U.S. energy consumption. Milling and finishing of aluminum adds a small additional energy to bring the total energy intensiveness to about 200 10^6 Btu per finished ton.

Only 5% of primary production energy is required to process scrap aluminum so a substantial saving is possible. At present about 18% of processed aluminum is recycled, corresponding to about 1 million tons per year. Fig. 4.5 is a material flow diagram for aluminum production. It is estimated that perhaps twice that amount is potentially available. Aside from problems of collection, transportation, and metals separation there have been problems during processing with oxidation of aluminum can scrap and with contamination by other metallic components such as magnesium. A very op-timistic estimate of the potential energy savings from recycled aluminum is 0.27 10^{15} Btu/yr or less than 0.4% of U.S. consumption.

A new smelting process under development by Alcoa promises to use 30% less energy than the most efficient cells now in use which produce aluminum at the rate of 6.5 kWh per pound. This will have a dramatic effect on energy use in the aluminum industry but not such a large impact on total U.S. consumption.

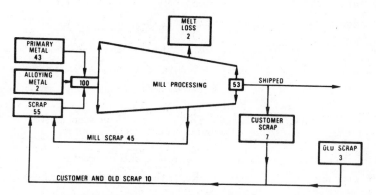

U.S. ALUMINUM INDUSTRY FLOW DIAGRAM

Fig. 4.5.   Material flow in the Aluminum processing industry.
            Of particular interest is the fraction of scrap that
            is recycled. Mill scrap is generated within the plant
            from rejects, trimmings, etc. Old scrap includes
            aluminum cans, automobile parts, etc. Source: Ref.(31).

2.   Cement

Energy consumption in the cement industry accounts for about $0.60 \ 10^{15}$ Btu/yr
or about 0.8% of total U.S. energy consumption. There are two processes in
use in the U.S. with slightly different energy intensiveness. The "wet"
process is used for about 61% of production and requires $8 \ 10^{6}$ Btu per ton
while the dry process uses $7.3 \ 10^{6}$ Btu per ton and accounts for the remainder
of the total U.S. production of about $85 \ 10^{6}$ tons (1972). It has been
estimated that a reduction of energy intensiveness of about 25% should be
possible. Heat transfer efficiency has been improved in at least one installation[26] yielding production at only $4.9 \ 10^{6}$ Btu/tons. Waste heat and
air infiltration can be reduced with a savings of perhaps $0.2 \ 10^{6}$ Btu/ton. A
much larger saving would be possible if the amount of water in the input
"slurry" could be reduced since between 20% and 25% of the process heat is
used to evaporate water in the feed. In the longer term energy savings will

be obtained with new types of more efficient kilns which are already in use in Europe and Japan and require only ∿3 10^6 Btu/ton. Progress in energy conservation has been slow up till now in this industry, as in many others, because of the relatively high cost of labor and low cost of energy. Although energy conservation progress in the industry is now expected to be substantial due to the increasing cost of fuel, it will not have a big effect on total U.S. consumption.

3. Food

The direct energy use in farming was about 1.2 10^{15} Btu in 1970[32] while the total energy that could be attributed to farming was about 2.1 Quads. This latter figure includes energy required for farm machinery, fertilizer, etc. In the same year the food processing industry directly consumed about 1.3 Quads and commercial and home cooking and refrigeration accounted for another 3 Quads. Commercial transport of food products account for 1 Quad. It has been estimated[32] that nearly 8.7 Quads or 13% of U.S. energy consumption go directly or indirectly into food production and preparation. The food processing industries themselves use about 2% of U.S. energy directly.

Some of the larger components of the U.S. food processing industry are listed in Table 4.6 along with their annual energy consumption.

TABLE 4.6. Energy Consumption of major sectors of the food
processing industry (1973).

Sector	Energy used 10^{15} Btu
Meat packing	0.11
Milk	0.08
Canned fruits, vegetables	0.05
Frozen fruits, vegetables	0.05
Animal feed	0.08
Beer	0.07
Bread, cake	0.07

Source: Ref. (33).

About half the energy is consumed in the form of natural gas and about 28% in the form of purchased electricity, a component which is growing.

The energy input to the entire food system has grown by about a factor of three since 1940 to the point that about 10 calories are required to prepare each calorie consumed. At the same time the farm labor input has dropped sharply; at present about 2% of the population is engaged directly in farm work but many others are in support and processing industries. Since food processing is mainly a consumer of low-grade (low temperature) heat it seems clear that energy savings are possible through better monitoring and perhaps utilization of waste heat from other processes.

    4.   The automobile industry.

Two   estimates[34,35]   have been made of the energy input to the manufacture of an automobile. The estimates range from about 100 to 150 10^6 Btu per automobile. Although one obviously has to look in detail at the manufacturing process to identify potential savings, this number is interesting since it bears on the energy savings if the number of automobiles per person were to decrease. Even more important in understanding the impact of the automobile on the American economy is the estimate[34] of the total energy per year consumed on all automobile related products and services including fuel, highways, etc.; this turns out to be 10.5 10^{15} Btu or nearly 15% of the total U.S. energy consumption.

5.   ELECTRICAL UTILITIES

The electrical utility sector is a major user of fossil fuels. In 1974 this sector used 19.6 10^{15} Btu of energy or nearly 27% of total U.S. energy consumption. Table 5.1 shows the energy used in electrical generation according to source. Hydropower has been converted at the average efficiency of fossil fuel plants (10,389 Btu/kWh$_e$) and nuclear-generated electricity at the rate of 10,660 Btu per kWh$_e$ according to the Bureau of Mines[1] procedures.

TABLE 5.1. Source of energy used in Electricity Generation in the U.S.

(1974)

source	energy used 10^{15} Btu	% of total	electricity generated 10^{12} kWh$_e$
Coal	8.67	44	0.834
Gas	3.33	17	0.321
Oil	3.45	17	0.332
Hydropower	3.02	15	0.291
Nuclear Power	1.17	6	0.110
TOTAL	19.64	100.	1.888

Source: Ref. (1).

In Table 5.2 is shown the installed capacity of U.S. generating plants and also the quantity of electricity generated in 1973.

TABLE 5.2. Installed capacity of generating plants. (1973)

type	Installed Capacity 1,000 MW	Electricity Generated 10^{12} kWh$_e$	Fraction of capacity actually generated %
Nuclear	21.07	0.083	45.13
Conventional Steam	351.23	1.489	48.38
Internal Comb. + gas turbine	4.91	0.0062	14.43
Hydropower	61.28	0.271	50.49
TOTAL	438.49	1.849	48.14

Source: Ref. (36)

One of the serious problems that faces the utilities is the growth of severe peaks in the demand, especially during the summer. Depending on location some utilities have peak demands during the winter but others have recently experienced increased severity in summer peaking due to air-conditioners. The average national load factor is 62%; this refers to the ratio of average

load to peak load. During summer peaks in demand with about 6% of capacity
unavailable there remains a margin of reserve of only 20%; in certain
localities the margin of reserve is nil and the supply voltage is reduced
deliberately. An example of the effect of poor load factors on overall
consumption can be deduced from Table 5.3, where generating efficiency for
several types of generating plants is listed.

TABLE 5.3.   Generating Efficiency for Several Types of Plants.

type	fuel required Btu/kWh$_e$	Efficiency %	Fuel cost $/kWh$_e$
Nuclear	11,000	31	0.0031
Base load-coal	9,100	37	0.0084
Intermediate load-coal	14,900	23	0.0137
Peaking unit-oil	14,200	24	0.0267

Source: Ref. (36).

Peak power plants are less efficient and produce electricity at a higher
cost. Residential consumers in the U.S. are not charged a premium for
electricity during peak demand periods although production costs are higher.
It has been estimated that an improvement in load factor from 62% to 70%
would result in a saving of about 0.5 Quads due to decreased utilization
of low efficiency plants. An important benefit would be reduced capital
costs and energy expenditures for plant construction. Both realistic peak
pricing and thermal storage or other types of energy storage could result
in this saving.

6.   CONCLUSION

In many areas of significant energy consumption the potential savings on each consuming unit are quite large.   The most important examples are given in Table 6.1.

TABLE 6.1.   Potential Conservation Targets

	Fraction of total U.S. Consumption %	Potential savings Per Unit by 1985 %	Is new Technology Required?
Space Heating	16	∿30	no see Table 2.9
Automobile Use	14	∿40	very little see Table 3.13
Industrial Low Temp. Heat and Utilities	∿30	∿20	no
TOTAL	60	25	

These three categories together account for nearly 60% of total U.S. consumption of energy so the potential savings are enormous.   The catch is that in only a very small fraction of the houses, automobiles, and factories will conservation measures be implemented, at least within the next few years.   The reasons for this are that: (a) the stock of units turns over rather slowly; in the case of cars about 12 million are produced each year (about 12% of the stock) while in the case of dwelling units about 2 million new ones (3% of stock) are added each year; (b) Retrofitting may involve an initial outlay of money which private individuals are unwilling to spend because they either do not have it or do not perceive the long term savings that are possible; (c) builders and manufacturers may be unwilling to incorporate conserving features because of the added expense or increased sale price; (d) energy conserving measures in industry may entail additional capital equipment purchases and labor costs; (e) at this point energy costs are still not high enough to provide an incentive to conserve; gasoline consumption, for example, has not displayed a very large elasticity in the past year while its price was rising.

The present Administration has proposed a number of energy conservation
measures as part of a set of actions[12] aimed at making the U.S. "in-
vulnerable to import disruption by 1985".

Fuel economy standards. Standards for new automobiles will probably be
established by 1980 with the goal of improving performance by 40%. This is
contingent on compatibility with auto emissions standards but does not require
any very radical technological developments. This will save about 2 Quads in
1980 if implemented.

National Mandatory Thermal Efficiency Standards for Buildings. A Building
Energy Conservation Standards Act (1975) is proposed to set both prescriptive
and performance standards for new residential and commercial structures. The
plan is to set up prescriptive standards to be effective for an initial period
followed by carefully developed performance standards which would take effect
no later than 1980. For commercial buildings performance standards would
come into effect about two years sooner. It is predicted[12] that by 1985
this legislation will have affected 11% of the stock of houses and 26% of
commercial buildings then standing. The expected savings of oil imports in
1985 is 0.75 Quad from the residential sector and 0.49 Quad from the commercial sector for a total of 1.24 Quad or about 1% of the "Historical Growth"[2]
prediction for demand in 1985.

Winterization Assistance Act (1975). This administration proposal would
provide money to retrofit the homes of low-income families with insulation,
weather stripping, and plastic sheet for windows (not a very attractive
solution to the heat loss problem). The amount of money per family would
range from $40. to $175 depending on the climate zone. Sufficient funding
was requested to retrofit a total of 1.7×10^6 low-income dwellings over a four-
year period. This will clearly be of some benefit to the families whose home
is being insulated but the annual energy saving in 1985 will be about 0.04
Quad or about 0.03% of the pessimistic prediction of demand for 1985. Since
this measure would reach only about 2% of the residences the extremely small
saving is not a surprise.

Residential Conservation Tax Credit. This would provide a 15% tax credit for
home improvements leading to increased thermal efficiency. The maximum credit
over a three-year period would be $150. The savings anticipated from this
incentive would be about 0.6 Quad by 1985.

These examples demonstrate that incomplete or voluntary measures can save
little in comparison with legislation which attacks the whole problem at the
source; the manufacturing or construction stage. Taken by themselves these
conservation measures do not seem to be at all adequate to reduce consumption
from the "historical growth" curve to levels which we can comfortably hope to
supply with sources that could be available in 1985.

It is evident that government pressures, both economic and legal, are
essential in order to take advantage of the conservation potential which
already exists with current or imminent technology. The strain on resources
and on the environment imposed by the "historical growth" rate of consumption
would be impossibly great. An effective conservation program notwithstanding,
new energy sources will have to be developed and research on them will have to
be encouraged. Along with vigorous energy conservation programs and a search
for new sources there will need to be plans for substituting other energy
sources for the rapidly diminishing supply of oil and natural gas. For many
applications this means increased use of electricity and for some applications
it should mean the consideration of solar energy. One hopes that a policy of
strong conservation measures together with the development of new energy
sources and supplies will enable the U.S. to balance its energy budget
gracefully by 1985 or soon thereafter.

APPENDIX

CONVERSION FACTORS

Energy

1 calorie = 4.184 J

1 Btu = 1054.4 J

1 kWh = 3,414 Btu

1 kWh = 3.6×10^6 J

1 hp hr = 2.686×10^6 J

1 therm = 10^5 Btu

1 Quad = 10^{15} Btu

Power

1 Btu/hr = 0.293 W

1 kWh/yr = 0.114 W

1 hp = 746 W

1 kW = 3,414 Btu/hr

1 kcal/sec = 4,184 W

1 Quad/yr = 3.34×10^{10} W

1,000 therms/yr = 1.142×10^4 Btu/hr

Time

1 yr = 8,760 hr

1 yr = 3.16×10^7 sec.

1 day = 86,400 sec.

Volume

1 gal = 3.785 liters (U.S. gal)

1 ft^3 = 0.0283 m^3

1 barrel crude oil = 42 gallons

Area

1 ft^2 = 0.0929 m^2

1 acre = 4,047 m^2

1 sq. mile = 2.589×10^6 m^2

Length

1 mile = 1,609 m

1 foot = 0.3048 m

1 inch = 2.54 cm

Energy Flux

1 Btu/ft^2 hr = 3.152 W/m^2

Mass

1 ton = 907.2 kg

1 lb = 0.4536 kg

U value

1 Btu/ft^2 hr $^\circ$F = 5.674 W/m^2 $^\circ$C

Fuel Economy

1 mile/gal (mpg) = 0.425 km/liter

Fuel Heats of Combustion

(These are approximate values which depend on
the type and origin of fuel)

1 barrel crude oil (42 gal)	$5.8 \ 10^6$ Btu
1 ton coal	$2.2 \ 10^7$ Btu
1 ft^3 natural gas	1,021 Btu
1 gallon gasoline	$1.25 \ 10^5$ Btu
1 gallon distillate fuel oil	$1.39 \ 10^5$ Btu

Fuel Consumption Rates

10^6 barrels/day = 2.1 Quads/yr

10^9 tons coal/yr = 25 Quads/yr

Electricity Generation

1 kWh$_e$ = 10,400 Btu$_t$ (approx.)

Solar Influx

1 kW/m^2 (peak)

REFERENCES

1. Department of the Interior, Bureau of Mines; News Release, "U.S. Energy Use Down in 1974 after Two Decades of Increases", Washington, D.C., April 3, 1975.

2. Ford Foundation Energy Policy Project, <u>A Time to Choose</u> , Ballinger Publishing Co., Cambridge, Mass. (1974).

3. American Physical Society, "Efficient Use of Energy: A Physics Perspective", New York, N.Y. (Jan. 1975).

4. Stanford Research Institute, Patterns of Energy Consumption in the United States, Menlo Park, Calif. (Jan. 1972).

5. Federal Energy Administration, "Project Independence Report", (Nov. 1974).

6. Department of the Interior, "Energy Perspectives, Washington, D.C., (Feb. 1975).

7. Federal Energy Administration, "Project Independence, A Summary", Washington D.C., (Nov. 1974).

8. Bureau of Mines, Department of the Interior; various news releases and circulars.

9. Federal Energy Administration, "Residential and Commercial Energy Use Patterns, 1970-1990", Washington, D.C.,(Nov. 1974).

10. Bureau of the Census, Department of Commerce, <u>The Statistical Abstract of the United States,</u>(also published as the U.S. Fact Book, the American Almanac), Grosset and Dunlap, New York (1975).

11. American Society of Heating, Refrigerating and Air Conditioning Engineers (ASHRAE) , <u>Handbook of Fundamentals</u>, New York, N.Y. (1972).

12. Federal Energy Administration, "Energy Independence Act of 1975 and Related Tax Proposals", (March 1975).

13. Pilati, D., "The Energy Conservation Potential of Thermostat Setback", Oak Ridge National Laboratory, ORNL-NSF-EP-80.(Feb. 1975).

14. Herendon, Robert A., "Appliance Energy Use", Energy Research Group, University of Illinois, (Jan. 1975).

15. Hirst, E., Energy Intensiveness of Passenger and Freight Transportation Modes", Oak Ridge National Laboratory, ORNL-NSF-EP-44 (April 1973).

16. Federal Highway Administration, "Highway Statistics", (various years).

17. Hirst, E., "Total Energy Use for Commercial Aviation", Oak Ridge National Laboratory, ORNL-NSF-EP-68 (April 1974).

18. U.S. Civil Aeronautics Board, "Handbook of Airline Statistics", Washington, D.C. (various years).

19. Pilati, D.A., "Conservation Options for Commercial Air Transport", to
    be published in Energy Systems and Policy.
    Pilati, D,A., "Airplane Energy Use and Conservation Strategies", Oak
    Ridge National Laboratory, ORNL-NSF-EP-69 (May 1974).

20. Federal Energy Administration, "One day a week driving ban on Private
    Autos", Internal Report (Feb. 1975).
    Federal Energy Administration, "Sunday Service Station Closings",
    Internal Report (March 1975).

21. Federal Energy Administration, "Analysis of Gas Rationing", Internal
    Report (Jan. 1975).

22. Federal Energy Administration, "Gasoline Consumption", Internal Report,
    (March 1975).

23. Mitre Corporation, "Policy Assessment of the 55 mph Speed Limit",
    McLean, Virginia (Oct. 1974).

24. Department of Transportation and Environmental Protection Agency,
    "Potential for Motor Vehicle Fuel Economy Improvement", Washington, D.C.,
    (Oct. 1974).

25. Kummer, J.T., "The Automobile as an Energy Converter", Technology Review
    77, 26(1975).

26. Garrett, H.M., and Murray, J.A., "Energy Conservation in the Cement
    Industry", in Energy Delta, G. W. Morgenthaler, A.N. Silver, eds.,
    AAS Publications, Tarzana, Calif. 1975.

27. Hannon, Bruce, "Options for Energy Conservation", Technology Review
    76, 24 (1974).

28. Dow Chemical Co., The Word is Save, Midland,Michigan. 1975.

29. Gordian Associates, "Industrial Energy Conservation - Pilot Study",
    New York, N.Y. (June 1974).

30. Meyerson, M., "Industrial Energy Conservation in Overview", in Energy
    Delta, G. W. Morgenthaler, A.N. Silver, eds., AAS Publications, Tarzana,
    Calif. (1975).

31. Russell, A.S., "Energy Conservation in Primary Metals Processing", in
    Energy Delta, G.W. Morgenthaler, A.N. Silver, eds., AAS Publications,
    Tarzana, Calif. (1975).

32. Steinhart, J.S., and Steinhart, C.E., "Energy Use in the U.S. Food System",
    Science 184, 307(1974).

33. Development Planning and Research Associates, "Industrial Energy Study of
    Selected Food Industries", Manhattan, Kansas (1974).

34. Herendeen, R.A., and Bullard, C.W., Energy Cost of Goods and Services,
    1963 and 1967, Center for Advanced Computation, University of Illinois,
    (Nov. 1974).

35. Berry, R.S., and Fels, M., "The Energy Cost of Automobiles", Science and
    Public Affairs, The Bulletin of the Atomic Scientists (Dec. 1973).

36. Gordian Associates, "The Opportunities and Incentives for Electric Utility
    Land Management", New York, N.Y. (April 1975).

DISCUSSION

Question: Would you comment on the figure of 50% mill scrap shown for the
            aluminium industry? As a metallurgist, I find the figure very
            high.

Answer: I can't comment on the figure, but the loss does not affect the
            energy use greatly. The aluminium simply has to be re-smelted,
            which comes after the energy-intensive electrolytic stage.

Comment: Lots of examples of small savings were given. This is in the
            nature of energy conservation studies, i.e. many small savings
            can be made which together are significant. The important ques-
            tion is how we can get all of these small changes to occur with-
            out giving each one individual attention. Energy pricing policy
            can achieve this. In the US the average and not the marginal
            price of energy is paid, which tends to reduce saving incentive.

Response: In my personal view, pricing policy is a definite case for
            government intervention.

Question: On the subject of commuter travel, why not work a 4 day 10 hour
            week, rather than a 5 day 8 hour week, and cut transport energy?

Answer: Statistics show the car mileage distribution by day-of-week to
            be flat, and so people will simply drive elsewhere on free days.

Question: What is the average journey to work by car in the US?
Answer: About 8 miles.
Comment: It is important for the US to take a lead in conservation, partly
            because of the larger wastage in the US, and as an example to the
            world. We cannot advocate thrift without pursuing it ourselves.

Response: The Ford foundation energy study tried to establish whether life-
            styles could be maintained, whilst using less energy. 1985 pro-
            jected energy use requires 3.5% p.a. growth. Indications were
            that this rate could be halved without affecting lifestyles or
            the quality of life.

Question: How much of the US oil is imported?
Answer: 35%, but it could easily grow to 50%.

Question: Regarding aircraft fuel use vs. journey lengths, aircraft seem
            comparatively efficient in level flight. How many miles of level
            flight are equivalent to a take off or landing?

Answer: In level flight about 6,000 Btu/passenger mile are consumed. I
            don't know the requirement for a take off or landing.

Comment: You use the phrase "small savings", but 1 Quad is about £1,000
            million saved in balance of payments and is not insignificant.

Response: It is only a small percentage, but I agree that these small
            percentages can add up to a significant figure. It takes a long
            time for conservation measures to take effect. Suppose $\frac{1}{2}$%/year
            is saved for 5 years. This is against the inertia of a 3% growth
            for the previous 5 years so has little effect.   The products of
            the society must be changed. A 4 day 8 hour week should be
            worked. The US would have 20% less GDP, and would use 20% less
            energy.

ENERGY USE IN INDUSTRY

G. E. H. Newton

Reed Engineering & Development Services Ltd., Maidstone

Once man ceased to be a nomad he learned the systematic use of the powers of water, wind and fire. By the end of the Middle Ages, advanced agricultural man in Western Europe not only had the water wheel, the windmill, wood and some coal as energy sources but was able to conquer distant parts of the world because he was able to build armed sailing vessels.

The biggest advance was man's discovery that fire could do work. From wood, he progressed to coal, the resource that fuelled the industrial revolution.

With the advent of electricity the way was open to industrial development of unprecedented flexibility.

Since that time, world energy consumption, particularly in the industrialised countries, has grown at a rate which is matched only by the increase in material wealth. Meanwhile, other fuels, oil at the beginning of this century and natural gas in the early thirties, first augmented and later began to supplant coal.

F.C. Fisher and R.H. Pry of General Electric, U.S.A. presented a simple substitution model of technological changes; they considered 17 examples such as synthetic fibre, plastics, organic insecticides etc. Marchetti of the International Institute of Applied Systems Analysis, applied this model to fuel substitution in the U.S. market during the last century. Fig. 1 shows the results. During the last century the prevailing fuel was wood. It was substituted by the advent of coal. In the early part of this century oil began to penetrate the market and to substitute coal. It has now reached its maximum and is being substituted by gas. If the model is extended by allowing for the advent of nuclear fission, one can see the substitution of gas taking place; likewise nuclear fission being substituted by some other energy source. (Fig. 2)

Fig. 3 shows the United Kingdom inland energy consumption for the period 1920 to 1974; the type of pattern of energy usage as depicted in Fig. 2 can be seen.

Table I gives the energy usage by main groups of users while Table II gives the breakdown of energy consumption by the major industrial groups, which is also shown in chart form in Fig. 4.

The Department of Energy has produced an energy flow chart for 1974 (Fig. 5) showing the energy supply utilization and losses in terms of million barrels per day of Oil Equivalent. The Americans have produced a similar diagram for 1970 and another covering the projections for 1980. (Figs 6 and 7)

U.S. MARKET PENETRATION OF SUBSEQUENT FUELS

FIG. 1.

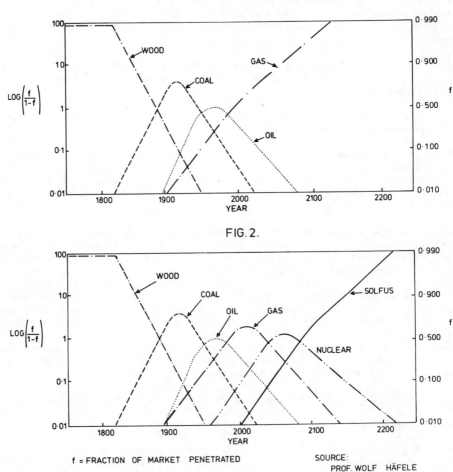

FIG. 2.

f = FRACTION OF MARKET PENETRATED

SOURCE:
PROF. WOLF HÄFELE
II ASA

TOTAL U.K. INLAND CONSUMPTION PRIMARY
FUELS (DEPARTMENT OF ENERGY STATISTICS)
FIG. 3.

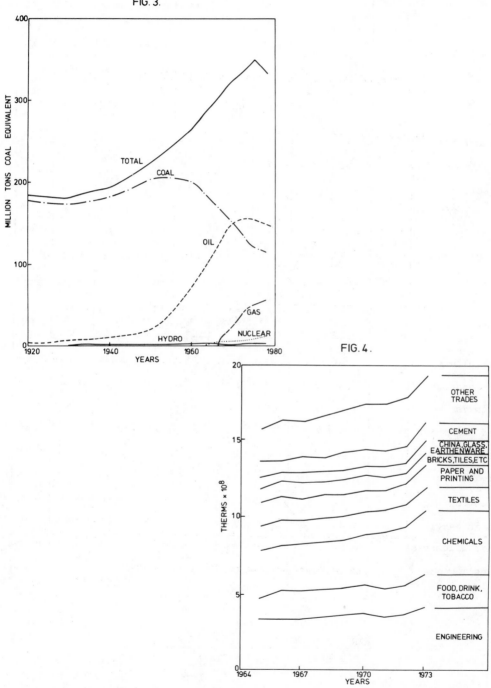

FIG. 4.

	1956	1957	1958	1959	1960	1961	1962	1963	1964
Agriculture	625	587	605	589	601	640	654	692	658
Iron and Steel	7 227	7 348	6 513	6 281	7 378	7 063	6 593	6 867	7 299
Railways	3 695	3 551	3 309	3 016	2 858	2 630	2 255	2 016	1 649
Road Transport	3 577	3 309	3 815	4 134	4 489	4 830	5 105	5 442	6 119
Water Transport	570	566	559	555	570	568	593	575	536
Air Transport	795	739	710	737	798	976	954	1 011	1 039
Domestic	14 507	13 879	14 489	13 704	14 414	14 109	14 588	15 305	14 095
Public Services	2 326	2 193	2 372	2 221	2 370	2 180	2 361	2 536	2 606
Other Industries	16 496	15 876	15 396	15 134	15 713	15 755	16 157	16 251	14 259
Miscellaneous	2 877	2 804	2 764	2 764	3 082	2 705	2 906	3 191	1 899

	1965	1966	1967	1968	1969	1970	1971	1972	1973
Agriculture	663	643	663	721	745	750	772	849	882
Iron and Steel	7 497	6 849	6 490	6 871	7 087	7 193	6 561	6 312	6 523
Railways	1.280	1 021	762	625	619	640	597	559	558
Road Transport	6 537	6 886	7 341	7 795	8 080	8 505	8 897	9 343	9 974
Water Transport	544	533	516	469	487	505	454	391	436
Air Transport	1 070	1 164	1 309	1 424	1 480	1 536	1 686	1 792	1 903
Domestic	14 520	14 407	14 154	14 529	14 720	14 643	14 141	14 395	14 925
Public Services	2 781	2 944	3 079	3 182	3 354	3 452	3 422	3 488	3 555
Other Industries	15 798	16 230	16 270	16 647	17 053	17 496	17 056	17 944	19 254
Miscellaneous	2 991	2 888	2 769	2 906	3 143	3 231	2 946	2 967	2 983

TABLE 1

ENERGY CONSUMPTION BY FINAL USERS

HEAT SUPPLIED BASIS

Department Energy Statistics
Million Therms

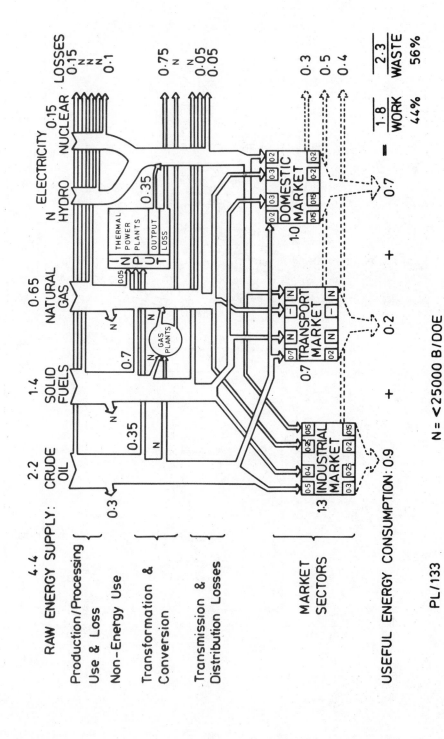

FIG. 5. ENERGY FLOWCHART FOR U.K. IN 1974 (MILLION B/D OIL EQUIVALENT)

N = <25000 B/DOE

PL/133

Figure 6
UNITED STATES ENERGY FLOW PATTERN
ACTUAL – 1970

1970

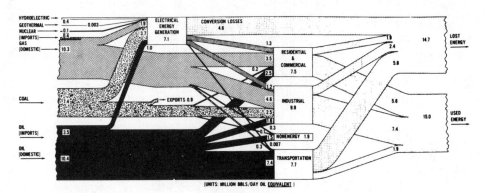

(UNITS: MILLION BBLS/DAY OIL EQUIVALENT)

Figure 7
UNITED STATES ENERGY FLOW PATTERN
PROJECTED – 1980

1980

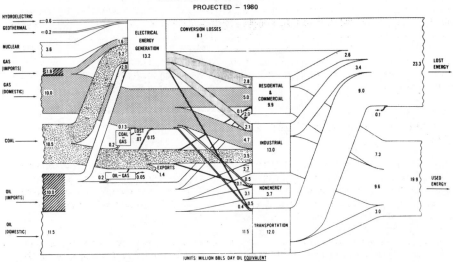

(UNITS: MILLION BBLS/DAY OIL EQUIVALENT)

SOURCE: "UNDERSTANDING THE 'NATIONAL ENERGY DILEMMA'," JCAE, 1973

Such diagrams must be further extended if the energy input and usage in industry is to be considered and Fig. 8 shows the type of breakdown from raw energy to the type of energy at the end use.

The total industrial concerns using energy in one form or another number some 189,000 according to the Electricity Supply Statistics 1974. Manufacturing concerns total some 82,000 as given in the Census of Production Tables 1968.

An indication of the relative numbers of concerns in the major energy consuming manufacturing industries is as follows:

Metal Manufacture	2,800
Ship building, marine and vehicle manufacture	2,500
Other metal goods	12,000
Chemicals	2,800
Food, drink and tobacco	6,000
Textiles	7,000
Paper, printing and manufacture	10,300

Industry consumes nearly 60% of the primary energy consumption, the manufacturing industries (excluding iron and steel) consume some 22%.

Of the manufacturing concerns about 75% have an energy requirement equal to about 1 to 3% of total product cost and in the majority some 60% of the energy is used for heating and ventilating and 12% for lighting.

In the process industries energy represents from 5 to 20% of the total product cost and in this category heating, ventilating and lighting will range upwards of 10%.

From the foregoing it is not difficult to realise the number of permutations and combinations that are possible when considering energy for industry.

Data on energy use in manufacturing is difficult to obtain.   Historical records of plant are generally not studied in detail within the company; frequently they are not retained.   Often they are expressed in monetary terms based on easily measurable values such as kilograms of steam per hour, having no indication of the heat value.

It is only since the oil crisis that interest has been aroused in energy data, audits and accounting.   Although pioneers such as Sir Oliver Lyle were preaching Energy Conservation some 40 years ago.

There is now an increasing use of heat unit accounting in manufacturing. In this system careful track is kept of all energy-bearing materials that enter a plant in order to determine where energy is used.   This approach permits the detection of problem areas where energy savings can be instituted.

FIG.8.

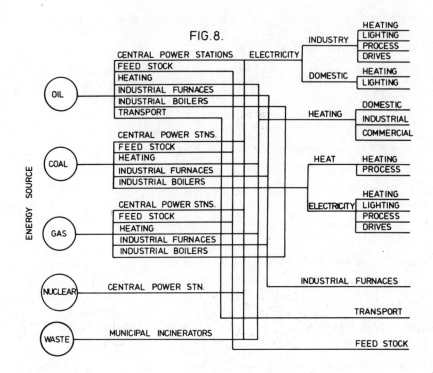

FIG.9.

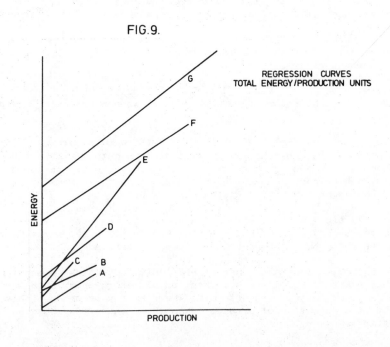

REGRESSION CURVES
TOTAL ENERGY/PRODUCTION UNITS

An important factor in determining energy use per unit of output within
industry or firm is the product mix.   This is an application of the
same principles that underlies the shift from energy intensive industries
to other manufacturing industries.   Within the range of products of a
single plant, energy use per unit for the entire plant can rise or fall
according to whether energy intensive products become more or less
important in the product mix.

There is also a cyclical effect on energy use per unit of output.   When
capacity utilization is low, energy use per unit is high.   Similarly,
when capacity utilization is very high, energy use per unit of output is
often high.   There is an optimum level where energy use per unit of output is minimal.

There is, of course, the effect of ambient conditions on energy use and
often the energy used for heating and ventilating is not segregated from
that used for production.

One has the dilemma, therefore, that to use aggregate figures is unlikely
to give the accuracy required to give meaningful data.   While, on the
other hand, to consider every variable would result in an uneconomic
exercise.   Circumstances and available information will finally determine
the approach.

In general the available information will be aggregate figures, it is
politic to use these in the first instance in such a way as to promote
the next stage of questions which usually result in more information being
made available, or obtained.   The operation is a progressive one up to
the stage where the effort to obtain the figures is greater than the value
of the information.

As an illustration of this approach, Fig. 9 shows plots for individual
factories producing the same type of product but of different quality from
similar types of raw material.   Fig. 10 shows the variation between
production units in the same factory, again with variations in machine
design, raw material and product.

These are now being developed into specific products and specific areas
of plant and equipment.

Fig. 11 shows plots of total energy against raw material input and it can
be seen how the high energy requirement of the heating and ventilating
system affects the results.   Fig. 12 shows the total energy input measured
against degree days.   The values are from a factory where the energy for
production is only 30% of the total energy demand.

These results are only part of a total material balance;   there is
obviously an interaction between analysis of energy and economy, the total
material balance as in Fig. 13 in which energy, water, raw materials and
pollution have to be dealt with.

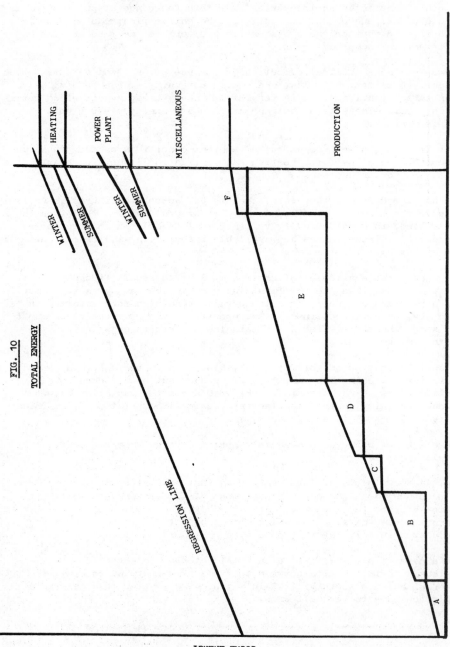

FIG. 10

TOTAL ENERGY

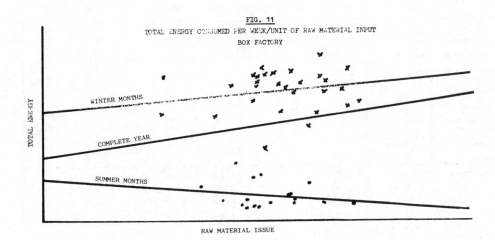

FIG. 11

TOTAL ENERGY CONSUMED PER WEEK/UNIT OF RAW MATERIAL INPUT

BOX FACTORY

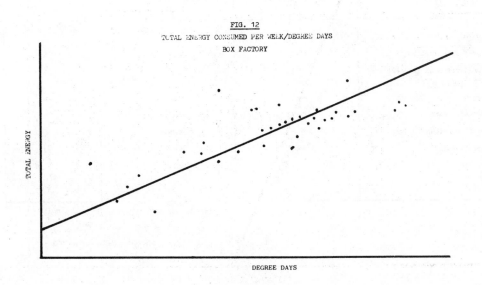

FIG. 12

TOTAL ENERGY CONSUMED PER WEEK/DEGREE DAYS

BOX FACTORY

TABLE II   ENERGY CONSUMPTION BY MAIN INDUSTRIAL GROUPS   DEPARTMENT ENERGY STATISTICS
                                                                 MILLION THERMS

HEAT SUPPLIED BASIS

	1965	1966	1967	1968	1969	1970	1971	1972	1973
Engineering and other Metal Trades	3 485	3 498	3 496	3 617	3 748	3 857	3 640	3 702	4 235
Food Drink and Tobacco	1 819	1 872	1 876	1 883	1 880	1 951	1 968	2 027	2 104
Chemicals and Allied Trades	2 640	2 730	2 842	2 876	2 797	3 016	3 463	3 662	4 234
Textiles	1 616	1 637	1 583	1 592	1 584	1 540	1 364	1 356	1 419
Paper and Printing	1 397	1 469	1 435	1 469	1 491	1 521	1 460	1 464	1 465
Bricks, Tiles, Building Materials	966	911	892	885	836	797	744	602	675
China, Earthenware and Glass	670	677	662	671	675	695	716	693	701
Cement	1 127	1 179	1 155	1 209	1 236	1 124	1 077	1 213	1 274
Other Trades	2 078	2 257	2 329	2 445	2 806	2 995	3 074	3 225	3 147
TOTAL	15 798	16 230	16 270	16 647	17 053	17 496	17 506	17 944	19 254

(NOTE
  SHOWN AS OTHER INDUSTRIES IN TABLE 1)

TABLE III   ENERGY INTENSIVENESS AND PRICE INCREASES OF BUILDING MATERIALS

Material	Energy usage as percentage of delivered price 1973	Price increases 1973-77 from higher fuel costs %	Proportion of total materials %
Cement	25-30	35-50	8
Non-fletton bricks	10-15	20-25	4
Constructional iron and steel	7-10	10-15	14
Glass	7-9	10-13	2
Sanitary ware	7-9	10-13	1
Fletton bricks	5-8	7-10	4
Aggregates	4-6	6-9	10
Timber	1	–	–

SOURCE: N.E.D.O. REPORT

FLOW DIAGRAM FOR MATERIALS ENERGY, WATER, WASTE
AND RE-CYCLING
FIG. 13.

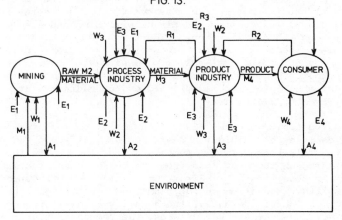

M - FLOW OF MATERIAL   E - FLOW OF ENERGY
R - FLOW OF RE-CYCLING   W - FLOW OF WATER
A - FLOW OF WASTE AND POLLUTANTS

PRINCIPLE OF ANALYSIS FOR THE CONSUMPTION OF
ENERGY IN BUILDINGS
FIG. 14.

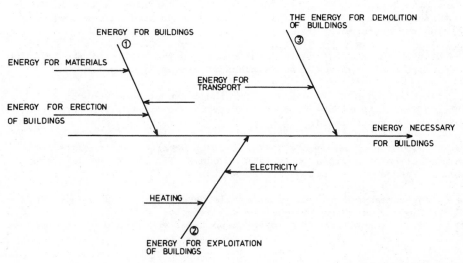

BY PROF. P.C. KREIJGER

The Building

Basically, a building is required for weather protection for personnel
and plant; it can be a simple shed of corrugated asbestos or iron or
it can be the most sophisticated form of structure.   The energy input
is in the form of fuel for heating and air conditioning and electricity
for ventilation, cooling and lighting.   There will also be an energy
input for equipment processes and for feedstock and, in certain instances,
an increased energy input for anti-pollution measures.   Desirable factors
for comfort are that air movement should be provided throughout the
building and that the higher the temperature and/or radiant heat, the
higher the air movement desirable.

In any building containing process plant using heat, boiler, turbines,
motors, piping carrying hot fluids; quantities of heat are given off
from the plant.   This increases the temperature of the air in the building
to a value depending on the heat loss from the building and the scheme of
ventilation or air conditioning.   First, the heat given off depends on
the efficiency of the insulation on the plant and the type of finish of
the surface of the plant.   Secondly, the quantity of heat given off
depends on the air movement across surfaces, the temperature of the air
drops with increased throughput but the total heat given off rises.
Also heat is liberated into the building by leakages from the plant,
leakages of the heating medium or the vapour from a drying process.

If the environment is not comfortable this can be for any one or combination of a number of reasons – the air temperature may be too high,
the relative humidity may be too high, radiant heat may be excessive or
air movement may not be sufficient.

In a building containing a process requiring heat it is a simple matter
to keep the temperature as high as desired, the problem is to make the
temperature lower or to alter the air movement or humidity to compensate
for high temperatures.

In achieving the correct environment condition the design of the building
or envelope enclosing the process plant is extremely important.

As a first approach to an analysis for the consumption of energy in
buildings, Fig. 14 illustrates the energy input, arrows 1, 2 and 3 can,
of course, be sub-divided.

The N.E.D.O. report on energy states that the energy requirements for
the production of building materials accounted for 5% of the delivered
price.   The range is wide and is shown in Table III.

The energy contents of building materials have been calculated by Prof.
P.C. Kreijger, following the chain – raw material → material → product →
structure and quantifying the energy aspects of each phase of the processes.

The breakdown approximates to:

Raw Materials	16%
Manufacture of industrial raw materials	8%
Manufacture of product	16%
Transport of products and people	20%
Human energy consumption (mainly food)	10%
Heating	30%
	100%

In Sweden it was calculated that the energy content of buildings was:

Building materials for flats	160 kWh/m^3 of building volume
Building materials for offices	200 " " "
Building materials for detached houses	63 " " "
Materials for water, gas and electrical installations	24 " " "

Transport of bricks represented some 10% of the energy content of the brick and transport of reinforcement represented 2% of the energy content of the reinforcement. Expressed in another way the transport of bricks represented 10 to 15% of the delivered price.

Assuming a depreciation period of building of 40 years, the energy consumption for buildings can be expressed approximately:

Materials	4.0%
Transport	0.5%
Building Process	3.0%
Heating of Buildings	92.5%
	100 %

This means that the energy content of completed new buildings is about double the energy content of the building materials.

The Process Industries

The paper industry may be taken as an example of the process industries because it is concerned with the basic elements of Fire, Wind and Water.

To produce about 4 million tons of paper per year, something like 2 million tons of oil equivalent are burned, about 8 million tons of water are evaporated, 320 million tons of air are handled and 4500 gigawatts of electricity are used.

Like most industries, it has old, modern, small, large, efficient and
inefficient, well designed and not so well designed mills, making a wide
range of paper from fine tissues to heavy board.   Add to this list
winter, summer and any point of the compass and we finish up with a
variety of energy inputs.   But energy has, in any process, three prime
functions:

> To assist in producing the product
> In protecting buildings and equipment
> In improving working conditions

Fig. 15 shows a typical cross section of an average type paper mill, while
Fig. 16 shows the difference in systems when using a totally enclosed
hood and an open canopy hood.   Fig. 17 shows an elevation of a typical
fourdrinier type of paper machine.

Fig. 18 illustrates the energy distribution in a paper machine house;
the energy sources are usually steam and electricity but they are often
supplemented by direct firing of oil or gas in the drying process.

Some losses from the process will affect the working condition and an
indication of the condition of humidity and temperature in a paper mill
is shown in Figs 19 and 20.

What must be realised is that in any process where heat or motive power
is used, energy is liberated ultimately into the atmosphere or the sea.

If we consider Fig. 18, the heat supplied to the dryers, 75% will be
liberated to the atmosphere in the form of vapour, 20% will be returned
to the boiler house in the condensate and 5% will be liberated into the
machine through leakage, radiation etc.   80% of the heat supplied to
velocity air system, air replacement, roof heating and annexe heating
will be liberated in the machine house and ultimately exhausted to
atmosphere.   About 20% of the heat supplied will be returned to the
boiler house.

Stock heating is heat supplied to process water and fibre to improve
drainage.   Some of this heat will be liberated into the machine house in
the form of vapour from the wire part and radiation from pipes;   the bulk
of it will be discharged in the effluent and thence to river and sea.
Much has been done in recent years to close up the water systems and
reduce the quantity of effluent, thus reducing the energy requirement.

Heat from prime movers will be dissipated into the atmosphere or into
the fluid being handled and once again into the atmosphere or river.
Some will be liberated into the building by radiation and convection from
the equipment.   Large installations will have motor cooling where air
is directed to and from the equipment in order to reduce the amount of
heat released to the machine house and, more important, to keep the
equipment within specified operating temperatures.

Of the total heat supplied in the form of steam, only about 15% is
returned to the boiler house in the form of hot condensate.

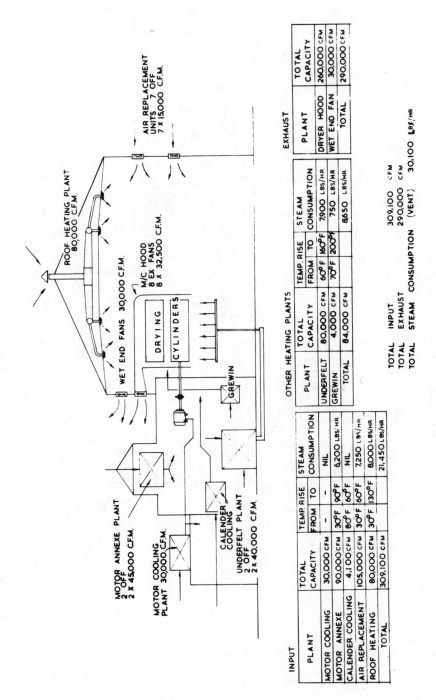

FIG. 15. TYPICAL AIR DISTRIBUTION IN MACHINE HOUSE

ROOF HEATING PLANT
80,000 C.F.M.

AIR REPLACEMENT
UNITS 7 OFF
7 X 15,000 C.F.M.

M/C HOOD
8 EX FANS
8 X 32,500 C.F.M.

WET END FANS 30,000 C.F.M.

DRYING
CYLINDERS

GREWIN

MOTOR ANNEXE PLANT
2 OFF
2 X 45,000 C.F.M.

MOTOR COOLING
PLANT 30,000 C.F.M.

CALENDER
COOLING

UNDERFELT PLANT
2 OFF
2 X 40,000 C.F.M.

INPUT

PLANT	TOTAL CAPACITY	TEMP RISE FROM	TEMP RISE TO	STEAM CONSUMPTION
MOTOR COOLING	30,000 CFM	—	—	NIL
MOTOR ANNEXE	90,000 CFM	30°F	90°F	6,200 LBS/HR
CALENDER COOLING	4,100 CFM	80°F	60°F	NIL
AIR REPLACEMENT	105,000 CFM	30°F	60°F	7,250 LBS/HR
ROOF HEATING	80,000 CFM	30°F	130°F	8,000 LBS/HR
TOTAL	309,100 CFM			21,450 LBS/HR

OTHER HEATING PLANTS

PLANT	TOTAL CAPACITY	TEMP RISE FROM	TEMP RISE TO	STEAM CONSUMPTION
UNDERFELT	80,000 CFM	60°F	160°F	7900 LBS/HR
GREWIN	4,000 CFM	70°F	200°F	750 LBS/HR
TOTAL	84,000 CFM			8650 LBS/HR

EXHAUST

PLANT	TOTAL CAPACITY
DRYER HOOD	260,000 CFM
WET END FAN	30,000 CFM
TOTAL	290,000 CFM

TOTAL INPUT 309,100 CFM
TOTAL EXHAUST 290,000 CFM
TOTAL STEAM CONSUMPTION (VENT) 30,100 LBS/HR

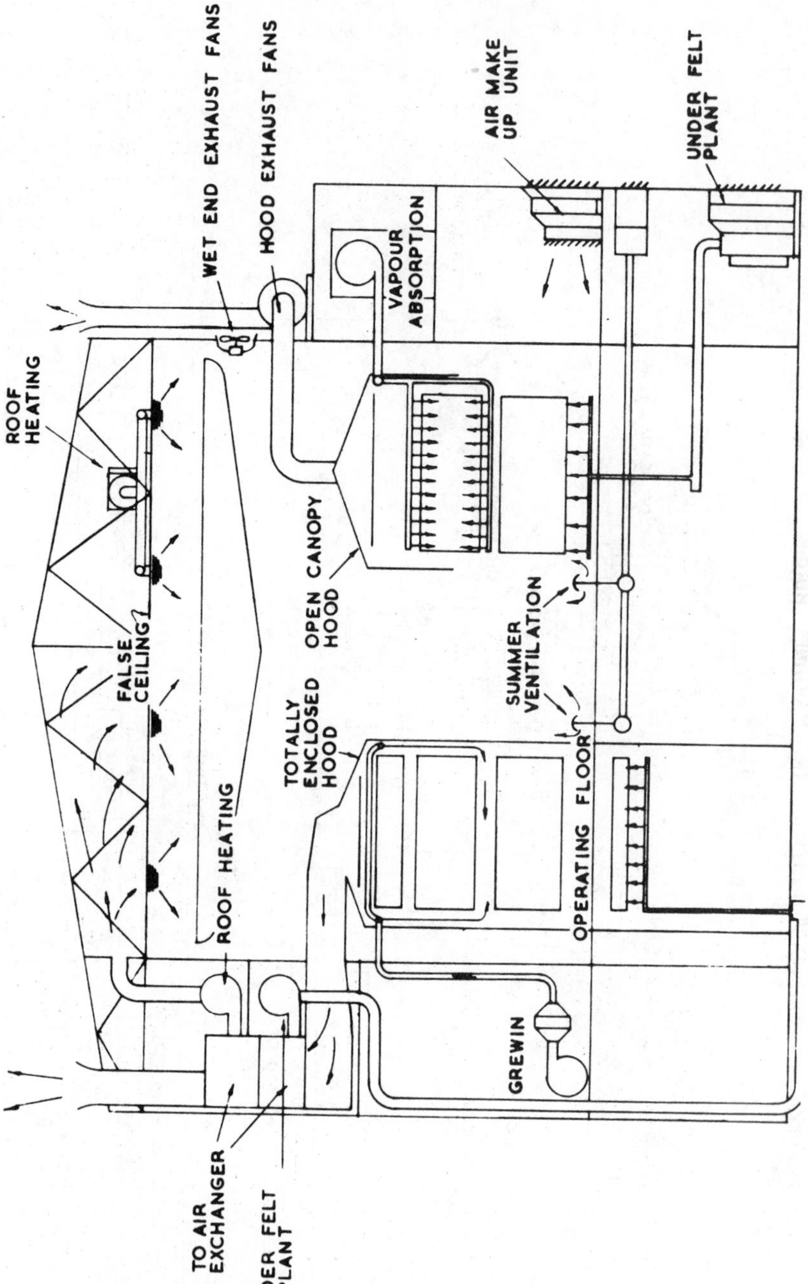

FIG. 16. AIR HANDLING UNITS IN MACHINE HOUSE

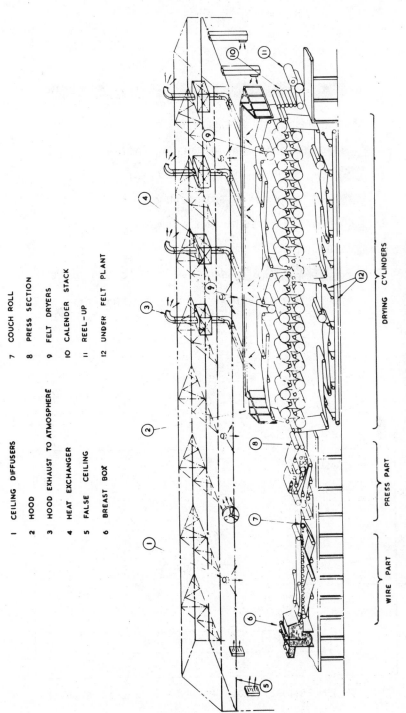

1 CEILING DIFFUSERS
2 HOOD
3 HOOD EXHAUST TO ATMOSPHERE
4 HEAT EXCHANGER
5 FALSE CEILING
6 BREAST BOX
7 COUCH ROLL
8 PRESS SECTION
9 FELT DRYERS
10 CALENDER STACK
11 REEL-UP
12 UNDER FELT PLANT

WIRE PART   PRESS PART   DRYING CYLINDERS

Fig. 17. PICTORIAL LAYOUT OF A FOURDRINIER PAPER MACHINE

FIG. 18.

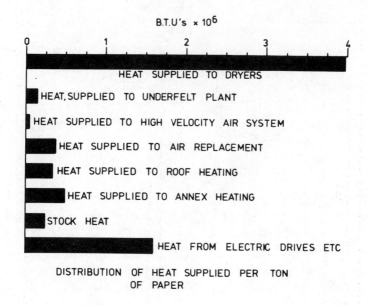

DISTRIBUTION OF HEAT SUPPLIED PER TON
OF PAPER

FIG. 19.

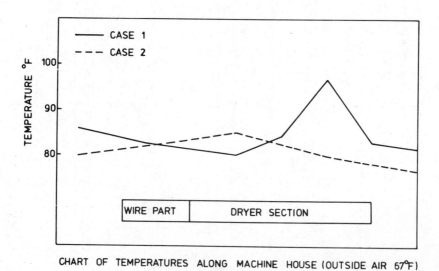

CHART OF TEMPERATURES ALONG MACHINE HOUSE (OUTSIDE AIR 67°F)

FIG. 20.

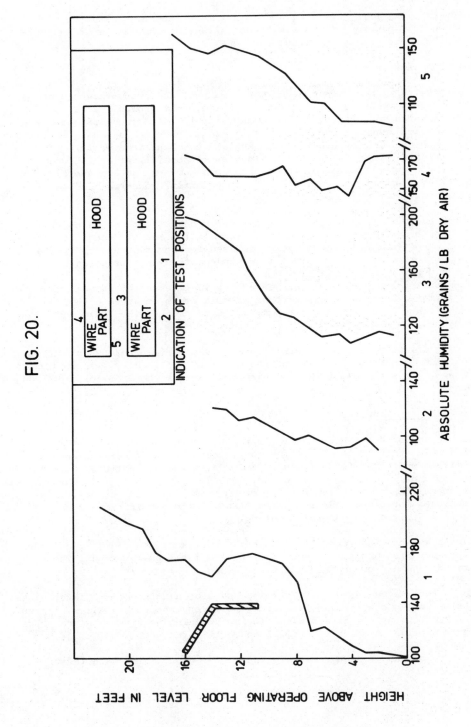

Recovery of heat is practical and is now more important with the high
fuel costs; however, it is a subject warranting its own paper and outside the scope of this particular paper.

The Food Industry

The industry is based on convenience; many of the products could be,
and at one time were, made by the housewife, most probably less
sophisticated and with a greater energy input.   The products which are
processed are natural materials;   this means there are several limitations
as to temperature, mechanical handling and chemical treatments.   The
industy's main efforts are those directed to sophistication in food
preparation, shelf life, distance, (between growers and consumers) which
basically means it is concerned with preparation, conservation and distribution.

Energy accounts for about 5% of the total product cost, of which 2% is
used for process and 3% for heating, ventilating and air conditioning,
which means that 60% of the energy is used for environmental purposes
and protection of the structure from condensation and corrosion.

The industry is not energy intensive; it is interesting to note that it
is estimated that to grow one tonne of soya beans requires 4,140,000
B.t.u./tonne of gross energy input, while to process them into margarine
takes 730,000 B.t.u./tonne.

Cooking of the products is mainly carried out at temperatures of 100 to
120°C and the energy input is of the order of 80/120 B.t.u./lb in terms
of heat, while the electrical requirements are of the order of 5/300 kWh/
tonne.

The Packaging Industry

In the manufacture of Corrugated Board the energy requirement of the
process is about 55 to 65% of the total energy demand, heating and
ventilating account for between 10 and 20%, as does lighting.   The total
energy input is between 100,000 and 150,000 B.t.u./1000 m^2 of board.
The energy input is in the form of steam from oil fired boilers, this
being used for process drying and heating, while electricity is used for
machinery drives and general lighting.

Where boxes are made from solid board, (i.e. fancy chocolate boxes) the
energy used for process is about 30% of the total energy input, the
remainder being for heating and ventilating.

Many materials are used for packaging, such as aluminium, tinplate, glass
bottles, cellulose film, polystyrene, polypropylene etc.   All will have
differing energy requirements, differing building requirements, ranging
from a chemical complex to a small converting factory.

The type of energy requirement for some of the packaging materials is
listed below:

1 million m^2 of packaging fibre	Polypropylene Cellulose film	110 tonnes oil equivalent 155 " " "
1 million fertilizer sacks	Polyethylene sacks Paper sacks	470 " " " 700 " " "
1 million 1 l. bottles	P.V.C. bottles Glass bottles	97 " " " 230 " " "

The above figures include oil required as feedstock and to provide energy for the manufacture of the article.

Newspaper Printing

The types of installation in use by this industry are many and varied and range from the nationals in Fleet Street to the local paper produced in the High Streets.

The level of energy input for the printing of the paper, the offices for editorial and other staff, is of the order of $\frac{1}{2}$ to 2 million B.t.u. per tonne of newsprint processed.   In this instance it is difficult to relate the energy input to newspapers produced because some of the work is only partly done and then transferred to another part of the country.   There is also a very large difference between the relationship of office space to production area, depending on building and organisations under consideration.

Future

There is a growing activity in seeking information of where the energy is used, how it is used and what are the gross and nett energy demands for a given product.   The search for information on energy usage has its spin offs which could ultimately result in the optimization of all the resources.

The additional information will assist in creating better designs of buildings housing process plant; it will lead to better designs of equipment that will lead to less stringent requirements on the building. It should also lead to the improvement of existing processes and systems and the creation and design of new ones.

The future can be classified as:

Short Term	Improvement of existing plant operation
Medium Term	Improvement of existing plant by adaption and modification
Long Term	Development and installation of new plant, equipment, processes and systems

It is evident that in the future energy will be derived from sources other than fossil fuels as these will be considered as more valuable feedstock for chemical industry and alike.   By inference it can be deduced that in the future there will be an electrical or hydrogen economy or even a combination.   If so, then it is possible that radical changes to existing process industries will be necessary.   For instance, in the paper industry

with an electrical economy, electrode boilers could be used to supply
steam, resulting in no necessity to change the existing process equipment or building.   However, it may be that the drying of paper could
be achieved by the use of infra red, di-electric or micro-wave, or
one might even speculate on the dry laid method of making paper.   Whatever is used would have serious repercussions on the equipment and
buildings as well as men, materials and money.

It would seem necessary for all interested parties, Government,
professional bodies, researchers, inventors, manufacturers and fuel
suppliers to work towards a common long term objective - that is to use
energy - money - materials and men as effectively and efficiently as
possible.

FURTHER READING

1   P C Kreijger, "Building Materials versus Energy", S.C.I.E.N.C.E.
Conference, Brussels, 1975.

2   C P Gerhardt, "Energy Consideration for the Food Industry", S.C.I.E.N.C.E.
Conference, Brussels, 1975.

3   "Competitiveness of Low-Density Polyethylene, Polypropylene and
Polyvinyl Chloride after the 1973 Oil Crisis", Imperial Chemical Industries.

4   "A Simple Substitution Model of Technological Change", General Electric
Research and Developers, Schenectady, New York.

5   W Häfele, "Future Energy Resources", World Power Conference, Detroit, 1974.

6   P F Chapman, "Energy Costs of Materials", Energy Policy, March 1975.

7   G E H Newton, "Heating and Ventilating the Pulp and Paper Industry",
Institution of Heating and Ventilating Engineers.

8   "Energy Prospects to 1985", O.E.C.D., Paris.

9   "A Time to Choose - America's Energy Future", Energy Policy Project of
the Ford Foundation.

10   "Energy Consumption in Manufacturing", Energy Policy Project of the
Ford Foundation.

11   "Financing the Energy Industry". Economic Policy Project of the Ford
Foundation.

12   "Fuel Conservation or Crisis", The Combustion Engineering Association
Conference Proceedings.

13   "The Nation's Energy Future" - A Report to Richard M Nixon, President of
the United States By Dr Dixy Lee Ray.

14 "Increased Cost of Energy; Implications for UK Industry", National
Economic Development Office.

15 "Energy Conservation in the United Kingdom", National Economic
Development Office.

16 W Häfele, A S Manne, "Strategies for a Transition from Fossil to Nuclear
Fuels", International Institute for Applied Systems Analysis, Vienna.

17 R J Deam, "World Energy Supply Analysis", Queen Mary College, London.

DISCUSSION

Question: What is the efficiency of the heated-roller drying process, i.e.
heat used to evaporate the water divided by the heat supplied to
the rollers?

Answer: 50-60%.

Question: Can the use of mechanical drying (squeezing) reduce the amount of
energy required for drying?

Answer: Mechanical drying requires the use of electricity which can upset
the steam-electricity balance. If it becomes necessary to buy
electricity from the grid, any possible energy gains may be lost.

Question: Would you comment on the economics of electric heating?

Answer: It can be used for levelling the moisture profile in the roll as
normally the roll is overdried. Dielectric, microwave, infra red,
and fluid beds could be considered but again there is the problem
of upsetting the steam-electricity balance.

Question: The high cycle efficiencies of the energy plant were mentioned but
is it possible that the very short life and wasteful nature of
some of the applications of the product, e.g. packaging, might
make these figures look over-impressive?

Answer: The paper industry has millions of pounds invested in machinery
and cannot stop production overnight; production must continue as
efficiently as possible.

Question: Does the paper industry perpetuate a product that may not be useful?

Answer: Do we want newspapers and books, or just T.V. and tapes? We need
some paper packaging and bags for carrying potatoes home.

Question: What about recycling the packaging?

Answer: Paper is recycled already, e.g. the paper industry supplies the
Daily Mirror with newsprint which has a large recycled fraction;
it is also used for hand towels and corrugated cardboard. Recycling used energy however, e.g. for removing the printing ink.
The economics depend on whose energy it is, ours in the case of
recycling or the exporters' in the case of new pulp.

Question: The only incentive for industry to conserve energy seems to be
that energy costs money and making profit is what industry is
about. Should industry not respond in a more "humanitarian"
manner to energy conservation?

Answer: Government control is required as shown by the cleaner air and
rivers after Government anti-pollution legislation. Anti-pollution
machinery gives no return on capital, so industry will not install
the machinery unless bound to.

Question: How do you set reasonable standards considering that, for example,
             some of the U.S. anti-pollution laws are technically unachievable
             at the present time, while some pollutants have been overlooked?
Answer: Legislation on SO_2 has been implemented too quickly; at one time
             when the sulphur content of a fuel oil was 2.7% a certain chimney
             was high enough to comply with anti-pollution requirements. Now
             the sulphur content is more than 3% so a higher chimney will be
             required. The industry will have to spend large amounts of money
             to reach some seemingly arbitrary safe SO_2 level.

Question: Should the sulphur be removed from the fuel oil at the refinery?
Answer: An E.E.C. directive will limit the sulphur in fuel oil to 2% and
             eventually 1%. Why, as in the case above, should an industry
             spend money on a higher chimney when eventually it might become
             unnecessary and the fuel oil prices will include an extra amount
             for the removal of the sulphur at the refinery? Given that SO_2
             must be reduced to a certain level there are two possible methods:
             remove the SO_2 after combustion or remove the sulphur from the
             fuel oil. A large user of fuel oil may be able to install large,
             efficient machinery and build high chimneys in order to comply
             with regulations, while the smaller user cannot possibly afford
             this and must pay for low sulphur fuel.

Question: What caused the recent paper shortage, was it a fuel shortage?
Answer: No, a raw material shortage, viz. pulp. The large pulp exporting
             countries also manufacture paper, so are reluctant to export pulp
             when it is in short supply. Until recently, cyclic variations in
             demand for paper were long term (7 years) but now are short term,
             too short to allow large machinery to follow demand. I suggest
             two means by which rapidly fluctuating demand may be met:
             1) storing waste paper for pulping during a shortage, though
                storage can be expensive,
             2) use of smaller more flexible units rather than the present
                large machines.

DOMESTIC ENERGY USE AND ENERGY CONSERVATION IN BUILDINGS

S. J. Leach

Building Services & Energy Division, Building Research Establishment

This lecture is drawn from a Current Paper of the Building Research Establishment "Energy conservation: A study of consumption in buildings and possible means of saving energy in housing" CP 56/75 and copies are available from the Distribution Unit, Building Research Establishment, Garston, Watford WD2 7JR.

Some main conclusions are summarised below

ENERGY CONSUMPTION

All the energy used in buildings has been analysed in terms of the primary energy buildings are accountable for, which ultimately includes the energy overheads incurred in the processing and the distribution of the fuels received by consumers. In terms of primary energy buildings consumed at least 40 per cent and possibly as much as half of all the country's energy in 1972 compared with 41 per cent by industry and 16 per cent by transport. The significance of the large building consumption is in part the important potential for energy saving and in part the possible impact of building energy conservation on the operation of the energy supply industries and on the planning of future energy supply.

The bulk of the building energy is accounted for by the use of the services rather than the materials and construction. Within buildings housing is by far the most important category. Most energy in housing is required for space heating and of the average net annual consumption per dwelling of 81 GJ, 52 GJ is for space heating, 18 GJ for water heating, 8 GJ for cooking and the remaining 3 GJ for lighting, TV, refrigerators etc.

ENERGY CONSERVATION

The report examines a variety of technologies and estimates their potential as a proportion of the national energy consumption. All estimated energy savings are based on maintaining existing thermal comfort and other environmental standards.

The ultimate aggregate potential for energy savings in buildings is estimated to be about 15 per cent of the national consumption which is about as much as used by transport altogether. This total could only be achieved over a long period of time since it depends on the use of technologies which are more fitted to new constructions designed for the purpose. The estimated potential of possible measures applied in existing housing is about 6 per cent of the national consumption. Some of the measures discussed in the report are as follows.

COMBINED HEAT AND POWER: Power stations in which the heat content of the cooling part of the steam cycle is fed to consumers for space and water heating through district mains saves energy by utilising heat which is otherwise lost. It is an established practice in some other countries. The BRE estimation of the energy saving potential is about 10 per cent of the national consumption. The feasibility of combined heating and power in UK conditions is a complex issue. If implemented it would require a long time before its full potential would be realised.

HEAT PUMPS: These operate on the same principles as a domestic refrigerator. They can remove heat such as from the air outside and upgrade it to a higher temperature useful for space heating. Their significance is that they will always supply more heat than the energy required for the operation of the machine. In UK conditions they would supply at least twice and possibly three times as much heat as the electric energy they consume. Their estimated energy saving potential for domestic space and water heating is about 9 per cent of the national consumption. This may become even larger with further development of heat pumps in conjunction with ventilation heat recovery. Heat pumps lend themselves more readily to new constructions designed for the purpose but their scope in existing housing is also a subject for the BRE research programme.

THERMAL INSULATION: Improved thermal insulation saves energy by reducing the fuel consumption required for space heating altogether. The Building Regulations in force from January 1975 constitute a significant advance of the mandatory requirements applicable to new housing. However, some improvement of thermal insulation of most of the existing housing would also be technically feasible. The estimated potential from upgrading the thermal insulation of existing housing is 3-4 per cent of the national energy consumption.

CHOICE OF FUELS: Electricity is the most efficient at the point of use but it will nevertheless save primary energy to use the fossil fuels such as oil, coal and gas directly for certain purposes rather than to convert them first into electricity and then back into heat again. The estimated potential from the direct use of fossil fuels instead of electricity in existing use for domestic purposes such as space and water heating is 3-4 per cent of the national energy consumption. When seen in relation to a particular house the direct use of fossil fuels instead of electricity amounts to a saving of about 50 per cent of the primary energy compared to about 35 per cent from increasing the thermal insulation to a high standard. Heat pumps driven by electricity generated from fossil fuels are as good as, or better than, the direct use of fossil fuels.

ENERGY CONSERVATION ECONOMICS

The report contains an economic assessment of most of the technologies discussed in the paper. The emphasis is on the national viewpoint rather than on the costs and benefits as seen by an individual householder. Working in terms of resource costs rather than prices charged to consumers and using standard criteria for public investment appraisal the economic value of an energy saving has been compared with the total capital and running costs of the measure to assess its cost-effectiveness.

A particular hypothetical case study is that of a package of upgrading the thermal insulation and changing from electricity to fossil fuels in existing housing. The estimated energy saving is about 6 per cent of the national consumption. The estimated total value of the energy saving is about £3,300m and the estimated total cost is marginally less.

The report also considers the situation for individual householders but the issues are so complex that the analysis is little more than one particular illustration of a possible approach. Households differ widely in their energy consumption and one important question for an owner-occupier is how much his particular fuel expenditure is now. A second consideration is how long he expects to remain in the particular house. And last but not least how he would propose to finance any expenditure which might have to be incurred.

DISCUSSION

Question: Is the cost effectiveness of heat pumps in domestic applications as expressed in your lecture too optimistic if the pumps cost ∿ £1,000?

Answer: If a proper calculation is done with, say a coefficient of performance of 3.0, then it is worthwhile.

Question: Are larger heat pumps more efficient?

Answer: At present, yes, as shown by the installation in Nuffield College.

Question: Are solar collectors better than converters of solar energy bearing in mind the cost of materials?

Answer: In new buildings it is possible that converters are better but in pre-existing houses collectors might be preferable.

Question: Have calculations been made of how quickly heat pumps could penetrate the market?

Answer: No. When they are more reliable, this could be done.

Question: Has the potential for well-orientated community collectors been examined?

Answer: No, rather difficult problems are involved but there is potential for capital savings to occur if it is socially acceptable.

Question: Are there any potential savings to be achieved in insulation and
             design of domestic appliances?
Answer: Certainly by lagging the hot water tank but energy consumption of
             the appliances is too small at present to be significant. House-
             hold kitchens are often hotter than they need be due to hot water
             tanks being lagged as an afterthought if energised by a boiler.
             Also pilot lights and unlagged boilers contribute significantly to
             the heating. As estimate shows that pilot lights consume as much
             as 10% of the gas consumed and electric ignition is preferable here.
             Gas ovens are less efficient than the corresponding electric ovens.

Question: Returning to heat pumps, their efficiency may not increase with
             size as it rather depends on the type. Could there be social and
             environmental problems e.g. from noise? With absorption systems
             toxic problems from leaks may exist.
Answer: Most noise comes from the fan distributing the heat and the heat
             exchanger. There is no technical reason why this could not be
             reduced.

Question: The Building Research Association is to build three experimental
             houses with different heating systems. Will the 'solar house'
             have a steerable collector system?
Answer: There will be no steering so that the house is recognisable as such
             and will be acceptable to people and to the Building Industry.
             Thus the systems will not be optimised.

Question: In houses without cavity walls cannot insulation be achieved with
             say 2" thick polystyrene covering the inside surfaces of the walls?
Answer: Lining walls in this manner causes generally unacceptable problems
             around doors and window sills and a reduction in size of rooms as
             well as fire risks.

Question: Do fluorescent lights provide sufficient heat to compensate for
             the removal of skylights in factories?
Answer: We have no information on this problem.

Question: Are large office blocks economical from an energy point of view?
Answer: Generally these follow the regulations of ~35% glass but here the
             problem is often one of prestige, hence the larger glass areas of
             many buildings.

ENERGY USE IN AGRICULTURE

D. J. White

Ministry of Agriculture, Fisheries and Food

SYNOPSIS

An analysis is given of energy use in agriculture and of the benefits of
using fossil energy to supplement solar energy. It is argued that the expenditure of fossil energy in the form of mechanisation and fertilisers has
given us varied foods to satisfy all our needs and enabled us to maximise
food production in relation to land area and men employed. However, fossil
fuels will ultimately prove to be finite and while there is promise of abundant energy at some time in the future, there are formidable technological
problems to be overcome before the promise becomes reality. Thus, the paper
also discusses the prospects for improvements to efficiency of energy use in
agriculture that may be possible through the use of new techniques and
practices without impairing agriculture's high productivity.

INTRODUCTION

Agriculture is thought of as an activity which depends on solar energy to
grow plants. However, in the so-called developed countries of the world,
additional energy is used in the form of fertilisers to increase crop yields
and mechanisation to decrease labour requirements. This supplementary energy,
largely in the form of fossil fuels, is as essential to UK agriculture as
solar energy but whereas solar energy is abundant, fossil fuels are a finite
and diminishing resource.

It is true that there is no shortage of energy at the present time but we
cannot be sure how long fossil fuels will last or if and when alternative
sources of energy will become available to replace them. Estimates of the
lifetime of world energy reserves are notoriously uncertain because of the
unpredictability of such factors as the size of the energy resources, the proportion of these that may practicably be recovered, world population trends
and the growth of energy consumption in both the developed and under-developed
countries of the world. Rocks and Runyon[1] suggest that world oil and gas
reserves will be near exhaustion by 2020 and to satisfy world energy require-

ments then existing, world reserves of coal would have a constant consumption
life span of about 3 to 4 centuries. These authors suggest that there is even
greater uncertainty in relation to resources of nuclear fissile materials but
that the USA may be able to satisfy her present total power requirements for
nearly 300 years. However, on a worldwide basis these estimates may well be
optimistic because they imply little improvement in living standards for much
of the world's population.

One thing is clear and that is that within the lifetime of many now living,
we are faced with depletion and possibly exhaustion of our most versatile and
convenient fuel – petroleum. When that happens, the supply of hydrocarbon
fuels will presumably be continued by synthesis from coal. In spite of this
note of optimism, it is nevertheless sensible to examine our use of energy in
all spheres of activity so that we may plan strategies to meet possible energy
shortages in both the short and longer term for whatever reason. To do so is
prudent, it is by no means a panic measure. And there is another reason too,
so far as agriculture is concerned. With some exceptions, the monetary cost
of energy itself is not the major cost in producing food. However, five-fold
increases in basic oil prices since June 1973 have already shifted the balance
and we must consider the possibility that further adverse shifts are likely in
the future.

In this paper an attempt is made to cover four main topics:

(1) An analysis of energy use in agriculture

(2) Some specific commodities are examined to review the energy inputs
     to their production and the corresponding outputs obtained in the
     form of products useful to man

(3) The benefits to agriculture of fossil fuel to supplement solar
     energy

(4) The prospects for greater efficiency of energy use i.e. greater
     productivity per unit of supplementary energy employed

Throughout this paper the International System of Units is used in which the
unit of energy is the joule (J). Some useful definitions and conversion
factors are given in Appendix 1.

ANALYSIS OF ENERGY USE IN AGRICULTURE

National energy consumption

To give perspective to agriculture's use of energy, in the national context,
it is first useful to look at the national energy consumption[2] and this is
shown in Table 1 for 1973. Our overwhelming dependence on fossil fuels is
illustrated by the fact that 97% of the energy was provided by coal, petroleum and natural gas and only 3% by nuclear and hydro-electricity. The slow
growth of energy from nuclear fission is particularly to be noted; 18 years
after the first nuclear power station, Calder Hall, supplied electricity to
the national grid, nuclear energy accounts for only 2.6% of national energy
used.

Table 1   National energy consumption in the UK 1973

Resource	Consumption Mt	Energy equivalent TkJ	Energy per cent
Coal	134	3,500	37.8
Petroleum	95	4,300	46.5
Natural gas	–	1,170	12.6
Nuclear electricity	–	240	2.6
Hydro-electricity	–	50	0.5
Total		9,260	100.0

Agricultural use of energy

A breakdown of the main items contributing to energy usage in UK agriculture
is given for 1973 in Table 2. This shows that at 23.6% petroleum fuel is the
largest single user of energy, closely followed by fertilisers (23.1%), with
off-farm feedstuff processing (14.2%), machinery (14.4%) and electricity
(9.2%) also having substantial inputs. In the following paragraphs, a little
more is said about the derivation of some of these values and further details
are given in references 3, 4 and 5.

The largest use of energy in agriculture is in the form of petroleum fuels and
in 1973[2] amounted to 1.67 Mt, constituting 1.7% of total petroleum fuels
used in the UK, with an energy equivalent of 75 TkJ. However, in order to
make petroleum fuels available for use, energy has to be expended in refining

and transport and it has been established[6] that for every litre of petroleum produced, an additional 0.13 litre must be put in. Thus the energy consumed (75 TkJ) when multiplied by 1.13 gives the primary energy input (85 TkJ).

The consumption of electricity[2] on UK farms in 1973 was 3.95 TWh which is equivalent to 14.2 TkJ and amounted to 1.8% of national consumption. There is reason to believe[7] that 41% of this electricity is used for domestic purposes and that only 59%, that is, 8.4 TkJ is actually used for agricultural purposes. However, this is the energy used on the farm and it is necessary to convert to its primary energy input. The generating efficiency of UK power stations[2] during 1973 was 30%; 8% of the power generated is consumed at the station and there is a further loss of 8% of that sent out in transmission and distribution. The result is that the amount of energy available for consumption is only 25.4% of the primary energy input at the power station. Thus the power consumed on the farm (8.4 TkJ) when multiplied by the reciprocal of 0.254 (3.94) gives primary energy consumed (33.1 TkJ).

Table 2   Primary energy consumed in UK agriculture, 1973

Item	TkJ	per cent
Solid fuel	4.1	1.1
Petroleum	85.0	23.6
Electricity	33.1	9.2
Fertiliser	83.5	23.1
Machinery	52.0	14.4
Feedstuff processing (off-farm)	51.3	14.2
Chemicals	8.5	2.4
Buildings	22.8	6.3
Transport, services	16.3	4.5
Miscellaneous	4.3	1.2
Total	360.9	100.0

While petroleum fuel and electric power are the most obvious energy inputs used on the farm, of equal importance are the complex network of commodities and services behind the provision of fertilisers, agricultural chemicals, machinery, farm buildings, water, animal feedstuffs and many others too

numerous to mention. All of these inputs demand the use of energy to make
them available, as was shown for fuels and power themselves and may be considered to have primary energy equivalents. The values shown in Table 2 for
primary energy consumed for off-farm feedstuff processing, chemicals,
buildings, transport, distribution services and miscellaneous were assessed
by Leach[8] for 1968 and for the purposes of Table 2 it has been assumed that
these values also apply in 1973. The energy values for fertilisers and machinery for 1973 have been assessed as explained in the following paragraphs.

Table 2 showed that fertilisers constituted the second largest user of energy
in agriculture and considerable energy is expended in their manufacture.
Table 3 shows the amounts of fertilisers used annually[9], the energy equivalents[10], that is, the energy expended in manufacture taking account of that
in the basic materials plus that demanded by the manufacturing process and in
transport, and the resulting calculated energies. The total energy involved
in the manufacture of all fertilisers is 83.5 TkJ and 87% of this is due to
nitrogen. Nitrogen is the most energy intensive nutrient because its manufacture is based on natural gas which is used as the feedstock for the synthesis of ammonia which is used to produce ammonium nitrate, the most common
nitrogenous fertiliser. The elements phosphorus and potassium are included
in many compound fertilisers and both require energy to mine and to transport.

Table 3 Energy involved in fertiliser manufacture, UK 1972/73 (June to May)

Nutrient	Consumption kt	Energy equivalent GJ/t	Energy consumed TkJ
Nitrogen (N)	946	77.0	73.0
Phosphate (P_2O_5)	481	14.3	6.9
Potash (K_2O)	435	8.3	3.6
Total			83.5

Whenever machinery is purchased an expenditure of energy is necessarily incurred which is the energy that was involved both in the machine's manufacture
and in the production of materials from which the machine was made. Thus the
idea arises of capital (money spent on machines) having an energy equivalent
and it is possible to make an approximate estimate for a particular industry
if both total energy consumption and product value are known. For the manu-

facturing industries of the UK as a whole in 1973, the total energy used[11] was £2.1 x 10^{10}. Thus the energy equivalent of capital for manufacturing industry was 3740 TkJ/£2.1 x 10^{10} = 178 MJ/£. The monetary value of new plant, machinery, vehicles and spares used to support UK agriculture in 1973 has been assessed as £292 M from updated information[12] and involved the assumption that spares and labour each constitute half of the sum spent under the heading 'repairs'. Thus the total energy involved in the production of new machines is £292 x 10^6 x 178 MJ/£ = 52 TkJ.

Agricultural energy use in the national context

It was shown in Table 1 that the total consumption of energy in the UK in 1973 was 9260 TkJ and when this is divided into the total energy consumption of agriculture of 360.9 TkJ, agriculture's share of national consumption is 3.90%, that is, about 4%. It should be appreciated that this is the proportion of energy used to produce food at the farm gate and that further substantial inputs of energy are involved before the food reaches the consumer in such things as processing, transport, packaging, distribution, retailing and household preparation. Leach[8] has made estimates for 1968 of the primary energy consumed in the fishing industry, the food processing industries including any accompanying transport and packaging, and in food distribution. Assuming these figures still to apply in 1973, Table 4 shows the primary energy consumed in the UK in making all foods (indigenous and imported) available to the consumer. It will be noted that food production uses 11% of national energy consumption and that the food processing industries use over 5%, a little more than agriculture's 4%. Table 4 does not consider the energy consumed in food preparation in the home but it is useful to refer to some figures for the USA[13] which suggest that the total energy in bringing food to the plate accounts for up to 15% of the USA's energy consumption and that as much as one-third of this occurs in the home. No assessments appear to be available for the UK at the present time.

The matter of imported foods and animal feedstuffs has not been considered so far because in production these represent no charge to the UK energy budget. However, it is salutary to remind ourselves that in one sense, they may also be regarded as supplementary energy in much the same way as fuel, for if they were not available to us to maintain our production we would have to grow substitutes, assuming we had the land resources to do so, and that would involve an additional demand on our national energy supplies. Leach[8] has

Table 4 Primary energy consumed in the UK in producing food

Function	Energy TkJ	per cent	per cent of national consumption
Agriculture (to farm gate)	361	35.7	3.90
Fisheries	33	3.3	0.36
Food processing industries	476	47.2	5.15
Food distribution	139	13.8	1.50
Total	1009	100.0	10.91

estimated that in 1968, the energy for producing and transporting these imports totalled some 326 TkJ made up of food (260 TkJ), fish (13 TkJ) and animal feedstuffs (53 TkJ). If the energy equivalent of imports is added to the total in Table 4, the total energy involved in bringing food to the consumer in the UK is 1335 TkJ, which is equivalent to 14.4% of national energy consumption. As explained earlier, the actual expenditure of UK national energy consumption is about 11% but of course domestic consumption is excluded from both of these figures and on the basis of American experience may well be expected to add one-third as much again. It becomes clear at this point that while agriculture's use of energy in the national context appears to be relatively small, the total amount of energy used in the overall food production cycle represents a significant proportion of national usage.

While digression on the energy involved in food production was justified to give perspective to agriculture, which is the primary production stage of food, this paper is concerned with agriculture rather than food and it is pertinent to enquire what we get in exchange for this expenditure of energy. In 1971/72, 53.6% of our unprocessed food was produced in the UK[14], this figure being the monetary value of food moving into manufacture and distribution derived from home agriculture from all sources. Thus, we may say that 4% of national energy consumption produces a little more than half of our unprocessed food and in the light of these figures it must be concluded that agriculture has a claim to receive high priority as an energy user and that even in times of dire energy shortages its modest demands should continue to be met. It follows also that there is little scope for energy economies to have any effect in reducing national consumption significantly although such economies are clearly desirable where they contribute to minimising the rising costs of food production without putting productivity at risk.

ENERGY FLOWS IN AGRICULTURAL PRODUCTION SYSTEMS

It is instructive to look at systems for particular commodities to study the
energy flows, to see what goes in and what is obtained in return. Such calculations as these have been done by a number of authors, namely, Slesser[15],
Pimental et al[16] and Leach[8, 17, 18]. Following the general methodology
of these authors, the present author has made some assessments[3, 19] for
selected arable and horticultural crops and for some animal products and the
principal results are summarised in Table 5. In addition, some results of
Leach[8] are shown for comparison. In the following section, an explanation
is given of the approach adopted and some examples of energy budgets for
wheat and dairy cattle (milk and meat) are given in Appendix 2. The publication by Leach[8] presents some detailed energy budgets for a range of crops,
animal products and farming systems.

The methodology of energy accounting

To estimate the energy input, support energy or energy subsidy as it is var-
iously called all items involving an expenditure of energy from expendable
resources must be included. Solar energy is not included as an input because
it is abundant. The results given in this paper have been derived taking
into account wherever possible the energy inputs due to petroleum fuels,
electricity, fertilisers, agricultural chemicals and machinery using the
following sources. The energy inputs due to petroleum fuels and electricity,
for a selection of cropping systems, were given by Rutherford[20, 21] and for
poultry production by Hann[22]. Rutherford[20] also gave machinery depreciation values in terms of £/ha for selected enterprises and so energy values may
be calculated using the energy equivalent of capital given earlier of 178 MJ/£.
Fertiliser values for the principal arable crops are the average amounts
supplied (total amount of fertiliser used on that crop divided by the total
acreage of crop, including any acreage receiving no fertiliser) as determined
by the Rothamsted Surveys[23, 24] supplemented where necessary and for other
crops by MAFF recommendations[25]. Fertiliser energy equivalents used were
those given in Table 3, amounts of agricultural chemicals were obtained from
various sources and the energy equivalent used was 106 MJ/kg from reference
26. Data relating to various systems of animal production were supplied by
Frost[27]. It may be taken that energy inputs are mostly underestimated since
it has not been possible to assess all items involving energy expenditure.

The energy output of the system is generally the metabolisable energy of the food and in some systems the output may be of more than one kind, for example, eggs and meat from poultry and milk and meat from cattle. Crop yields and animal populations have been obtained from various sources[12,22,28,29] and nutritional data relating to metabolisable energy and protein yields from MAFF sources[30,31]. Apart from the outputs consumable by man as food there are other products such as crop residues, animal wastes, manures, wool and hides all of which have a use to man either directly or indirectly by re-cycling through the crop-animal food chain. However, the by-products are not included in the present analysis, not because they are unimportant, but be-cause the prime purpose of agriculture is the production of food. Those by-products which are recycled within the agricultural system are of course taken care of automatically.

Energy in the production of particular commodities

The results presented in Table 5 may be explained as follows. The first col-umn gives the product or commodity and the second gives the energy input or support energy on a per annum basis in relation to land area employed to raise the crop (GJ/ha year). The third column gives the output energy of the crop per annum in relation to land area (GJ/ha year) and provides a measure of the effectiveness with which land area is used to produce energy from food. For animal products, the relationship to land is established through that required to grow all the feed the animals need. This is easily established for rumin-ants where the feed consists of combinations of grass, hay, silage and cereals but less readily for pigs and poultry. For pigs, feed may be mainly barley with fish meal and for poultry a compounded feed consisting of grains such as maize and barley with fish meal and soya. For these animals, an all-barley diet has been assumed to assess the 'energy cost' of the foods they eat and to relate their requirements to land area. The fourth column gives the ratio energy output to energy input and is a measure of the 'efficiency' of a food conversion process. The higher the value of E, the greater is the energy out-put for a given energy input. The fifth column gives the protein output per annum on a basis of land area (kg/ha year) and the sixth column the cost of the protein in energy terms (MJ/kg).

It is necessary to add the cautionary note that unique values of E for parti-cular commodities are not to be expected since these estimates depend on particular agricultural practices and mechanisation systems. However, an

Table 5 <u>Estimates of agricultural use of support energy</u>

Commodity or product	Energy input or support energy GJ/ha year	Energy output or metabolis- able energy GJ/ha year	E = Column 3 / Column 2	Protein output kg/ha year	Energy input to produce protein MJ/kg
Wheat	19.6	61.0	3.11	435	45
*Wheat	17.8	56.2	3.35	400	42
Barley	18.1	60.6	3.36	310	58
Oats	18.8	66.4	3.52	480	39
White bread	31.7	47.1	1.48	368	86
Potatoes	52.0	69.3	1.33	460	113
*Potatoes	36.2	56.9	1.57	376	96
Sugar beet (at farm gate)	25.2	82.5	3.28	Not applicable	
Sugar from beet	109	82.5	0.76	"	"
*Sugar from beet	124.4	82.9	0.67	"	"
Carrots	25.1	32.5	1.30	234	107
*Carrots	27.6	30.0	1.10	219	126
Brussels sprouts	32.4	10.9	0.34	296	109
*Brussels sprouts	47.9	9.1	0.19	244	196
Onions, dry bulb	93.4	27.7	0.30	276	338
Tomatoes (glasshouse)	1300	62.0	0.05	945	1360
Milk	17.0	12.0	0.70	145	118
*Milk	26.9	10.0	0.37	129	208
Beef (from dairy herd)	10.4	3.2	0.31	40	257
Beef (from beef herd)	10.6	2.4	0.23	31	348
Pigs (pork and bacon)	18.0	11.4	0.63	76	238
Sheep (lamb and mutton)	10.1	2.5	0.25	22	465
Poultry (eggs)	22.5	6.0	0.26	113	200
*Poultry (eggs)	48.5	7.0	0.14	137	353
Poultry (broilers)	29.4	4.3	0.15	145	203
*Poultry (broilers)	58.9	5.9	0.10	203	290
Poultry (turkeys)	23.6	7.1	0.30	129	184

* Results of Leach[8]; all other values by author[3,19]

attempt has been made throughout to use average values, that is, average fertiliser inputs and average crop yields. It will be noted that the values of Leach[8] and of the present author sometimes differ by a factor of 2 but this does not affect the general conclusions which are drawn in the succeeding paragraphs.

It is apparent from Table 5 that in terms of both energy and protein some commodities are produced more efficiently than others. For example, the arable crops cereals and potatoes have E values ranging from 1.3 to 3.5, while animal products have lower E values generally in the range 0.15 to 0.31 with pig products and milk being somewhat higher. More energy is required to produce protein from animal products, generally in the range 184 to 465 MJ/kg, than from arable crops for which energy values range from 39 to 113 MJ/kg of protein. In relation to land area employed, both the energy output and protein output were significantly greater for arable crops than for animal products. None of these conclusions should cause any surprise since animals feed on plants and are bound to produce less energy as meat than that contained in the plants eaten. It is of course as well to remember also that some plants cannot be eaten by man directly and have to be processed through ruminants to produce an edible product.

The horticultural crops in Table 5 showed wide contrasts but the various values generally lay intermediately between those for arable crops and animal products. The exception was glasshouse grown tomatoes where the energy supplied to maintain the required growing temperature was by far the dominant factor and resulted in an E value of 0.05.

While estimates based on both usable energy and on protein serve to illustrate our very considerable dependence on energy subsidies to capture the solar energy that goes into producing our food, results such as those in Table 5 must be treated with circumspection. Important though they are, energy and protein are not the only things that we get from food or even in some cases the most important. Food also supplies minerals and vitamins[31] and, of course, there is the pleasure that is derived from eating varied foods. These are benefits which are not readily quantifiable but it should be accepted that it is as valid to use energy to produce pleasure foods as it is to use energy for pleasure motoring.

It is not suggested that the results in Table 5 should be used to argue that

we should cease to eat animal and horticultural products and live instead on
a diet of cereals and potatoes. However, we cannot ignore the fact that we
are given an alternative dietary strategy that could be invoked in a situation of dire energy shortage. But it is not suggested that such a situation
exists today and the fact is that in affluent societies there is an increasing demand for meat, milk and eggs which, if satisfied, can tend only to reduce our efficiency of energy use in producing metabolisable energy and protein. The question is as Leach[8] puts it 'not whether animal products are a
wicked luxury, but when is enough enough?'. Leach[8] shows that of all farm-
land in the UK, only 8.3% provides food directly to man, nearly all the remaining 92% being devoted to feeding livestock, which also consume around
7 million tonnes of imported feedstuffs each year. Since this 8% provides half the energy output of UK farms and animals the other half (as can
be seen from Table 6), the vegetable system is much more efficient in the use
of land resources than is the animal system. It is not true of course that
all land used to raise animals could be used for cropping but it certainly
would appear feasible to feed the UK population for a lower energy input
with a shift away from animal products in favour of cereals and vegetables.
It must be emphasised again that such a shift is practical politics only under
conditions of dire emergency.

Energy budget for UK agriculture

The presentation of energy budgets for particular commodities leads to the
idea of presenting an overall budget for UK agriculture and this is given in
Table 6. The inputs are those in Table 2 with the addition of energy associated with feedingstuffs imported to aid production of animal products in the
UK[8].

The outputs in Table 6 are the energy equivalents of the crops passing into
human consumption. Thus for crops such as cereals which are used to feed
both humans and animals, only that portion used directly for humans is included. Similarly, other intermediate outputs such as animal feedingstuffs,
wastes and plant residues are not shown because they are recycled (albeit
some wastefully) within the system to feed animals or returned to the land to
raise more crops. The output values for the plant and animal products were
assembled[19] using various sources[22,29,30]. Human food was chosen as the
basis for determining output because food production is the primary purpose of

agriculture but it should not be overlooked that energy is also involved in the by-products that man uses directly such as hides and wool just as energy must be expended in the production, say, of man-made fibres. Thus, in that sense, it may be said that the usable energy is greater than that based on food alone.

Table 6 <u>Overall energy budget for UK agriculture up to the farm gate,</u>
<u>TkJ per annum</u>

Input or energy subsidy		Output or energy available to man			
Solid fuels	4.1	Cereals	56.8		
Petroleum	85.0	Potatoes	13.0	Arable crops	84.4
Electricity	33.1	Sugar beet	14.6		
Fertilisers	83.5				
Machinery	52.0	Vegetables	2.4	Horticulture	3.2
Feedstuff processing (off-farm)	51.3	Fruit	0.8		
Chemicals	8.5	Milk	38.0		
Buildings	22.8	Beef	13.8		
Transport,distribution and services	16.3	Pigs (pork and bacon)	17.3		
Miscellaneous	4.3	Sheep (lamb and mutton)	4.5	Livestock	80.6
Imported feedstuffs (including transport)	53.2	Poultry (eggs)	4.5		
		Poultry (chicken)	1.9		
		Poultry (turkey)	0.6		
Total	414.1		168.2		168.2

$$E = \frac{\text{Output energy}}{\text{Input energy}} = \frac{168.2}{414.1} = 0.4$$

Table 6 shows that the overall E ratio for UK agriculture is 0.4. Other authors have derived similar but lower values e.g. Leach[8] (0.34) and Blaxter[32,33] (0.36). Thus to obtain the energy we consume in the 50% of our unprocessed food that is produced in the UK, we put in $2\frac{1}{2}$ times as much energy in the form of fossil fuels and imported feedstuffs. While this is one way of expressing the use that agriculture makes of the energy that it consumes, it is unfortunate that no realistic common basis exists to compare the efficiency with which various industries or activities use energy. Thus, we are not able

to assess agriculture's performance in relation to other users of energy but
the author would risk making the judgment that it is energy well spent in
producing such a major necessity of life.

BENEFITS TO AGRICULTURE OF SUPPLEMENTARY ENERGY USE

Trends in energy use and manpower and tractors employed

Table 7 illustrates some trends over two decades from 1950 to 1970 of the increase in primary energy used, in the form of fossil fuels and electricity[2,34]
the number of all types of worker employed on agricultural holdings[29,35,36];
and the number of tractors employed. Over the two decades considered, energy
use increased by a factor of 1.7 while the labour force was reduced to less
than one-half. These figures clearly contain an indication of the increasing
role played by mechanisation and this is additionally borne out by the increase in the number of tractors by a factor of 1.5 between 1950 and 1960. In
the subsequent decade, the number of tractors has tended to stabilise, never
exceeding 400,000, and the reason for this is that the average horsepower of
a tractor has increased very considerably during that time. If figures for
total horsepower employed could be shown, instead of number of tractors, a
continued increase in the power used would be revealed. It may seem surprising for an activity that takes place away from industry, but Leach[8] states
that in terms of energy use per man employed, UK agriculture is in the same
bracket as heavy engineering. More pertinent however, is the fact that diminishing returns are now being experienced in that the energy required to
substitute for labour is rising substantially.

Table 7 Trends in primary energy usage in the form of petroleum fuels, solid
fuels and electricity, number of workers employed on agricultural
holdings (full-time, part-time and temporary) and number of tractors

Year	1950	1960	1970
Primary energy used TkJ	67	80	104
Number of workers, thousands	918	693	430
Number of tractors, thousands	240	360	350

Table 8   <u>Trends in primary energy usage in the form of fertilisers, yields of
some arable crops and corresponding energy outputs</u>

Commodity	Year	1950	1960	1970
Wheat	Energy input GJ/ha	2.77	4.83	7.69
	Crop yield t/ha	2.72	3.64	4.08
	Energy output GJ/ha	39.2	52.4	58.7
Barley	Energy input GJ/ha	2.76	4.06	6.58
	Crop yield t/ha	2.52	3.25	3.63
	Energy output GJ/ha	37.9	48.9	54.6
Potatoes	Energy input GJ/ha	9.27	13.4	17.2
	Crop yield t/ha	19.1	20.5	26.7
	Energy output GJ/ha	48.1	51.6	67.3
Sugar beet	Energy input GJ/ha	9.16	13.4	16.2
	Crop yield t/ha	26.8	34.8	36.0
	Energy output GJ/ha	59.6	77.4	80.1
Primary energy of all fertilisers used in agriculture TkJ		22	42	72

Energy use and output

Table 8 shows for four different arable crops, primary energy usage on the
farm in the form of fertilisers[24,37] (nitrogen, phosphate and potash), crop
yields[29,35,36] and the corresponding energy outputs calculated as metabolisable energy. Because of the vagaries of the UK climate, crop yields vary
significantly from year to year and to compare one year with another may produce misleading conclusions. To reduce this possibility the crop yields
given are 5 year averages of the central year named and the 2 years on either
side. For all four arable crops in Table 8, increasing fertiliser energy inputs are accompanied by increased crop yields and correspondingly higher
energy outputs. By taking differences between adjacent columns for different
years (see Table 9), it will be noted that increased energy inputs in the
form of fertilisers produced, in general, a greater return in output energy.
An exception to this was sugar beet over the decade from 1960 to 1970 and it
seems likely that this crop may already be fertilised to achieve near maximum

yields. It would no doubt be erroneous to link improvements in crop yields
solely with increased use of fertilisers but this achievement can be attributed to the adoption of improved seed varieties coupled with enlightened use
of fertilisers and better crop protection chemicals.

Table 9 Increase in energy usage in the form of fertilisers and increases
         in yields and corresponding energy outputs of some arable crops

Commodity		Decade 1950 to 1960	Decade 1960 to 1970
Wheat	Energy input GJ/ha	2.06	2.83
	Crop yield t/ha	0.92	0.44
	Energy output GJ/ha	13.20	6.30
Barley	Energy input GJ/ha	1.30	2.52
	Crop yield t/ha	0.73	0.38
	Energy output GJ/ha	11.00	5.70
Potatoes	Energy input GJ/ha	4.13	3.80
	Crop yield t/ha	1.40	6.20
	Energy output GJ/ha	3.50	15.70
Sugar beet	Energy input GJ/ha	4.24	2.80
	Crop yield t/ha	8.00	1.20
	Energy output GJ/ha	17.80	2.70

PROSPECTS FOR IMPROVEMENT IN EFFICIENCY OF ENERGY USE

Petroleum fuels

The largest use of energy in agriculture is in the form of petroleum fuels
(Table 2, 23.6%) and Table 10 gives the breakdown of usage in various
sectors[21]. The energy is in terms of energy consumed, not primary energy
input, and the total (86.7 TkJ) is greater than that given earlier of 75 TkJ.
The difference is more than accounted for by the item listed as vehicles,
lorries, vans and cars in Table 10 which was not included before because it
did not appear in the relevant national statistics[2] quoted. Table 10 shows
that nearly one-half of petroleum fuels are used to power tractors and other
machines, one-quarter for glasshouse heating and one-tenth for heating and

drying.

Table 10 Use of petroleum fuels in agriculture, UK 1972/73 (May to June)

Sector	Consumption kt	Energy equivalent TkJ	Energy per cent
Tractors and self-powered machines	925	42.1	48.5
Vehicles, lorries, vans, cars	293	13.7	15.8
Glasshouse heating	496	21.8	25.2
Heating, drying, lighting	198	9.1	10.5
Total	1912	86.7	100.0

Cultivations

The largest use of petroleum fuels is in connection with tractors and other self-powered machines and much of this is no doubt consumed in field operations comprising cultivations, seed drilling, spraying, harvesting and general farm transport. However, for reasons quite unconnected with the energy situation, there has been a move towards minimising cultivation operations with the object of decreasing manpower requirements and reducing the damage done to soil structure by compaction and smearing due to the repeated passage over the soil of heavy machines. It just so happens that there is a further bonus to be derived in that fuel may be saved due to the reduced amount of work done on the soil. Where soil conditions allow it, seed may be directly drilled into the ground without prior cultivation and the area of land so treated is increasing annually. A disadvantage with direct drilling is that greater use of herbicides may be necessary to control weeds than when control is by cultivations and the use of herbicides introduces an additional energy expenditure. However, calculations suggest[38] that direct drilling with the application of herbicides still results in a significant saving of energy over conventional systems of cereal establishment.

Drying of crops

Table 10 shows that heating and drying operations, apart from glasshouse heating, consume a little more than 10% of the petroleum fuels used in agriculture. Where artificial drying of crops is practised, this can be a considerable user of energy in relation to the field operations that the crops require[21].

Unfortunately, in the UK climate cereals are frequently harvested at a
moisture content in the range of 18-22% and this must be reduced to nearer
15% for purposes of storage. With respect to conservation of grass, minimum
energy is used if natural drying methods are employed or if the grass is
made into silage (put into a clamp or tower silo, from which air is excluded)
but other methods of preservation such as barn drying of hay and high temperature drying of forage crops, with conversion into cobs, pellets or wafers
are also still economic practices although they are substantial energy users.
Some possible developments that may lead to improved efficiency of energy use
are considered in the following paragraphs.

Forage, as harvested and dried, often has a protein content of 18-25% in the
dry matter and this is higher than ruminants can make use of with the result
that the excess is wastefully excreted. However, by mechanically squeezing
the harvested crop some of the proteinaceous dry matter can be removed in
the liquid[39] which is squeezed out and a fibrous residue is left behind,
which is little degraded, and can be fed directly to cattle in that form,
made into silage or can be dried and made into pellets. The liquid represents a very suitable feed for non-ruminants such as pigs. Thus much better
use is made of the available protein and research and development work is
aimed at improving the storage qualities of the liquid and at extracting from
it a protein containing powder which is a suitable ingredient for animal
feeds. With respect to the pressed fibre, the advantages are two-fold for
the enterprise which normally makes pellets from high temperature dried forage crops. First, there is a saving on the drying load because there is less
water to be removed from the crop and second it is possible to achieve an increased output of dried crop per unit time from a drier of a given size. As
a particular example, it can be shown that if an incoming crop of 80%
moisture content is squeezed to a fibre of 67% moisture content, the amount
of water to be removed by drying is more than halved and so is the fuel required. Notwithstanding, high temperature drying is still a costly business
in energy terms and the real advantages for energy conservation must be in
ensiling the pressed residue. While it seems possible to link wet fractionation of forage crops to the activities of the large grass driers, who tend
to be farmers' cooperatives more akin to an industrial enterprise, it is
somewhat more difficult to work wet fractionation into the agricultural
system allied to ensiling the pressed residue, except perhaps on very large
farms. However, these aspects are receiving attention.

With respect to cereals, post-harvest drying is normally essential in the UK, the moisture removal necessary depending on seasonal weather conditions. One investigation which is under way by Wilton[40] envisages using the heat of combustion of the straw to dry the whole crop (straw and grain). An average hectare of barley in the UK yields roughly 3.5 tonne of straw[41] and with an assumed moisture content in the field of 30% and a gross energy content of 18.4 MJ/kg, the gross energy value is about 45 GJ/ha. To dry the whole crop, under the assumed conditions, could well take up to 10 GJ/ha so that energy considerations alone are favourable to the idea of burning straw to dry grain and there could be a surplus left over to dry, for example, forage crops. However, the question is can it be done in practice and be economic in monetary terms. Usually, when combine harvesting grain, the straw is left behind on the field and it is collected only if it is needed for some purpose such as animal bedding or feed. It is doubtful if it would be collected purely for the purpose of drying the grain but there is another circumstance in which the whole crop, both grain and straw, could be harvested together where the grain is to be used, as is most of the cereal crop, for animal feed.

The cereal crop is generally harvested when the grain is ripe and it is well known that losses occur by shedding of the grain before the combine enters the crop, from the action of the combine header and cutter bar and through failure of the combine to extract all the grain from the straw. Some authorities put these losses as high as 10%. Unfortunately, the combine harvester is not capable of harvesting grain at the point when maximum potential yield occurs, this being some 2-3 weeks earlier than the stage at which a combine would normally be set to work. However, Wilton[40] has harvested the whole crop by means of a precision chop forage harvester at the point at which the grain is held well in the ear. The chopped whole crop is taken to a rotary drier and then passes to a separator where the grain is removed and the straw is fed to a straw burning furnace which supplies heat to the drier. It may well be that harvesting the crop in this manner in an immature state is only practicable if the straw can be used to dry the crop, as otherwise very substantial inputs of fuel are required.

Glasshouse heating

Glasshouse heating, as shown in Table 10, accounts for 25% of the petroleum fuel used in agriculture and is perhaps the sector which has been hardest hit by rising fuel prices, simply because heating costs form such a large proportion of the cost of producing protected crops. Nevertheless, there are prospects for making improvements[42] though admittedly they are marginal in relation to the cost increases that have taken place in recent years in fuels for heating purposes. The situation today is that 90% of the heated glasshouse acreage of the UK is burning oil and although such boilers are capable of high efficiency this can only be maintained by regular checks on the air/oil mixture and this is readily done by simple measurements of carbon-dioxide concentration and the temperature of the flue gases. Irrespective of the grade of oil burnt, heating an excessive amount of air through the boiler is the largest source of heat loss. It is elementary but necessary to say that things such as unlagged or inadequately lagged pipes and flanges are to be avoided yet they do occur.

The ability to control the environment within glasshouses at precisely defined levels has resulted in the collection of sound experimental data and the publication of what are called blueprint methods of growing by which specified minimum night and day temperatures are maintained. There is further scope here to devise programmes for crop growing which ensure optimum use of supplementary heating (fossil fuel energy) to complement the available solar energy. There is clearly little point in maintaining a higher temperature than is necessary in the absence of solar energy to facilitate growth and plant performance may be optimised by applying positive heating to increase temperature with increasing light intensity. Such systems must of course be judged not necessarily on the amount of fuel saved but on the relationship between fuel used and crop output. It is in fact unwise to be attracted by the fuel savings possible by lowering set temperatures. While fuel may be saved, all the evidence seems to be quite clear that reducing blueprint temperatures results in a greater loss of yield and earliness of the crop than can be compensated for in monetary terms of fuel saved. Equally, it is important to adhere to the recommended temperatures since maintaining for example $17^{o}C$ instead of $16^{o}C$ could result in an extra 13000 gallons of oil being used per hectare per year. Even with an accurately calibrated thermostat, if this is exposed to radiation losses to the cold glass the temperature will fall below that of the surrounding air. The use of an aspirated screen avoids such errors and improves the response of the thermostat by reducing the cyclic fluctuations in on/off control systems.

Glass has a high rate of heat loss and it is tempting to consider double

glazing as a method of conserving heat. Quite apart from the high capital
cost, the main drawback is that light transmission is significantly reduced
with a resultant loss of crop. This was brought out in some experiments[42]
where internal lining of the glasshouse with polyethylene film reduced heat
loss by as much as 38% but this was offset by a reduction of as much as 14%
in light transmission. It is unfortunate that the critical months of low
light intensity when most heat would be saved are likely to result in a
greater monetary loss of crop than is saved in fuel. Inevitably also, the
film collects dirt and the light loss would increase with time to the detriment of crop production. Dirt on glass is a particularly intractable problem
and it is worthwhile saying that much of the dirt comes from boilers with
faulty combustion. Another approach that has been examined[42] to the reduction of heat losses is by the use of blinds, normally used to control day
length, drawn only during the hours of darkness. Measurements on a commercial nursery growing all-the-year-round chrysanthemums in an 0.2 ha block
fitted with black polyethylene blinds has shown that when the covers are in
position the heat required is reduced by one-third. Because the blinds are
used only for fuel saving, rather than for modifying day length, savings can
only be made at night, and it is estimated that the annual saving is about
20%. The installation of the blackout blinds for crops not requiring day
length control could perhaps be justified but there are practical problems
to be overcome, such as crop supports and the effect of the blackout on the
plant environment. Because the blinds prevent the plant foliage losing heat
by radiation to the cold glass it is likely that the plants would be at higher temperature for the same air temperature, compared with the glasshouse
without an opaque cover. There is then a realistic possibility of maintaining the same plant temperature with a lower air temperature and if this is
the case then further heat savings are also possible. It has also been
found[42] that under average night time conditions of partial cloud and an
8 km/h wind, a greenhouse with a double skinned inflated plastic film roof
required 20% less heat than a glasshouse. The light transmission of this
novel house is, if anything, slightly better than the modern light alloy
framed glasshouse because of the reduction in structural members required.
With this type of house, the heat losses do not increase so markedly as the
losses i : a traditional house as the wind speed rises. Because increased
losses due to wind arise from decreased resistance to convective heat transfer at the outer glass surface and also from increased air leakage, it
follows that some form of artificial windbreak could be worth considering and

indicates that care should be taken in the siting of greenhouses.

Since the object of heating the glasshouse is to maintain the correct plant
temperature, rather than controlling the air temperature, it may be possible
to use radiant heating to achieve the required leaf temperature for a reduced
air temperature. At present there is a lack of data on the plant response to
such a regime and clearly there are difficulties in arranging a radiant heating system inside a glasshouse that would not interfere with light transmission. Some work on radiant heating has nevertheless been done abroad in
which the heating was obtained by circulating oil at $400^{o}C$ in pipes and by
the burning of gas within steel pipes.

A further process which is often inefficient is that of sterilising the soil
in glasshouses by means of steam heating. Sheet steaming as it is called may
be cheap in labour but it is expensive in energy because of the length of
time needed to achieve any depth of soil treated. To raise the temperature
adequately at a depth of 15 to 23 cm, 40 to 50 kg of steam per square metre
of soil are needed but when the soil is heated by buried perforated pipes
only half the amount of steam is needed for the same depth. For every single
kg of steam per square metre of soil treated that is wasted, 1000 gallons of
oil per hectare are needed.

Fertilisers

Fertilisers are a major user of energy (Table 2) at 83.5 TkJ per annum representing 23.1% of energy usage in agriculture. Much of the increase in
agricultural output is directly related to the increased use of fertilisers
but this does not imply that present supplies are being used in the optimum
manner or that present patterns should be continued in the future. More
effective systems for predicting fertiliser requirements in relation to such
things as soil type, previous weather, cropping sequence and crop needs
should lead to their more effective use[43]. In addition, there may well be
reason to grow more leguminous crops and to choose rotations which decrease
the need for artificial fertilisers.

An exciting possibility is the development of biological methods of nitrogen
fixation[44,45] from the atmosphere by means of plant and soil microbes, the
transfer of this fixation ability to plants that do not possess it, or of re-
producing the process on a commercial scale. Nitrogen is fixed in nature by

bacteria, which employ enzymes, called nitrogenases. Because of the presence of transition metals (iron and molybdenum) in the nitrogen fixing enzyme, complex compounds in which transition metals combine with the nitrogen molecule have been prepared and their reactions studied[45]. Of particular relevance to the biological processes is a recent success in preparing complex molecules based on tungsten and molybdenum in which a bound nitrogen molecule became reduced to ammonia. If such a process could be harnessed on an industrial scale it would be possible to make nitrogen fertiliser at a small fraction of the present energy cost. It is possible though that there are more direct biological solutions in that legumes harbour colonies of nitrogen fixing bacteria and have the ability to absorb their output directly. Legumes have diminished in agriculture as the result of cheap nitrogen fertiliser, but a more exciting prospect than increasing their use would be to develop cereals and root crops capable of coming to the same comfortable symbiotic arrangements with colonies of nitrifying bacteria as do peas, beans and clover. An outstanding advance has been the recent discovery that a third type of nitrogen fixing bacterium, different from both the free living soil types and the module forming rhizobia exist in a tropical grass, a Digiteria. The bacteria live in the cortex of the root and they not only fix nitrogen almost as well as rhizobia but they are prepared to do so in a test tube. This gives rise to prospects of a type of industrial brewing in which we might be preparing nitrates instead of alcohol. An even more far reaching and exciting possibility is the actual transferring of the genes which confer a capacity for nitrogen fixation from a nitrogen fixing bacterium to one of a different species. The ultimate triumph of molecular biology could be to transfer the nitrogen fixing capacity to the cells of crop plants themselves so that, for example, wheat would manufacture its own nitrogen compounds.

Animal manures are a source of plant nutrients that could be exploited more effectively than is the case at present. An estimate of the quantities of manures produced by housed livestock and poultry[46] shows that the quantities of nitrogen, phosphorus and potassium in the manures correspond to two-fifths of the amount of nitrogen, one-third of the phosphorus and two-thirds of the potassium which are purchased annually[21]. While it is not suggested that all of these manures go to waste at present, there are several factors which reduce the effectiveness with which nutrients are recycled through crops. For example, some large producers of pigs and poultry do not own sufficient land to be able to use all the nutrients in the manures they produce while in other cases manures are badly stored so that there are losses by leaching and

dilution by rain water. It is essential to time the application of manures
to crops to suit the plant requirements and this implies the necessity of
having sufficient storage capacity to retain all the manure produced. It is
also necessary to have a nutrient analysis of the manures so that they may be
balanced with artificial fertiliser to meet the requirements of different
crops.

Machinery and spares

The concept of an energy equivalent of machinery and spares was shown in
Table 2 to account for 52 TkJ, that is, 14.4% of agriculture's energy usage.
Although there is some evidence to suggest that existing machinery is not
being used to full capacity, the scope for economy in this sector is somewhat
problematical. The use of larger machines than are really necessary, has advantages in ensuring that cultivation and harvesting operations are carried
out in good time and it may well be that an energy penalty would be incurred
through using obsolescent and deteriorating machines, particularly if increased maintenance was required, which could nullify the gains that are indicated by extending machinery life. On the other hand, expenditure on machinery attracts tax relief and it may well be that this exerts a strong influence on the machinery replacement policies followed by many farmers.

Solar energy

So far, consideration has been given to ways in which fossil fuel energy is
used in agriculture and possible ways of increasing the efficiency of this
usage. But equally, there is a challenge to increase the capture of solar
energy since of the 100 J/m^2s, which reaches the earth's surface in temperate
zones, plants normally convert less than 1% by photosynthesis into dry matter.
Clearly, there is an incentive here to increase the photosynthetic efficiency
of plants. Because they are slightly but fundamentally different in both
structure and chemistry, tropical crops photosynthesis more efficiently than
temperate crops, can take up more carbon dioxide and are less wasteful in
their respiration. These C_4 species as they are known, of which maize is one,
do not grow well in the UK but some research workers believe that it may be
possible to breed their desirable characteristics into temperate species.
In addition, the efficiency of solar energy conversion by a crop will depend
on factors such as the plant spacing and structure, temperature, availability
of water and nutrients and of diseases and pests which may affect the photo-

synthetic process.   The parts of the crop used as human food, may also contain only a minor part of the total energy fixed in the whole crop and it is therefore important to recycle the parts that humans are not able to use directly.

It has been suggested[47] that some crops may be grown which can serve the dual function of being food or fibre sources as well as a source of cellulose and that the cellulosic material may be converted into hydrocarbon fuels by fermentation, pyrolysis or hydrogenation.   The paper estimates that 1 tonne of dry organic material can be converted into 2 barrels of oil which is equivalent to 275 kg of oil and an energy equivalent of 12.4 GJ.   It further assumes that crop dry matter yields of at least 25 tonne per ha per annum may be expected and at this rate of production, goes on to suggest that 4 million ha devoted to a fuel crop could yield about 3% of the USA's current consumption of crude oil.   Such calculations as these are probably wildly optimistic in relation to UK conditions.   To take cereals as an example, in the UK we would expect to get say not more than about 5 tonne/ha per year of dry organic material after the grain has been removed.   This is equivalent to 5 t/ha x 275 kg/t = 1.375 t/ha of oil.   The present oil consumption of the UK is 95 million tonne per annum, so the land area required to produce this oil equivalent would be 95 x 10^6 t/1.375 t/ha = 69 million ha. This is greater than the area of arable land (7.2 x 10^6 ha), the total area devoted to crops and grasses (12 x 10^6 ha) and of the total area of the UK (19 x 10^6 ha). Even if we were to devote the whole of the area down to crops and grass to the raising of fuel crops, it would still be possible to produce only (12 x 10^6/69 x 10^6) x 100 = 17% of our oil requirement.   Viewed another way, we should need to use all our arable acreage to produce only 10% of our petroleum fuels so that even if we were to use 10% of our arable land in this way, we should then only make a modest contribution of 1% to our petroleum fuel demands.   Under UK conditions, the use of land to raise crops to produce fuel would appear to be a non-starter but it is possible that there could be something in it for tropical countries and a particular crop mentioned is Napier grass which can produce over 100 tonnes of dry matter per ha per year in warm climates.   This would give 20 times the yields per ha that have been considered in the calculations given here.

The prospects for converting solar energy to power, by-passing the photosynthetic process, are formidable and, in the UK climate, quite economically

impracticable at the present time. Of course, consideration of solar energy
in the design of buildings and in the orientation of glasshouses is essential
to ensure that best use is made of it and it will continue to be very important in the natural drying of crops. Except in small sizes and specialised locations the use of wind as a source of energy has not been a viable
proposition except perhaps where none other is available and power is required in small amounts discontinuously. However, even here there are developments which may well grow to challenge this view. One company with a
history of helicopter rotor design[48] believes it can apply this experience
to the design of cheaper and more efficient power producing rotors than
hitherto. They envisage a twin bladed rotor of 18 m diameter which would
generate a maximum power of 150 kJ/s (150 kW). The essence of the design
is a simple built-in blade pitch control, which enables the rotor to start
up in a light breeze, increase its speed proportionally up to a predetermined wind speed and then limits it to constant speed and power for any
higher wind speeds. This company have a number of proposals for making use
of the power generated but in its simplest form, the rotational energy of
the rotor would be converted into hydraulic power and thence into heat
energy which could, for example, be used to heat a glasshouse. It has been
calculated that a 150 kW rotor could provide sufficient power to replace 55%
of the fuel required annually to heat a glasshouse of 0.04 ha. Because of
the uncertainty of availability of wind power it would clearly be sensible
to retain the existing heating installation and only replace part of the demand by wind power. The viability of such a system hinges critically on the
capital cost involved and rational assessment of the scheme is not possible
until such a system has actually been built and tested.

Waste products

Agriculture must also seek to make better use of its waste products both
from crops and from animals. Animal wastes are a valuable source of nutrients and the most sensible course is to return these wastes to the land.
This matter has already been fully discussed and will not be considered
further.

A second use for animal wastes is by anaerobic fermentation to produce
methane gas[49] which can be used as a direct fuel. Its production has
been a feature of municipal sewage works for many years but what is feasible

on a large scale, is less attractive on a smaller scale and there are many
drawbacks to farm operation. The first of these is capital cost of the plant.
Anaerobic bacteria are organisms that thrive in the absence of free oxygen
and so the digestion process must be carried out in an airtight vessel, a
digester in which raw wastes are added daily and sludges withdrawn simultaneously. To sustain the operation it is essential to ensure that the methane
fermenting bacteria are in balance with other types and since optimum digestion is obtained between 35 and 37°C, this means that a proportion of the gas
which can be as much as 50% of the total gas produced, must be used to heat
the plant to sustain operation. It is also unfortunately true that supplementary heating is needed most when ambient temperatures are lowest and this
is the time when the gas would be needed, say, for heating of livestock
houses. While the gas can be used in gas burning appliances, to power
mobile equipment its use in engines requires plant for the removal of carbon-
dioxide and hydrogen sulphide, it must be compressed into cylinders and this
requires special compression equipment, the use of power of a substantial
order and presents safety hazards. The prospects for methane are probably
best in relation to controlled environment houses but considerable gas storage facilities would be required since the heat demand would fluctuate
throughout the year. For example, 16% of the energy used in glasshouse heating is required in February.

Perhaps the most important crop residue is straw. About 3.8×10^6 ha of cereals are grown annually[12] and assuming a gross energy value of 45 GJ/ha
(derived earlier) gives a total energy content of 170 TkJ, that is, about
one-half of agriculture's total energy usage. There is little doubt that if
straw is to be disposed of, the most economical method is to burn it and a
Committee has reported[50] to that effect. At the present time, it appears
that about 38% of the straw produced is surplus to requirements and is burnt.
However, the above figures suggest that straw should be regarded as a resource, not as an inconvenience to be got rid of in the easiest possible
manner. The energy cost of baling, handling and storing straw[20] is only
about 0.2 GJ/ha and this is trivial compared with the energy yield of 45
GJ/ha. Admittedly, there are problems in that straw collection can be costly
in money terms and has to be done at a time of high labour demands and of
critical harvesting and cultivation operations. Critical then to the matter
of making use of straw, is the collection, packaging, storage and transport
so that an economic return can be obtained. Consideration has already been

given in an earlier section to the use of straw as a fuel in drying the
whole cereal crop. Assuming that straw could be stored, preferably in a high
density form, it could be used on the farm to fire furnaces to dry forage
crops or to provide heating for controlled environment houses. It has already been shown[41] that straw can be stably compacted to a density relative
to water of approximately 1.4 by mechanical means and a mild heat treatment
and that the total energy consumption of the process is less than 5% of the
energy that would be liberated by burning the product as a fuel. Weight for
weight, the compacted material has a calorific value of about 45% of that of
high grade coal.

Other present uses for straw include animal feed, animal bedding, and as an
aid to animal waste disposal. The main problem with straw as an animal feed
is that it has low digestibility, is low in protein and cattle will eat
little of it. To increase the digestibility, BOCM–Silcock have developed a
method of delignification[41] in which chopped or ground straw is soaked in
strong caustic soda and the mix forced under pressure through dies. The combination of strong alkali, temperature and pressure gives almost instantan-
eous delignification and the digestibility of the organic matter in the straw
is increased. BOCM have made a decision to go ahead with a plant to produce
20,000 tonne per year and if this is successful it seems likely that other
plants will follow. One other use of straw is also under active consideration[41] and that is as a substitute for hard wood pulp in paper making. For
present purposes, this is noted only because it is a competitor to the uses
of straw that rank as conservers of energy.

Waste heat from power stations

The possibility of utilising waste heat from power stations is one that is
frequently raised in the agricultural context. Unfortunately, the heat from
condensation of the steam in the condenser is available typically at a temperature of 25°C and this is only a few degrees above that which may be required in growing some glasshouse crops. Thus, the most appropriate large-
scale application of heat available from the conventional design of power
stations may be to fish farming of marine and freshwater species. To use
power station waste heat to heat glasshouses either very extensive heat transfer surfaces will be required or it will be necessary to raise the temperature of the heat by the use of heat pumps or by extraction of steam in an

earlier part of the cycle as is done in industries using process steam.
There may well be possibilities in relation to commercial glasshouse heating,
within close proximity to power stations, and while it is encouraging that
these problems are receiving attention it is hardly to be expected that any
solutions are likely to arise that will have widespread applicability in the
near future. A more fundamental modification to power stations has been proposed to make use of waste heat by replacing the conventional cooling tower
and condenser arrangement by a system of pipes buried in the top soil of the
ground area surrounding the station used for growing crops. It is assumed
the waste heat from steam condensation would raise the crop-going potential
of the warmed area but there are few data available on which to make rational
assessment.

CONCLUSIONS

It has been shown that agriculture (up to the farm gate) uses less than 4% of
national energy and produces a little more than half of the nation's unprocessed food. When food processing and distribution are taken into account,
11% of national energy consumption is involved in making indigenous and imported foods available to the consumer. Agriculture has a claim to receive
high priority as an energy user and even in times of dire shortages, its modest demands should continue to be met. It is clear that there is no scope
for energy economies to have any effect in reducing national consumption significantly but such economies are desirable where they contribute to minimising the rising costs of food production without putting productivity at risk.

The principal contributors to use of energy in agriculture are direct use of
petroleum (23.6%),the manufacture of fertilisers (23.1%), off-farm processing
of feedstuffs (14.2%), the manufacture of machinery (14.4%) and direct use of
electricity (9.2%). These inputs of fossil energy, in the form of mechanisation and fertilisers, have enabled us to substitute for manual labour and for
land respectively.

It has been shown that some farm products are more expensive in energy terms
than others. For example, on a basis of energy input compared with metabolisable energy or protein in the commodity, animal products and some horticultural crops are more costly than those based on arable crops. Similarly, the
vegetable system is much more efficient in the use of land resources than is
the animal system. It is not suggested that these facts provide any basis

for determining food policy at the present time. But they must be borne in mind in a climate of finite and diminishing energy reserves although a shift from animal products in favour of cereals and vegetables could be regarded as practical politics only under conditions of dire energy shortage. For the present, monetary economics and consumer demand not energy economics dictate the situation.

In summary, it may be said that the expenditure of energy should be recognised as an admissible means of giving us varied foods to satisfy all our needs and to maximise food production in relation to men and land area employed. In using supplementary energy to produce the kinds of food that we want, in minimising the number of working hours that have to be expended and in minimising the land area employed, we are merely expressing a preference for receiving the benefits of energy in these particular forms as opposed to some other. Clearly, our continued ability to do so will depend upon the availability of energy and the pressures of other conflicting demands upon it. The truth of the matter is that in spite of all that has happened, oil is at the present time still cheap and abundant but ultimately it will certainly prove to be finite. Our present dependence on fossil fuels and their exhaustibility cannot be ignored and it would be wise to control the rate at which they should be consumed in relation to the various calls made upon them.

An indication has been given of the prospects that exist for making more efficient use of energy in agriculture. In the case of plant nutrients, there is scope for optimising use of manufactured fertilisers in conjunction with improved management of animal manures and, in the longer term, the hope that biological methods of nitrogen fixation may be exploited at will. There is also the prospect of more efficient methods of conserving and feeding forage crops to animals and of further reductions in the mechanical cultivations of land to grow crops. In the protected crops sector, there are prospects for reducing heat losses from greenhouses and optimisation of the use of supplementary heat to complement available solar energy. Considerable challenges are presented by the need to improve the photosynthetic ability of growing crops, to make use of the calorific value of cereal straws and, more generally, to put to useful purposes the waste products of both agriculture and industry.

References

1. Rocks, L and Runyon, R P. 'The Energy Crisis', 1972, Crown Publishers Inc, (New York)

2. Digest of United Kingdom energy statistics 1974, HMSO

3. White, D J. 'Energy in agricultural systems', Conference on the Use of Energy in Agriculture, Institution of Agricultural Engineers, 13 May, 1975. **'The Agricultural Engineer', Vol 30, No 3, Autumn 1975.**

4. White, D J. 'The use of energy in agriculture', Farm Management, Vol 2, No 12, Summer 1975, p 637

5. White, D J. 'Agriculture and Energy', Agricultural Progress, Vol 50, 1975, p 39

6. Chapman, P F, Leach, G A and Slesser, M. 'The energy cost of fuels', Energy Policy, September 1974, p 231

7. Bayetto, R A. Central Electricity Council, private communication

8. Leach, G. 'Energy and Food Production', International Institute for Environment and Development, June 1975

9. 'Fertiliser Statistics 1973', Fertiliser Manufacturers' Association Ltd.

10. 'The energy input to a bag of fertiliser'. Imperial Chemical Industries Ltd, Agricultural Division, Billingham, Cleveland

11. National income and expenditure 1963-1973, Central Statistical Office 1974

12. Annual review of agriculture 1974, HMSO

13. 'The Cast Report' Agricultural Engineering, March 1974, p 19, April 1974, p 37, May 1974, p 21

14. Annual abstract of statistics 1973, HMSO

15. Slesser, M. 'Energy subsidy as a criterion in food policy planning', Journal of Science and Food in Agriculture 1973, Vol 24, p 1193

16. Pimental, D, Hurd, L E, Bellotti, A C, Forster, M J, Oak, I N, Sholes, O D and Whitman, R J. 'Food production and the energy crisis', Science 1973, Vol 182, p 443

17. Leach, G A 1974. 'The energy cost of food production'. Chapter in Bourne, A. The Man Food Equation, Academic Press, London

18. Leach, G A 1974. 'Industrial energy in human food chains', Chapter in Duckham, A N and Jones, J G W. Human Food Chains and Nutrient Cycles, Elsevier

19. White, D J. Estimates of energy inputs and outputs for various agri-
     cultural products. Unpublished work 1975

20. Rutherford, I. Private communication, February 1974. Agricultural
     Development and Advisory Service, Ministry of Agriculture, Fisheries
     and Food, Silsoe, Bedford

21. Report of the Energy Working Party, December 1974, Report No 1. Joint
     Consultative Organisation for Research and Development in Agriculture
     and Food

22. Hann, C M. Private communication, December 1974. Agricultural
     Development and Advisory Service, Ministry of Agriculture, Fisheries
     and Food, London

23. Church, B M (1974). Use of fertilisers in England and Wales, 1973.
     Mimeographed report prepared for MAFF, ADAS 'Closed' Conference of
     Advisory Soil Scientists Soil Analysis and Fertiliser Committee
     (SS/SAF/9)

24. 'Fertiliser Statistics 1972', Fertiliser Manufacturers' Association Ltd.

25. Fertiliser Recommendations, MAFF Bulletin 209, (1973), HMSO

26. Leach, G A and Slesser, M 1973. 'Energy equivalents of network inputs
     to food producing processes', Strathclyde University, Glasgow, private
     circulation

27. Frost, B. Private communication, January 1974. Agricultural Develop-
     ment and Advisory Service, Ministry of Agriculture, Fisheries and Food,
     Reading

28. Output and utilisation of farm produce in the United Kingdom 1967/68
     to 1972/73, Ministry of Agriculture, Fisheries and Food

29. Agricultural Statistics, United Kingdom 1972, HMSO 1974

30. Food Science Division, Ministry of Agriculture, Fisheries and Food

31. Manual of Nutrition, HMSO

32. Blaxter, K L. 'The energetics of British Agriculture', J Sci Food and
     Agric 1975

33. Spedding, C R W and Walsingham, J M. 'The production and use of energy
     in agriculture' Annual Conference of The Agricultural Economics Society,
     14 July 1975. To be published in Jl of Agricultural Economics, Vol 26

34. Department of Energy, private communication

35. Agricultural Statistics, United Kingdom 1962/63, HMSO 1965

36. Agricultural Statistics, United Kingdom 1953, HMSO 1955

37. Fertiliser Statistics 1965, Fertiliser Manufacturers' Association Ltd.

38. Jones, D P. 'Energy considerations in crop protection', Outlook on Agriculture, Vol 8, No 3, 1974, p 141

39. Pirie, N W. 'Leaf Protein; its Agronomy, Preparation, Quality and Use' Blackwell Scientific Publications, 1971

40. Wilton, B. 'Heat from burn-up can dry corn' Farmers' Weekly, 22 February, 1974

41. Staniforth, A R (editor). Report on Straw Utilisation Conference, Agricultural Development and Advisory Service, Oxford, 5 December, 1974

42. Winspear, K. 'Glasshouse engineering research and the energy crisis', National Glasshouse Energy Conference, Glasshouse Crops Research Institute, Littlehampton, 15 October 1974

43. Austin, R B. 'Economy in the use of manufactured fertilisers' Conference on The Use of Energy in Agriculture, Institution of Agricultural Engineers, 13 May 1975

44. Pereira, H C. Address to Conference on The Use of Energy in Agriculture, Institution of Agricultural Engineers, 13 May 1975

45. Chat, J. 'The work of the unit of nitrogen fixation' Agricultural Research Council, Paper 185/74, 1974

46. O'Callaghan, J R, Dodd, V A and Pollock, K A. 'The long term management of animal manures', J Agric Engng Res (1973), 18, 1-12

47. Graham, R W. 'Fuels from crops: renewable and clean' Mechanical Engineering, May 1975

48. Shapiro, J S. Servotech Ltd, private communications

49. Jones, K B C and Brown, R T. 'Anaerobic digestion production of methane gas from livestock manures' Ministry of Agriculture, Fisheries and Food, Farm Waste Unit, Report FWU 2, 14 January 1974

50. Report on straw disposal. Advisory Council for Agriculture and Horticulture in England and Wales, October, 1973

APPENDIX 1

DEFINITIONS AND CONVERSION FACTORS

The joule (J) is a very small unit of energy and in more familiar units, 1 therm = 10^5 Btu = 105.5 MJ; 1 kWh = 3.6 MJ; 1 calorie = 4.187J. The unit of mass is the kilogramme (kg) and tonne (t) where 1 t = 1000 kg = 0.984 ton. The unit of length is the metre (m) and area the hectare (ha) where 1 ha = 10000 m^2 = 2.471 acre. Crop yields are expressed as tonne/hectare (t/ha) where 1 t/ha = 0.398 ton/acre. The following prefixes are used to denote the multiples indicated.

Prefix	kilo	mega	giga	tera	terakilo
Symbol	k	M	G	T	Tk
Factor by which unit is multiplied	10^3	10^6	10^9	10^{12}	10^{15}

APPENDIX 2

ENERGY BUDGETS FOR SELECTED COMMODITIES

All energy values are given in terms of primary energy consumed unless otherwise indicated. Some energy inputs that are likely to be of significance have not been accounted and these are mentioned where this applies.

WHEAT

Fertilisers

Nitrogen, 96.5 kg/ha x 77 MJ/kg	= 7430 MJ/ha
Phosphate (P_2O_5), 46.5 kg/ha x 14.3 MJ/kg	= 665 MJ/ha
Potash (K_2O), 39 kg/ha x 8.3 MJ/kg	= 322 MJ/ha
Herbicides, 2 kg/ha x 106 MJ/kg	= 212 MJ/ha
Seed, 188 kg/ha x 4.65 MJ/kg (result of calculation)	= 875 MJ/ha
Fuel used in field operations (cultivations, harvesting etc)	= 2331 MJ/ha
Grain drying, 6% moisture reduction	= 5760 MJ/ha
Machine depreciation and repairs, 11.4 £/ha x 178 MJ/£	= 2030 MJ/ha
Energy input	19625 MJ/ha

Yield of wheat at 15% moisture content   = 4220 kg/ha

Energy subsidy for wheat = (19625 MJ/ha)/(4220 kg/ha)   = 4.65 MJ/kg

Energy output = 4220 kg/ha x 14.4 MJ/kg   = 61000 MJ/ha

E = (61000 MJ/ha)/(19400 MJ/ha)   = 3.11

Protein content of wheat, 10.3% = 4220 kg/ha x 0.103   = 435 kg/ha

Energy to produce protein = (19625 MJ/ha)/(435 kg/ha)   = 45.0 MJ/kg

DAIRY CATTLE (MILK AND MEAT)

The following figures are given on a per cow per annum basis and it is
assumed that the average cow feed is divided between hay and silage in the
proportion 65% to 35%

Concentrates (assumed to be barley), 1200 kg x 4.48 MJ/kg   = 5360 MJ

Grass (14 kg d.m.* for 185 days) = 14 kg x 185 x 1.37 MJ/kg   = 3550 MJ

Silage (8 kg d.m. for 180 days) = 8 kg x 180 x 0.35 x
1.82 MJ/kg   = 917 MJ

Hay (7.7 kg d.m. for 180 days) = 7.7 kg x 180 x 0.65 x
1.81 MJ/kg   = 1630 MJ

Total forage and concentrates   = 11457 MJ

A separate calculation (Ref.19) gave an input of 8360 MJ to
  raise a heifer and if it is assumed that the cow is
  replaced after 5 years, then the portion to be accounted
  per annum is 8360 MJ/5   = 1672 MJ

Energy used in the milking parlour (vacuum pump, milk cooling,
  hot water etc) is estimated to be about 0.3 kWh per gallon
  of milk produced = 0.3 kWh/gallon x 888 gallons x 3.6 MJ/kWh
  x 3.94   = 3780 MJ

Energy input (total)   = 16909 MJ

The area required to produce the forage and concentrates has
  been assessed (Ref.19) as   = 0.993 ha

Energy input = 16909 MJ/0.993 ha   = 17000 MJ/ha

Average milk yield per cow per annum   = 888 gallon

* d.m. means dry matter

Energy output of milk = 888 gallons x 4.54 kg/gallon

    x 2.75 MJ/kg   = 11100 MJ

Energy output of calf assumed culled at birth (carcase

    weight, 55% of liveweight of 45 kg = 24.8 kg) = 24.8 kg x

    10.17 MJ/kg   =   250 MJ

Energy output of culled cow per annum (assumed culled after

    5 years with carcass weight 55% of liveweight of 545 kg =

    300 kg) = 300 kg x 10.17 MJ/kg x 0.2   =   610 MJ

Energy output (total)   = 11960 MJ

Energy output = 11960 MJ/0.993 ha   = 12000 MJ/ha

E = (12000 MJ/ha)/(17000 MJ/ha)   = 0.70

Protein content of milk (3.3%) and beef (12.9%) = 888 gallons

    x 4.54 kg/gallon x 0.033 + (24.8 + 300 x 0.2) kg x 0.129   =   144 kg

Protein output = 144 kg/0.993 ha   =   145 kg/ha

Energy to produce protein = (17000 MJ/ha)/(145 kg/ha)   =   118 MJ/kg

Not accounted above but likely to be significant in the production of milk are energy inputs in feed preparation, delivery of feed from store to animals, removal of animal wastes and buildings to house animals.

DISCUSSION

Question: Are there essential elements in animal products other than protein?
Answer: There are amino acids, which are also available in peanuts. The main point about the continuation of the eating of meat and other foods is that people derive pleasure from it. As long as pleasure motoring is tolerated, so should pleasure eating be tolerated.

Question: When assessing the efficiency of animals as converters of energy, were items such as wool, manure and other by-products taken into account?
Answer: No, these would be bonuses to the system. Some energy analysts have taken into account, for example, the value of animal excrement as a potential source of methane power.

Question: When considering the energy input of crops compared with the yields, is there not a law of diminishing returns?
Answer: Yes, but that does not necessarily mean that the optimim point has been reached. It has been with Soya beans, but grass still has a long way to go, for example.

Question: Would a rise in the price of oil affect the proportional energy cost of agrochemicals?
Answer: No more than it would affect the energy costs of fertilisers or transport which are also oil-related.

Question: What about the 'environmental' costs, for example with the use of DDT and Dieldrin?
Answer: The argument that long-term harm is being done is contestable. Nevertheless DDT and Dieldrin have not been used in the UK or the USA for some time.

THERMODYNAMICS AND ENERGY

N. Kurti

Clarendon Laboratory, University of Oxford

INTRODUCTION

Thermodynamics is a vast subject and I shall restrict myself to only a few aspects, viz. those which are important when discussing energy questions. I shall begin by recapitulating the laws of thermodynamics.

The first law can be paraphrased: You can't have something for nothing, i.e. in order to obtain heat, work or some other form of energy must be converted and vice versa.

The second law can be stated: You can't have something for very little. It qualifies the first law and makes it less attractive. If only the first law applied, one could imagine a tanker sailing from the Persian Gulf to Milford Haven without using any oil - it would simply draw heat from the ocean and, perhaps, leave a few ice floes in its wake. This is not possible according to the second law which may be re-stated: It is not possible for a system to produce work if all that the system does is to drain heat from a single reservoir with no other changes occurring in the surroundings. The 'drinking duck' which you have all seen endlessly bobbing its beak in a beaker of water seemingly violates the second law since its only source of heat is the atmosphere. Its functioning is based, however, on the fact that the relative humidity of the surrounding air is less than unity and water can therefore evaporate from the duck's beak to produce the necessary temperature difference, i.e. the physical state of the surroundings changes. Indeed, if one places a bell-jar over the contraption, its movement will stop as soon as the partial pressure of the water vapour in the air reaches saturation value.

The third law of thermodynamics is not relevant to what you will be considering at this School but can be stated briefly: You cannot create perfect molecular

<u>order, or, you cannot reach absolute zero.</u>

We can formulate the second law by analogy with mechanics. The work produced by a mass m descending from a height h_1 to h_2 is $W = mg(h_1-h_2) = m(V_1-V_2)$, V being the gravitational potential. Similarly, the work produced by a heat engine performing between two reservoirs at temperatures T_1 and T_2 (measured on the thermodynamic Kelvin scale) is given by $W = S(T_1-T_2)$, where $S = \frac{Q}{T}$ is the entropy. You see that the work is produced, <u>not</u> by a quantity of heat Q falling through a temperature difference, but by the amount of entropy S, which, if the processes are reversible, remains constant, i.e.

$$\frac{Q_1}{T_1} = \frac{Q_2}{T_2}$$

where Q_1 and Q_2 are the quantities of heat exchanged with the two reservoirs.

Incidentally, this relation shows clearly that thermodynamic temperatures are defined in terms of <u>ratios</u> only and that the numerical value of -273°C for the absolute zero has no physical significance. It is simply the result of the convention of dividing the temperature interval between the ice point and the boiling point of water into 100 equal degrees[*]. It is appropriate to mention this and the following true story will show why I do so. In the early 1930's, there was much speculation about the significance of Eddington's number 137. Three young physicists (G. Beck, H. Bethe and W. Riesler) working in Cambridge sent a letter to the highly respected German scientific periodical Die Naturwissenschaften. Under the title "The Quantum Theory of Absolute Zero" they calculated the value of absolute zero in the following way. Cooling a substance to absolute zero means "freezing out" all the degrees of freedom of of its constituent atoms. Now, according to Eddington, every electron has 137 degrees of freedom and since matter is made up of an equal number of positive and negative particles, the number of degrees of freedom is 274. The zero point motion of the electron round the nucleus persists even at absolute zero and this leaves 273 in agreement with experiment. This obviously spoof article was actually published in Naturwissenschaften (<u>19</u> 39 (1931)).

If we combine the first law of thermodynamics (Q + W = 0 for a cycle) with the

---

[*] The international Kelvin temperature scale is now defined by giving the <u>triple</u> point of water the value 273.16000....

second we find that

$$\frac{W}{Q_1} = 1 - \frac{T_2}{T_1} \, ,$$

where T_1 and T_2 are the temperatures of the steam in the boiler and in the condenser, respectively, and Q_1 the heat absorbed in the boiler. If we consider one of the Central Electricity Generating Board's most efficient power stations, the one at Ratcliffe on Soar, with $T_1 = 840^{\circ}K$ and $T_2 \backsim 310^{\circ}K$, the maximum thermodynamic efficiency $\frac{W}{Q_1} = 0.63$.

This assumes an ideal Carnot cycle and complete reversibility; the actual efficiency is only about 0.35. It may well be that trying to bridge this large gap does not make economic sense but its existence does represent a challenge to both physicist and engineer.

APPLICATION TO HOUSE HEATING

As an example of the application of thermodynamics, let us consider the various ways in which a house may be heated. We shall calculate for the various cases the heating efficiency R defined as the ratio of the amount of heat (Q_d) delivered to the water (maintained at a temperature T_d) of the heating system and the quantity of heat Q provided by some chemical fuel (coal, gas, oil, paraffin, etc.). We shall further assume that we can run a heat engine between a "boiler" temperature T_1 and some "condenser" temperature and that an inexhaustible heat reservoir at a temperature T_o is available as a source for a heat pump. For the numerical examples we shall take $T_1 = 700^{\circ}K$, $T_d = 350^{\circ}K$ and $T_o = 270^{\circ}K$. We shall further assume that all processes are reversible, viz. are Carnot cycles, that we have ideal heat exchangers and that there are no combustion losses through chimneys, etc. The calculated efficiencies will therefore be larger than one could expect in practice but they will at least provide comparative figures.

1. Direct use of fuel, e.g. in a central heating installation $\frac{Q_d}{Q} = 1$.
2. Heat pump run by a heat engine. Fig. 1 shows the arrangement. The heat pump draws heat from the reservoir at T_o and delivers Q_d to the dwelling.

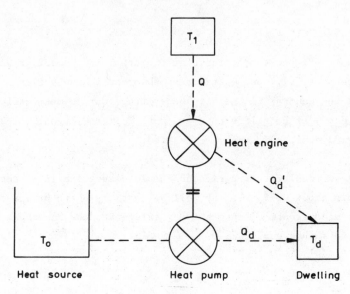

Fig. 1

If W is the work performed by the engine driving the heat pump

$$Q_d = W \frac{T_d}{T_d - T_o}$$

The heat engine will be run between T_1 and condenser temperature T_d so that the exhaust heat Q_d' may be used directly to warm the dwelling. We then have

$$W = Q \frac{T_1 - T_d}{T_1} \quad \text{and} \quad Q_d' = Q \frac{T_d}{T_1} \quad \text{and hence the heating}$$

efficiency becomes

$$R = \frac{Q_d + Q_d'}{Q} = \frac{T_d}{T_1} \cdot \frac{T_1 - T_o}{T_d - T_o} = 2.7$$

3. Heating by electricity obtained from a power station run between T_1 and condenser temperature T_c ($= 300^\circ K$, say) so that

$$W = Q \frac{T_1 - T_c}{T_1}$$

a) Resistive heating: $Q_d = W$, hence

$$R = \frac{Q_d}{Q} = \frac{T_1 - T_c}{T_1} = 0.6$$

b) Use a heat pump powered by electricity:

$$Q_d = W \frac{T_d}{T_d - T_o} \quad \text{and} \quad W = Q \frac{T_1 - T_c}{T_1} \text{ , hence}$$

$$R = \frac{Q_d}{Q} = \frac{T_d}{T_1} \cdot \frac{T_1 - T_c}{T_d - T_o} = 2.5$$

c) Take T_d as a power station condenser temperature instead of T_c so that
the "cooling" water may be used directly for heating the dwelling as in
a district heating scheme. This means a reduction in power station
efficiency and a loss of electrical energy output given by

$$W = Q_d \cdot \frac{T_d - T_c}{T_d}$$

While the additional amount of heat Q the boiler must receive to compensate for this loss is

$$Q = W \frac{T_1}{T_1 - T_c} \text{ , and hence}$$

$$R = \frac{T_d}{T_1} \cdot \frac{T_1 - T_c}{T_d - T_c} = 4.0 \text{, i.e. the highest value of R.}$$

Two points must be emphasized. First, the figures for the heating efficiency
R are given for comparison only; they were calculated throughout on the assumption of ideal, i.e. reversible, conditions. Second, the heating efficiency is
only one of the several factors which determine the choice of any particular
heating method. It may well be that if one considers cost, e.g. capital cost
of heat pumps, heat transmission and heat distribution system, or the cost of
running power stations under variable conditions, the balance may tilt against
systems which save primary energy. It all depends on the way that the cost of
primary energy and that the sources of primary energy will develop during the
next decades and one will have to rely on the prophetical wisdom of economists,
sociologists and politicians for the best solution.

Before leaving the question of house heating, here are a few more remarks about
the use of heat pumps. Since the efficiency of a heat pump is inversely proportional to the temperature difference $T_d - T_o$, it is advantageous to have a
heat source at as high a temperature T_o as possible. One might therefore consider the use, in certain favourable locations, of the low grade (say $20^{\circ}C$)
heat discharged by high-efficiency power stations. Another possibility worth

mentioning is the upgrading by means of a heat pump of heat made available in
a power station district heating scheme (case (c) above) at about $350^\circ K$ to
$500^\circ K$ or so, thereby making it suitable for some industrial processes.

INTERNAL ENERGY VERSUS FREE ENERGY

In the foregoing section I assumed that the chemical energy in the fuel is
first converted into heat irreversibly, e.g. by combustion, and then, with the
help of a heat engine, into mechanical or electrical energy. We have also
seen that even if the heat engine is run under ideal conditions the efficiency
of this conversion barely exceeds 60%.

If, however, the reaction is carried out reversibly then the mechanical or
electrical energy that can be gained from the process could considerably exceed
that obtainable from combustion in a heat engine. It follows from the second
law of thermodynamics that for a reversible reaction (at constant pressure)

$$\Delta G = \Delta H - T\Delta S$$

where G is the Gibbs free energy or free enthalpy, H is the enthalpy and S the
entropy, and

$$\Delta G = G_i - G_f, \qquad H = H_i - H_f, \qquad \Delta S = S_i - S_f,$$

are the changes in G, H and S, respectively, during the chemical reaction.
ΔH is the heat of reaction, e.g. the heat of combustion of carbon or of hydrogen in oxygen, while ΔG, the change in free enthalpy represents the maximum
amount of work (or electrical or mechanical energy) that may be obtained from
the reaction. We will see that, depending on the sign of ΔS, the entropy
change during the reaction, the amount of work obtainable can actually be
larger than the heat of reaction.

If W_{rev} is the maximum amount of work obtainable from a reversible reaction,
we have

$$\frac{W_{rev}}{\Delta H} = \frac{\Delta G}{\Delta H} = 1 - T \cdot \frac{\Delta S}{\Delta H}$$

For this ratio to exceed unity, ΔS must be negative ($S_f > S_i$), or, since entropy
is a measure of molecular disorder, the final state of the chemical reaction

must be more disordered than the initial state.

Let us consider as an example the reversible combustion of a mole of carbon
in a "fuel-cell". Two processes are possible, namely

$$\text{a)} \quad C + O_2 = CO_2$$
$$\text{b)} \quad C + \tfrac{1}{2}O_2 = CO$$

For (a) at $300^{\circ}K$ there is very little difference between ΔH (393.5kJ) and
ΔG (394.2kJ), i.e. the entropy change is small. This is plausible because the
entropy of a solid (C in this case) is small compared with that of a gas and
we have the same amounts of gas of similar complexities (i.e. similar entropies)
before and after the reaction.

For (b) there is an increase in entropy ($S_i - S_f < 0$) because $\tfrac{1}{2}$ a mole of gas
in the initial state gives rise to one mole in the final state.   Indeed, we have

$$\Delta H = 110.5kJ \quad \text{and} \quad \Delta G = 139.0kJ$$

and $\quad \dfrac{W_{rev}}{\Delta H} = \dfrac{\Delta G}{\Delta H} = 1.25$

i.e. the amount of work obtainable is <u>more</u> than the heat of reaction.

A case of great practical importance is the burning of hydrogen since there
is a possibility that H_2 produced by nuclear power will become a widely used
fluid fuel in a hydrogen economy.

For the reaction $H_2 + \tfrac{1}{2}O_2 = H_2O$, clearly $\Delta S > 0$ and, indeed, $\Delta H = 285kJ$,
$\Delta G = 240kJ$ giving

$$\dfrac{W_{rev}}{\Delta H} = \dfrac{\Delta G}{\Delta H} = 0.84$$

Let us now compare W_{rev}, obtainable under ideal conditions from a $H_2 - O_2$ fuel
cell, with W_{therm}, the work obtainable from a Carnot engine based on the combustion of H_2, again under ideal conditions,

$$\dfrac{W_{therm}}{\Delta H} = 1 - \dfrac{T_c}{T_h} \quad \text{where } T_h \text{ and } T_c \text{ are the boiler and condenser}$$

temperatures respectively.

With $\dfrac{T_c}{T_h} = 0.4$, $\quad \dfrac{W_{therm}}{\Delta H} = 0.6$, and we find $\dfrac{W_{rev}}{W_{therm}} = 1.4$.

This refers to ideal conditions but, since it seems that a fuel cell can get
closer to ideal conditions than a heat engine, the fuel cell should become
even more favourable.

If light and compact H_2-O_2 fuel cells could be produced cheaply, they would certainly play an important part in hydrogen economy.

To sum up: Thermodynamics tells us what is the maximum efficiency that may be realised under ideal conditions in any particular energy conversion. It does not tell us about costs, engineering complications and other difficulties. The decision to choose any particular scheme rests ultimately on WHETHER SAVING OF ENERGY RESOURCES OR SAVING IN COST IS THE PRIME CONSIDERATION OR WHETHER A QUANTIFIED COMBINATION OF THE TWO SHOULD BE ATTEMPTED.

DISCUSSION

Question: Which is the better heat source - the heat pump or the power station?

Answer: Straight-forward joule heating has no thermodynamic merit overall and should not be used. Heating systems using bled steam or waste heat, although sacrificing some power station efficiency, are best in terms of overall thermodynamic efficiency. Heat pump systems although thermodynamically somewhat less efficient, are still good and have the advantage that they can be operated as small units locally. One system was developed for domestic use in which a heat pump was used to extract heat from the larder to heat the domestic hot water supply, the two functions being compatible in terms of heat flow. Unfortunately, this system was developed before its time, viz. in the 1950's era of cheap power and it did not prove sufficiently attractive. With the rapid rise in energy costs, however, this system may once more be offered to the public. In brief, if one must use electricity for heating, it should be indirectly used via a heat pump.

Comment: In Denmark and in Germany, heat is bled from power stations to heat homes and for industrial use and this leads to an effective overall thermodynamic efficiency of about 70%, i.e. double that of a good modern conventional thermal power station.

Question: Are domestic heat flow meters available?
Answer: Those available are neither cheap nor reliable and some development is needed in this area.

Question: Is it necessary to meter the heat flow into a consumer's premises in an area supplied by a district heating scheme?
Answer: If heat is not metered, consumers tend to waste up to 20% of the available energy, so metering is justified.

Question: What is the relationship between heat and electricity demand?
Answer: It is relatively easy to store heat energy in bulk and it has been shown that the thermal inertia of a district heating system is such that the heat supply is virtually unaffected by electrical load fluctuations, a very important feature.

Question: How far can heat be transmitted from power stations when the location of power stations near load centres is often not acceptable for many reasons?
Answer: Systems for the transmission of piped hot water, at 80-100°C, over distances of up to 40 km are already in operation outside the United Kingdom and longer distances are thought to be feasible although the attractiveness of the system decreased with increasing distance, because of increasing capital costs and heat losses.

SECTION 2

FOSSIL FUELS

ENERGY AND THE COAL INDUSTRY

J. S. Harrison

Project Assessment and Development Branch, National Coal Board, Harrow

Introduction

The recent explosion of interest and research into new ways of exploiting coal offers coal the opportunity for a much larger contribution to future energy supplies than had generally been considered feasible. The era of cheap oil in the 1950's and 60's led to a rapid expansion of this industry and an almost universal decline in coal production. Oil had advantages in many fields and, in particular, it was ideally suited to supporting the very rapid expansion of the transport and chemical feedstock industries.

The increases in oil price, although surprising in their suddenness, were inevitable in the face of ever increasing demand and diminishing reserves. Their effect was, however, to catalyse a re-appraisal by the industrialised nations of their energy needs, particularly with considerations of independence and security of supply in mind. The development of indigenous resources has naturally become a priority item.

World wide reserves of coal are several times greater than those of conventional crude oil and several times larger than oil shales and tar sands. Furthermore, for historical reasons, large coal reserves are close to the heavily industrialised regions which developed when coal was almost the only energy source.

Fundamental changes in the pattern of energy supplies will become necessary as reserves of premium fossil fuels are depleted. Although reserves of coal are considerably greater than those of other fossil fuels, this is not reflected in the present pattern of energy consumption, which is heavily dependent on petroleum as illustrated in Fig.1.

In response to the mounting pressures on energy resources, a world wide expansion for coal production is planned. The United States proposes to expand its coal industry by a factor of three, and in the United Kingdom, where production is currently about 113 million tons a year, the National Coal Board proposes a capital investment of some £600 million to attain an output of 150 million tons a year by 1985. The NCB proposals are discussed in the Report of the Tripartite Steering Committee (Government, NCB and the Unions), which records the Government's acceptance and support of the NCB's strategy for developing the industry as outlined in the 'Plan for Coal'.

In order to realise its full potential for contributing to future energy supplies, the technology must be developed to enable coal to be used in ways appropriate to the second half of the 20th century. In particular increasingly stringent pollution control regulations can be expected and the higher cost of energy will encourage the development of high efficiency processes.

RESERVES AND PRODUCTION OF FOSSIL FUELS

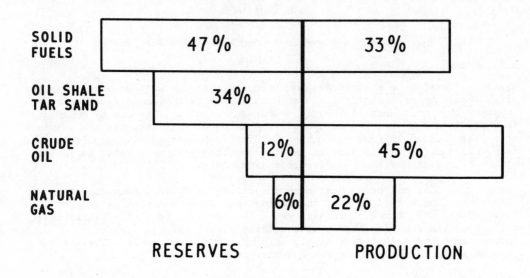

FIG. 1

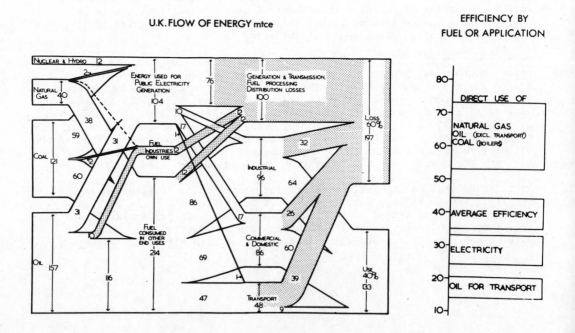

FIG. 2

Energy resources in the U.K.

The present energy supply system in the United Kingdom is fairly typical of an industrialised nation, and is illustrated in Figure 2. In round figures, the total energy input to the UK is 330 million tonnes of coal equivalent (mtce) of which 230 mtce is supplied to final consumers who obtain useful energy equivalent to approximately 130 mtce. The energy supply and conversion industries therefore have an efficiency of about 70%. About three quarters of the loss by the energy supply industries can be attributed to electricity generation. The overall efficiency from primary energy input to useful energy obtained by consumers is approximately 40%. The above average growth rates in transport and power generation have been responsible for a downward trend in the efficiencies of supply and overall use during the last decade.

The general experience within this country and others, through changing economic circumstances, supports the view that there is a close relationship between energy demand and general economic growth. Thus it is clear that to maintain or improve our economic position requires a long term strategy for increasing energy supplies and the efficiency of energy utilisation.

Nuclear power should play an increasingly important role in power generation. At present, nuclear stations provide 10% of the electricity generated in the United Kingdom but in terms of the total energy supply to consumers this represents only a little over 1%. The expansion of nuclear power eventually to take over the base load electricity generation duty will enable coal and oil to be reserved for markets where their special properties as hydrocarbon fuels are valuable. A moderate and balanced programme of development would retain the flexibility necessary in today's rapidly changing conditions. The need for co-ordination of the policies pursued by the various energy industries will be highlighted by the future energy situation.

Figure 3 shows the projected distribution of energy supplies towards the end of the century if a 2% growth rate in useful energy requirements and a 4% growth rate in power generation is maintained. It seems likely that the difficult materials problems involved will prevent nuclear power from contributing significantly to industrial process heating in the near future. The use of nuclear heat sources in group heating schemes may also be restricted by resistance to the siting of reactors near centres of population. If nuclear power takes over base load power generation duty, as assumed in Figure 2, it appears unlikely that it will contribute more than a quarter of the useful energy (excluding transport) required by the end of the century. Furthermore, the requirements for fossil fuel will continue to grow well into next century.

The United Kingdom reserves of coal are substantial, and could support 100 years use at present rates of consumption. In order for coal to be able to supplement the diminishing supplies of oil and gas towards the end of the century, it is vital that investment in coal production, and the development of new technologies for coal conversion are given a high priority. The availability of North Sea gas and oil will give the United Kingdom a valuable breathing space at a time when other countries are experiencing increasing difficulties and expense in obtaining adequate energy supplies. This should be used to reorientate the energy producing and consuming industries to the new energy supply and demand situation.

FUTURE PATTERN OF CONVERSION AND
DISTRIBUTION OF ENERGY FROM COAL

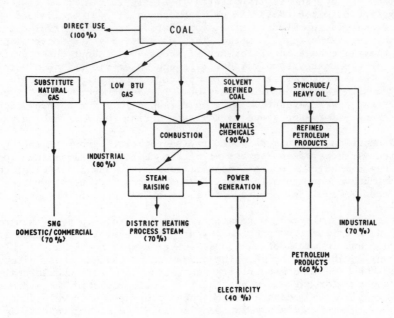

FIG. 3

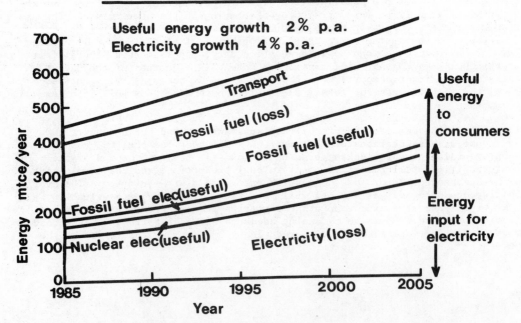

FIG. 4

Coal Utilisation Processes

The substitution of oil and natural gas by coal will occur initially in the bulk steam raising and power generation markets, and will enable oil and natural gas to be reserved for premium uses, such as transport and inter-mittent heating, where their special qualities are more valuable. This change will be aided by the introduction of new combustion techniques, such as fluid-ised bed combustion and low-Btu gasification, which are expected to have lower capital costs and environmental impact than existing combustion methods.

Although direct combustion will remain preferable where considerations of amenity and environment permit, in the longer term it will become necessary to manufacture substitute liquid and gaseous fuels from coal. The timescale for the introduction of these conversion processes is difficult to estimate. More stringent pollution control and the rate of introduction of nuclear energy will clearly be influential. Strong incentives, arising from the shortage of natural gas and the requirement for clean fuels for power generation, already exist in the USA for the development of coal gasification and liquefaction technology. In the UK, however, the discoveries of off-shore oil and gas reserves have made the introduction of coal conversion technology less urgent. In particular, it seems likely that coal liquefaction will be exploited here in order to produce chemical feedstocks and materials rather than fuels in the first instance.

The main conversion and distribution routes which are expected to become avail-able for coal in the future are shown in Fig.4, which illustrates the consider-able versatility of coal as an energy source. This greater sophistication in the way in which energy is used will emphasise the desirability of greater co-ordination and integration of the various energy industries.

New Combustion Processes

The new coal processes fall naturally into two categories, those which improve the efficiency or amenity of combustion itself, and those which upgrade coal to a cleaner, more convenient liquid or gaseous fuel. In the first category there are two main contenders, fluidised bed combustion and low-Btu gasific-ation.

Fluidised bed combustion is a technique in which combustible material is burn-ed in a bed of inert particles through which air is passed. Its application to coal has many advantages. The system provides a high intensity of heat release and can be used with coals of high and variable ash content. The bed can contain tubes for steam raising and because good heat transfer is obtained the resulting design is compact. Fluidised bed boilers suitable for industrial applications are currently being tested. The reduced size and cost of fluidised bed boilers also make their use for power generation attractive. In addition, pollution arising from the emission of sulphur dioxide to the atmosphere can be considerably reduced by adding limestone or dolomite to the fluidised bed. Recent design studies for a full-scale power station indicate capital cost savings of 14% for an atmospheric pressure design and 23% for a pressurised (combined cycle) unit, compared with a conventional p.f. coal-fired station.

An alternative approach involves converting the coal into a low-Btu fuel gas by reacting it with a mixture of air and steam. The resulting gas can then

be cleaned and would normally be used immediately and on site. Storage and
transport over distances greater than a few miles seems likely to be uneconomic, although distribution on the scale of, for example, an industrial
estate may be attractive. A clean low-Btu gas can also be used in combined
cycles for power generation. In this application it could enable full
advantage to be taken of the higher cycle efficiencies which will follow the
anticipated increases in gas turbine inlet temperatures.

New Conversion Processes

The most striking difference in the chemical composition of coal compared with
oil is in the proportion of hydrogen which it contains (Table 1).

TABLE 1.   APPROXIMATE CHEMICAL COMPOSITION OF FUELS

	Carbon	Hydrogen	Oxygen	Nitrogen	Sulphur
Wood	50.0	6.0	43.0	(------1.0-------)	
Peat	57.5	5.5	35.0	(------2.0-------)	
Lignite	70.0	5.0	23.0	(------2.0-------)	
Low-rank coal	81.0	5.1	11.6	1.6	0.7
Steam coal	92.4	4.0	1.3	1.4	0.9
Anthracite	94.4	2.9	0.9	1.1	0.7
Low-temp. tar	82–84	8.0–8.5	7.0–9.0	0.5–0.7	0.7–0.9
High-temp. tar	90–93.5	5.0–6.0	0.1–3.0	0.7–1.2	0.7–0.9
Crude oil	84–88	11.5–14.5	(--------0.5–4.5----------)		
Fuel oil	85–85.5	11–11.5			3.0–4.0
Kerosine	85.9	14.0			0.1
Naphtha	84–85.7	14.3–16.0			

In general the older the coal, i.e. the higher the rank, the less hydrogen it
contains. Thus when replacing oil for the production of petro-chemicals,
either hydrogen has to be added to the coal substance, or carbon has to be
removed. In addition, coal is much more complex in its chemical structure
than is oil; the carbon can occur in various proportions as aliphatic and
aromatic structures, oxygen can be present in ether linkages, hydroxyl groups,
or in ring structures and other elements such as nitrogen and sulphur can
also be present in varying proportions. Much work has been done on elucid-
ating the chemical structure of coals in the hope that this would lead to
improved methods of isolating particular chemicals of commercial interest.
Unfortunately, the methods that have to be employed on commercial scale tend
to destroy the original structures and replace them with simpler groupings.
At the same time that the coal is being broken down into smaller fractions,
other reactions take place leading to repolymerisation of the molecular fragments into larger molecules of much different structure from the original
coal; and, at the temperatures involved in these reactions, gas is also
formed.

In the laboratory, coal can be hydrogenated in the presence of suitable catalysts to produce a range of aromatic compounds, or it can be oxidised, for
instance, by nitric acid, potassium permanganate, ozone, or oxygen to form
organic acids, or it can be fluorinated or chlorinated to give a range of

products.   However, none of these laboratory techniques has so far shown
promise for large-scale operation, and they are never likely to become the
basis of large-scale commercial processes.

On the other hand, a sizeable chemical industry was built in the past on the
production of acetylene from calcium carbide made from coke and lime reacted
in electric furnaces.   Acetylene can be used as the precursor for an enormous
range of chemicals. Today it has been largely replaced by processes based on
petroleum feedstocks, particularly ethylene and propylene. Whether the rising
prices of petroleum-based feed-stocks will be such as to revive interest in
the acetylene route remains to be seen.

The main practicable processes for the conversion of coal into premium fuels
or chemical precursors are:

(a)   Gasification with steam to produce synthesis gas;
(b)   Carbonisation, either in traditional coke ovens or in newer processes,but
      always involving the production of liquid, gaseous and solid products;
      and
(c)   Liquefaction using solvents either alone or in the presence of hydrogen
      and catalysts to dissolve the coal substance.

Gasification

Compared with the gasification of petroleum fractions, gasification of coal
requires more hydrogen to be introduced, and hence more steam to be reacted.
As a consequence, more CO_2 is produced during the gasification of coal than
in the case of oil.   The presence of mineral matter in coal makes it
impossible to use catalysts, and thus the various reactions involved in the
gasification process, some of which produce heat and some absorb it, can
be optimised only by carrying them out in separate stages. This, of course,
complicates the problem from the engineering point of view, and can involve
quite difficult problems of heat transfer.

Various new methods of gasifying coal are being studied;   they have in common
the objectives of:

(a)   encouraging direct hydrogenation in order to reduce the amount of methane
      which has to be produced by synthesis from carbon monoxide and hydrogen;
(b)   reducing or avoiding altogether the use of oxygen which in conventional
      processes is used to provide the heat by burning part of the coal;
(c)   extending the range of coals which can be used.

Of the processes under development, the following are the most important:

(a)   Lurgi
The Lurgi process is an established system using oxygen.   It is the only
commercial pressurised system operating on a large scale to produce a synthesis gas.   The plant at Westfield in Scotland, which is still operated as a
basis for town gas, has recently attracted interest from America as a convenient site on which to demonstrate a commercial process for the production of
methane from synthesis gas.

Disadvantages of the Lurgi system are that it is restricted in the type of
coal which can be treated, and in the size range of coal which can be accepted.

The Westfield Plant has been used as a test bed by the Office of Coal Research in an attempt to widen the range of coal types and sizes.

(b) Bi-gas
This is a process in which coal is gasified when it is entrained in a hydrogen-rich gas. The residue from the gasification stage is used in a slagging plant to produce hydrogen-rich gas.

A pilot plant on a scale of 120 tons/day is being built at Homer City and is due to be commissioned. The Bi-gas process relies on combustion with oxygen to generate the necessary heat for the gasification. It is hoped that it will accept a wide range of coals.

(c) Hygas
In the Hygas process the gasification is carried out in a series of fluidised beds in the presence of a hydrogen-rich gas. Several schemes are under investigation for the production of the hydrogen-rich gas, avoiding the use of oxygen, but the pilot plant operated at Chicago on a scale of 75 ton/day, uses oxygen. A pre-treatment stage can be incorporated in the process to enable a wide range of coals to be accepted.

(d) CO_2 Acceptor process
In this process, heat for gasification is supplied by the reaction between carbon dioxide and lime, the lime being regenerated in a separate vessel using air for combustion. A 50 ton/day pilot plant has been built in Rapid City and has been operated successfully. The CO_2 Acceptor process appears to require a reactive coal; lignite is being used in the pilot plant.

Carbonisation

The process of carbonisation is essentially the heating of coal in the absence of air, resulting in the direct decomposition of the coal into liquid, gaseous and solid products. The actual nature of the three products is of course important to the overall economics of the process. In conventional practice, in which coal is heated in coke ovens, it is the nature of the coke which dominates the overall economics of the process. Much work has been done in recent years by the NCB and BSC together in improving the properties of metallurgical coke, while avoiding the use of those coals which are in short supply. It is interesting to note that the tonnage of coke produced for use as a chemical reductant for the production of iron far exceeds the tonnage of oil used in the production of petrochemicals.

In conventional coke oven practice, the yield of tar is about 5-7% of the coal carbonised. Yields from low-temperature processes, such as those involved in the production of reactive smokeless fuels, can be somewhat greater; but much higher yields can be obtained under laboratory conditions in which coal is heated very rapidly. It has been shown in work carried out at the Coal Research Establishment and at the BCURA laboratories that, under experimental conditions or under conditions similar to those which obtain in a pulverised-fuel burner, yields of tar as high as 30% can be achieved. The yields of tar do not seem to be directly associated with the 'Volatile Matter' as measured under standard conditions; highest yields seem to be obtained from coals of intermediate rank.

The tar which is formed under conditions of rapid heating is chemically un-stable and rapidly re-associates into larger molecules, at the same time shedd-ing smaller fragments into the gas phase. In commercial practice, therefore, the yields of tar and gas depend not only on the rate of heating, but also on the residence time, and temperature history of the initial products of the decomposition reaction.

New processes are being developed which attempt to produce much higher yields of liquid and gaseous products, for example a pilot plant in California has been built to explore the Garrett process in which coal is heated and the products removed from the reactor in a matter of milliseconds. The COED process developed by the FMC Corporation, operates on the principle of progressively higher temperatures, the volatile product being removed at each stage. This multi-stage pyrolysis system is now being used as the basis of the COGAS process, some aspects of which are under investigation at the NCB's Leatherhead Laboratory.

Carbonisation processes produce substantial quantities of solid residue in the form of char which can be used in various ways including gasification to produce hydrogen-rich gas for recycle to the refinery section of the plant, or to produce synthesis gas and hence, liquid products or SNG for export from the plant. Alternatively, the char can be used for the generation of electricity in which case the process of fluidised combustion is particularly suitable as it can cope with residual materials of a wide range of properties such as are likely to be produced in carbonisation processes.

Liquefaction

The basic concept behind coal liquefaction processes is that coal can be digested with suitable solvents at temperatures about $400^{\circ}C$, i.e. just below the temperatures at which coal begins to decompose. The coal extract can then be separated from the undissolved residue to give an ash-free liquid for further treatment. The two main variants of the process, both of which were first investigated in Europe in the 1930's are:

(a) the use of hydrogen-donating solvents which, after recovery from the digestion process, can be re-hydrogenated before being re-cycled;

(b) the use of catalysts with hydrogen gas in the digestion stage.

The amount of coal that can be extracted by either route depends, ultimately, on the amount of hydrogen that can be introduced. Overall economics are thus strongly affected by the costs associated with hydrogen production, and the efficiency of the overall hydrogenation reactions.

Two pilot plants are currently under construction in the US, both aimed at the production of ash-free coal extract of low-sulphur content, suitable for combustion for the production of electric power. Funds have also recently been approved for the continued operation of a large pilot plant at Cresap, with the objective of producing synthetic crude oil for up-grading to petrol and other grades of liquid fuels.

The new processes are capable of producing much higher yields of liquid products from coal than were the original processes from which they have been

FLEXIBLE COALPLEX USING DIURNAL GAS STORAGE

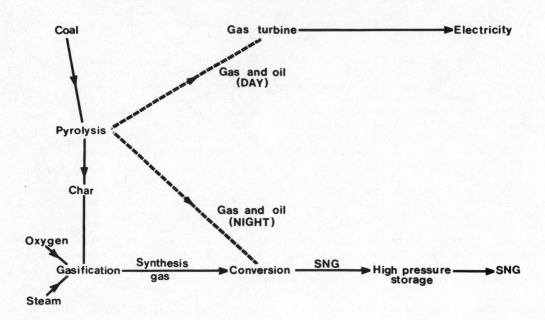

FIG. 5

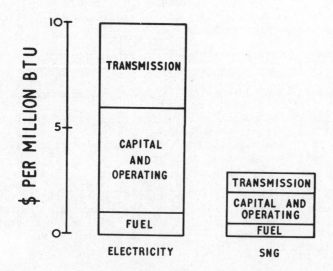

COST OF ELECTRICITY AND SNG
(LINDEN—USA COST)

FIG. 6

derived.   The improvement of catalysts, for example, has decreased the
proportion of insoluble residue from 40 to 10%, even though the pressures
were somewhat less than in the original process.

At the Coal Research Establishment, work on liquefaction has concentrated on
the production of ash-free liquids for the preparation of special carbon
products such as electrode coke and carbon fibre. Work has also been carried
out on the use of gases above their critical temperatures and pressures as
solvents on coal liquefaction processes. One advantage of this technique is
that the problem of separating undissolved material from the extract is
greatly simplified.

Optimisation of Coal Utilisation Processes

An optimised scheme for manufacturing substitute fuels from coal will probably involve a combination of the individual processes discussed. Some of
these processes result in products which are suitable for further conversion,
e.g. char, and others require heat, steam and hydrogen.   The word 'Coalplex'
has been coined to describe a complex of coal based processes which have such
complementary requirements.   Several schemes have been proposed, particularly in the U.S.A. The COG refinery is a typical example. A gasification stre
stream is used to manufacture hydrogen (for a hydrogenation stream) and also
SNG. The carbon-rich residue from hydrogenation is used as a feedstock for
gasification.

This type of coalplex is best suited to base-load operation, however, since
oil production is lost when SNG is not required. Although not possible with
the present separation of the energy supply industries, coalplexes can be
conceived which would operate flexibly to meet varying demands for their
products.   Figure 5 shows an example which makes use of diurnal gas storage.
The pyrolysis unit is operated continuously and the gas and oil produced
is used to supplement the manufacture of SNG during the night and for
electricity generating during the daytime.

For the full advantages of these combined processes to be realised, greater
co-operation between the energy industries will become necessary. For
example, it will be preferable on grounds of efficiency and cost to meet
demands for high amenity low-grade heating with substitute fuels from coal
rather than electricity even allowing for the higher efficiencies of power
generation which can be achieved using combined cycle systems.   This
contention has recently been supported by a study of energy supplies in the
USA, the results of which are illustrated in Fig.6. A higher cost is
associated with electricity for each of the three components fuel, capital
and transmission.

The efficiency of using electricity for heating purposes can be improved by
a factor of two or three times by using a heat pump. At present the investment required is large and can only be justified under special circumstances.
The use of SNG in total energy schemes could, however, still provide a more
efficient heating system than electricity when heat pumps become widespread.
The generation of the mechanical energy required for the heat pump from SNG
by the consumer could enable the waste heat to be recovered and to contribute to the heat supply.

Conclusions

Efforts to maintain energy supplies in the future will require a greater
dependence on coal, but not necessarily in the traditional ways.   Coal as a
source of energy for the future offers considerable versatility.   The direct
combustion of coal will remain preferable whenever considerations of amenity
and environment permit.   However, in order to meet the more specialised
markets for clean liquid and gaseous fuels, the conversion of coal into these
forms will become necessary.   Such change in the pattern of energy supply
could accompany a significant improvement in the overall efficiency of energy
use.   It will also become desirable to integrate several of the energy supply
and conversion processes.   For example, high overall efficiencies are possible
by the recovery of the waste heat from electricity generation for process
steam or district heating.   Significant benefits are also expected to result
from the combination of coal conversion processes in coalplexes.   These
developments will result in a considerably more complicated energy economy
than exists at present, but one in which our limited resources of energy and
capital are exploited to the greatest effect.

Any views expressed are those of the author and not necessarily those of the
NCB.

Further Reading

General

L. Grainger, "Coal into the Twenty-First century", Robens Coal Science Lecture,
London, 7th October, 1974.

L. Grainger, "The Role of Coal in an Integrated Energy Policy".

L. Grainger, "Coal Conversion Processes for the Future", Paper to Glukauf, 1975.

Fluidised Bed Combustion

D. G. Skinner, "The Fluidised Combustion of Coal", Mills and Boon Monograph
CE/3, London.

H. R. Hoy and H. B. Locke, Paper to the meeting of the Swedish Institute of
Engineers, Stockholm, October, 1972.

H. B. Locke, Paper to Achema, Frankfurt, June, 1973.

Gasification.

K. H. Krieb, Chemical Economy & Engineering Review, June 1974, Vol. 6 No. 6.

Chemical Engineering Progress Vol. 69, No. 3 p 31 to 66.

Chemical Engineering Progress, April 1975, Vol. 71, No. 4, p63, p85, p87, p89.

Liquefaction

M. D. Gray and J. Owen, Paper presented at Round-Table Discussion "Chemical
and Physical Valorisation of Coal", Rome, September, 1973.

Chemical Engineering Progress, April 1975, Vol. 71, No. 4, p61 - 80.

DISCUSSION

Question: What is the N.C.B. doing about underground gasification of coal?
Answer: The N.C.B. is reviewing the position but, at the present time,
            the prospects in the U.K. are not hopeful, mainly because of U.K.
            seam characteristics.

Question: What are the operating temperatures of fluid bed combustion and
            system capital costs?
Answer: Operating temperatures are about $900^{o}C$; projected capital cost
            savings are up to 20%.

Question: What solvent would be used in liquefaction?
Answer: The solvent would have to be regenerated in the process and would
            probably resemble anthracene oil. Super critical extraction gives
            a higher recovery of solvent, which would then probably be toluene
            or a similar material.

Question: What is the status of the coalplex project?
Answer: This is still a pen and paper exercise.

Question: How long will it take to develop the new gasification technologies?
Answer: They are available now, for example, the Lurgi system, but the
            existing processes have efficiencies of only about 40%. Develop-
            ments are taking place to increase this figure, to widen the range
            of coal which could be used, and to improve the overall economics
            of gasification.

Question: What route will the N.C.B. choose to make basic feedstocks such
            as ethylene for the chemical process industry?
Answer: The probable route to ethylene will be via synthesis gas, although
            some attention will be paid to the question of whether ethylene
            provides the optimum route to the final products.

Question: What are the economics of the production of syncrude relative to
            crude oil?
Answer: On a pertherm basis, the cost of petrol would be 2 or 3 times
            that of the cost of coal; however, the value of coal should be
            based on the future price of crude oil less the cost of converting
            coal to a syncrude, rather than on the thermal value of oil, which
            tends to be the present basis of comparison.

Question: Is coal really suitable for the domestic market?
Answer: The convenience of coal has to be taken into account, but coal is
            expensive to deliver in small quantities to domestic consumers,
            and the CEGB could use all the coal the NCB produces. Nevertheless,
            the NCB consider it important to continue the development of
            improved appliances for domestic and small industrial consumers.

Question: What is your progress in formed coke development?
Answer: In theory, formed coke could perform better than metallurgical
            coke in blast furnaces, but this remains to be proved. Plant
            studies are required and these would be expensive, needing about
            20,000 tonnes per experimental point. The British Steel Corpora-
            tion are building a unit using the Bergbauforschung process and
            the construction phase will be finished in the near future. The
            product will be used to establish experimentally the virtues of
            formed coke compared with metallurgical coke.

Question: What is the role of the International Energy Agency?
Answer: It has been arranged that Britain will take the lead in coal
             research, there being five proposed collaborative projects:
             1) an intelligence service,
             2) a survey of coal resources including accessibility,
             3) a clearinghouse on mining technology,
             4) an economics service,
             5) a facility for research on fluid bed combustion to collect
                 performance data which will improve understanding of the
                 basic phenomena underlying the process. The facility will
                 complement engineering plant being developed in various parts
                 of the world.

ENERGY AND THE OIL INDUSTRY

P. J. Garner

Chemical Engineering Department, University of Birmingham

At the present time petroleum provides about 70 per cent of the world's energy requirements. It is also the source of 90 per cent of the world production of organic chemicals and plastics.

It is clear that any consideration of the availability of energy and material resources must pay particular attention to petroleum and to the activities of the petroleum industry, which range from geological exploration to the development of advanced technology for the efficient exploitation of petroleum products as convenient sources of energy and as raw materials for industrial and individual consumer use.

The development of petroleum has been rapid and spectacular in its scale and in the range of science and technology involved. While the occurrence of natural seepage of petroleum had been known and expoited from time immemorial (e.g. Egyptian mummies preserved in bitumen) the first oil well was drilled in 1859 to provide additional supplies of a crude oil which was valued for its medicinal purposes. It was soon recognised that crude oil could provide a source of illuminating oil for lamps which up to that time had been supplied by vegetable oils and by the distillation of special coals and shale. The more volatile components which we would now describe as gasoline were a disposal problem and the heavier fractions were used as heating fuel. There was a steady increase in the consumption of petroleum up to about 1900 with no indication that the established progression of energy sources was about to be overturned.

From earliest times wood had been the sole fuel for domestic and industrial purposes such as iron making. By the year 1700 wood for charcoal manufacture

was in short supply. This stimulated efforts to use coal instead and in 1750
Abraham Darby was successful in the development of coke from coal and its use
for the manufacture of iron in his blast furnaces. With ample fuel the steel
industry could go forward and support the industrial revolution which transformed the standard of living in the western world.

From time to time there were alarms about the availability of coal reserves
but these proved unfounded and by the middle of the nineteenth century economists had come to postulate that coal would not be superseded as an energy
source until supplies became inadequate, by which time "natural laws" would have
ensured that a replacement source of energy would be available. The early 20th
century saw the spectacular rise in petroleum as a source of energy at a time
when coal supplies and reserves were ample for the foreseeable future and the
coal-based town gas industry was already highly developed.

In order to be able to understand the present position of petroleum as our
major energy source and its importance in the future it is necessary to follow
the course of the history of the development of the demand for petroleum and
its products and how these have been satisfied by the petroleum industry. The
availability of gasoline as a byproduct of lamp oil manufacture permitted the
development of the internal combustion engine in a form which could be used
for the propulsion of vehicles to replace horse-drawn transport. The automobile developed as the rich man's toy until Henry Ford realised its potential
as a means of mass transportation and started mass production of what turned
out to be the people's car. The effect of this development on oil production
in the USA is shown in Fig. 1 over the years 1870 to 1960. Gasoline demand for
the same period is also shown. There are several reasons for using statistics
from the USA; one is that they are available in a comparable form over a very
long period of time and, furthermore, they refer to a closed system. The USA
has always been a large self-contained area from the petroleum point of view.
It has been self-sufficient to recent times and the figures are not distorted
by exports to any significant extent, nor by imports which until recent years
have been non-existent. It will be seen that up to the year 1890, there was a
steady growth in oil production, about 20 per cent per annum, which is a very
respectable figure. There are no figures for gasoline demand because this ante-
dates the development of the car as a consumer of gasoline. However, after the
introduction of the automobile had ushered in the age of gasoline, the growing

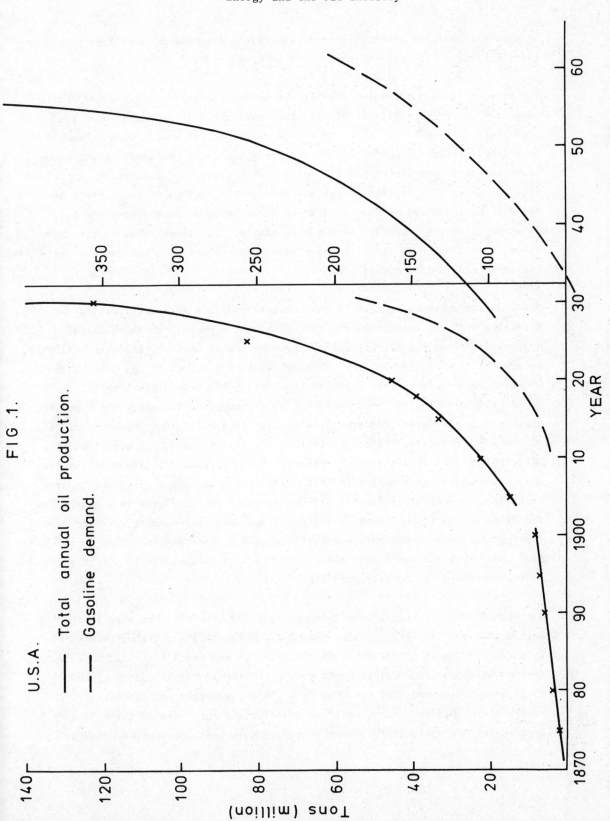

FIG .1.

demands for gasoline soon forced the production of unneeded quantities of heavier
fractions and their disposal was a very serious problem.

The problem is illustrated in Fig. 2. While the proportion of gasoline obtainable from crude oil by simple distillation varies widely depending upon the
source, it is possible to speak of an average crude oil which, when processed
in a simple refinery, would give a gasoline yield of 21 per cent. Since gasoline (and other petroleum product) specifications are not absolute by composition but determined by fitness for use in gasoline engines it is possible by
suitable manipulation of the blending of products to produce a yield of, say,
27 per cent of gasoline but, as one might expect, fractions obtained directly
by the distillation of crude oil have never been produced in the exact quantities required to meet market demands.

One solution would be to convert the heavier fractions into gasoline and the
first means of achieving this was by "cracking", a process in which large
hydrocarbon molecules are broken down by exposure to high temperatures. Already,
before the First World War, the cracking of heavier materials had been introduced to supplement the yield of gasoline obtainable from a given barrel of
crude oil. Cracking has been described as a fortuitous discovery. A fortuitous
discovery in this case undoubtedly means that it arose because somebody did not
do their job properly. In the early 1900's crude oil was still distilled in
large boilers and it was a batch process. I expect that what happened was that
when the plant manager went home in the evening, he told the operators to stop
distilling when they had removed all the gasoline but they went to sleep leaving
the fires on. This overheated the residue and, when the manager returned next
morning, he found to his surprise that he had much more gasoline than he should
have had. With the increasing demand for gasoline he was able to put this
fortuitous discovery to very good use.

The graphs (Fig. 1) show a very interesting trend, you will see that in the
period 1910 to 1930, the gasoline demand is an increasingly higher proportion
of the total annual oil production. By 1918, 25 per cent of the oil marketed
in the USA appeared as gasoline and this had risen to 45 per cent by 1940.
There are good grounds for assuming that the average gasoline content of USA
crude oil in 1918 was 21 per cent and calculations show that of the 9.8M tons
of gasoline consumed, about 10 per cent must have been produced by cracking

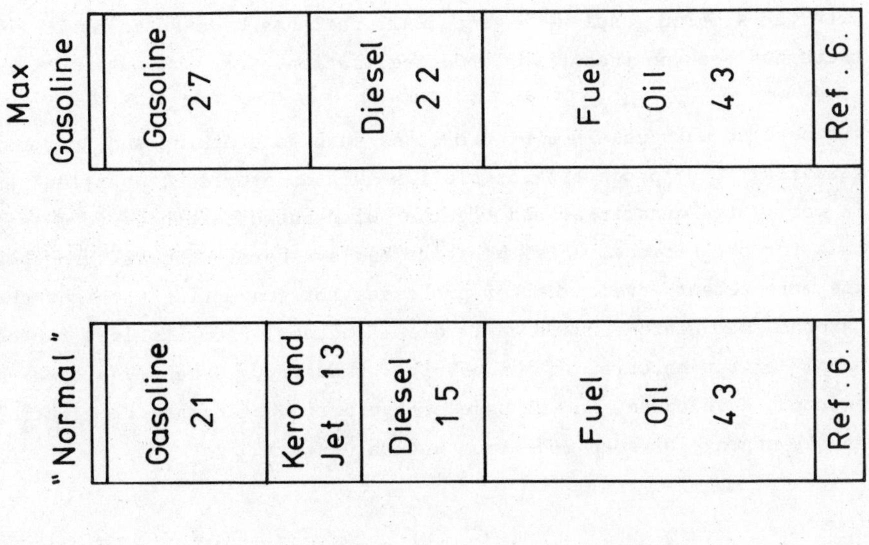

FIG .3.

Demand (Ex. Soviet)

1960 % by weight

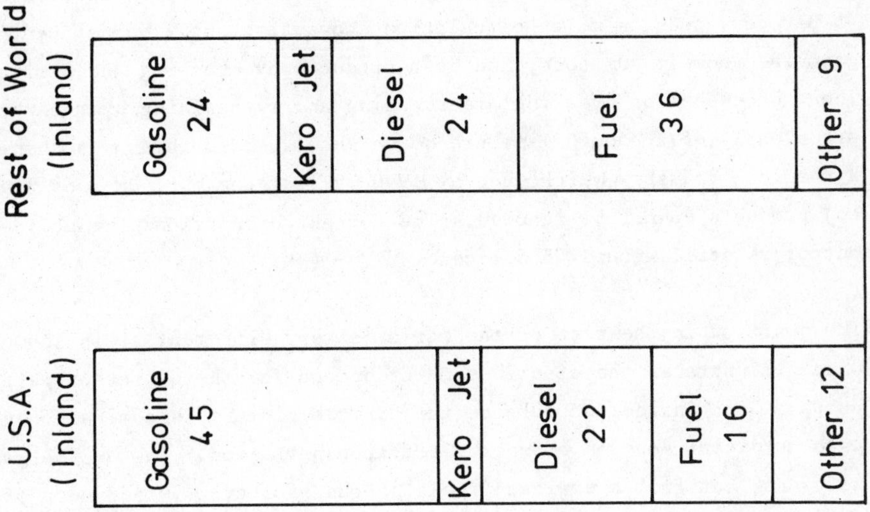

U.S.A.   Rest of World

(Inland)   (Inland)

Rest of World (Inland)
Gasoline 24
Kero Jet
Diesel 24
Fuel 36
Other 9

U.S.A. (Inland)
Gasoline 45
Kero Jet
Diesel 22
Fuel 16
Other 12

FIG .2.

Product Yield

Simple Refinery

Max Gasoline

Max Gasoline
Gasoline 27
Diesel 22
Fuel Oil 43
Ref. 6.

"Normal"

"Normal"
Gasoline 21
Kero and Jet 13
Diesel 15
Fuel Oil 43
Ref. 6.

heavier fractions. By 1940 the proportion of gasoline to crude had risen to
45 per cent and a similar calculation shows that about 55 per cent of this
gasoline, namely 40M tons, had been produced by cracking and similar processes.
These processes of breaking, rearranging and sythesising hydrocarbon molecules
are clearly of a chemical nature so we should class this as a chemical industry;
a chemical industry which even by 1940 was turning out 40M tons a year of chemical products cannot be ignored. The current proportion of gasoline in the USA
output is still around 45 per cent.

The position in the rest of the world is very different as is shown in Fig. 3
which illustrates the distribution of demand for the different petroleum products. Both inside and outside the USA the picture is dominated by products
used as direct sources of energy for transportation. In the USA gasoline,
kerosene, jet fuel and diesel fuels account for over 70 per cent of petroleum
use and of this over two thirds is gasoline. In the rest of the world only
55 per cent of petroleum is used as transport fuel and the consumption of
diesel fuel equals that of gasoline. This remarkable difference is due to
contrasting taxation policies. The USA, because of its dependence on the automobile for almost all its transport requirements, has always been a low-cost
fuel society and taxation on fuel has been kept at a very low level. In contrast, governments in Europe have welcomed petroleum fuels as a source of
unlimited revenue from taxes collected for them free of charge by the oil companies. The resulting high cost of fuel has stimulated interest in the development of the diesel engine for road transport because its more efficient use of
fuel can more than compensate for the extra cost of the diesel engine compared
with the simpler gasoline engine. All fuel has been so cheap in the USA that
there has been no incentive for dieselisation.

The need to make gasoline by processes such as cracking and "reforming" also
resulted in the production of olefins such as ethylene, propylene and butylenes
in very large quantities and together with butadiene these form the base materials for organic chemicals, plastics and synthetic rubbers throughout the world.
The more recent development of processes for converting straight chain hydrocarbons into benzene has finally displaced coal from its last foothold in the
supply of raw material to the chemical industry. Since 1945 there has been an
enormous development in the sale of synthetic detergents from petroleum sources
at the expense of soap made from natural fats.

While the out-of-balance demand for gasoline has been satisfied by the conversion of heavier petroleum fractions which would previously have been available
as fuel oils for heating in the domestic, industrial and steam generating
fields, the fuel oil market is still very large indeed. However, in the early
days of the petroleum industry fuel oil was almost as much of an embarrassment
to the lamp oil producer as gasoline had been before the development of the
automobile. The world was well satisfied with plenty of cheap coal but by the
time of the first World War fuel oil was beginning to displace coal in naval
ships and soon after in the merchant navy. The stimulus was the ease of transport, storage and handling of liquid fuels which among other things resulted
in major reductions in manpower needs. Further unexpected advantages were the
ease of control and the rapid accommodation to load changes which this control
made possible.

The case for oil fuel in land installations was not so obvious but the oil
companies set out to develop equipment which made the burning of oil more efficient than that of any other fuel. The elucidation of the fundamentals of
combustion phenomena has been almost entirely due to the efforts of the oil
companies and this has enabled them to supply fuels for gasoline and diesel
engines and gas turbines of such quality that their thermal and thermodynamic
efficiency is limited only by materials of construction.

Apart from energy and materials, petroleum is the only significant source of
another ingredient of civilisation without which life as we know it would
literally grind to a halt. This ingredient is lubricating oil, the production
of which, in the UK, accounts for about 1 per cent of the total consumption of
petroleum.

The breakdown of consumption of petroleum products in the UK for the year 1970
is shown in Table 1. The petroleum industry is probably unique in that losses
in manufacture are negligible since any byproducts can be utilised as refinery
fuel and there are no ash disposal problems.

The contribution of petroleum to our needs and comforts has gone largely un-
noticed, not only because of its efficiency and convenience but also because
there have in the past been very few interruptions in the supply. Outside the
oil industry itself very little thought has been given to the maintenance of

Table 1

U.K. Consumption of Petroleum Products

(1970 thousands of long tons)

(a)   Transportation

    Aviation spirit   74

        "   turbine fuel   3,353

    Motor spirit   14,070

    Diesel engine road vehicle fuel   4,955

(b)   Industrial power (diesel)   11,918

(c)   Heating, domestic, industrial and power
    stations

    Kerosene   2,494

    Fuel oil   37,975

(d)   Lubricating oils   1,155

(e)   Bitumen   2,036

(f)   Chemical feedstock

    (naptha etc.)   9,700

(g)   Liquefied petroleum gases

    Propane and butane   1,155

(h)   Other products   579

(i)   Refinery consumption (fuel)   5,933

                    Grand total   95,337

resources and the long term prospects for the development of new resources to
meet ever increasing demands.

Between 1950 and 1970 the total energy demand of the world increased at an
annual rate of 4.5 per cent. The annual average growth rate of both oil and
natural gas was 7 per cent, (a rate which doubles annual production every ten
years). The contribution of oil to the world energy demand increased from
33 per cent in 1950 to 54 per cent in 1970. Over the same period the consumption of solid fuels stayed constant and their share dropped from 53 per cent
to 23 per cent. A recent world energy demand forecast for 1985 (excluding the
USSR, Eastern Europe and China) assumes that oil will be able to contribute
60 per cent. The quantity of oil required to meet these forecasts between now
and 1985 is about 250×10^9 barrels which is about 12 per cent of current estimates of world ultimate recoverable reserves of crude oil. If oil consumption
were to continue to increase at rates envisaged in the forecast for 1985,
(i.e. 5.5 per cent per annum) recoverable reserves would be completely depleted
in about 35 to 40 years from now. The conclusion that petroleum reserves will
be dissipated soon after the year 2000 is in good agreement with early estimates
made by the oil companies in the early 1950's when world oil discoveries first
showed a dramatic fall (see Table 11).

The irregular distribution of petroleum deposits makes the estimation of ultimate recoverable reserves difficult but informed opinion is that no major oil
fields comparable with those in the Middle East remain to be discovered. There
are undoubtedly large numbers of small oil fields remaining to be discovered
but it is unlikely that many of these could be exploited economically. The
largest factor in determining the size of the world's oil reserves is taxation
policy, price and control and the demand of governments for "participation" in
oil operations. Already, the estimates of reserves in the North Sea have been
downgraded in the light of proposals for "participation" which would make it
very difficult to borrow the enormous amount of capital which must be found
(borrowed) for the exploitation of existing discoveries.

The recent increases in the posted price of crude oil have come as a shock to
western governments but should have been expected. In 1970 the posted price
of Arabian crude was $1.80 a barrel of which the Saudi government received
$0.92. At the same time consumer government's taxes in Western Europe on

Table II

Estimates of World Oil Reserves and New Discoveries

(10^9 barrels) 1950 - 1973

Year	Reserves	Discoveries
1949	1000 to 1500	-
1950	-	27
1951	-	15
1952	-	14
1953	1000	15
1954	-	12
1955	-	16
1956	1250	15
1957	-	16
1958	-	16
1959	2000	17
1960	-	13
1961	-	17
1962	-	19
1963	-	23
1964	-	22
1965	2480	25
1966	-	19
1967	2090	18
1968	1800 to 2200	14
1969	1350 to 2100	14
1970	1800	10
1971	1200 to 2000	8
1972	1900	7

petroleum products were around $5.0 per barrel, and OPEC decided that they
should be able to exact the same and so increase their oil revenues, which
now stand at about eight times what they were before 1970.

The large increase in the price of crude oil has already drawn attention to
the importance of efficiency in the use of oil as a means of prolonging the
life of the world's oil reserves and of providing additional time for the construction of facilities for the manufacture of substitute fuels from the coal
which will be available in sufficient quantities for several hundred years.
In the short term oil should not be used as a fuel where coal can do the job.
The same should apply to natural gas which as premium product should be reserved for premium applications such as chemical manufacture. The use of
natural gas for domestic heating where half of the heat goes up the chimney
seems unjustifiable. One unpublished advantage of electricity for domestic
heating is that none of the energy delivered to the home is wasted in this way.

The fact that the major use of liquid hydrocarbons is as fuel for transport
poses many problems. Immediate steps need to be taken to restore and improve
public transport systems; in particular the railways, tramcars and electric
trolley buses. The battery-powered electric car will become much more attractive for private transport when gasoline is no longer available.

The reservation of available oil and natural gas for chemical synthesis can
prolong the time available for the building of synthetic oil and gas plants
using the still abundant coal as raw material. The necessary large scale in
Germany during the war and further developed since then as in South Africa.
The technology of the use of coal as a source of all organic chemicals as well
as plastics and rubbers is also available and, in fact, hardly differs from
that used when oil is the base material.

Looking ahead to the time when coal reserves are running low one visualises
that most of our energy needs will be supplied by nuclear power stations. With
abundant nuclear power it will also be possible to synthesise such hydrocarbons
as may be essential (e.g. for lubricants) by reacting hydrogen produced from
the electrolysis of water with carbon dioxide produced by the dissociation of
limestone by nuclear heat. Here again the necessary technology is already
available and only needs the incentive to be applied.

There remain conditions when a mobile fuel source beyond the reach of electric power lines or battery charging points would be extremely useful, for instance, in the development of undeveloped regions and for military purposes. Thanks to atomic energy we can always have electric power available at our command. Some people have already suggested that, based on space flight technology, hydrogen from the electrolysis of water should be distributed in liquid form as a substitute for gasoline. The problems associated with the development of containers for distribution of liquid hydrogen and of handling equipment on a significant scale would appear to be severe and the cost probably astronomical.

Let us hope that for safety reasons and the benefit of the long-suffering tax-payer an already existing solution will be preferred. This is based solely on air, water and electrical energy from nuclear power plants and the scheme is shown in Fig. 4. No new technology is involved. The generation of electric power, the electrolysis of water, the separation of air and the synthesis of ammonia on the required scale are well established. The transport of liquid ammonia by ocean-going tanker, pipe-line, tank car and individual cylinder is practised world-wide. It has been known for a hundred years that ammonia is combustible and the possibilities of its use as a fuel in both standard spark ignition or compression ignition (diesel) internal combustion engines have been established. Minor modifications to the diesel engine result in thermal efficiencies which cannot be matched by hydrocarbon fuels because of combustion problems[1]. Ammonia is also superior to hydrocarbon as a fuel for gas turbines on the basis of power output and thermal efficiency, but poorer on the basis of fuel economy[2]. The specific fuel consumption for ammonia (kg per kilowatt hr) is about twice that for diesel oil and would be quite uneconomic at present 1975 prices. However, when diesel fuel also becomes a synthetic product, ammonia is likely to be the cheaper of the two. Published information indicates quite intensive study of ammonia as a fuel for internal combustion engines for military application. Dissociated ammonia can be 'burned' as an impure fuel in any hydrogen-consuming fuel cell with a slight loss in efficiency. Ammonia itself can be 'burned' in a hydrazine-consuming cell without any of the difficulties arising from the high toxicity of hydrazine. It remains to be seen whether anyone can afford the catalytic electrodes required by any fuel cell system[3].

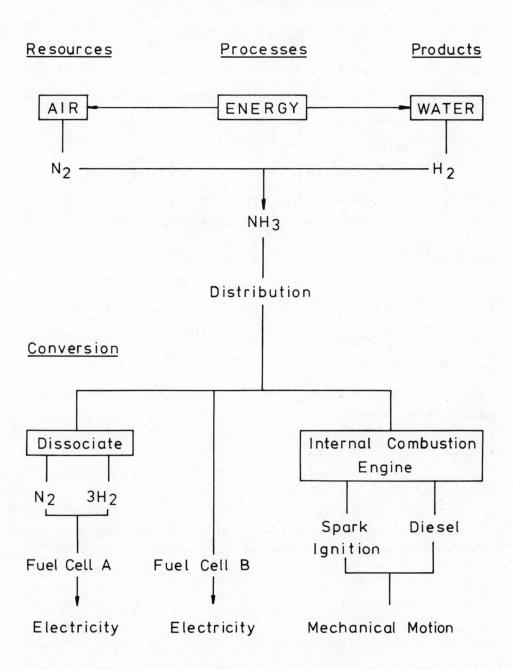

FIG .4.

Mobile fuel for the future

P. J. GARNER

Sources and References

Fig. 1   U.S. Government Statistics on Petroleum Production and Use.

Figs. 2 and 3 Data from "Modern Petroleum Technology 3rd Edition". Institute

                of Petroleum, London 1962.

Fig. 4   Author

Table 1   "Petroleum Statistics 1970" Institute of Petroleum, London.

Table 2   Data from Sir Eric Drake, "Oil Reserves and Production".

                Phil. Trans. R.Soc. Lond. A $\underline{276}$ 453-462 (1974).

References

1.   Pearsall, T.J. et al., $\underline{S.A.E. Jl}$ 76 (11) 49 (1969).

2.   Newhall, H.K. and Starkman, E.S., $\underline{S.A.E. Jl}$ 75 (6) 71 (1967).

3.   Anon. $\underline{S.A.E. Jl}$ 76 (12) 72 (1968).

Recommended General Reading

"Modern Petroleum Technology", 4th Edition.

Hobson, G.D. and Pohl, W.

Applied Science Publishers, 1973.

DISCUSSION

Question: How would you distribute ammonia?
Answer: It can be bottled at low pressure and can be distributed by road
            tanker or in pipes. Substantial quantities are imported to
            Europe from Puerto Rico at the present time by large-capacity
            chemical tankers.

Question: Have you any information on the composition of North Sea oil?
Answer: The oil companies have not revealed much information. However it
            is a light crude which contains paraffins. The sulphur content
            is less than 1%.

Question: Is not hydrogen a safer fuel than L.P.G. because its light density
            ensures that it disperses rapidly upwards?
Answer: The flammability limits of hydrogen are wider than LPG, so hydrogen
            is more likely to ignite and the high flame speed of hydrogen
            means that it explodes. Both LPG and hydrogen have to be handled
            carefully.

Question: You have suggested that nuclear power could be used to generate
            ammonia from water and nitrogen. The World Energy Conference
            survey of energy resources says that uranium is only 2% of the
            world's resources of non-renewable fuels. An OECD survey suggests
            that nuclear power stations built by 1980 will, during their life,
            consume all the known reserves of uranium costing less than $15/lb.
            Does this not cast doubts on an "ammonia economy"?
Answer: The sizes of reserves are closely related to the market values of
            the fuels. If the value of a fuel rises the reserves will also
            increase.

Question: Will not the extent of fossil fuel reserves also increase with
            fuel price?
Answer: Nuclear power at present is competitive with coal and oil elec-
            tricity generation. Perhaps we should build more nuclear power
            stations so that we can export our oil.

Question: How much energy is consumed in the production of ammonia?
Answer: The production process is very efficient and is a well developed
            technology.

Question: Can ammonia be used as a transport fuel?
Answer: Possibly - it has a high calorific value, is readily transportable,
            does not require much greater tank volumes (for comparable vehicle
            performances), and is readily obtained from electrolytic hydrogen
            and air (nitrogen). However, it has probably got a better future
            as one of the components of a fuel cell.

Question: Should petroleum be conserved for lubricating oils, greases, etc.?
Answer: No - but lubricating oils should be used more efficiently and care-
            fully, as in recycling or closed systems. Lubricating oils are
            also now being synthesised from non-petroleum constituents, such
            as vegetable oils.

ENERGY AND THE GAS INDUSTRY

J. A. Gray

Research and Development Division, British Gas Corporation

1.   INTRODUCTION

British Gas is the largest single gas utility in the world in terms of
number of customers, total supply and turnover. It has control of all the
piped supply in Great Britain although there is an independent supply of
bottled gas. In national terms it is the sixth largest business in the
country based on capital invested, and in the domestic market it is
comparable with the next two European gas industries put together.

The Industry operates as a primary supplier rather than, as formerly, an
energy converter, providing 90% of the gas to customers as natural gas and
the rest as town gas, most of which is reformed from natural gas. The
growth of gas supplied is shown in Figure 1, and its proportion of the total
fuel market in the U.K. indicated in Figures 2 and 3.

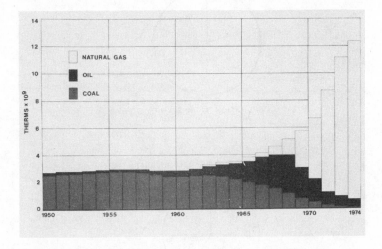

Figure 1. Growth of gas supply

In 1974 the 11,500M therms supplied constituted 16% of the total fuel
market. If reckoned in terms of useful heat supplied this is more like
25% of the total consumption excluding transport. This is likely to rise
towards 40% by the early 1980s.

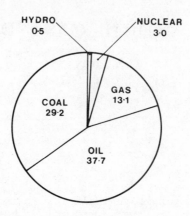

TOTAL PRIMARY INPUT = 83·5 x 10⁹ THERMS

(≡ 331·0 M.T.C.E)

Figure 2.   <u>United Kingdom primary fuel supplies for 1974</u> (Therms x 10⁹)

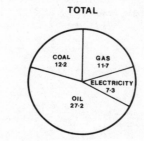

TOTAL CONSUMPTION = 58·5 x 10⁹ THERMS

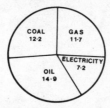

TOTAL CONSUMPTION LESS TRANSPORT = 46·0 x 10⁹ THERMS

Figure 3.   <u>United Kingdom energy supplies for 1974 – thermal basis</u>
<u>(Therms x 10⁹)</u>

Despite this preponderance of direct supply the industry is profoundly concerned with energy conversion in two senses:

(a) Ensuring that the fuel is used in a way commensurate with its premium qualities which may be summarised as follows:

- controllable,
- clean combustion,
- no on-site storage required,
- heat supplied where it is wanted,
- low pollution.

These qualities justify an emphasis on utilisation, and an extensive part of the organisation of British Gas is concerned with this, not least on the research side where utilisation accounts for 36% of the budget. The R&D budget amounts to £12M per year and a total of about 1,400 people are employed in the R&D function.

(b) In looking to the future there is an interest in being able to make gas from any feedstock that may be available. Initially this will be as a supplement to natural gas to enable us to meet in the most economic way a load which varies widely from winter to summer and from day to night. Ultimately sometime in the fairly distant future SNG will take over as a replacement for natural gas. This latter aspect requires some background explanation so the next four sections will outline the natural gas supply system, the characteristics of the load and the reserves position.

2. NATURAL GAS SUPPLY SYSTEM

Natural gas is received at four major terminals on the East Coast and transmitted to all regions of the country through a high pressure transmission system of welded steel pipe up to 36" diameter, designed to operate at 1,000 psi pressure. The system which is shown in Figure 4 has been developed in its entirety in the last ten years.

Gas passes from the national system to the regional systems at 550 psi and is successively broken down to 350 psi, 100 psi, 30 psi and eventually to 12-20 inches water gauge for supply to domestic consumers.

This extensive system of pipes - 120,000 miles of distribution mains, 2,500 miles of transmission pipes with another 550 miles being built for gas from Frigg and another 250 miles for associated gas from the Brent oilfield - constitutes the industry's major asset. The distribution system alone far outweighed the production facilities in value even when all gas had to be manufactured.

The doubling of the capacity of the existing pipe systems, arising from the high calorific value of natural gas (about 1,000 Btu/cu.ft. compared with 500 Btu/cu.ft. for town gas) was one of the main reasons for deciding to supply natural gas direct rather than converting it to town gas. The penalty of this was the enormous conversion operation, now reaching its end.

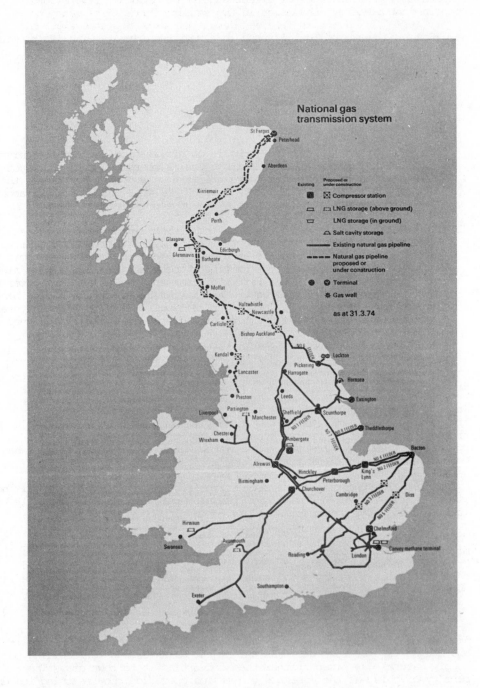

Figure 4. U.K. national gas transmission system

The load supplied is, of course, primarily a heating load. Figure 5 shows
that it is made up of a large and steadily growing domestic element (about
45% of the total), a large industrial element of about the same size which
has multiplied over five fold in the last five years, and a commercial
sector taking about 10%.

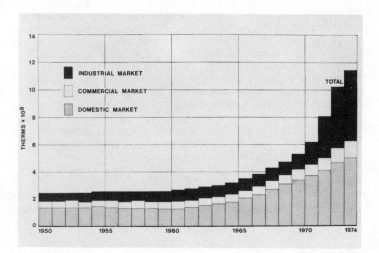

Figure 5. <u>Growth in the markets for gas since nationalisation</u>

The domestic load - which formerly was dominated by cooking, water heaters
and gas fires - is now mainly a space heating load, with a substantial
contribution from domestic water heating. As such it is increasingly
temperature sensitive and subject to large seasonal change and to diurnal
swings.

The industrial load is to meet heating requirements, particularly where
controllability and consistency are important, and to meet a fairly large
long-term commitment to feedstock supply. The industrial load is much less
susceptible to seasonal variation although much of it is confined to the
daytime.

The total result is a winter load that can be of the order of four times
the summer load under normal conditions, and a daily load that can itself
vary by a factor of 4 between day and night.

3. <u>LOAD MATCHING</u>

The industry has a number of means at its disposal to cope with seasonal and
diurnal variations.

<u>Seasonal load</u>

(a) The take from the North Sea can be varied. Thus the contracts allow
     for a 60% load factor (defined as the ratio of average daily take for

the year to the maximum take rate) in the southern North Sea, but 80%
is likely in the northern North Sea, where the pipelines are longer,
and the figure could be even higher where the gas is associated with
oil.

(b) Storage, which can take several forms.

    (i) Bulk storage in underground formations. But very large
       quantities of gas are involved, and the opportunities are
       limited by geology. However, few suitable sites have been
       identified and depleted wells seem the best prospect.

    (ii) Peak shaving storage provided by liquefied natural gas (LNG)
       tanks at or near extremities of the system to cover needle-
       peak demands and to ensure security of supply.

    (iii) Storage in artificial salt caverns. The function of gas
       stored thus is similar to that of LNG, but here again
       opportunities are limited by geology.

(c) Manufacture.
Substitute Natural Gas (SNG) made from stored and easily gasifiable
feedstock, e.g. naphtha, liquefied petroleum gases, and methanol,
is suitable for meeting high demands lasting several weeks.

(d) Interruption.
Some industrial customers, particularly those converted from oil, are
willing to accept the risk of interruption of supply, under well
defined conditions, in return for a favourable price. They have, of
course, to maintain an alternative supply and the means of using it.

(e) Sellers option.
During the load building phase, there may be a surplus of gas which
cannot be placed in the normal market. Arrangements are made to sell
this gas to customers who are prepared to take it when it is available,
i.e. at the option of the seller. As the premium load grows this kind
of use might be expected to become fairly exceptional.

Diurnal variations

Daily swings in demand can be covered by:

(a) conventional low pressure holders;

(b) high pressure storage in pressure vessels, pipe arrays or storage
mains;

(c) line pack, which uses the spare capacity of the transmission system.

4. GAS RESERVES

The future availability of natural gas dominates the longer term planning of
the gas industry. Presently contracted supplies account for 24 TCF (1 TCF ≡
10^{12} standard cubic feet) of reserves remaining at the beginning of this

year if the Norwegian part of the Frigg field is included. Quite a lot of
extra gas is known to exist (e.g. 600M cu.ft./day of associated gas from the
Brent field which has just been contracted to British Gas), and there is a
good chance of substantial further discoveries. Total proven reserves are
26.9 TCF as shown in Figure 6.

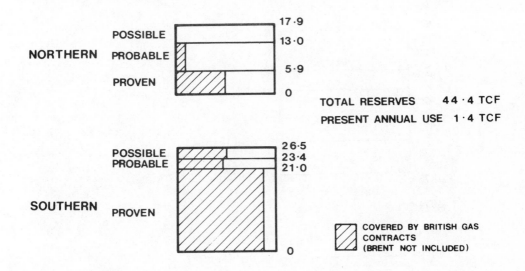

TOTAL RESERVES   44·4 TCF

PRESENT ANNUAL USE 1·4 TCF

COVERED BY BRITISH GAS
CONTRACTS
(BRENT NOT INCLUDED)

Figure 6. Reserves of natural gas

If probable and possible reserves are included the figure rises to 44.4 TCF.
Even higher figures are possible and must be considered for planning
purposes. Present consumption is about 1.4 TCF per annum. The more
immediate reinforcements in supply are likely to be from the north of
Scotland – not exactly where one would choose since long transmission lines
add to cost and operational limitations. The most recent addition is
associated gas from the Brent field. There are hopes of discoveries off
the west coast – for example British Gas has made a gas strike, as yet
unquantified, off the Lancashire coast. Such a supply would be useful in
allowing back-feeding into the system, thereby easing the transmission
problem and reducing the amount of compression needed.

The size and duration of the supply from these reserves is determined by the
depletion policy followed. Under our present system of licensing, the only
return recovered by the licensees/operators is from selling oil and gas
discovered. Having put the investment into discovering and proving the
presence of hydrocarbons, there is naturally a desire to reap the financial
rewards as quickly as possible. A large number of factors enter into
consideration – such as the length of pipes, the choice of size, the number
and type of drilling platforms for production, the nature of the formation
– but it generally turns out that the optimum rate of depletion for dry gas
fields from the producers viewpoint is 5-7% of recoverable reserves per
annum. The supply builds up to a plateau over a year or two when the field
is first opened up, and declines as the pressure in the reservoir drops, so

that the producing life is perhaps 25-30 years. In terms of ensuring a
continuing supply of indigenous energy for the U.K. this is not a long
period. British Gas might well prefer slower rather than faster depletion,
since this reduces the investment in transmission, allows steadier growth
and more orderly marketing. To maintain supplies one would like a steady
succession of discoveries located so as to minimise the transmission problem
and maximise security of supply. The ideal picture is too much to expect,
but is met in part by the way our supplies have built up over the last eight
years (Figure 7).

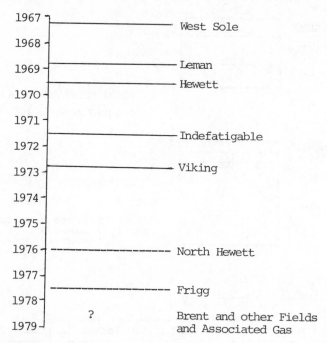

Figure 7. Build-up of natural gas supplies

New gas supplies - Frigg

The next major increment in supplies will be from the Frigg field. This is
situated east of the Shetland Islands and exemplifies a number of the
factors which affect natural gas supplies:

(a) It comes in large increments: within two years of coming onstream Frigg
     alone will provide more gas than the entire gas industry of ten years
     ago. This rapid build up is desirable because it is only worth laying
     fairly large pipes and because the large capital expenditure required to
     develop an off-shore field is only justified for large supplies.

(b) The suppliers' desire for a high load factor - in the southern part of
     the North Sea the contracts allow for about 60% (ratio of average/peak
     day) but because of the distances involved and the greater investment
     much higher load factors are likely to be necessary in the north.

(c) Gas has to be taken where it can be found.

(d) The composition can vary between fields and sometimes during the life
     of a field. Fortunately this has not yet occurred to any significant
     extent in the North Sea. There are minor differences but overall the
     gases from widely separated areas and different formations are
     remarkably compatible. But anything as different as the Dutch Groningen
     gas with its high nitrogen content would pose serious problems.

Associated gas bring further problems of load factor. The rate of supply is
linked to the oil production which will generally be at a much more uniform
rate than is desired for gas supplies and is also subject to interruptions
determined by the oil operations.

Size and growth of supplies

The current rate of supply is about 14,000M therms per annum - about 54 MTCE
equivalent to an average of 3,800 MSCFD. The reserves are capable of
supporting production of 5,000 MSCFD or 6,000 MSCFD well into the 1980's
if Norwegian Frigg is included, i.e. 22,000M therms per annum or 85 MTCE.

To sustain even the present level of supplies into the next century, further
reserves are required. It seems likely that more gas will be found since
there are many areas so far unexplored. Furthermore, the intense activity
in the northern sector is yielding further oil discoveries, many of them
with substantial quantities of associated gas.

Longer term future

While the long term continuation of the gas industry as a major supplier of
primary energy clearly depends on further discoveries of natural gas, the
market supplied by gas could, in large part, be supplied by a manufactured
gas if necessary. This is because the marketing policy of British Gas has
been to concentrate on the premium end of the market to sell gas for its
special qualities. As a long term insurance, British Gas aims to be able
to make a substitute natural gas (SNG) from any carbonaceous feedstock
available. Even before there is a requirement for SNG as a base load supply,
there is likely to be a major role for it for supplementary and/or peak load
supply as part of our armoury in combating the load factor problem. However,
before looking at the future possibilities for supplying this premium load,
perhaps we had better see how it is made up.

5. THE MARKET FOR GAS

In this country gas has long been a major fuel but for some purposes a
relatively expensive one. When it had to be made from feedstocks provided
by its competitors, coal and oil, it could only survive by ensuring that
appliances were available which used gas cleanly and efficiently. This
applied in both domestic and the industrial markets where full advantage
had to be taken of the premium qualities of gas, especially its controllability. The upsurge of interest in conservation and efficiency-in-use
has found British Gas well placed technically to respond to these demands
because of its past activities.

Domestic market

One of the significant factors in the resurgence of the gas industry in the
1960's was the development of the convector fire which combined attractive
appearance with high efficiency. This was only the first of a whole series
of developments which has transformed the domestic market.

A whole range of appliances is available with efficiencies in the region of
70% and sometimes higher. Where heat is required in the home, gas is
undoubtedly the most thermally effective way of providing it. This point
was brought out strongly in the NEDO (National Economic Development Office)
report on Energy Conservation. Only by introducing condensing appliances
– which would give cooler combustion products and extract the latent heat in
the water vapour produced by combustion – could the appliance efficiency be
significantly increased. This has been examined several times in the past
and rejected because of installation problems and the formation of vapour
clouds at the appliance flue terminal, but a new appraisal is being made.

As an example of the improvements made, British Gas, in association with
appliance manufacturers, has developed wall-hung, low-thermal-capacity
central heating boilers. These have improved the overall thermal efficiency
for space and water heating.

Although the scope for further improvement of appliance operating efficiency
is limited, there are opportunities for improving the matching of heating
appliances to the building served. Mathematical modelling of the dynamic
response of a building to external and internal heating and cooling should
help. Greater knowledge is required of how a building is used and what are
the required comfort conditions that would allow the right degree of
sophistication of controls to be specified.

With the improved insulation standards that are now being brought into the
Building Regulations, and emphasis on further improvement, there will be a
need for appliances of smaller output and these are currently being developed.

Industrial market

In the industrial market gas has to make its way in competition with oil and
electricity. That it has been able to match the lower price of the former
and the sophistication of the latter, with overall economy, is a tribute to
its own qualities and the ingenuity of our combustion engineers. The recent
changes in price levels have given a great boost to gas in this area.

Examples of current industrial utilisation work aimed at more efficient
conversion are:

(a) Self-recuperative burners.
     In large scale processes requiring high temperature heat it is well
     established practice to use the hot flue gases to preheat the combustion
     air and/or fuel. This is difficult to achieve economically on smaller
     batch or non-continuous processes. British Gas has developed a series
     of self-recuperative burners which greatly facilitate this approach.
     The heat exchanger is integral with the burner and allows considerable
     simplification in furnace design. Fuel savings of up to 45% are
     possible. These burners are now being manufactured under licence and
     used in pottery firing, steel reheating and non-ferrous melting.

(b)   Vat and tank heating.
      Heating of liquids in tanks is a common process which represents a very
      large heating load - about 8 MTCE p.a.   It is usually done by passing
      steam through coils or directly into the liquid with typical
      efficiencies of less than 50%.   British Gas has developed a direct-fired,
      forced-draught, immersion-tube combustion system which is achieving 85%
      efficiency.

(c)   Rapid heating furnaces.
      Rapid heating furnaces for metal have been developed which can be
      tailored to the billet size and which can heat up and cool down quickly
      with minimum heat wastage.   They reduce scale and decarburisation by
      minimising the time during which the material is held at high tempera-
      tures - thus reducing metal loss by a factor of 4 (2% to 0.5%).

(d)   Gas fuelled stationary engine.
      Significant overall primary fuel savings are possible using gas engines
      to produce shaft horsepower.   In comparison with electricity centrally
      generated and supplied at an overall efficiency of about 29%, a gas
      engine can achieve efficiencies of 30-40% which, with waste heat
      recovery, may be lifted to 55-60%.   In using shaft power, the variable
      speed capability of a gas engine can lead to further energy savings,
      for example, when driving compressors which often run less efficiently
      at fixed speeds under part load conditions.   Work is being carried out
      on automotive type, spark ignition engines adapted to operate on natural
      gas;   these can be especially attractive because of their low capital
      cost.   Such engines have been operating successfully in industry and
      commerce for driving air conditioning systems and air compressors.

(e)   Technical service.
      The greatest short-term potential for fuel saving lies in the proper
      application of existing knowledge.   To promote this in industry the
      British Gas School of Fuel Management has been set up.   This will run
      courses for senior management, for fuel managers and for lower levels
      of management.   These courses are being well supported and are part of
      a wider advisory service available to industry.

The foregoing indicates that gas has a market in its own right and is not
just an alternative fuel that is used only because it is readily available.

In most applications the advantages of gas over coal or oil would justify
conversion of these fossil fuels to gas.   To cover the situation when natural
gas can no longer supply the market, substitute natural gas (SNG) will be
made.

6.   SUBSTITUTE NATURAL GAS

The provision of processes for making SNG is a large element in the R&D
programme - forming 15% of the budget.   This work is based on considerable
past achievement, recognized by the election to the Royal Society of the late
Dr. Dent in 1967, and by the receipt of the MacRobert Award by Dr. Dent and
his team in 1971.

Dr. Dent's work was originally aimed at coal as the feedstock but with the
shift in the balance of cost to oil the results of his efforts which have

received widest application have been processes for gasification of oil fractions.

Before mentioning the individual processes, the general problem of gas making from fossil fuel feedstocks will be outlined.

All current thoughts on gasmaking relate to production of methane or something of comparable calorific value and burning properties – as distinct from town gas which has half the calorific value and a higher burning velocity. Methane CH_4, has a C/H mass ratio of 3. All other hydrocarbons have a higher value, ranging from 4-6 for LPG and light petroleum fractions to 14-18 for coal as shown in Figure 8. Conversion to methane, therefore, needs addition of hydrogen and/or rejection of carbon. Both usually occur, the hydrogen being made from water by the conversion of carbon to carbon dioxide, which can then be removed.

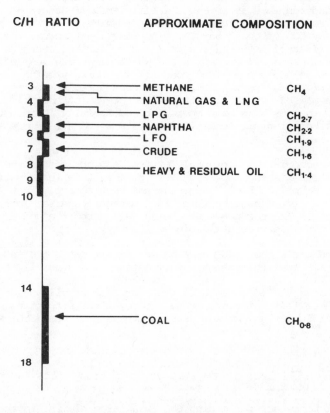

Figure 8. Characteristics of hydrocarbon fuels

As figure 9 shows, the further the feedstock is from the required C:H ratio, the more process work will be required to effect the change and the more capital intensive will be the process plant.

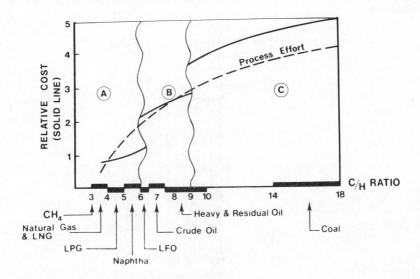

Figure 9. **Process effort as a function of C/H ratio**

The overall reaction is:

$$CH_n + (1 - \frac{n}{4})\, H_2O \longrightarrow (\tfrac{1}{2} + \frac{n}{8})\, CH_4 + (\tfrac{1}{2} - \frac{n}{8})\, CO_2$$

or for oil
n = 2

$$CH_2 + \tfrac{1}{2}H_2O \longrightarrow \tfrac{3}{4}CH_4 + \tfrac{1}{4}CO_2$$

but for coal
n = 0.8

$$CH_{0.8} + 0.8H_2O \longrightarrow 0.6CH_4 + 0.4CO_2$$

7. OIL GASIFICATION

Catalytic Rich Gas

The way in which these changes are achieved depends very much on the nature of the feedstock. For a light feedstock such as LPG or naphtha it can be readily accomplished in a simple catalytic reaction. This is the basis of the Catalytic Rich Gas (CRG) process developed by Dent. No hydrogen supply is needed and the process is carried out over a highly active nickel catalyst in an adiabatic reactor, i.e. there is no requirement to supply or remove heat. The process requirements are relatively simple and plant costs are the cheapest of all the routes to SNG. The process can be applied to any feedstock that can be vaporized and desulphurized, and the application is limited only by the availability of suitable feedstock. There are a number of competing processes, but the British Gas process has taken nearly all the business in the USA and Japan.

Gas Recycle Hydrogenator (GRH)

Where the feedstock cannot be desulphurized, the catalytic process cannot be used. If, however, the feedstock can be vaporized it is possible to hydrogenate directly using another process developed by Dent - the Gas Recycle Hydrogenator.

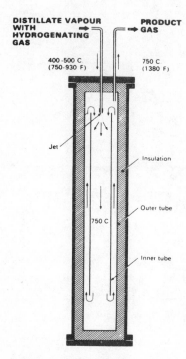

Figure 10. Gas recycle hydrogenator reactor

As its name implies, this requires a supply of hydrogen. The hydrogen and hydrocarbon vapour react in an empty refractory lined vessel where mixing and temperature control are achieved by rapid internal recirculation. The extent of reaction is determined by the chemical kinetics, and equilibrium is not reached as it is in the catalytic processes.

$$-CH_2 + H_2 \longrightarrow CH_4$$

This process was used extensively to produce a rich gas to mix with lean gas from ICI reformers in town gas days. Its value in making SNG has been demonstrated recently in a series of full-scale trials at Avonmouth Works. The hydrogen can be produced by reforming part of the methane:

$$\tfrac{1}{4} CH_4 + \tfrac{1}{2}H_2O \longrightarrow H_2 + \tfrac{1}{4}CO_2$$

or by partial oxidation of oil followed by shift:

$$\tfrac{1}{2}CH_2 + \tfrac{1}{4}O_2 \longrightarrow \tfrac{1}{2}CO + \tfrac{1}{2}H_2$$

$$\tfrac{1}{2}CO + \tfrac{1}{2}H_2O \longrightarrow \tfrac{1}{2}CO_2 + \tfrac{1}{2}H_2$$

Fluidised Bed Hydrogenator (FBH)

The heaviest fractions of oil are difficult to vaporize and have the highest
C:H ratio. To cope with them a fluidized bed hydrogenator is being developed.

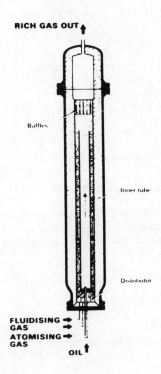

Figure 11. Fluidized bed hydrogenator reactor

In this a fluidized bed of coke acts as the mixing and temperature control
medium. Heavy oil is sprayed in together with hydrogen. Some carbon is
formed which deposits on the coke and a whole series of volatile hydrocarbons
comes out in the product gas which consists of methane, ethane and hydrogen.

This process is less well developed than the others previously mentioned. A
semi-commercial plant has been built in Japan by Osaka Gas, with whom British
Gas has a development agreement.

There are, therefore, a range of processes, some fully developed others
rather less so, which can cope with any likely liquid feedstock.

8.   COAL GASIFICATION

Coal is the most abundant fossil fuel and must eventually be the primary feedstock for gasmaking - indeed the situation is already imminent in the US where
great efforts are being made to develop efficient and cheap complete

gasification processes. The thermal efficiency of coal gasification is
about 60%. It could, therefore, be argued that gasifying coal is preferable
to using coal in power stations for electricity. Coal has the highest C/H
ratio of fossil fuels and poses severe handling problems particularly in high
pressure processes.

The basic reaction we would like to perform is:

$$C + 2H_2 \longrightarrow CH_4 \qquad\qquad - 17.88 \text{ kcals}$$

Given the right conditions of temperature and pressure (900°C and 500 atm)
coal can be completely hydrogenated, but these conditions are too severe for
an economic process. However, 30-40% of coal can be hydrogenated fairly
easily leaving an inactive char, which can then be used to provide hydrogen:

$$C + H_2O(g) \longrightarrow CO + H_2 \qquad\qquad + 31.34 \text{ kcals}$$

$$CO + H_2O(g) \longrightarrow CO_2 + H_2 \qquad\qquad - 9.83 \text{ kcals}$$

$$\overline{}$$

$$C + 2H_2O(g) \longrightarrow CO_2 + 2H_2 \qquad\qquad + 21.51 \text{ kcals}$$

The endothermic heat of reaction can be provided by burning more coal. For
this purpose oxygen is required since the use of air would introduce inert
nitrogen which is inadmissible for SNG although it is acceptable for low Btu
gas.

The different ways of achieving these reactions have been the subject of
much process development.

Lurgi process

The only fully developed high pressure process for gasification of coal to
SNG is the LURGI process.

This uses a steam/oxygen mixture to gasify coal in a rotary grate gasifier
at 300 psi. Coal is admitted and ash is removed through lock hoppers.
Temperatures are kept down to about 900°C by use of excess steam to prevent
fusion of the ash. The product is largely CO + H$_2$ with some methane.

The process has been used for many years for making town gas at Westfield
Works and elsewhere but in recent years it has been extended by British Gas
to produce SNG. The key step has been the catalytic synthesis of methane
from the CO and H$_2$ which has been demonstrated on a large scale at Westfield.
It has also been shown that the Lurgi gasifier can accept a much wider range
of coals than hitherto thought.

There is thus a fully proven process available - but with one principal
disadvantage: rather small throughput per reactor.

The small throughput is associated with the need to avoid ash fusion which:

(a) keeps temperatures down and limits reaction rates;

(b) requires large volumes of excess steam which tend to blow fines out of
     the bed.

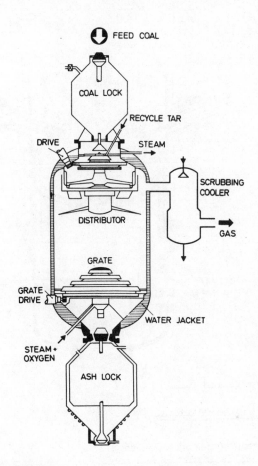

Figure 12. Lurgi gasifier

Slagging gasifier

These disadvantages are overcome in the British Gas Slagging Gasifier.

This process was operated at pilot plant scale some years ago and is now the subject of a $10M development at the Westfield Development Centre.

The reactions are carried out at a temperature high enough to melt the ash completely and allow it to be removed as a liquid slag. This increases the reaction rate and cuts down steam requirements to stoichiometric. It is expected that the throughput of a gasifier of given size will be increased four-fold by this means.

The first runs have been made successfully. Assuming all goes well, this process will greatly advance the method of gasmaking from coal but the apparent simplicity of the idea should not mask the mechanical difficulties inherent in slag tapping at high pressures.

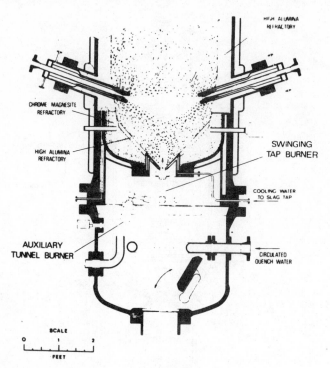

Figure 13. Slagging gasifier

U.S. Development

Coal gasification research and development has recently been very active in
the U.S.A. There are many separate approaches to the problem, several based
on Dent's pioneering work. They generally seek to maximise direct hydrogenation and to minimise or eliminate the requirement for oxygen. They are
outstanding examples of technical ingenuity but none has yet operated on a
scale comparable to the Lurgi or the slagging gasifiers.

9.   CONCLUDING REMARKS

From what has been said it is clear that there are ways of gasifying coal
on a large scale but that there is a great scope for development work. So
long as there is a carbonaceous fuel to act as a feedstock, British Gas aims
to be able to convert it to methane.

But if carbonaceous fuels become scarce very expensive supplementary therms
will be required. The most likely source seems to be nuclear energy. This
summer school is discussing the question of hydrogen from nuclear energy and
it is unnecessary to cover the same ground again. But if hydrogen were
available under the right economic conditions, it would be preferable from

many points of view to use it for making methane from fossil fuel by hydrogenation rather than supplying hydrogen direct. This doubles the amount of methane per ton of coal, thus effectively stretching fossil fuel reserves.

There is no doubt that gas is an important element in the energy economy of Great Britain, that it has a great deal to contribute in terms of increasing the efficiency of fuel use, and that its qualities are such that much of its present market will require us to make gas from oil or coal when natural gas can no longer meet the load.

BIBLIOGRAPHY

1.   Annual reports. British Gas Corporation.

2.   Digest of United Kingdom energy statistics 1974. HMSO.

3.   Development of the natural gas supply system. W.J. Walters. Paper presented to Institution of Gas Engineers, London and Southern Section, 1970.

4.   Further development of the natural gas supply system. W.J. Walters and R.S. Hackett. Paper presented to the Institution of Gas Engineers, London and Southern Section, 1975.

5.   The philosophy of gas storage. D.J. Clarke, G.S. Cribb and W.J. Wallis. Journal of the Institution of Gas Engineers 11 748 (1971).

6.   Development of the oil and gas resources of the United Kingdom. HMSO.

7.   Marketing implications of recent North Sea gas discoveries. G.F. Claxton. Paper presented to the Institution of Gas Engineers, London and Southern Section, 1974.

8.   Marketing vitality in our vital industry. J.A. Buckley. Gas Marketing 18 (7) 23-27 (1974).

9.   Marketing gas central heating. G.A.J.D. McMillan. Domestic Heating and Air Conditioning 7 (3) 2-21 (1974).

10.   Energy from gas in tomorrow's homes and buildings. Paper presented to Public Works and Municipal Services Congress, London, 1968.

11.   Some current topics affecting domestic gas utilization. C.H. Purkis. Paper presented to Scottish Junior Gas Association, Institution of Gas Engineers, Scottish Section, 1975.

12.   Some aspects of commercial gas marketing. C.H. Watson. Gas Marketing 19 (1) 29-30 (1975).

13.   A review of industrial and commercial gas burner developments.
       J.M. Downie and M.L. Hoggarth. Journal of the Institute of Fuel,
           124-129 (1974).

14.   Applications of recuperative burners in gas-fired furnaces.
       D.J. Bryan, J. Masters and R.L. Webb. Paper presented at the 40th
       Autumn Research Meeting of the Institution of Gas Engineers, 1974.
       IGE Communication 952.

15.   Utilization research and development at the Midlands Research
       Station. W.E. Francis. Paper presented to the Institution of
       Gas Engineers, Scottish Section, 1970.

16.   New horizons for pressure gasification. The production of clean
       energy. D. Hebden. 1971 MacRobert Lecture. Journal of the
       Institution of Gas Engineers, 12 229 (1972).

17.   Production of SNG from coal. D. Hebden. Paper presented to the
       Institution of Gas Engineers, North of England Section, and the
       Coke Oven Managers Association, 1973.

DISCUSSION

Question: At what pressure is gas transmitted?
Answer: The pressure is 1000 psig. Compression losses amount to 0.25% in
           the present transmission system.

Question: Who has first option on any gas found in the northern North Sea?
Answer: If it is found in the British sector, it must be first offered to
           British Gas.

Question: If natural gas becomes scarce, why could we not revert to the old
           town gas?
Answer: The old town gas processes are uneconomic today; they are inef-
           ficient and labour intensive. Modern methods of gas manufacture
           are efficient and capital intensive. To go back to town gas would
           mean that we would be halving the capacity of the transmission and
           distribution system since town gas has half the calorific value of
           natural gas.

Question: How much helium is there in North Sea gas?
Answer: Usually about 0.1%. If there were gas with levels of 0.3%, it
           would be economic to extract the helium. In Dutch natural gas,
           cryogenic processes are used to remove the 14% nitrogen and under
           these circumstances the helium is also extracted.

Question: What is the origin of the helium?
Answer: Radioactive decay.

Question: Does one get carbon fouling of catalysts for SNG manufacture or
           in the Gas Recycle Hydrogenerator?
Answer: Carbon laydown on high surface area nickel-in-alumina catalysts
           is a problem to which we are devoting considerable research with
           the objective of preserving reactivity. Carbon deposition is not
           a problem with the GRH, but if one had heavy oil feedstocks, with
           which carbon would be a problem, one would use the Fluidised Bed
           Hydrogenerator in which solid carbon would be absorbed into the
           bed.

Question: What are the prospects for gas or oil in the Irish Sea?
Answer: British Gas has found gas off the Lancashire coast. There has
            been an oil strike in the Eire part of the Irish Sea. There has
            been drilling in the Western Approaches, so far with little
            success. There is at present no international agreement on dril-
            ling rights in the English Channel. Oil production in the Wareham
            area, oil shale deposits in Kimmeridge Bay and the British Gas oil
            find at Wytch Farm suggest prospects for oil or gas in this area.

Question: How much gas associated with oil is not recoverable?
Answer: Where one has gas associated with oil, it is a practice in opera-
            tion to re-inject the gas into the well to maximise oil production.
            In principle, the oil company might have the option of flaring
            excess gas, but the Department of Energy is at present formulating
            rules on this point.

Question: Is the limitation on the rate of North Sea gas supply economic or
            technical?
Answer: The rate of supply is a compromise, reflected in the price paid
            for the gas, between a load factor acceptable to British Gas and
            the cost to the oil company of drilling wells. The limitation is
            both economic and technical.

Question: How efficient are storage techniques which use salt cavities?
Answer: All gas is recoverable but the cost is substantially the monetary
            interest on the gas stored.

Question: What are the gas transmission losses?
Answer: There is no loss of gas in the high pressure transmission system.
            In the low pressure distribution system, the losses are about 5%,
            i.e. it is a problem of old joints in the 120,000 miles of dis-
            tribution main. The Dutch initially had a loss of 30% in their
            mains after conversion from town gas to natural gas, and at this
            level one might imagine some hazard involved. In this country,
            since the problem is one of substituting dry natural gas for wet
            town gas (which contained aromatic compounds that swelled jointing
            materials and maintained sealing), one can try to swell joints by
            glycol injection as an aerosol into the main or, in bad cases, by
            excavating the joint and encapsulating it.

Question: Is liquified petroleum gas injected into SNG to maintain the
            calorific values and is there thus a condensation problem?
Answer: Small amounts of l.p.g. can be added to adjust the calorific
            value but there is no condensation problem.

Question: Do we still buy LNG from Algeria?
Answer: Yes. This is still a valuable contract and accounts for 2.5% of
            output.

SULPHUR POLLUTION AND EMISSION CHARGES*

R. Wilson

Physics Department, Harvard University

INTRODUCTION

Edward I of England banned the burning of coal in his kingdom because of the
noxious fumes produced. At the time there was enough wood in the country,
but when the population increased and the woodlands had been cleared, the
resultant energy crisis swept environmental considerations aside and coal
burning began on a large scale. The court physician to Queen Elizabeth I
took out a patent to remove sulphur[1] but, as we all know, the problems
still exist today, four centuries later.

Sulphur is, of course, not the only problem with burning coal; the particulate matter in the smoke is at least as serious. Coal burning also emits
other trace substances – it is one of the main ways mercury and other heavy
elements enter the biosphere, and their health hazard is unknown. The amount
of radioactive material, radium and thorium, moreover, in coal is enough to
make the problem of long-lived (greater than 500 years) radioactive waste
from coal burning comparable to the waste from a nuclear power plant. However, in this talk, I shall discuss the sulphur dilemma alone.

In the last fifty years there have been major attempts to reduce pollution.
First and foremost, many cities have banned the burning of soft coal, and
coal is now burned primarily in large factories and power stations. These
have electrostatic devices to remove up to 99.5% (by weight) of the particulate matter that causes the smoke. Another important step has been the
spreading by the use of tall chimney stacks. Tall stacks have been more

* Work assisted by the Energy Research and Development Agency through a sub-
contract from Brookhaven National Laboratory No. 33-542-9042-2.

widely used in England than in the U.S., but they are being installed in the
U.S. also.

In Glasgow as late as 1950, three tons of soot fell per acre per year, but
when this is cleaned up, and 99.5% removal of particulates (by weight) is
possible and usually mandatory, we still have sulphur.

Until recently this was all the pollution control that most experts thought
was necessary, but new assessments suggest that there may be further
problems.

EPIDEMIOLOGY OF SULPHUR POLLUTION

Before we discuss abatement procedures, it is necessary to get some idea of
the effect of sulphur emissions on man. There are two excellent survey
reports[2][3] and I will give a brief summary here.

There have been laboratory tests of exposing people to sulphur dioxide in
concentrations of 1 part per million and above. The effects are marked and,
for long times, lethal. At 10 parts per million (30,000 micrograms per cubic
meter), it is lethal for shorter times. The early English work in which the
name of Lawther stands out[4] was mostly concerned with effect on up to 100
people of "large" concentrations, but we are now talking about exposures of
millions of people. These are much more difficult studies.

In Figure 1, I show the result of a Norwegian survey of death rates in 152
winter weeks in Oslo[5], and in Figures 2 and 3 the results of a Japanese
comparison of bronchitis incidence in several cities[6]. There appear to be
effects at low concentrations - 0.03 parts per million (100 μg per m^3).

There are important reservations and restrictions to be considered with these
graphs. As can be seen clearly in the Japanese studies, the effects at the
present average levels are about 10 times smaller than the effects of cigarette smoking, and we all know how hard it has been to convince people of
that problem. It will be 100 times harder to convince people that

the effect exists, although a rough estimate shows that about 10,000 people in the U.S. could have their lives shortened by existing levels of sulphur. An examination of Chapter 4 of reference 2 shows these reservations clearly.

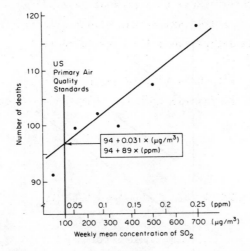

Fig. 1 - Total number of deaths for 156 winter weeks in Oslo, Norway (1958/9 to 1964/5).

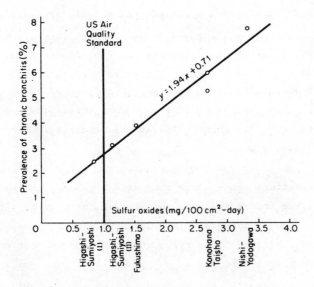

Fig. 2 - Correlation between the prevalence of chronic bronchitis and sulphur-oxide precipitation in Japan without separation of smokers and non-smokers.

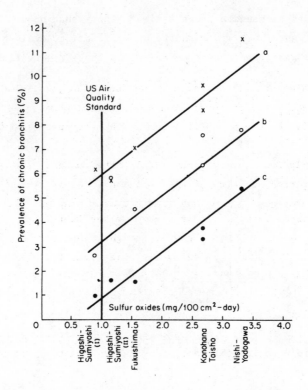

Fig. 3 - Correlation between the prevalence of chronic bronchitis and
          sulphur-oxide precipitation for several Japanese cities. Curve a,
          11-20 cigarettes/day; curve b, 1-10 cigarettes/day; curve c,
          non-smokers.

We therefore have a question - do we proceed as if there is no problem, since
it is hard to prove? (i.e. is the polluter innocent until proven guilty?) or
do we insist that there is a problem until proved otherwise? Clearly we must
choose a rational middle ground.

There are many reservations about these studies. Both have chosen the SO_2
concentration to be the independent variable (along the x axis) although
different cities (in the Japanese studies) were involved. Since SO_2 concentration is higher in the center of a city, might not the correct correlation
be to the strains of city life?

Nonetheless, SO_2 has dominated everyone's thinking. Moreover, pathologists
always believe in a repair mechanism; that leads to some threshold of

pollutant below which the body can repair itself, although figures 1 and 2
show none.

When the health hazard was believed to be due to SO_2, all that seemed necessary was to <u>reduce</u> the concentration to below the threshold, which, if our
regulatory bodies have done their jobs, is the air quality standard.

There is a simple procedure for calculating the concentration of a pollutant
at ground level from emission of the pollutant. We are interested in the
concentration averages over a long time and over many wind directions.
As we look at smoke from a chimney stack, we see that it rises above the top
of a stack, to 1½ to 2½ times the stack height, then stretches in a long
plume downwind. The plume gradually disperses into the rest of the air, and
eventually reaches ground level. There is almost no concentration at ground
level, near the stack, and at 2 miles from the stack there is space for it
to spread widely. Simple calculations along these lines are performed in
references 7 and 8. The accuracy is quite good – about a factor of 2 – and
has been checked by using the same model for radioactive materials.

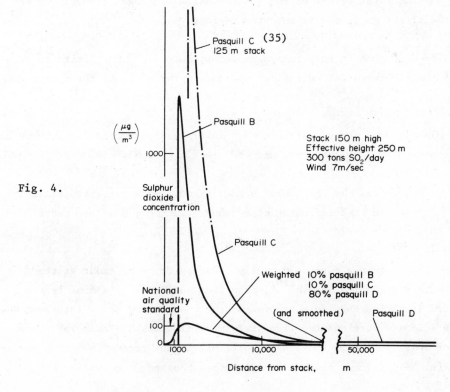

Fig. 4.

Figure 4 shows a crude calculation for average concentration for a 1000 mega-watt (electrical) power station burning 3% sulphur fuel and with a 500 ft high stack (plume rising to an "effective stack height" of 800 ft, or 250 metres).

What are the effects on health? At short distances of up to 30 miles, the concentration decreases as the square of the reciprocal of the distance from the stack ($1/r^2$). At distances greater than 30 miles, the pollutant is often trapped between the ground and an inversion layer at 3000 feet; it no longer disperses vertically and the concentration falls as ($1/r$) because the pollutant no longer disperses vertically, but only horizontally. If there is a large metropolitan area (50 miles x 50 miles), the number of people in increasing circles around the power station increases as r. If we believe, as suggested before, that there is zero medical effect on people at concentrations below the threshold of 100 $\mu g/m^3$ (or 0.03ppm) [9], this dispersion procedure is all we need. But the data we cited on large populations suggest the medical effect is proportional to dose. Then the two terms ($1/r$) and (r) cancel, and the total effect of one power station increases linearly as we add up the effect on people in outlying regions.

For smaller distances it only increases logarithmically. The limits to this are only the limits on the metropolitan area and the ultimate absorption of the sulphur.

This simple argument shows that sooner or later we should consider sulphur as a global and not a local problem, and the time is probably now.

Another reason arises why the global effect is now important. It is likely that SO_2 is not the best indicator of sulphur pollution; sulphate particulate levels, SO_4, are better.

Experiments with guinea pigs, especially those by Dr. Mary Amdur at the Harvard School of Public Health[10], show that sulphuric acid mist is ten times more effective than sulphur dioxide in constricting bronchial tubes, and some sulphate particulates five times more effective (Table 1a). The amount of constriction is roughly proportional to the concentration of pollutant (Fig. 5). Sulphur dioxide readily gets absorbed onto small smoke

Table 1a

Ranking of Sulphates for Irritant Potency

(Constant size particle 0.3μm)

Compound	% Increase Resistance $/(\mu g\ SO_4.m^{-3})$
H_2SO_4	0.410
$Zn(NH_4)_2(SO_4)_2$ [+]	0.135
$Fe_2(SO_4)_3$ [*]	0.106
$ZnSO_4$ [+]	0.079
$(NH_4)_2SO_4$ [+]	0.038
NH_4HSO_4	0.013
$CuSO_4$	0.009
$FeSO_4$ [*]	0.003
$MnSO_4$ [*]	-0.004

[+] Data of Amdur and Corn (1963)

[*] Data of Amdur and Underhill (1968)

Table 1b

Irritant Potency of Sulphuric Acid

Size μm	Number of Animals	$\mu g/SO_4/m^3$	% Incr. Resistance	% Incr. $/(\mu g\ SO_4.m^{-3})$
0.1	10	74	32 [*]	0.432
0.3	23	100	41 [*]	0.410
0.7	10	120	43 [*]	0.358
1.0	9	68	14 [*]	0.205
2.5	10	215	18 [*]	0.084

[*] $p < 0.01$ or better

particles to form sulphate aerosols. Moreover, small particulates, the ones
that escape the electrostatic suppressors, are found to be most irritant and
remain longest in the air   (Table 1b).

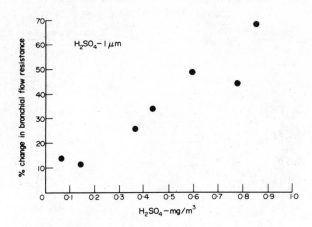

Fig. 5.

Several epidemiological studies confirm this laboratory view that sulphate
particulates are more important than sulphur dioxide. I want to remind you,
however, of the difficulty of these studies. The first statistical study
was done by an economist[11] who had no reputation in the field to lose and
could be bold. Lave used a regression analysis and gave a formula for the
death rate in the U.S.A in 1961:

    Death rate per 10,000 persons per year =
    $20 + 0.04C + 0.7S + 0.001 (P/mile^2) + 40 NW + 7000 (>65)$
where

       C is the concentration of particulates
       S is the minimum concentration of <u>sulphates</u>
       $(P/mile^2)$ is the population density
       NW is the proportion of non-whites in the population
       >65 is the proportion of old people in the population

Noteworthy features are again that a <u>linear</u> effect is found, and that a
better correlation is found to <u>sulphate</u> concentrations than to sulphur
dioxide concentrations.

A correlation with total suspended particulate matter (TSP) has been observed

by Winkelstein[12] and the data has been re-examined, and the conclusions confirmed, by a Brookhaven group[13]. A large EPA study (Community Health and
Environmental Surveillance System or CHESS, reference 14) shows a correlation
of various medical effects to sulphate levels, and I present them here as
figures 6, 7a, 7b. You will note in the solid lines on this study, the prejudice of the medical man (Dr. Finklea) in charge of this study for a threshold below which medical effects are zero although the data does not really
prove it.

This SO_2/sulphate issue is very important. While sulphate pollution remained
a local effect, it seemed likely that SO_2 was the main way of ingesting sulphur. SO_2 is absorbed out on plants and vegetation[15] before conversion to
the irritant H_2SO_4 or sulphates. Now we disperse SO_2 with high stacks, we
guarantee that a large fraction gets converted to sulphates before reaching
the ground, and becomes 10 times more irritant. In this respect, the past is
not a guide to the future - we have dispersed our sulphur sources - put them
in the suburbs - but emit more sulphur, in high stacks. We may be killing
many more people than figures 1-7, based on the past, suggest.

I would like to highlight this: all European abatement practices are based on
a discussion of sulphur dioxide pollution. Perhaps they are inadequate.

Although the health hazard may not be proportional to concentration, a growing body of opinion believes that it is wise to assume that the health hazard
is proportional to concentration for the purpose of setting pollution standards. For example, the committees that recommend standards for exposure to
radiation have recommended such a "linear theory" for many years. A low concentration of pollutant can be very important if it is widespread and many
people are exposed.

There is further evidence that sulphates are a regional or global problem.
This comes from sulphate measurements in the eastern U.S.A. Concentrations
($10\mu g/m^3$) are 10 times higher than in the western U.S.A. Let me make a
rough calculation of their magnitude, to show the physical effects involved.

In figure 8 I show the SO_2 emission density in the U.S.A. showing the large
emissions in the area.

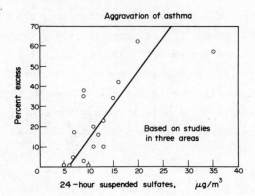

Fig. 6.

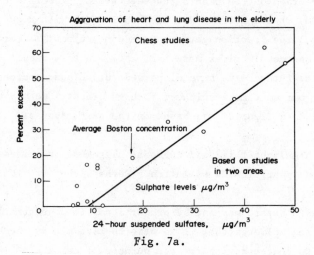

Fig. 7a.

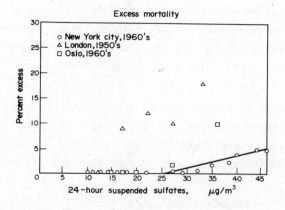

Fig. 7b.

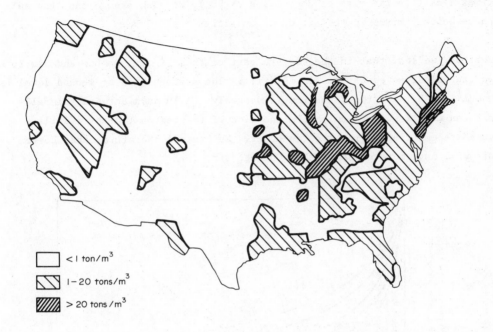

Fig. 8 - Nationwide Geographic Variation in SO_2 Emission Density.

Legend:
- < 1 ton/m^3
- 1 - 20 tons/m^3
- > 20 tons/m^3

ROUGH ESTIMATE OF THE HAZARD

We will calculate the sulphate concentration due to sulphur emissions from sources in the northeastern U.S. under very simple assumptions. This, we believe, demonstrates a prima facie case that industrial sources are the cause of the high sulphate concentrations in the northeastern U.S. and hence are a health hazard.

In 1968, the total U.S. emissions of SO_2 were 33 million tons[15]. We here assume that 15,000,000 tons are emitted yearly over an area 1000 km north to south and 2000 km east to west covering the Chicago to New York region. (Figure 8 shows the actual density of such emissions)[16]. We will assume that all this sulphur dioxide converts to sulphate, and that these sulphates stay in the air till the next rainfall (the rate of deposition on land is 50 times less than the rate of deposition of SO_2). We also assume that the sulphate quickly mixes in the air up to an altitude of 10 km.

Further, we show that in the eastern part of this region, Boston and New York, we can calculate as if the air is still. The average wind at ground level is about 1 km/hr and the direction is westerly; in 10 days between rainfalls air can move 2400 km - only from one side of the region (which is still a sulphate emitting region) to the other. We expect, therefore, that Boston air will have more sulphate than Chicago air.

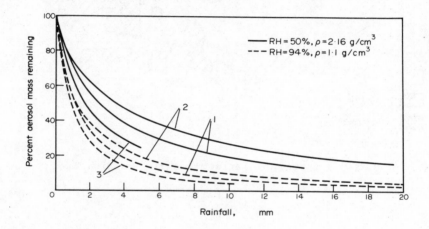

Fig. 9.

Rainfall of greater than 2mm comes roughly every 10 days (reference 17). Here we take an estimate from reference 18 (figure 9) that 2mm of rain is enough to cause a rainout. The frequency depends upon the amount of deposition. Thus we expect the concentration to rise between rainfalls and then fall to zero. If 15 million tons are emitted in this area in one year, then 500,000 tons are emitted per 10 days, or 5×10^{11} grams of SO_2, in a volume $10^6 m \times 2 \times 10^6 m \times 10^4 m = 2 \times 10^{16} m^3$. The concentration then rises to 25 $\mu g/m^3$, and averages $12\frac{1}{2}$ $\mu g/m^3$; since sulphates are a bit heavier, this would be about 18 $\mu g/m^3$ of sulphate. The distribution of sulphates over the U.S. is shown in figure 10 (from reference 14). This shows that there is indeed a high concentration of sulphate in this heavily industrialized region.

1970 Sulfate concentrations, urban ($\mu g/m^3$)

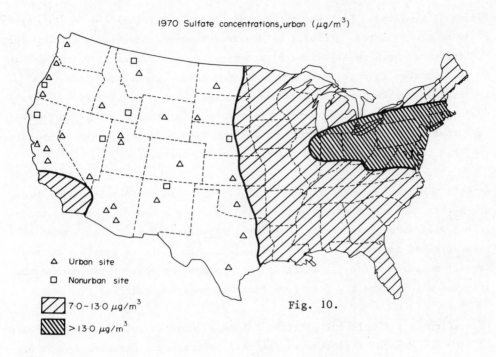

△ Urban site

□ Nonurban site

7·0–13·0 $\mu g/m^3$

>13·0 $\mu g/m^3$

Fig. 10.

Sulfate rainout map

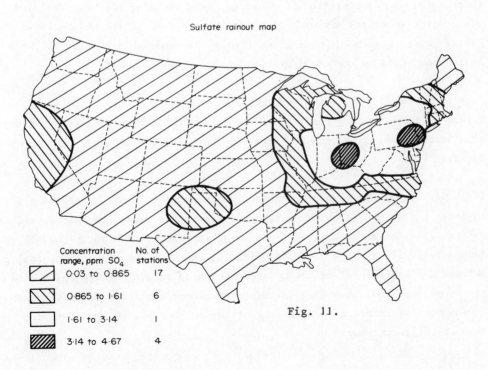

Concentration range, ppm SO$_4$	No. of stations
0·03 to 0·865	17
0·865 to 1·61	6
1·61 to 3·14	1
3·14 to 4·67	4

Fig. 11.

Figure 11 confirms a high washout of sulphates in rainfall[19,20]. Unfortunately no one measured carefully before coal burning commenced a century ago,
and there has been no experimental stopping of coal burning for long enough
to tell whether this sulphate concentration indeed comes from industrial
sources. Since we have no volcanoes emitting sulphur in the East, natural
sources seem unlikely. I believe, therefore, that there is a _prima facie_
case that these high concentrations _do_ come from industrial sulphur emissions,
and that the onus is on any polluter to prove otherwise.

We note that the concentrations of sulphates _now_ in the north-eastern U.S.
are higher than the thresholds suggested in figures 6 and 7 and that the oil
crisis will make these rise. Therefore, the question of the existence of a
threshold is, for those of us in Boston, academic. It is sensible to consider all measures which can mitigate the problem, so that we can choose the
solution which gives the best comparison of cost and benefit.

To confirm this simple calculation, I show a computer calculation in figure
12 from Ron Meyers of Brookhaven National Laboratory. He takes a puff of
sulphur from a power station in Chicago and assuming the measured winds
and complete vertical mixing, calculate the progress of the sulphur over the
U.S.A. Averaging these puffs gives figure 13, remarkably similar to the
measurements and my crude calculation.

Such transfrontier pollution problems are being studied now by Ottar in
Norway[21]; preliminary results confirm that sulphur indeed crosses national
boundaries, passing on occasion from the Ruhr over Ireland, and the Faroes to
Norway!

If it were cheap and simple to remove the sulphur, it would be everyone's
choice , but scrubbers to remove sulphur from stack gases are expensive (the
capital cost for an electricity generating station is $100 per kilowatt),
reduce reliability and add to the operating cost. Conversion of the coal to
methane gas and removal of the hydrogen sulphide before burning would also
remove mercury and other heavy elements mentioned earlier; coal gasification
is the most desirable method of controlling sulphur but is more expensive
than stack gas scrubbers.

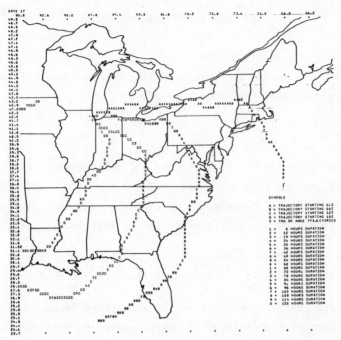

Fig. 12 - Meyars/Cederwall trajectories for the layer 0-1000 meters.
            Brookhaven National Laboratory.

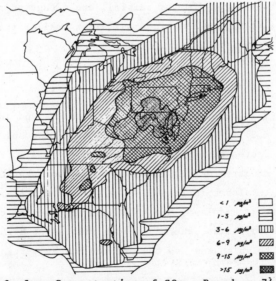

Fig. 13 - Mean Surface Concentration of SO_4. Based on 74 fossil-fuel power
            plants east of the Mississippi River with capacities greater than
            1000 MW (median SO_2 emission; 275 tons/day). Actual NWS winds
            (0-1000 meters) for 9/26-10/31/74 used with trajectory model
            (Heffter, ARL), simple chemistry and deposition added.
            Meyars and Cederwall, Brookhaven National Laboratory.

DISCUSSION OF ASSUMPTIONS

We believe it to be useful to summarize those assumptions where we may have
over- or under-estimated the health hazard.

Over-estimate

1) The epidemiology may be all wrong and the correlations observed between
    mortality rate and sulphate concentrations may have a third cause. This
    we consider unlikely because the evidence by now comes from several
    countries and studies, and is supported by laboratory studies.

2) There may be a threshold concentration below which sulphates cause no
    health hazard, and this may be higher than present evidence suggests.

3) The sulphates may be removed from the atmosphere faster than we have
    assumed, and the high concentrations of sulphates observed in the eastern
    U.S. may have natural causes.

4) We have used figures that 2mm of rainfall are enough to rain out the sul-
    phates. Much of them may be rained out after smaller rainfalls which
    therefore occur more frequently.

Under-estimate

1) The epidemiology is based upon levels of sulphates and sulphur dioxide
    present in the U.S. in the 1960's. During that time sulphur burning in
    the cities has decreased and has been confined to tall stacks, but sulphur
    burning in the countryside has increased, as power stations are driven
    there, and the total has increased. The sulphur dioxide emitted in many
    sources close to ground level may have been absorbed by plants before
    they convert to sulphates, whereas oxides emitted at high levels may
    always convert to sulphates, so the problem may be getting worse.

2) In the 1960's the sulphates may have been attracted to large particulates
    which are removed from the air quickly and in any case do not easily pass
    the nasal passages. With the improved particulate control of the 1970's
    we may be attaching sulphates to smaller particulates which stay in the

air and more easily pass the nasal passages to cause a health hazard.
This agrees with the evidence from laboratory animal studies[10] that
smaller aerosols are a greater health hazard.   Thus again, sulphates in
the 1970's may be several times worse than sulphates in the 1960's.

3)   The small aerosols may not be washed out by the rain as we assumed, and
     this therefore gives one more reason why the sulphate problems of the
     1970's may be worse than the epidemiology of the 1960's suggests.

4)   In this study we omit effects of the acid rainfall on human health.   The
     increased acidic nature of the water supplies may be serious.   However,
     these can presumably be dealt with by appropriate chemical treatment of
     water supplies (at some expense).

FUEL SWITCHING

There have been many suggestions that fuel with a low sulphur content be
burnt during unfavourable meteorological conditions, and that with high sulphur content during favourable conditions, thereby stretching the supply of
expensive, sparse, low polluting fuel.   Most of these discussions are based
on the premise that the public health problem arises from sulphur dioxide,
which is a local problem, whereas (as shown in this paper) the regional problem of sulphate aerosols is more important.

If SO_2 were the problem, we could consider good conditions for the seaboard
cities to be a westerly wind with Pasquill D,E,F, or G weather conditions[35],
where sulphur dioxide from a tall (300 metre effective height) chimney stack
will not hit the ground for 20 kilometres (see Table II from reference 22).
The prevailing wind is westerly, and these weather conditions occur 75% of
the time.

This simple prescription is not quite so simple for an inland source, because
many people live downwind.   The basic procedure still works, however, if we
rely upon dispersion, and neglect possible effects of very low doses of sulphur dioxide.   For example, in a long report prepared for the electric

utility companies of Massachusetts by Environmental Research and Technology[23]
it is claimed that all power sources with stacks over 100 ft in height in
Massachusetts could burn high sulphur fuel all the time without violating the
National Air Quality Standard, and if fuel switching were adopted there would
be a large margin of safety. We do not necessarily accept this claim, for
there are assumptions and over-simplifications in the report. In the context
of this report we are not concerned with the legalism of meeting an air
quality standard, but in protecting human health. The assumption of a
threshold at the air quality standard is unproved, and the neglect of the
sulphate problems is almost certainly wrong.

Table II
Meteorological Data for Atmospheric Release Calculations[*]

Weather type	A	B	C	D	E	F	G
Probability of weather condition.	0.019	0.081	0.136	0.44	0.121	0.122	0.08
Wind velocity u_x (m/sec)	2	3	5	7	3	2	1
Distance, x (m)				σ_z (m)			
200	28.8	20.3	14	8.4	6.3	4.05	2.63
500	100	51	32	18	13	8.4	5.5
1000	470	110	59	32	21.5	14	9.2
2000	3000	350	111	51	34	21.5	13.7
5000	–	1900	230	90	57	35	23
10,000	–	–	400	140	80	47	31
20,000	–	–	650	200	110	58	37
50,000	–	–	1200	310	150	75	48
100,000	–	–	1800	420	180	90	55

[*] Table adapted from (22).

When we consider sulphates, and if we do not admit a threshold, the procedure
must be considerably modified, and it becomes complex when it is recognized
that sulphates are the major hazard, and that the problem is not merely local.

For the seaboard polluters, the presumption is still about the same. For
example, we consider the winds in Boston; data from Logan Airport shows that

winds from the south predominate, and this is more nearly true for heavy
winds. Therefore, burning of high sulphur fuels might be permitted during a
wind from southwest to northwest, and Pasquill conditions D, E, F, or G,
where a major wind reversal is not expected for two days, or until after a
rainfall has removed the sulphates. This might only be allowed for stacks of
100 metres or more in height to accentuate the effect of carrying the sulphate
out to sea. (Table III).

Table III

Month and Winds	Absorption on sea (%/hr)	Improvement over uniform emission on land %
April, 1974		
Logan Airport ground level	0%	19.8
winds, continuous emission	5%	27.8
of pollutants	10%	31.7
Emission only when wind	0%	58.2
blowing offshore	5%	77.5
September, 1974		
Logan Airport ground level	0%	8.3
winds, continuous emission	5%	16.8
of pollutants	10%	21.5
Emission only when wind	0%	28.0
blowing offshore	5%	43.6
Logan Airport 1000 feet winds	0%	–
	5%	–
	10%	–
September, 1974		
Logan Airport winds assumed	0%	2.9
exactly opposite	5%	4.6
	10%	5.6
September, 1974		
Logan Airport winds; sulphate	0%	6.0
only emitted during offshore	5%	9.4
wind		

(Calculations by Mr. B. Chang, Harvard University).

Note that even with 0% absorption, air quality is improved because the pollutant is being blown to uninhabited regions.

FORECASTING AND ASSESSING

The proposals for fuel switching run into a major psychological and administrative hurdle. Who is to forecast the weather, and what happens if the
forecast is wrong? As we proceed from a discussion of SO_2 pollution (local
and short term) to one of sulphates (regional and longer term) this problem
gets worse. The cost advantage of fuel switching is so great that the fore-
caster may well be under improper political pressure.

One way to avoid this problem is by means of the pollution charge or tax (to
be discussed later). The polluter would be entitled at almost any time to
burn high sulphur fuel; the restrictions could be few and well-defined -
burning only with stacks of 100 metres or more, for example, or no burning
during specially announced air pollution emergencies. But the pollution
charge would be directly related to the health effect calculated (by a well-
defined method) from the measured weather as it actually occurred.

Then the incentive for correct forecasting will fall squarely on the polluter; if he forecasts correctly, he will save money; more than enough to pay
for a good forecast. Even improper political pressure cannot easily change
definite weather measurements.

POLLUTION CHARGE OR TAX

There have been several suggestions [24-28] that a tax be charged on emissions
of sulphur in order to discourage the emission of sulphur, and possibly to
compensate its victims. In most cases, a simple tax of 10 cents per pound of
sulphur has been suggested. This low a tax will not discourage much the
emissions of sulphur; at \$100/kw installed capacity, the cost of a stack gas
scrubber is about 20 cents/pound of sulphur.

I have suggested [28] that the tax be more closely related to the effect on

public health. The procedure suggested was to calculate the average SO_2
concentration caused by each emission source and to tax the product of the
concentration and population exposed thereto. With the recent improved
knowledge, the SO_2 concentration in this formula must be replaced by the
sulphate concentration. Once this is done, we revert closely to the original
simple tax per pound of sulphur emitted. I prefer the use of the charge
instead of tax because this removes all element of guilt and the procedure
is simpler.

I estimate the health hazard as follows. From SO_2 concentrations in the U.S.,
we find about 10,000 people might have early deaths. In another public
health matter[29] on radiation standards, NRC have suggested that $1000 be
used per man-rem and this corresponds to $10,000,000 per cancer. If we take
the same cautious figure, we should charge $100,000,000,000 to the U.S.A. for
the 30,000,000 tons of sulphur emitted yearly; this is $3,000/ton or $1.50
per pound.

Using a less cautious figure of $1,000,000 per life saved, and doubling to
allow for property damage, we find $.30 per pound as a rough guess of a fair
charge - enough to encourage stack gas scrubbing for all new plants.

We reiterate here some of the advantages claimed for a pollution tax, and
group them according to the type of charge. Whatever the form of tax, we
have several advantages:

1) The total cost of coal burning would be correctly assigned and people may
   then, by ordinary economic (market) considerations, choose the best balance between environmental and public health concerns and cost, and perhaps choose alternate fuels or even just save on fuel.

2) A pollution tax is more flexible than a strict ban. A ban on burning
   high sulphur fuel is all too easily waived, in times of crises, on insufficient data. However, if the option to burn high sulphur fuel automatically entails a considerable cost, then the utility will spend considerable effort to find low sulphur oil, to make extraction systems work,
   and so forth.

3) Industry tells us[30] that present   (closed cycle) SO_2 scrubbers are
   expensive and unreliable. Yet if they work 75% of the time, simple

arithmetic shows that the charge is reduced to $\frac{1}{4}$ and so is the pollution.
Yet more complex arguments show that the tax will be reduced still further. The high charge during the 25% off time will have several salutary
effects. Spare parts will be rushed to repair the system, and even company presidents will drive to the airport to pick up air freight. Moreover, a utility system runs with 10-25% "reserve" capacity. A power station with the SO_2 scrubber temporarily inoperative can be switched off to
save the charge. The public will then be protected more than $\frac{3}{4}$ of the
time. If we relate charge to public health effect, we obtain further
advantages.

4) Industry has the incentive to build plants in such a way as to reduce
    these effects. Tall stacks can help in towns; remote and downwind siting
    - perhaps offshore coal burning - may be encouraged.

5) Industries, who have more money and incentive than the Environmental
    Protection Agency or public health departments, will spend more money on
    research, both basic and applied, to understand these public health
    effects and their abatement properly.

There have been problems in the few cases where pollution taxes have been
assessed, primarily on public water treatment plants. The main offenders
have been public authorities, and they have often been assessing taxes
against themselves. The incentive effect of taxation to govern actions by
ordinary economic incentives has therefore been absent.

This does not apply to sulphur pollution, which is mostly industrial. Moreover, by relating the charge to health effects, it is possible to pay the
charge to the people who suffer these health effects; the State can collect
the charges to pay calculated fractions to local communities for equal distribution among the residents. "Closing the feedback loop" in this way
should be an important aspect of any pollution charge or tax[25,31].

INTERSTATE PROBLEMS

If the large-scale effects of sulphate pollution are indeed as important as
suggested in this report, (and we reiterate that we believe there is a _prima_

facie case for this), midwestern states are sending pollution to the East
Coast. If regulation is to be effective, it must be federal regulation;
presumably Massachusetts, for example, has a right to sue Illinois or Ohio
for inadequate regulation. If a pollution charge related to health effects
is adopted, and the charge is returned to the citizens affected, then Illinois
and Ohio would be paying sums to the residents of downwind states - New York
and Massachusetts - and England might by paying to Norway! These pose interesting jurisdictional problems for the legal experts to solve.

RECOMMENDATIONS FOR RESEARCH AND ACTION

We regard the sulphate pollution issue as sufficiently important that action
is needed now, both to prevent further expansion of polluting facilities,
and to reduce the emissions from existing ones. However, we fully recognize
the uncertainties not only in our approach, but in the epidemiology and the
meteorology that we have explicitly accepted.

Since the onus of proof is now on those who wish to continue and expand the
pollution to demonstrate, if they can, that it is innocuous, we suggest some
possible avenues of approach:

1) One of the major unknowns is whether the sulphates in the north-eastern
    U.S. come from natural or man-made sources, although circumstantial evi-
    dence suggests the latter. Two major experiments suggest themselves.
    All coal burning and other man-made sulphur production could be stopped
    east of the Mississippi for one month in April (after the winter and
    before the summer electricity peaks). Industries concerned could take a
    vacation. In this time a massive programme of monitoring of ground and
    air concentrations of SO_2 and sulphates could be undertaken with emphasis
    on the size of the aerosols, followed in due course by epidemiology. In
    a month all man-made concentrations should be rained or washed out.

2) For a week all sulphur polluters could be required to add to their stacks
    a proportion of radioactive sulphur dioxide[32] (S^{35} is a
    beta ray emitter with half-life of 86-87 days); this could then be traced

throughout the area. Experiments so far with sulphur tracers have been
on a small scale, and concentrations have only been traced over large
distances. To detect sulphur in Boston might require 10^{-3} microcurie per
100 meter or 10^{-6} curie per gram of sulphate; the radioactive element
would be 10^{-10} be weight. 5×10^{5} curies - a huge quantity of sulphur -
would have to be emitted in ten days. Since this would come from many
sources, it can easily be shown to be acceptable from a standpoint of
radiating the public. However, emission of this much sulphur from a
single source might be unacceptable.

3) The model is only tentative. We think that several studies would be useful to establish it or improve it:

   (i)   There should be a search for correlations between sulphate concen-
         tration and time from the last rainfall. This can also help esta-
         blish the amount of rainfall necessary for rain out.

   (ii) The usual weather data give wind frequency, but does not give cor-
         relations in wind from one day to the next. Our assumption of a
         steady weather condition for 50 km and then an average drift with
         the wind may not be adequate. Our assumption suggests that re-
         lease of sulphur with a steady west wind, on the Eastern seaboard
         where the prevailing wind is westerly, would not give sulphate
         pollution[33].

   (iii) Other studies of the sulphur cycle can be made[34] to see whether
         or not the sulphates observed in the eastern U.S. are man-made or
         natural.

We are therefore led to the environmental dilemma of the year with which our
politicians must grapple. The rise in price of oil is forcing industry to
return to coal burning; should they be forced to install stack gas scrubbers?
The total cost in the U.S. could be $20 billion capital expense amortised at
20% to give $4 billion a year; increase in operating cost can be another $2
billion a year. If we do not remove the sulphur, 10,000 people a year may
lose their lives. Is it worth spending $6 billion on the possibility that
we are saving the lives of 10,000 people?
It seems absurd that after six centuries of burning coal we have not devised
an effective way of burning it cleanly, are still not sure of the adverse
health effects in any detail, and therefore cannot answer this important
question easily.

REFERENCES

(1) A Social History of Engineering, W. H. Armytage, MIT Press, p.69.

(2) An excellent general (and cheap) reference is Air Quality and Station-
     ary Source Emission Control Report by Commission on Natural Resources,
     National Academy of Sciences, National Research Council, prepared for
     Committee on Public Works, U.S. Senate, March 1975, Serial 94-4, avail-
     able from Superintendent of Documents USGPO $8.60.

(3) A good summary is Air Quality Control Criteria for Sulphur Oxides.
     U.S. Department of Health, Education and Welfare, Environmental Protec-
     tion Agency Report AP-50 (1970) USGPO.

(4) e.g. Some Recent Trends in Pollution and Health in London. P.J. Lawther
     and J. A. Bonnell, International Air Pollution Conference, Washington
     DC, December, 1970.

(5) General Air Pollution in Norway, report by W. Lindberg to the Smoke
     Damage Council, Norway (1968).

(6) Atmospheric Contamination of Industrial Areas Including Fossil Fuel
     Stations and the Method of Evaluating Possible Effects on Inhabitants.
     Y. Nichiwaki et al. report to Conference on Environmental Effects of
     Nuclear Power Stations IAE-SM-146/16 International Atomic Energy
     Agency, Vienna.

(7) Energy, Ecology and the Environment. R. Wilson and W. Jones. Academic
     Press (1974), Chapter 8.

(8) Handbook of Atmospheric Dispersion Estimates. D. Bruce Turner, Environ-
     mental Protection Agency AP-26 (1972) USGPO.

(9) "Tall Stacks - How Effective are They?" A. J. Clarke, D. H. Lucas,
     F. F. Ross, Second International Air Pollution Conference, Washington DC,
     December, 1970.

(10) M. O. Amdur, Animal Studies, for Conference on Health Effects of Air
     Pollution. National Academy of Science, October 3-5, 1973.

     M. O. Amdur, The Long Road from Donora, 1974, Cumming Memorial Lecture.

     M. O. Amdur, Arch. Environmental Health 23, 459 (1971).

     M. O. Amdur, Journal of the Air Pollution Control Association 14, 638
     (1968)

(11) "Air Pollution and Human Health", L. Lave and E. Seskin, Science 169,
     723 (1970).

(12) W. Winkelstein et al., Archives of Environmental Health, 14, 162 (1967).

(13) "Health Effects of Fossil Fuels Plants", L. Hamilton et al., in press.

(14) "Health Consequences of Sulphur Oxides", U.S. Environmental Protection
     Agency, March, 1975.

(15) "Nationwide Inventory of Air Pollutant Emissions" publication of U.S.
     Department of Health, Education and Welfare, 1974.

(16) A. P. Altsheller, Atmospheric Sulphur Dioxide and Sulphate, Environmental
     Science and Technology, 7 709 (1973).

(17) Climatology of the U.S. No. 82-19, 1951-60, Logan Airport, U.S. Department of Commerce, USGPO 100.

(18) "The Influence of Meteorological Parameters on the Concentration of Contaminants in Rainfall Due to the Washout of Hygroscopic Particles", G. J. Stensland and R. G. dePena, Penn. State Univ.

(19) C. V. Coghill and G. E. Likens, Acid Precipitation in Northeastern U.S.A. (1975).

(20) Anon. Acid Rain in the U.S.A., Scientific American 230 122, 1974.

(21) Dr. Ottar, Norwegian Institute for Air Research, P.O. Box 15, N2007, Kjeller, Oslo, Norway.

(22) Handbook of Atmospheric Dispersion Estimates, D. B. Turner, U.S. Environmental Protection Agency, AP-26 (1972).

F. Rogers and C. C. Gamertsfelder in Environmental Effects of Nuclear Power Stations, IAEA, Vienna (1970).

(23) "An Assessment of the Air Quality Impact of Increased Sulphur Dioxide and Particulate Emissions from Stacks taller than 100 feet in the Commonwealth of Massachusetts". Environmental Research and Technology Document, ER 1214, September, 1974.

(24) N. F. Ramsey, "We need a Pollution Tax!", Bulletin of Atomic Scientists 26, 3 (1970).

(25) D. Chapman, "A Sulphur Emission Tax and the Electric Utility Industry", Cornell Agricultural Economics Paper 73-17 (August, 1973).

D. Chapman et al., "Power Generation; Conservation, Health and Fuel Supply", Report to FPC (1973).

(26) Studies for Norway, Sweden and Denmark made by OECD available from OECD, Paris.

(27) Communication from L. Ruff, Ford Foundation.

(28) R. Wilson, "Tax the Integrated Pollution Exposure", Science 178, 182 (1972).

(29) "As Low as Practical", Hearing, RM-50-2, Nuclear Regulatory Commission, May, 1975.

(30) Consolidated Electric Utility Hearing Examiners Report, Ohio EPA, September, 1974.

(31) D. G. Wilson, "Social Cost Feedback Possibilities", MIT, 1974, (unpublished).

(32) C. E. Junge, "Air Chemistry and Radioactivity", Academic Press, New York, 1963.

(33) B. Chang has made some detailed studies of this using Logan Airport winds. B. Chang, private communication.

(34) W. W. Kellogg, R. D. Cadle, E. R. Allen, A. Z. Lazrus and E. A. Martell, "The Sulphur Cycle", Science 175, 587 (1972).

(35) F. Pasquill, "Atmospheric Diffusion", published Van Nostrand-Reinhold, Princeton, New Jersey, 1961.

DISCUSSION

Question: Could the pollution tax be applied to all pollutants?

Answer: Yes, in principle, but there is no reason why one should not start with one, i.e. sulphur. Sulphur is a good case for taxation since the clean-up measures are considerably cheaper than the health charges to society.

Question: The lecture implied that there was no threshold level for sulphur dioxide. Does not the CHESS report imply that there may be a long term exposure effect?

Answer: The CHESS report is primarily concerned with sulphates. It is difficult to separate the effects of sulphur dioxide from those of sulphates. There may be a threshold effect but it does not seem possible to prove it at present.

Question: What is the effect of car exhaust fumes on deaths from bronchitis?

Answer: An adequate study has not yet been done.

Question: The incidence of bronchitis in the U.K. correlates with standard of living. Is climate a dominant effect?

Answer: Probably, but it is difficult to unravel such complicated effects since the total problem is equivalent to one-tenth of the effect of smoking ten cigarettes per day.

Question: Is there synergism between particulates and sulphur in their effects on the incidence of bronchitis?

Answer: Such a possibility has been investigated but synergism has not been demonstrated.

Question: Under the conditions of the temperature inversion in London in December, 1952, there were 3000 deaths. Was there a higher concentration of sulphur dioxide or of particulates?

Answer: Both went up. There were no sulphate measurements at that time. Five years later there was a similar fog with low particulates but the same sulphur dioxide concentration. On this occasion the excess deaths ran at one third of the previous level, but the B.B.C. broadcast warnings which advised bronchial sufferers to stay indoors. Also by this time stack heights were generally greater, which dispersed the pollutants over a bigger area.

Question: In building a new power station, what would be the cost of sulphur removal measures?

Answer: The average capital cost of building a new power station is 500 dollars per kilowatt and the inclusion of sulphur scrubbing measures would add 100 dollars per kilowatt. This could imply about a 10% rise in the cost of electricity in the United States.

Question: Are there different levels for sulphates and particulates indoors and outdoors?

Answer: There has been insufficient study up to now.

Question: Battersea and South Bank power stations use lime scrubbing, but we are now approaching the limit of solubility of calcium sulphate in the Thames. Fulham power station used ammoniacal liquor in the Fulham/Simon-Carves process. This was an experimental project which has now closed down. The problem is: what do you do with the sulphates if you burn 50 million tons of coal containing 2% sulphur? Flue gas scrubbing must, in addition, be complete,

	otherwise the cold plume sits down on the immediate area around
	the power station. Desulphurisation of oil prior to combustion
	would be an easier process to manage in power generation.
Answer:	Scrubbing has indeed got its practical problems.

Question: Power generation based on fluidised bed combusters would be advantageous in that coal could be burnt with limestone in the bed?

Answer: Yes. A pollution charge would encourage new technological developments of this sort. The mere establishment of legally-enforced regulations would not provide this kind of encouragement.

Question: If a pollution charge were imposed, there would need to be a guarantee that the government would direct the proceeds to those plants willing to install scrubbing measures. Is it not true that power generation will always have a pollution problem? It is not clear that the present problem is catastrophic.

Answer: I would tend to agree, but more evaluation needs to be done.

Question: Rather than donate a pollution charge to the power industry would it not be better to feed it into community health organisations?

Answer: This would be one possibility.

PANEL DISCUSSION:
THE PROSPECTS FOR FOSSIL FUELS

Panel: Prof P D Dunn, University of Reading (Chairman)
        Prof P J Garner, University of Birmingham
        Mr J S Harrison, National Coal Board
        Dr C G James, British Gas
        Prof D C Leslie, Queen Mary College, London
        Prof R Wilson, Harvard University

Dunn: If we look at energy use over the last two decades, we see that it
has been met almost entirely by fossil fuels. The 1970 figures are, I think,
that about 2/5 of our primary energy was supplied by coal, about 2/5 by oil,
about 1/5 by natural gas and only several percent by hydro and nuclear power.
It is interesting to notice that the increase in energy use over these last
two decades has been almost entirely in oil and gas. One may assume that the
situation over the next two decades or so will be one where fossil fuels
will still be meeting most of our energy needs. One would expect that nuclear
fission will play an increasing role but will not become really important
until the third decade or so, ie it is not of immediate significance. The
long-term prospects, eg solar, fusion, wave, geothermal, are really only
twinkles in the eye at the moment, fission being more like a piece of grit,
perhaps.

We should bear in mind that the developing countries are expecting to make
very much more use of energy and so I hope our discussions will extend to
the world as a whole. We should discuss the extent, use and lifetime of
fossil fuel reserves and the environmental and safety aspects of their use.
Finally, we should consider why we use energy at all and whether we could
reduce our requirements by modifying our practices.

Question: What do you feel about the competition that exists between the
various energy suppliers? Is it wasteful or does it improve efficiency?

Answer: A recent chairman of the Electricity Council said that the greatest
factor in improving the efficiency of the electrical industry was competition
from natural gas. This competition is part of government policy; either
competition exists or the government takes over full control. Evidently
our politicians do not think our situation is so bad that they should adopt
such a totalitarian approach. The government could usefully change the rules
under which the Central Electricity Generating Board operates so that it
would be allowed to produce "waste" heat at a temperature high enough to be
used for district heating; although the sacred electrical output would then
be reduced, the overall efficiency of use of primary energy could be doubled
compared with present typical values.

Question: Why not change the rules? The CEGB cannot be allowed to persist
in "wasting" 70% of its primary energy.

Answer: Before the October 1973 war, the CEGB was totally uninterested in selling heat. The second law of thermodynamics was not so much overridden as obscured by the low cost of fuel but it is now realized that the second law is valid and the CEGB is now interested in selling heat at 30°C or so, ie its present output, to greenhouse operators, fish farmers, to anybody at the power station boundary. Raising the heat grade inevitably reduces the electricity output and that is the problem.

Question: Will not topping and bottoming cycles help?

Answer: MHD is the obvious example of a topping cycle but recovery of the seed appears to be an insuperable problem. One could visualize a sodium boiler but a gas turbine with a freon bottoming cycle seems more hopeful; it is regrettable that more is not being done in this direction as efficiencies of 55% or so could be achieved. If hydrogen becomes available, it can be burned in an internal combustion steam turbine to give an efficiency of about 60%. Fluidized combustion of coal would allow turbine inlet temperatures up to 900°C and low - BTU gas will produce much higher temperatures when suitable turbine materials become available; efficiencies of about 50% should be possible.

Question: Will this ever-increasing use of energy produce insoluble pollution problems?

Answer: We do not have an immediate thermal pollution problem; about 95% of the oil floating on the sea is due to seepage from the sea bed and the sea contains micro-organisms trained to eat the oil and enjoy it. The main problem seems to be that of sulphur and the fluidized bed is an admirable way of dealing with this. It is regrettable that so little is being done on this technique here.

Question: Surely for every environmental problem, there is a technical solution? We should not only improve the efficiencies of power stations but reduce the demand for electricity.

Answer: This demand could certainly be reduced by not using high-grade energy for space heating. The CEGB has the problem that many large power stations have been built that have many years to live and produce large quantities of too-low-grade heat too far from urban centres. Future power stations should be smaller, nearer urban centres and should produce heat at a more useful grade for district heating, for example. The CEGB is presently constrained by the government to produce electricity as cheaply as possible.

Question: About 20% of the electricity produced in this country is produced by industry as part of total energy systems when process heat is also used. Could this idea expand?

Answer: The problems are that private industry is not allowed to sell surplus electricity and CEGB standby electrical capacity, for the event of plant-failure, is very expensive.

Question: Surely if one firm asked the CEGB for standby electrical capacity, that would be expensive but if hundreds of firms did so, the cost would not scale as their breakdowns would not all occur together?

Answer: This is true but cold-weather standby would tend to scale.

Question: When will petrol produced from coal be competitive with that from oil?

Answer: There are developments in coal liquefaction projects which are at that stage now but there are problems in scaling to commercial production, viz in the materials handling, hydrorefining, ash disposal etc. If coal and crude oil were equally priced, the petrol from crude would be cheaper of course.

Question: How much is spent on coal research?

Answer: In the UK, the total is only about £3 M/yr while the nuclear total runs off the end of my pocket calculator. In the USA, it is significant that the people who urge more research on coal seem to be those working in the nuclear field. Now ERDA has taken over coal, the ratio is rapidly approaching unity (coal research budget now up to about $150M) while for solar, ERDA requested $70M which was doubled by Congress. The nuclear dominance seems to be being eroded somewhat in the USA. The major problem of coal use seems to be the environmental effects of strip mining and sulphur emission. Indeed, the main emphasis in the USA in coal liquefaction is to produce solvent refined coal for burning in power stations just to deal with sulphur.

Question: Zambia has coal with a very high ash content. Could we not export fluidized bed technology there?

Answer: The coal industry is becoming more international in its outlook now and the Coal Board has been advising Zambia on the development of its coal resources. Africa would seem to be well-suited to the use of solar-energy where many states have economies based on small villages but the capital cost would be high. Any thing that would reduce the capital cost of producing electricity from solar energy would greatly benefit the human race. Meanwhile, small coal-fired power stations would appear to be the answer. Africa has the largest undeveloped hydropower resources which will need to be developed with care because of silting and environmental problems. The Aswan dam, for example, has affected the centuries-old tradition of using for agriculture the silt brought down in the annual floods, the distribution of mosquitos and the diseases they carry has been altered and the sardine catch in the Mediterranean around the Nile Delta has been badly affected.

Question: How much is being spent in this country on energy conservation?

Answer: We should really be talking about efficient energy utilization and research on this has been proceeding for a long time at a rate determined by the cost of energy. The USA is the greatest user of energy because energy has always been cheap there. Now that the price of oil has increased, attitudes have changed. The cement industry has always been energy intensive and one company which makes twice as efficient use of energy as the others is now in a better position though the other companies will catch up by employing consultants to advise them. The problem is that in the domestic sector, individuals cannot afford consultants so charges will have to be introduced as the design and construction level which will take a long time to have an effect.

Question: Could an energy tax be used as an incentive?

Answer: The UK has high taxes already, though not specifically imposed to encourage energy saving. We will have to decide what situation we want in the year 2000 and impose the appropriate taxes now. Here, natural gas is sold at cost, which is below the costs of other fuels, and this encourages waste of a limited resource. I have just converted my heating boiler from oil to gas because of the price difference and it is absurd that this incentive should exist. Generally, the prices charged for fuels do not include the environmental and other costs of their use and a tax could be imposed to include this.

Question: Our present major energy vectors are coal, gas, electricity and oil. What are they likely to be in the future?

Answer: Methane would be very good. It could be made from all available fossil fuels and could readily take in a proportion of nuclear energy (nuclear hydrogen combined with methane). An extensive gas distribution grid exists and there are many millions of gas appliances. As long as coal exists, it will be cheaper to deliver its energy to consumers in the form of methane than as coal or electricity. The direct use of hydrogen is not very appealing, mainly because of storage (lower calorific value per unit volume than methane) and it seems that chemical dissociation of water will be more expensive than electrolysis which is only 60% efficient. When oil has become too expensive for general use (owing to its scarcity), coal will be needed for chemical feedstocks and there is no doubt that we shall have to rely on nuclear electricity for most of our energy needs. By that time, however, when we are considering using nuclear power to electrolyze the English Channel, we could build dams across the North Sea and excavate the vast quantities of coal available there!

Concluding remarks

Wilson: Different countries have different views on energy. In the USA one has great enthusiasms and, when environmental movements erupted about four years ago, we suddenly realized that we were not taking environmental aspects into our energy costs which are much too low and we were rightly described as extravagant. It was a great shock to discover two years ago that we were importing 10% of our energy and there is now a major change in attitudes of the people, if not yet of the government, towards energy saving. There will be a reduction of 20-30% in energy use per capita but then population and standard of living will rise so anybody who thinks this fall will be maintained is a little starry-eyed.

Harrison: Energy problems seem to be being discussed much more rationally today than even two years ago. The supply industries are realizing that some of their assumptions are quite reasonably being challenged but users are also realizing that these industries have problems, too. I find this better awareness of each other's attitudes encouraging.

<u>Leslie</u>: I would like to stress the great inelasticity of the energy industry; the investment is so enormous that all we can achieve on a time-scale of less than a generation is the avoidance of very painful restrictions by the avoidance of foolish waste.

<u>Garner</u>: I echo Professor Leslie's remark. I had hoped to say something about the contribution everybody can make by saving energy, information on this being readily available, but I am glad this was said for me and not omitted.

<u>Dunn</u>: I would like to make a succinct summary of this discussion but I do not have the hour or two to do this. You are all, no doubt, very clear about our points of view and I now close this meeting and thank the members of the panel on your behalf.

NUCLEAR FUELS

FISSION REACTORS

D. C. Leslie

Nuclear Engineering Department, Queen Mary College, London

THE FISSION PROCESS

The fission process is shown in Fig.1. It involves the fission products which contain almost all the energy (this energy is degraded into heat by elastic collisions with the rest of the crystal lattice), surplus neutrons which make a chain reaction possible and gamma rays to maintain the energy and momentum balance. Now fissile nuclei are needed to maintain a chain reaction. In case that seems a tautology, by fissile I mean the thermally fissile nuclei uranium-233, uranium 235 or plutonium 239. There is some fission in the so-called fertile nuclei even in thermal reactors. But even in the fast region you cannot maintain a chain reaction with uranium-238 alone; this is a fortunate circumstance, otherwise everybody would be blowing themselves to pieces. The crucial number for the chain reaction is η, the number of neutrons emitted per neutron absorbed. This is not the number of neutrons emitted per fission, for some absorptions result in capture rather than in fission. Obviously η must be greater than 1 to sustain a chain reaction. To breed, the number must be greater than 2; one neutron is needed to continue the chain reaction, and one more must be absorbed in a fertile atom to produce another fissile nucleus by radioactive decay.

The neutrons are born in fission at about 2 MeV, and one can get good η values either by keeping the neutron energy as high as possible, or by bringing the neutrons quickly into thermal equilibrium with the reactor materials. The first choice is the fast reactor which is described in the next paper, by Mr. Dale: his aim is to keep the neutron energy above about 100 keV. The thermal reactor man aims to bring the neutron energy down to about 0.1 eV as rapidly as possible, and that is done by elastic collisions with the light nuclei of the moderator; these may be hydrogen, deuterium or graphite (carbon). In the intermediate region, say 1 eV up to several tens of keV, the η values are small and no chain reaction with good economics is possible.

The quantity listed in Fig.2 is not η but the so called conversion or breeding ratio. This is the number of fissile nuclei generated by neutron capture and subsequent decay for each fission. It is called a conversion ratio if it is less than one, and a breeding ratio if it is greater than one, but both are the same number. You will see that in practice, breeding is only possible by using plutonium in a fast reactor. It is just about on the edge with U-233 in a thermal reactor and it is certainly not possible in practice in such reactors with either U-235 or Pu-239. In the thermal reactor the numbers in Fig2 are just the η value minus 1.25. The 1 is to maintain the chain reaction and the 0.25 is the inevitable loss by leakage and by capture in canning, control rods etc.

The cross sections of the fissile isotopes are very much greater than that of U-238 in the thermal region; they are all of the order of hundreds of barns, whereas the cross section of U-238 is only a few barns. Therefore, a self-sustaining reaction can be achieved in the thermal region with a very low percentage of U-235. Indeed, it can be done with natural uranium which

Fig. 1

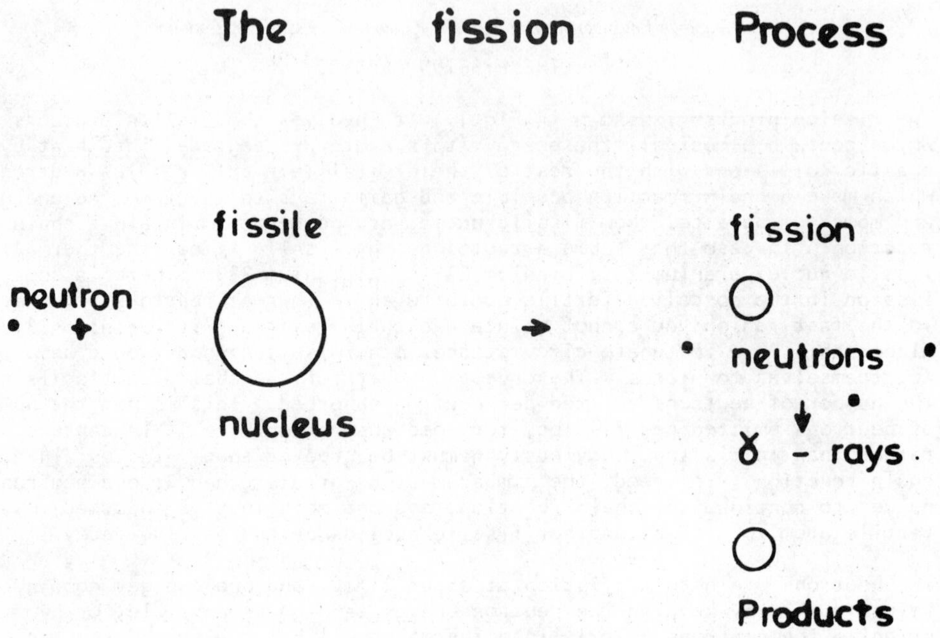

The   fission   Process

fissile

neutron

fission

nucleus

γ - rays

neutrons

Products

Fig. 2

Conversion/breeding ratios

Reactor	U - 233	U - 235	Pu - 239
Fast	?	0.8	1.2
Thermal	1.0	0.85	0.65

Losses - 0.1 neutron/fission in fast reactor

- 0.25 neutron/fission in thermal reactor

contains only 0.7% of U-235; however most reactors are enriched to between 2% and 3%, which gives more design flexibility. One has to strain rather hard to get a good reactor using natural uranium. In the fast region the cross section advantage of the fissiles is very much less and enrichments between 15% and 30% are needed.

REACTORS AND THEIR FUEL SUPPLY

Thermal reactors are much more highly developed than fast reactors; they are normal articles of commerce and are traded as such although they have rather special emotional overtones. Their disadvantage is that they can burn at most only a few per cent of all uranium nuclei. In fact all we burn at the moment is about $\frac{1}{2}$%, and this includes a little of the U-238 which has been converted into plutonium. These thermal reactors are also called convertors to remind us that some conversion does go on; however we do not, at the moment, pass back through the reactor the plutonium which is formed in this process. This matter of plutonium recycling is another emotional issue. It is also an economic one, because it is not at all clear that it is best to burn the plutonium in thermal reactors; it might be better to keep it for fast reactors.

Because of this low uranium utilization an early uranium crisis is almost inevitable. It will be staved off only by failure to reach the stated rates of nuclear construction. Personally, I think it is very likely that we will not achieve our current targets of nuclear construction and therefore that the uranium crisis will be later and less severe than present predictions suggest. Each GW = 10^9W electric of nuclear power requires about 5,000 tons of natural uranium over a 25 year working life, including the initial charge. This figure is valid for light water reactors (both boiling and pressurized); for AGR and SGHW the figure is slightly greater. For Magnox, which is relatively inefficient in utilizing uranium, it is considerably more. It is as little as 3,000 tons for the Canadian natural Uranium CANDU and for the High Temperature Reactor (HTR).

At the moment we have certain knowledge of the whereabouts of 1 million tons = 1 mte of natural uranium. We have a good idea where to find a further 1.3 million tons; with some effort, we might find another 2$\frac{1}{2}$ million of high grade ore, containing more than several hundred parts per million = ppm of uranium. The best ores contain as much as 3,000 ppm but these are exceptional. (I am indebted to Dr. S. H. U. Bowie, FRS, for these figures). Now the million tons of certainly known uranium will fuel 200 GW(e) of thermal reactors; this amount is already built, building or firmly committed. Every thermal nuclear power station committed from now on will rely for its fuel on uranium whose whereabouts are not known with any certainty.

When burnt in thermal reactors, uranium is a small resource; it amounts to between 10% and 20% of the known oil reserves. Therefore, there can be no long-term future for thermal reactors by themselves. We must have the fast reactor and we must have it pretty soon. Then thermal reactors will be useful for burning the surplus uranium which you may choose to make in fast reactors. The other remedy is, of course, to find more high grade uranium ore: people are getting very active about this as the facts given above become more generally appreciated. But once we have a fast reactor, the uranium we already have expands in energy worth about a hundred times, and the energy value of the known high grade uranium is then several times that of all the fossil fuel reserves. Moreover, the low grade uranium ores which

are not really burnable in thermal reactors become accessible to the fast
reactor which can convert them so much better.

ELEMENTS OF A THERMAL REACTOR SYSTEM

The primary choices are shown in Fig.3, and the first of these is the
moderator. In the literature of 25 years ago there are all sorts of rather
incredible choices but only three have survived. (Some not so incredible
choices have still not survived). In particular, organic fluids were
perfectly sensible; it has just not proved feasible to use them economically.
Today, the only three survivors are the light hydrogen atom, the heavy hydrogen atom and carbon. The hydrogen atom is invariably in the form of water,
while carbon is used in the form of graphite. The reactors which are now
either articles of commerce or have obvious commercial potential all use one
of those three.

Turning to coolants, the only liquid survivors are pressurised light water,
boiling light water, and pressurised heavy water. Among the gases, carbon
dioxide and helium have always been the only possibilities. When choosing
a coolant, you must make sure that it does not absorb too many neutrons. It
must also be capable of high temperature operation. Hydrogen is very much
the best coolant from a thermodynamic point of view but people have quite
rightly shrunk from even trying to engineer safely a hydrogen-cooled land-
based nuclear reactor. (The space rocket is hydrogen-cooled but the safety
considerations of this reactor were rather special). Another choice is
whether to use a pressure vessel or a pressure tube. In all cases you must
pressurise the coolant in order to get adequate thermodynamic conditions and
good rates of heat transfer; you can enclose the whole reactor in a pressure
vessel, or you may use individual pressure tubes. You must also choose the
nature of the canning. In almost all cases the fuel is uranium dioxide which
is capable of very high temperature operation and is singularly resistant to
radiation damage. Its chief disadvantage is its very low thermal conductivity, so the fuel has to be finely divided to get adequate heat rating.

Fig.4 lists the surviving thermal reactor types. The first two, PWR and BWR,
are American types and have been sold in a number of countries. In both
cases the combined moderator and coolant is light water. In the Pressurised
Water Reactor it is not allowed to boil. The operating pressure is about 150
bar and the heat is exchanged into a secondary coolant: this is also light
water, which is allowed to boil. In the BWR boiling is allowed, and the
steam is taken off directly. Next, we have the two heavy water moderated
reactors, which are both of pressure tube types. They differ in the coolant.
The British SGHWR or "steamer" is cooled by boiling light water. The Canadian
heavy water reactor (CANDU) is cooled by pressurised heavy water. In all
four cases the fuel is uranium dioxide and the canning alloy is predominantly
zirconium, chosen because it has low neutron absorption and good resistance
to water. Very few alloys are resistant to water at 300°C: happily zircalloy
is. It is also strong, having about two-thirds of the strength of steel.
Then we have the three gas-cooled reactors; the moderator is graphite in every
case. The first two are cooled by carbon dioxide which is cheap but has
higher chemical activity than helium. Helium could not have been used in the
first Calder Hall reactors; they leaked a ton of coolant a day when they
started work. (The amount of radioactivity released was very small). Smaller
and better engineered reactors were needed before helium cooling could be
considered. In each case the gas raises steam in a heat exchanger. In Magnox
the canning is magnesium alloy, magnesium also having low neutron absorption.

Fig. 3

Choices for thermal reactors

Moderator	Coolant
Light water (H_2O)	Pressurised H_2O
	Boiling H_2O
Heavy water (D_2O)	Pressurised D_2O
	Carbon dioxide
Graphite	Helium

Fig. 4

Classification of thermal reactors

Type	Moderator	Coolant	Cycle	Canning
PWR	H_2O	press: H_2O	indirect	Zr
BWR	H_2O	boiling H_2O	direct	Zr
SGHWR	D_2O	boiling H_2O	direct	Zr
CANDU/ Atucha	D_2O	press: D_2O	indirect	Zr
Magnox	Graphite	CO_2	indirect	Mg
AGR	Graphite	CO_2	indirect	S/S
HTR	Graphite	He	indirect	graphite

The fuel in this reactor is uranium metal which is, in many ways, a rather unsatisfactory fuel. Its conductivity is low for a metal and it suffers a phase change with volume expansion at $665^{\circ}C$, limiting the centre temperature to something below this figure; so AGR uses uranium dioxide fuel and stainless steel canning. Stainless steel is very strong and is a good high temperature material but it is a vigorous absorber of neutrons. Part of the difficulty and expense of AGR stems from that choice. In HTR, the fuel is again UO_2 in the form of coated particles about a millimetre in diameter. The canning is in effect graphite, and there is almost nothing in the core except graphite and fissile and fertile materials.

EXISTING REACTOR SYSTEMS

Magnox

Fig.5 shows the fuel element of Oldbury, which is a typical Magnox reactor. This figure shows the very elaborate finning. The gas enters the finning and is forced round in a spiral path until it comes to the splitters; it then rotates round the channel wall. This is a very high efficiency heat transfer surface and the success of Magnox is entirely dependent on it. With a plain can, Magnox would be wholly uneconomic instead of rather uneconomic.

Advanced Gas-cooled Reactor (AGR)

Fig. 6 shows a line diagram of a typical AGR. The reactor core is a hunk of graphite bricks about 9m each way with approxiamtely 400 channels passing through it. Each of these is 190 mm in diameter with vertical upflow of the coolant, which comes out at $675^{\circ}C$. The pressure vessel is fabricated of prestressed concrete, and it contains both the reactor core and the heat exchangers. Above is the refuelling machine which enables this reactor to be refuelled on load. This is an important operational advantage over the water reactor which must be shut down to refuel, and it is shared by the Magnox reactor. It is interesting how huge these power stations are, although the reactor itself is small. (This is seen most clearly in the case of the Prototype Fast Reactor. The reactor core itself is not much bigger than a dustbin, but the reactor building is tens of metres each way). Fig. 7 shows the standard AGR fuel element. The fuel is in the form of pellets of urnaium dioxide canned in stainless steel. They are arranged in rings of 6, 12, and 18 rods, with a central tie-rod which does not contain fuel. Grids hold the fuel "pins" in place inside a couple of graphite sleeves.

Magnox was a very worthy beginning to the nuclear power programme; the CEGB operators are devoted to these reactors because they work so reliably. But they are now out of date. They are too expensive, they give poor uranium utilization and there are safety worries because the fuel and the canning are not wholly compatible with the coolant.) AGR is very safe or, if you want to be neutral about it, it is particularly easy to make a satisfactory safety case for it. However, as we know to our cost, it is expensive and difficult to build, and we expect it to be difficult to maintain. I want to separate those general objections from the gory one-off difficulties of Dungeness B which will be known to all readers. These are to some extent "first of a kind" difficulties, but mainly they arose because the size of the task was underestimated and the contract was entrusted to an organisation that lacked resources to execute it.

Fig. 5

Oldbury fuel element

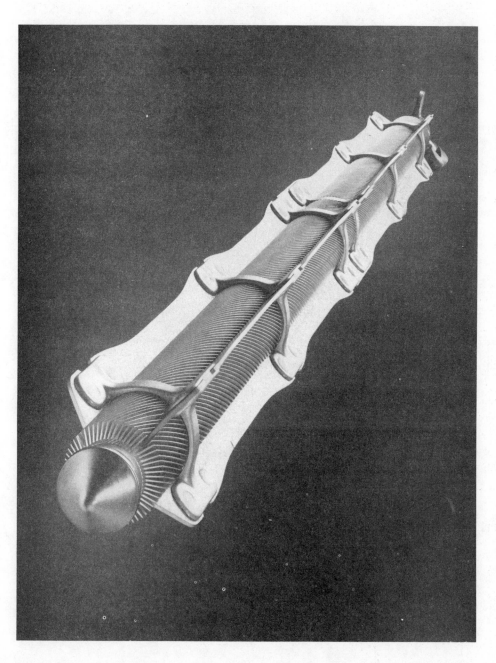

D. C. LESLIE

Fig. 6

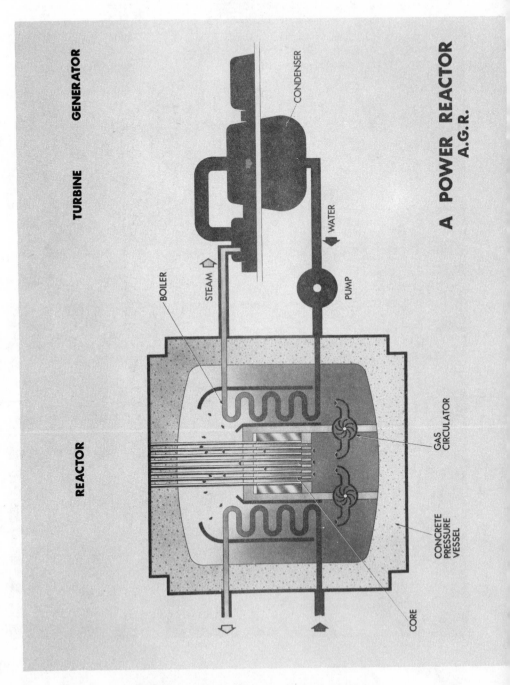

REACTOR

TURBINE GENERATOR

BOILER

STEAM

CONDENSER

WATER

PUMP

CONCRETE
PRESSURE
VESSEL

GAS
CIRCULATOR

CORE

A POWER REACTOR
A.G.R.

Fig. 7

AGR fuel element

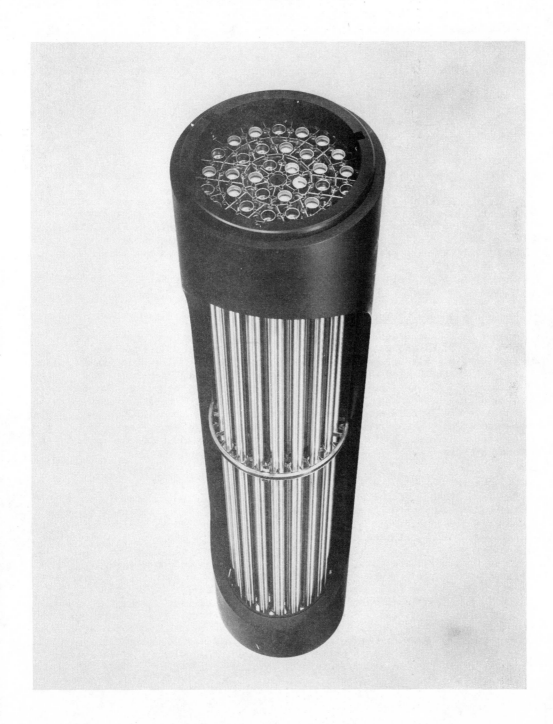

The AGR fuel pin has minute ribs on the fuel can surfaces to improve heat
transfer. Although they are less than a millimetre in cross section, these
ribs increase the heat transfer three times and the pressure drop eight times.

High Temperature Reactor (HTR)

There are a number of designs of HTR and Fig.8 is typical. The fuel element
itself is in the form of an annulus. The annulus contains large numbers of
little coated particles inside two graphite shells and the coolant flows both
inside and outside the annulus. The coolant is helium. The design is
generally similar to that of AGR but it is much more compact because of the
high temperature operation possible with helium and the high temperature
possibilities of graphite: these open the way to much higher ratings. Graphite gets stronger and stronger up to a temperature of about 2,500°C, whilst
all metals soften as the temperature increases.

This is potentially a very attractive reactor; there is no question about
that. It has good neutron economy, which means that its demands on natural
uranium are relatively small, and again it is easy to make a safety case for
it. But it still needs a fair amount of development. There is a commercial
prototype being built at Fort St. Vrain in Colorado: this is now three years
late and it is still (Aug 1975) not working. The view we have taken in
Europe, (I think correctly) is that this is still not an article of commerce,
but we expect it to become so during the next decade. I believe that it will
then start to compete with the water-cooled reactors which are described
below.

Pressurised Water Reactor (PWR)

Fig. 9 is a line diagram of a typical pressurised water reactor. Note the
pressure vessel and the reactor core, which consists of a number of pins of
uranium dioxide clad, not in stainless steel, but in zircalloy. The coolant
comes into the bottom of the vessel, passes up through the core, through the
primary side of the heat exchanger and back through the pump. The secondary
side of the heat exchanger raises steam and very poor quality steam it is.
The pressure is only 40 bar, and the temperature is about 270°C. The turbine
is thoroughly wet, with all the associated difficulties. The steam is, of
course, condensed in the condenser and returned to the secondary side of the
heat exchanger.

Boiling Water Reactor (BWR)

BWR is a direct cycle reactor (Fig.10), implying that the steam generated in
the core passes straight to the steam turbine. (This system used to be called a "nuclear kettle"). Water in the liquid phase enters the bottom of the
core and boils. It then passes up into the head of the vessel, where steam
separators as shown in the figure, separate the two phases. The liquid
phase flows down the outside of the core, being helped on its way by jet
pumps; the steam is taken out of the vessel into the turbine, condensed and
returned to the vessel. This reactor is mechanically simpler than the PWR
but it has other complexities. As the boiling starts the amount of moderator
available becomes less, so that there is an interaction between the power
rating and the neutronics. This interaction can be made favourable but it
took a lot of development to establish that this was so. Also more radioactivity passes out of the primary containment than with PWR. This radioactive contamination is of two kinds. Various activities are induced in the
canning, and some of these are dissolved by the coolant. Small amounts of

Fig. 8

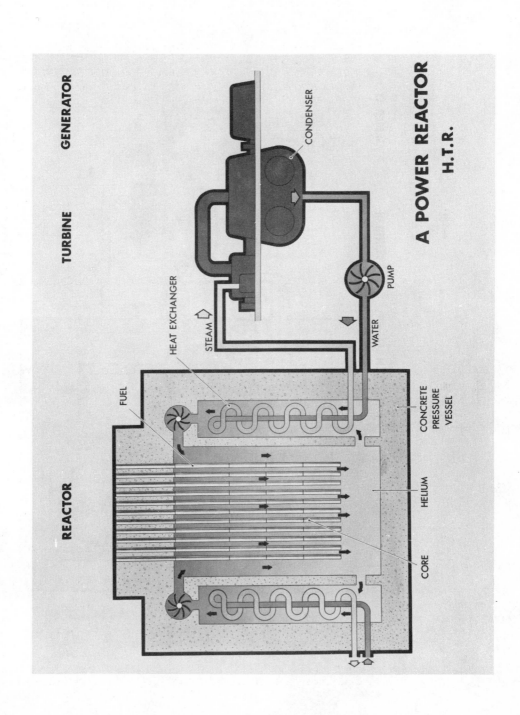

REACTOR   TURBINE   GENERATOR

FUEL

HEAT EXCHANGER

STEAM

CONDENSER

PUMP

WATER

CONCRETE PRESSURE VESSEL

HELIUM

CORE

A POWER REACTOR
H.T.R.

Fig. 9

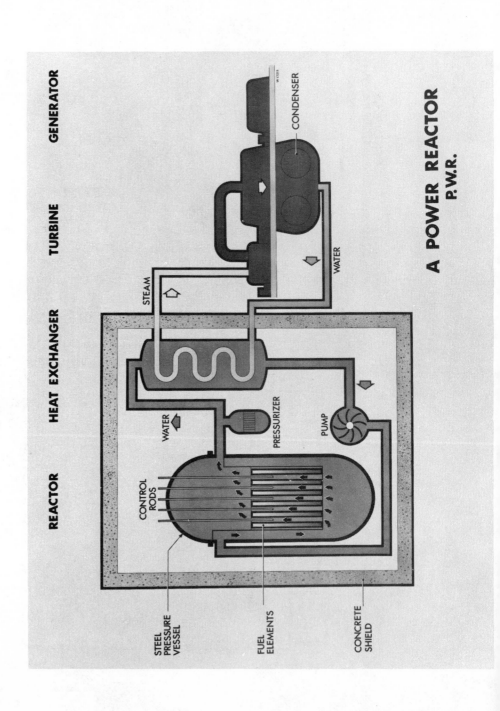

REACTOR   HEAT EXCHANGER   TURBINE   GENERATOR

A POWER REACTOR
P.W.R.

Fig. 10

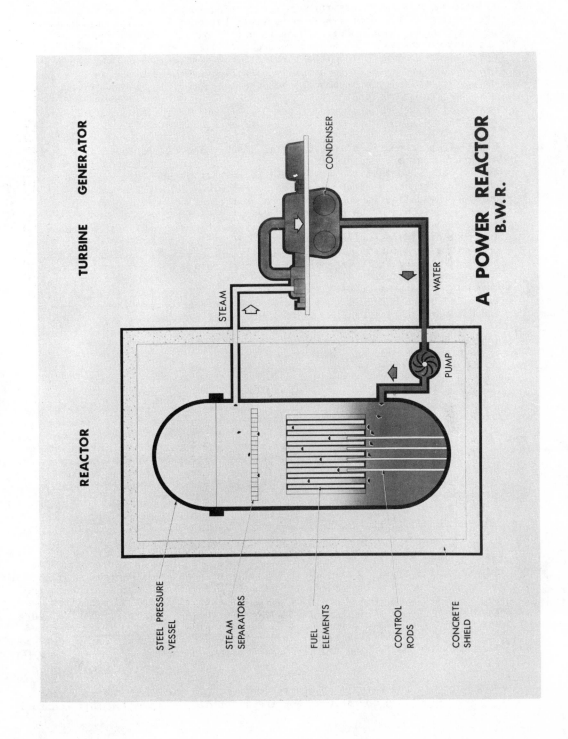

REACTOR   TURBINE   GENERATOR

STEEL PRESSURE
VESSEL

STEAM
SEPARATORS

FUEL
ELEMENTS

CONTROL
RODS

CONCRETE
SHIELD

STEAM

CONDENSER

WATER

PUMP

A POWER REACTOR
B.W.R.

fissile material may pass through the canning to produce fission products in
the coolant. This contamination is small but nontheless it must (and can)
be dealt with. Also, neutron bombardment of oxygen induces nitrogen-16
activity in the water with an 8 second half-life. In this reactor, some
nitrogen-16 does pass into the turbine. You must not go within about 8 feet
of the turbine while it is actually operating. Men can work on it as soon
as it is physically cool enough, because the activities die away very quickly
when the reactor is shut off.

CANDU

Fig.11 is a schematic of the Canadian CANDU. The calandria is a cylinder with
a large number of passages through it. The calandria tubes pass from the top
to the bottom so that the inside of the calandria tube is accessible from the
outside. The heavy water moderator is contained inside the calandria and
outside the calandria tubes, and through the calandria tubes pass pressure
tubes which contain the coolant at pressure. Referring again to figure 11,
the heavy water moderator is operated at atmospheric pressure and about 75°C.
The coolant is pressurized heavy water which flows from one end to the other
of the pressure tubes. Thermodynamically this is a pressurized water reactor,
but mechanically it is very different, particularly in the subdivision of
the pressure circuit. Also this is the only reactor with horizontal fuel
channels. Fuelling goes on from both ends. A clever fuel cycle is needed
to get adequate burnup with a natural uranium reactor; the fuel cycle used
demands more or less continuous fuel movements from both ends. The neutron
economy is very good; it produces almost twice as much energy per ton of
natural uranium as do the enriched reactors (other than HTR), with some penalty
in both capital cost and total cost. The gap in total cost will tend to
narrow as the prices of uranium and of separative work rise.

The Canadians have done a fine engineering job on this reactor. The prototype at Douglas Point finally reached full power some seven years late, and
those who did not know what was going on were rather scornful. However, by
the time this prototype had reached full power the Canadians had solved the
engineering problems of this reactor type. The construction and operating
record of the first four commercial reactors at Pickering has been splendid.
The last of these was built in less than four years (including a four month
strike) and No.3 reactor achieved an 80% load factor in its first year of
operation.

Steam Generating Heavy Water Reactor (SGHWR)

Fig.12 is a photograph of the SGHW prototype reactor at the Atomic Energy
Establishment at Winfrith in Dorset, while Fig. 13 is a line diagram. The
calandria is similar to that of CANDU but, is in a vertical configuration; again
there are pressure tubes rather than a pressure vessel. Coolant enters the
bottom of the core in the liquid phase, and starts to boil as it passes over
the fuel, which is contained in the pressure tubes. The steam-water mixture
goes to external steam drums where it is separated. The liquid phase comes
back down a pipe which is rather delightfully called a "downcomer". The steam
phase goes to the turbine and condenser and the condensed water goes back into
the drum.

At the top are the refuelling connections: these ports are closed except
during a refuelling operation. Not the least of the charms of this reactor

Fig. 11

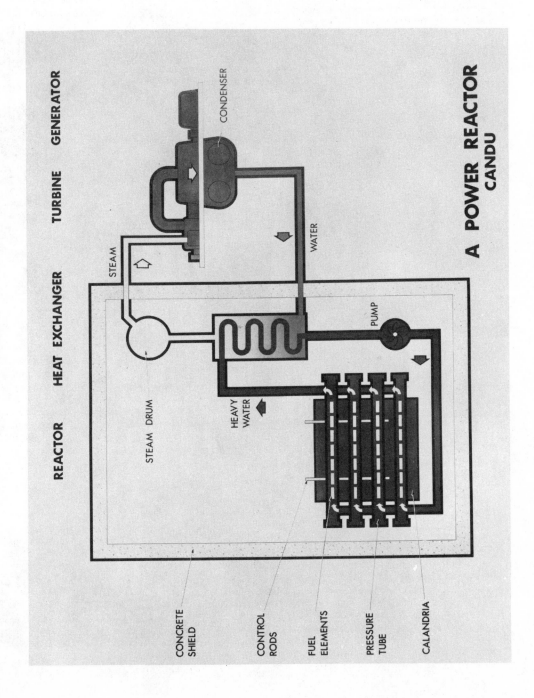

A POWER REACTOR
CANDU

REACTOR   HEAT EXCHANGER   TURBINE   GENERATOR

Fig. 12

The SGHW prototype reactor at AEE, Winfrith.

Fig. 13

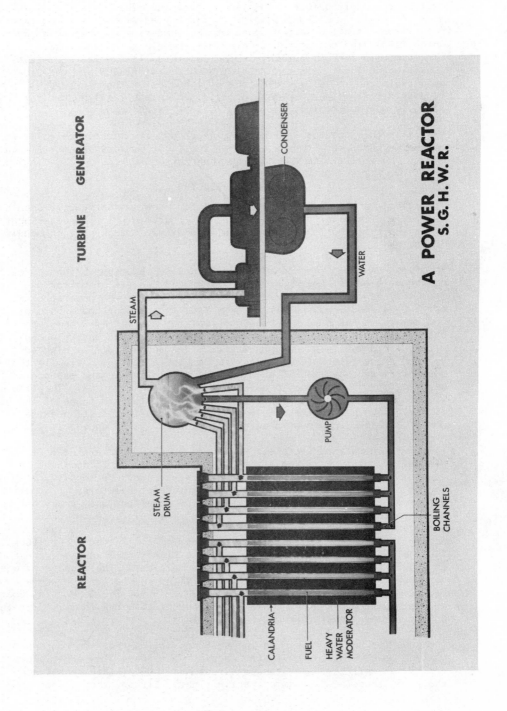

A POWER REACTOR
S. G. H. W. R.

is that refuelling can be done over a weekend when the electricity demand is lower than during the week.] On the American LWRs the pressure vessel head must be taken off before refuelling can begin. This operation takes some weeks, implying that LWRs should run for a year without fuel movement. The refuelling operation is best done at the same time as the annual maintenance, which is carried out at the time of minimum electricity demand. (This is in the summer in the UK, but may be in the winter in the USA because of the air-conditioning load). In contrast, the channels of SGHW are individually accessible and it is possible to fuel a number of them in a weekend, making it possible to maintain a more or less continuous fuelling cycle without actually moving the fuel while the reactor is on load.

[Thermally this reactor is like BWR; neutronically and mechanically it is much more like CANDU but it uses enriched fuel. It is less expensive than CANDU and may be a little more expensive than PWR and BWR.]

THE GREAT REACTOR DEBATE

I conclude by recalling the "great debate" about which reactor should form the basis of the third UK nuclear power programme. I will be as neutral as I can about it, but the reader should know that I took part on behalf of SGHWR.

The first UK programme was based on Magnox. The output of these reactors is now about 4 GW(e): they had to be down-rated because of corrosion. Apart from this down-rating they work very well. The second power programme calls for 6 GW of AGR, and the first power from this second phase is expected soon. The delays on AGR have not been abnormal for a first of a new type of reactor, but it has been excluded from this debate as being expensive and probably difficult to maintain. Magnox was excluded because of its high cost and because of certain safety worries. The general European view is that we are not yet ready to build HTR on a commercial scale. So the gas-cooled reactors were all out, though many people wanted to go on with AGR. They claimed (correctly) that we have expertise on this reactor, that one of the designs is very good, and that it could be duplicated with almost no extra design work. But they were overruled on the grounds stated above.

This left the water reactors. We looked at CANDU and excluded it mainly on grounds of cost. We liked it because of its uranium economy and this looks an even better argument than it did two years ago. Personally, I do not regret arguing against the American Light Water Reactors. I sometimes wonder about CANDU because it uses uranium so effectively and uranium is going to be short. But there are safety difficulties about CANDU. It is not clear that emergency spray water could be injected into the middle of those long channels in the event of a loss of coolant accident (LOCA). In the end the choice lay between one of the Light Water Reactors and SGHW, though it took some time for this to become clear. The assessment was done by comparing PWR with SGHW, on the understanding that if PWR won we would go back and see whether we really wanted PWR or BWR. The Generating Board and GEC both wanted PWR; the AEA has long thought that BWR is mildly preferable, but everybody agrees there is not much in it.

The case for PWR was based on the CEGB conviction that the construction of nuclear power stations must start again. The Board view of this question varies for entirely understandable reasons: indeed, they cannot hope to get their construction programme exactly right. Either they do not build

enough power stations and are then cursed for leaving us short of power, or
they build too many and are cursed for wasting national resources. The
question of whether enough have been built depends on things that nobody can
know, such as the future development of the national economy. Nobody would
have predicted seven or eight years ago that we would be in deep recession,
and that is the planning period for power stations. Recently we have had a
lot of spare capacity. The electric central heating market was expanding
very fast in the late fifties and early sixties, until it was cut off sharply
by the advent of natural gas. The resulting over-ordering to which the CEGB
was committed is with us yet, and until very recently the Board has been
taking the view that they do not need more power stations of any kind.
However their rolling five year projection then (end-1973) showed more power
being needed by the winter of 1978.

As the country moves deeper into recession the year in which one will need
more power stations tends to recede, but the Board would still like to see
more nuclear power. Since the October War, nuclear has become considerably
cheaper than fossil fuel power; the cost disparity is probably around 30%.
(Also the Board wants a certain amount of flexibility when it bargains for
coal). The position is changing; 19th Century economics is raising its head
and the producers and enrichers of uranium are starting to charge an "economic
rent". This implies that the price of nuclear fuel may rise to the point
where nuclear power is only just competitive with oil, although we are
nowhere near that point yet.

The reasons for restarting the nuclear power programme imply that it should
be restarted as quickly as possible. The CEGB argued that SGHW was ruled out
by this consideration alone, since SGHW experience was confined to one small
prototype while PWR is a developed article of commerce. It is also cheaper,
though the size of the margin is not clear. If trouble should arise there
are many other people working on PWR, and we would not have to solve the
problems by ourselves. Also there will be a chance to participate in a world-
wide trade in PWR components with possible benefits to the British balance of
payments.

The contrary case was set out most forcefully by David Fishlock of the
Financial Times. The first part of this case was that a crash programme was
not possible in this country, regardless of what reactor type was chosen,
because of the current disarray of the nuclear industry. It has been
demoralised by a long period without orders and by frequent changes in
organisation and government policy. Moreover, it is wishful thinking to
suppose that because PWR is a developed article in America, one can just send
across the Atlantic for a set of drawings and start building it. It is
emphatically not like an electric toaster. It is a very difficult piece of
technology which will take a lot of mastering. Those with experience of
large-scale computer programmes will know that to master somebody else's
computer programme sufficiently to be able to modify it, takes not less than
30% of the effort which went into writing it. If something goes wrong it can
easily be 100% or even 150%. Building nuclear reactors is exactly the same.
In my personal view, both the CEGB and General Electric deceived themselves
on this first point.

The second facet of the case against PWR is that it is not yet a fully
developed article of commerce. There are many reactors of this type, and a
good proportion of them work very well. However, breakdowns and faults are

frequent enough to suggest that the work of development is not yet complete. Finally, the export argument looks wrong too. We are not going to have any effort to spare for a long time. CEGB will want all the output of the nuclear industry for a good many years, though I hope we may develop an export trade in the eighties. Our share of the trade in PWR components would be whatever Westinghouse and Framatome are prepared to let us have. I reckon that, in the long term we have got a better chance of selling our own reactor rather than somebody else's.

Safety entered substantially into the argument, and there were many different points of view about it. However, I hope that the following statements will be generally agreed. The first is that the safety problems of PWR were hardly understood when the first sales were made. In particular, the need for an emergency core cooling system (ECCS) was not appreciated. It will take a few seconds to shut off the nuclear reactor after (e.g.) a break in one of the main coolant pipes. Both coolant and moderator are lost, so that the reactor shuts itself off, but heat generation continues by the radio-active decay of the fission products. It was understood early on that there would be a long-term problem; the reactor will be producing 0.3% of normal reactor power one month after shut-off. What was not understood was that there was an immediate problem and that emergency coolant must be injected before the vessel had depressurized. We in the UK realised this while we were developing SGHW, and the SGHW prototype started life with an ECCS: none of the early American water reactors had this. The Americans now have a very impressive programme to assure safety in this respect, but they have still to demonstrate that, following main coolant loss, emergency coolant will re-enter the core and cool it down in sufficient time. I believe that assurance of this will be provided over the next few years.

It is easier to demonstrate safety for SGHW, in which the pressure vessel is split into a large number of tubes: there will be about 600 in a commercial SGHW. It is only necessary to demonstrate the performance of the ECCS in one of these: this can be and has been done. In contrast, the protagonists of LWR have to demonstrate the performance of the entire contents of a pressure vessel, and that is another thing altogether. One has also to consider the possibility of pressure vessel fracture. The Americans have a programme on this too, and so far it is very reassuring. However the standard of integrity required is daunting; the tolerable failure rate is between 10^{-5} and 10^{-6} per year. One does wonder whether any single human activity can be assured to that sort of standard. In contrast, SGHW can probably be designed so that the fracture of one pressure tube can be tolerated: again, it is easier to demonstrate the safety of SGHW. Moreover, as the PWR builders strive for better safety they are likely to close the cost gap between their reactor and SGHWR.

Another part of the case for SGHW was that the Nuclear Installations Inspectorate had virtually approved the SGHW which the North of Scotland Hydro-Board were planning to buy for their Stake Ness site: in fact, the Inspectorate were within a few weeks of giving safety approval. But they (the Inspectorate) are still unconvinced about PWR; they are not saying it isn't safe but they want more information.

The initial phase of the third British nuclear power programme will consist of three two-reactor stations, one at Torness on the Lothian coast south-east of Edinburgh, and two at Sizewell on the Suffolk coast: each reactor will

produce 660 MW(e). The contracts for those are unlikely to be signed before
early 1977. This may not matter too much. What people are saying now is
"For heavens sake let's get this right; then we can build 20 more straight
off if we want to, without major design changes".

Finally we note the vast scale of the enterprise. The existing AGR
programme represents an investment of well over £1,000m at historic prices:
at today's prices, the investment will be between £2,000m and £3,000m. When
these stations are all working they will save the balance of payments over
£300m a year for oil purchase. The electricity produced by the existing
(Magnox) nuclear power stations saves about 9 mtce (million tons of coal
equivalent) a year. When the AGRs are all working the saving will be nearly
30 mtce. That figure, which should be achieved by 1980, should be set
against total UK energy consumption of 325 mtce. By 1990 the savings might
well be 60 or 70 mtce so that nuclear power will then be providing perhaps
20% of our primary energy input.

DISCUSSION

Question: Can reactors be used to load-follow economically?
Answer: SGHWR can load-follow at up to 30%/min.,   but very sharp load-
following is handled by gas turbines.

Question: Who owns the uranium ore that Great Britain uses?
Answer: We have long term contracts with Australia and Canada for the
supply of uranium ores.
Comment: The CANDU system seems superior than SGHWR to the layman as it can
be bought "off the shelf" and has a better fuel utilization.
Response: No reactor can be bought "off the shelf". Also the CANDU emergency
core cooling is suspect and the price (1973) was considerably
higher than SGHWR.

Question: Why is the Th^{232}/U^{233} fuel cycle not utilized in thermal reactor
systems?
Answer: Although Th^{232} is more abundant than U^{235} there exist at present
no chemical processing plants and no reactor systems capable of
handling Th^{232}.

Question: What happens to the spent fuel elements from Magnox and AGR?
Answer: There are no spent fuel rods from AGR at present. Magnox fuel is
reprocessed at Windscale   in the usual way after "cooling" in
ponds at reactor sites.

Question: How long will SGHWR last?
Answer: Well over 30 years, possibly indefinitely, as everything can
be easily replaced.

Question: Is the Canadian CANDU experience helpful to the British SGHWR
programme, and can SGHWR be modified to run on natural uranium?
Answer: The CANDU experience does help as the two systems are of a similar
type and the feasibility of a large system has been proven. It is
only possible to use natural uranium in the present SGHWR if the
D_2O content is increased but this unfortunately results in a large
($\sim$ 4%) positive void coefficient as found by the Canadians.

Question: Would you describe the safety aspects of the nuclear fuel cycle?
Answer: The spent fuel elements are transported in flasks which must withstand:
1) a fall from 30 ft.,

2)   a 55 mph impact with a bridge,
3)   2 hrs in a fire at 1,000°C.
The elements are reprocessed and major safety problems do not
arise until the Pu has been separated, at which stage a security
risk arises which may be avoided by diluting the Pu with uranium
for storage. Fission products can be stored in the form of glass
bricks which could be stored underground.

Question: Considering the possibility of large releases of fission products,
is it advisable to site reactors in densely populated areas?

Answer: It is believed that 2 miles is a sufficient distance from city
boundaries for the siting of a reactor when it is still economically
feasible to supply domestic heat to the city.

Question: Do you think it is reasonable for the U.K. to receive nuclear waste
from other countries?

Answer: Yes, it is a reasonably safe, lucrative trade.

Question: Can we obtain uranium economically from sea water?

Answer: No, it looks as if the pumping energy could exceed the energy
obtained from the uranium.

Question: What are the steam conditions and coolant pumping power for the
Winfrith SGHWR?

Answer: Steam conditions 285°C, 950 psi ($\sim$ 65 bars), pumping power 3.5 MW.

Question: Do nuclear fuel costs include reprocessing costs?

Answer: Yes. Operating reactor costs are approx. broken down as follows:
85% capital,
5% purchase of the original natural uranium,
5% enrichment,
5% fabrication and reprocessing.

Question: What is the feasibility of using process heat from HTR and what
is the maximum helium outlet temperature?

Answer: The maximum helium outlet temperature is 1100°C, but utilization
of this is difficult due to materials problems and the need for
secondary heat exchangers.

Question: People living in Kerala (India) are subject to a background radiation dose 12 times that of UK average background dose, but do not
ingest it, while for gaseous radioactive emission from a power
station, ingestion is inevitable. Would you comment?

Answer: Emissions from nuclear reactors are well below the limits set down
by ICRP and are also significantly less than those from many conventional coal burning stations.

Question: What is the life expectancy of Kerala residents?

Answer: About 40 years, which is not significantly different from a similar group living several miles away. Most deaths from radiation
effects (cancer) occur after the age of 50, however, so no real
conclusions can be drawn from the Kerala study.

FAST BREEDER REACTORS

W. B. Dale

Reactor Systems Directorate, UKAEA, Risley

Introduction

In considering the subject of fast breeder reactors the factors of greatest interest to the non-specialist are the possibility of increasing the total world energy resources by this means and whether there are undue safety risks attached. In this lecture, attention is focused on the understanding of these two issues, and general engineering design aspects of fast reactors have been mostly excluded.

When considering and comparing total energy resources many units have been used by different people including barrels of oil, tons of oil (equiv), tons of coal (equiv), BTU, therms, calories, MW year, kwh, and conversion and comparison of figures given in different papers becomes difficult and confusing. It is proposed to work here in joules which is the international standard unit of energy. However, the figures are also quoted in terms of barrels of oil as this is the commodity which nuclear power will eventually be required to displace.

Fuel Resources

It is by now commonplace knowledge that the world's reserves of oil and natural gas are of limited extent and at the present rate of expansion would be exhausted in a few tens of years (1). Energy conservation measures and a cutback in economic expansion can extend this time, but arguments have been advanced that all will be well because nuclear power will gradually take the place of oil. From a chemical raw material point of view this is extremely doubtful, and even from an energy point of view strenuous efforts will be required to ensure that adequate nuclear power is ready in time, and in particular, that the fast breeder reactor can be utilised quickly to extend the world's supplies of fissionable material.

The only naturally available fissile material that can be used to feed nuclear reactors is the isotope U235 which occurs as 0.7% of natural uranium, the remaining 99.3% being U238 which is not readily fissionable. Natural uranium is distributed fairly widely throughout the earth's crust and in the sea, but at very low concentrations. The total amount is of the order of 10^{10} tons in the top crust alone, but the majority of this would be very expensive to obtain in concentrated form. The amount that can be obtained in the western world at economical prices for fuelling in thermal reactors is estimated to be of the order of 8×10^6 tons (2) giving a readily available total world U235 content of 5.6×10^4 tons.

The fissioning of 1 kg of U235 produces 8.3×10^{13} J of heat. In converting this to equivalent barrels of oil it has to be remembered that where as oil can be burnt directly to provide heat, or indirectly to provide motive power or electricity; at present, the prospects of using nuclear power to provide process heat are small, and the almost exclusive use is to produce electricity. Where the production of motive power or electricity are the main aims, the heat content of nuclear fuel can probably be equated to that of oil, but where process heating is the main aim, the heat content of nuclear fuel has probably to be downgraded by a factor of two or three compared with oil to allow for the thermodynamic and other losses of having to produce a refined form of energy like electricity as an intermediate product. It is not easy to assess the total way in which nuclear power will replace oil and for simplicity the simple heat content of the two will be directly equated, bearing in mind that this tends to give a somewhat optimistic comparison. On this basis 1 kg U235 is equivalent to 13,500 barrels oil equivalent (boe) and the world stock of 5.6×10^4 tons U235 is equivalent to 46×10^{20} joules or 0.8×10^{12} barrels. This is about one half of the world's supply of crude oil and about one fifth of the total oil if oil from shale is included.

Thus, taken by itself U235 makes only a minimal contribution to overcoming the problem of oil scarcity. During the burning of U235 in thermal reactors some of the U238 is converted to Pu239 which is fissionable, and there is also a small component due to fast fission of U238. The effect of these is to increase the available heat content by 50% or so, and if the plutonium produced is recycled in thermal reactors, the total available heat is increased by a further 50% so that the total available heat is increased to about 100×10^{20} joules or 2×10^{12} boe and is equal to the total heat quantity of crude oil. This amount is useful but does not change the situation radically.

If the U235 or Pu can be burnt in fast breeder reactors, the cycle can be arranged so that more fissile material is produced than is consumed as explained later. In principle, therefore, the whole of the U238 can be converted to Pu and fissioned. In practice, however, small amounts of U238 are retained in the processing system during the frequent recycling and the total heat is increased by perhaps a further factor of about 50. This means the fuel ore, used in this way can provide a total nuclear heat equal to an amount 50 times the heat available in crude oil reserves. In addition, how-ever, the cost of mining ores can also be increased by a factor of 50 without suffering an economic penalty and this must increase by a very substantial factor the amount of low grade ore that can usefully be mined. Potentially, therefore, fast reactors can provide all the heat requirements for the fore-seeable future. If the uranium ore eventually becomes too expensive to obtain in sufficient quantity, thorium ores can be used as an alternative fertile material. Thorium, although non-fissionable itself, becomes converted to fissionable U233 under neutron irradiation and can thus act as an alternative fertile material to U238. It is not as efficient a material as U238 however and there seems little incentive to use it whilst U238 is still available.

A word of warning on the present situation is required, however, in that the general rate of build up of fast reactors appears to be too slow to be able to make an impact. The available time scale is extremely short and assuming only thermal reactors were to be built to fulfil the nuclear energy require-ments projected by OECD the world's ore supplies would be totally committed by 1985 (3) (4) and it would then not be worth building any further thermal reactors as they could not be fuelled for their full life.

If the first breeders are commissioned by 1985 and the manufacturing capacity then switched as far as possible to breeders, the total ore resources and rate of mining are still such that only about 60% (4) of the projected OECD nuclear programme can be achieved by the turn of the century. This OECD programme was aimed to fill the gap caused by increasing oil scarcity and it

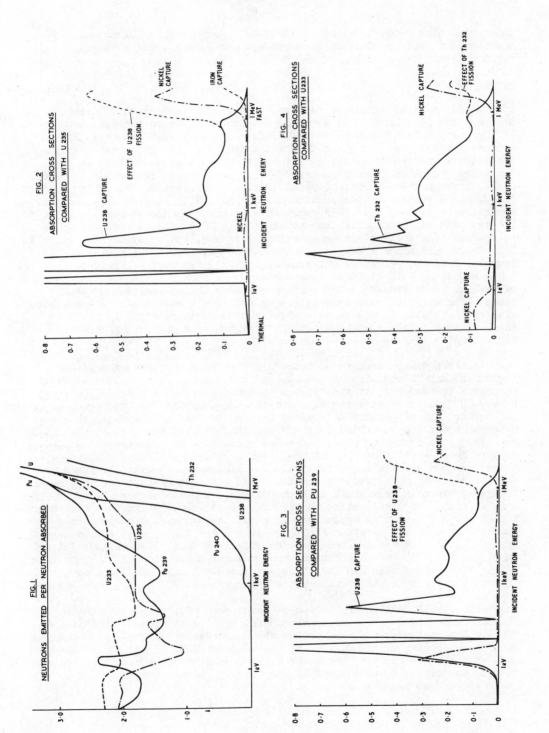

FIG. 2
ABSORPTION CROSS SECTIONS
COMPARED WITH U 235

FIG. 4
ABSORPTION CROSS SECTIONS
COMPARED WITH U233

FIG. 1
NEUTRONS EMITTED PER NEUTRON ABSORBED

FIG. 3
ABSORPTION CROSS SECTIONS
COMPARED WITH PU 239

thus seems that a continuation of adequate energy supplies cannot now be assumed, and that the further the breeders are delayed, the greater the world's energy gap will be.

Neutron Economy

If the moderator of a thermal reactor is partially removed so that the neutrons are not slowed down so much, the average neutron incident energy is increased and the absorption in U238 becomes very much greater. If each neutron absorbed in U235 at an incident energy of 1 kev produces on average two outgoing neutrons, then the enrichment of the U235 would have to be increased from 3 or 4% to at least 20% to cater for the absorption in U238 alone, irrespective of other losses. Figure 2 illustrates this point. At 1 kev, the absorption cross section of U238 is 20% of that of U235 and thus a mixture of 5:1, U238 to U235 would have an equal chance of capturing one of the two neutrons produced by the absorption in U235. As more moderator is removed and the neutron average energies increase, the relative absorption in U238 falls, and above 1 Mev it again becomes quite low. The number of neutrons produced per fission increases and U238 can now be fissioned. At 2 Mev about one half of the fissions can occur in U238 so that the presence of U238 in large amounts is not such a handicap to criticality and the reactor can again be made critical using relatively modest U235 enrichments. With Pu239, a similar situation applies and is further helped by the fact that Pu gives a larger average number of neutrons per neutron absorbed at high energies (Figure 1).

The situation can be seen to be extremely dependent on incident neutron energy and any degradation of this energy gives a significant loss of neutron economy. The majority of the flux is spread over the energies 20 kev to 2 Mev. U238 is generally placed round the edges of the core to form a blanket and capture the neutrons that would otherwise be lost and in this way a substantial amount is converted to Pu239 and this blanket material is a significant factor in producing a good breeding gain. Another factor is that the capture cross section of U238 remains high relative to that of fission products, but not so high as to preclude the attainment of criticality with an acceptable amount of fissile material. Thus the loss of neutrons to fission product capture is very small at high energies, and the capture by nickel and iron in stainless steel, although also very small could start to become significant above 2 Mev. The corresponding situation using Pu239 as a feed fissionable material is shown in Figure 3, and the Th232, U233 system in Figure 4. Here it can be seen that the effect of Th232 fission is relatively small and from Figure 1 the fission neutrons per absorption in Th232 is also small. The system thus has less breeding potential than one which converts U238 to Pu239 but is still quite satisfactory.

Thus either U233, U235 or Pu239 can be burnt very efficiently in fast reactors and from an ore conservation point of view the less U235 that is burnt in thermal reactors, the better the conservation of ore. Although the general enrichment of the fuel is of the order of 20%, or say seven times that of a thermal reactor, the power rating can also be increased by this amount, so that there is no overall increase in fissile inventory for a given electrical output.

As an example of the fate of the neutrons captured in a fast reactor, the following gives a breakdown for a typical CFR.

	Core	Blanket	
Pu239	10.9	1.6	Conversion to Pu240
Pu240	3.2	0.1	Conversion to Pu241
Pu241	0.5	0	Conversion to Pu242
U235	0.2	0.2	Conversion to U236
U238	32.3	27.3	Conversion to Pu239
Fission Products	3.1	0.3	Loss to System
Other	7.1	3.1	Loss to System
	57.3	32.6	
		57.3	
Capture in reflector		7.7	
Leakage from System		2.4	
		100%	

Neutrons producing fission were 52.4% compared with the above.

A short Pu doubling time does not increase the total world stocks of fuel that can ultimately be fissioned because providing the breeding ratio is greater than unity, every fertile U238 atom can eventually be converted into a fissionable Pu239 atom and fissioned, but it does enable Pu burning reactors to be built up more quickly. The doubling time depends not only on the breeding ratio, but also on the power rating and the time taken to reprocess the fuel after it has been taken out of the reactor. The processing of the fuel immediately after it is first removed from the reactor is difficult and costly because of the highly active fission products it contains and it is usually desired to allow a "cooling" period of several months so that the fission products can decay before reprocessing. For a present design of oxide fuel commercial LMFBR with a breeding gain of about 0.27 the doubling time would be about 25 years.

Coolant Fluid

Because of the high fissile inventory in the fuel, a high rating is essential and an efficient heat removal fluid is required which does not capture or slow down neutrons to a major extent. Possibilities are high pressure gas and liquid metals. Water or steam tend to moderate neutron energies so that the system would not be an efficient breeder.

For liquid metals, lead, mercury or sodium are possibilities. Sodium is cheap and plentiful and as its physical properties as a liquid are not too dissimilar from water, water can be used as a convenient transparent simulant fluid to check the behaviour of many of the components such as the pumps. In spite of sodium being active chemically it is generally safer than mercury or lead and any sodium vapour which escapes becomes fairly quickly converted to relatively innocuous sodium carbonate after release to atmosphere. There is, of course, an intermediate sodium oxide stage which is dangerous to breathe in high concentrations and eyes and lungs require substantial protection in the vicinity of a large sodium fire. Sodium, in spite of it being chemically active does not burn vigorously in air and

tends to smoulder giving off dense oxide fume rather than to have a
substantial flame.

A disadvantage of sodium as a coolant is that it tends to moderate the
neutron energies to some extent so that close spacings are desirable. The
use of sodium also means an intermediate heat exchanger is necessary to
eliminate the possibility of a sodium water reaction inside the reactor
vessel. A helium system does not suffer from either of these disadvantages.
On the other hand, the sodium system can be operated at atmospheric pressure
and can be made so that a loss of coolant accident is incredible. On the
whole, the two systems, liquid metal cooled and gas cooled, are finely
balanced and each have their firm advocates.

Fuel

For an efficient thermodynamic cycle, coolant temperatures of the order of
550°C are required. The fuel will be operating hotter than this and natural
uranium metal would be cycled through the alpha–beta phase change point and
would not be a suitable material. Fuel swelling and incompatibility with the
clad would also preclude a high burn-up. Oxide or carbide fuels are thus
required for commercial designs. UC, with a single carbon atom moderates
neutrons less than UO2 with two oxygen atoms and thus has a higher breeding
ratio and is likely to prove the long term fuel choice. At present, however,
there is very substantial experience with UO2, from its use in AGRs and
water reactors. Its manufacture, irradiation and reprocessing are not new
and as its breeding ratio is satisfactory it tends to be the first choice,
at least for the first series of breeder reactors. The relatively low
thermal conductivity of ceramic fuels coupled with high rating means that
the fuel diameter must be small to avoid high centre temperatures.

Although the above argument has been used in the context of UO2, the fuel
does, of course, contain about 20% PuO2 but the physical properties of UO2
and PuO2 are broadly similar.

Liquid Metal Cooled Reactor   *Sodium cooled fast reactor*

The generally favoured design of fast breeder reactor is at present the
sodium cooled fast reactor. Reactors of this type have been operating for
many years, BR5 in Russia was commissioned in 1959 and the Dounreay fast
reactor has been at power since 1963. The Dounreay fast reactor has metal
fuel and 24 heat exchanger circuits. Although it has proved a successful
irradiation facility, and has provided useful experience in dealing with
liquid metal coolants, considerable changes are required before it could be
extrapolated to a commercial design. The coolant is also a mixture of
sodium and potassium in order to lower the freezing point and prevent
solidification in cool circuits. This addition of potassium is not now
considered necessary and will not be adopted in commercial designs.

A prototype fast reactor has thus been designed and built at Dounreay and is
in the final stages of commissioning. The parameters are briefly as below.
A cross section of the vessel is shown in Figure 5.

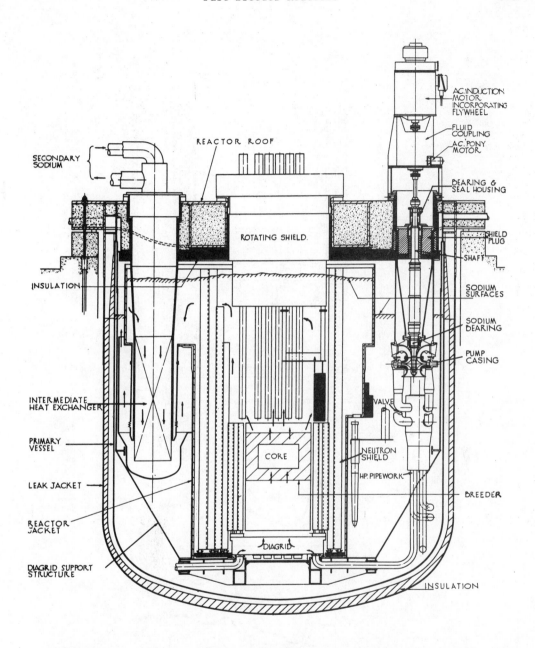

ELEVATION THROUGH REACTOR SHOWING PRIMARY CIRCUIT

FIG. 5.

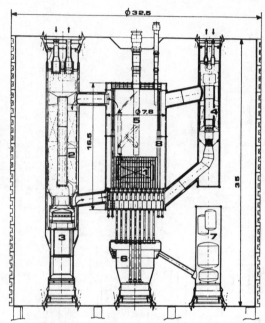

Fig. 6. **1200 MW gas cooled breeder reactor**

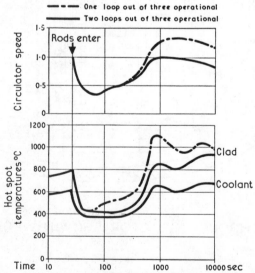

Fig. 7. **Transients for design basis depressurization**

PFR PARAMETERS

Thermal Output	560 MW Core
	40 MW Breeder
Gross Electrical Output	265 MW approx
Nett Electrical Output	250 MW
Weight of mixed oxide fuel in core	4 tonne
Weight of PuO2 in core	0.9 tonne
Outer diameter of pins	5.8 mm
Fuel pins per sub-assembly	325
Number of core sub-assemblies	78
Number of radial breeder sub-assemblies	42
Length of fuel in pin	915 mm
Length of axial breeder	229 mm
Diameter of primary vessel	12.2 m
Weight of sodium in primary vessel	900 tonne
Number of IHX	6
Number of primary pumps	3
Coolant inlet temperature	415°C
outlet temperature	580°C
Core Fissile Inventory	3.7 kg/MW(e)

The fuel consists of mixed PuO2 and natural UO2 and is in the form of long pins encased in stainless steel. Because of the high fuel temperatures, much of the fission product gases will be released from the fuel lattice and are accommodated in a plenum at the bottom of the pin to prevent excessive build up of pressure within the pin. 325 pins are accommodated in a stainless steel hexagon wrapper with a separate set of shorter pins at the top end containing natural UO2 to act as an upper axial blanket breeder material. There is also natural UO2 below the centre fuel portion to act as a lower axial breeder. The wrapper is freely supported by a spike at its lower end and is handled into and out of the reactor as a single unit. In order to obtain access to the individual core wrappers for refuelling, the control rods are disconnected from their top supports and a top concrete shield in the roof is rotated to position the refuelling machine above the desired part of the core. The core proper consists of pins in 78 wrappers but other wrappers containing pins of natural UO2 surround this core to form a radial blanket.

There are 3 pumps for the primary sodium circuit and 6 sodium to sodium intermediate heat exchangers (IHX). These IHXs are to avoid bringing high pressure water and steam into the reactor primary tank so that any boiler tube failure will not cause a vigorous Na/H2O reaction near to the core and will not pressurise the primary tank. It also ensures that any of the radioactive and possibly contaminated sodium is confined to the primary tank and in the unlikely event of fracture of a sodium pipe leading from the tank the tank itself is not drained and there is very little radioactive sodium about. Sodium itself becomes radioactive under irradiation, but the IHXs are at some distance from the core, within the tank, and protected by shield rods.

The primary tank containing the reactor core and sodium is a relatively thin stainless steel vessel, outside which there is a strong steel leak jacket and the interspace is monitored for sodium. There are no large apertures below sodium level so that loss of sodium coolant from the primary tank can be rendered incredible. The primary tank is unpressurised and the sodium at the surface is at atmospheric pressure, although covered with a blanket of inert argon gas.

The secondary sodium is pumped from the IHX to other vessels through which pass the tubes containing high pressure water and steam. The pipes between the reactor and the boilers through which this secondary sodium is pumped are also double walled with the interspace monitored for leakage sodium. Any leakage of water occurring from the boiler tubes into the sodium can be detected in very small amounts by the evolution of hydrogen and the sodium and water in the boilers can each be rapidly transferred to separate dump tanks. The arrangement of the turbines is conventional.

A secondary containment building surrounds the primary vessel and is designed to contain pressures slightly above atmospheric under emergency conditions. This is to prevent release of radioactivity in the event of a fuel handling mishap whilst transporting irradiated fuel from the reactor to the handling caves combined with a possible spillage of sodium from the transport flask.

The arrangement for the first commercial fast reactor will be broadly similar to the above except that because of the larger power output (1200 MW(E)) the number of wrappers and pumps will be greater. The strong outer steel vessel will be replaced by a strong concrete vessel. This strong vessel gives extra assurance of containing the contents of the primary tank in the very remote event of an accident occurring inside the core.

Gas Cooled Fast Reactor

The concentration on commercial gas cooled reactors in the UK might have been expected to lead to preference for a gas cooled fast breeder reactor. The pressures required to obtain an economical rating were, however, double what was then being used and the possibility of an accidental depressurisation and the resulting safety problems, although by no means insoluble led to a continuation of the sodium cooled fast reactor which had been successfully operating at Dounreay for many years.

The design has, however, been studied and developed in more detail by the European Gas Breeder Reactor Association who have produced a 1200 MWE design in a pre-stressed pressure vessel and re-examined the safety aspects (2). The general performance parameters are given in the table and the general reactor layout is as shown in Figure 6.

GAS COOLED FAST REACTOR

Main parameters

Reactor Heat Output	3600 MW	
Net Electrical Output	1200 MW	
Coolant Pressure	90 bar	
Inlet Temperature	260°C	
Outlet Temperature	560°C	
Fuel Pin Outer Diameter	7.7 mm	
Core Fuel In-pile Time	3 years	
Breeding Ratio	1.4	
Pu Linear Doubling Time	18 years	Low cost design
	15 years	Low inventory design
Core Fissile Inventory	4.06 kg/MW(e)	Low cost design
	3.2 kg/MW(e)	Low inventory design

It can be seen that the breeding ratio is rather better than the sodium cooled fast reactor. More advanced designs with higher gas pressure and ratings may ultimately reduce the doubling time to 12 years whilst maintaining the breeding ratio constant.

The coolant is helium at a pressure 90 atmospheres inside a pre-stressed concrete pressure vessel. All large penetrations are closed by concrete plugs held in position finally by the vessel pre-stress tendons. The fuel is in steel clad pins and can be either UO2 or UC. The high external pressure requires the pins to be operated near to reactor pressure and arrangements are thus made for them to be separately vented. The release of fission product gases is thus piped from the pins through pillars on steel penetrations in the concrete at the bottom of the reactor. Refuelling is performed annually with the reactor depressurised.

The gas pumps and the boilers are situated within the main vessel and no intermediate heat exchangers are required.

In order to overcome the depressurisation difficulties mentioned earlier there is an external leak tight outer vessel which can contain a pressure of $2\frac{1}{2}$ atmospheres preventing the escape of coolant to atmosphere and retaining an adequate pressure within the reactor to facilitate continued adequate removal of decay heat without the pins overheating too much and well below clad melting temperature Figure 7. The design is such that the reactor could still be safely cooled even at atmospheric pressure.

Molten Salt Breeder Reactor

For many years there have been advocates of a reactor in which the fuel has been in the form of a molten salt instead of fabricated into fuel elements. Such a scheme is superficially attractive in that it avoids fuel fabrication difficulties, avoids burn up limits, complex refuelling machinery and periodic shutdowns for refuelling.) The fuel can be dumped from the circuit to dump tanks in the event of trouble. However, like most reactor systems, the simplicity tends to disappear when detailed engineering and plumbing is started and there are considerable difficulties of containing very hot, highly radioactive, corrosive fluids. The high melting point of the salt (about 560°C) requires trace heating during shutdowns, and there is the possibility that local high concentrations of fissile material may give local criticality regions and hot spots. Although feasible, very considerable development is required and there appears to be little incentive towards this when two alternative systems (sodium cooled and gas cooled fast reactors) have already been carried to a high state of development.

Many conceptual proposals exist. One variant is where the salt is (U + Pu) Cl3: Na Cl circulated from the core vessel through high performance lead heat exchangers. By using salt temperatures between 800°C and 1,000°C the size of core and heat exchangers can be kept down but the fissile inventory required is still about twice that required for a sodium cooled fast reactor. The breeding gain is slightly better than an oxide fuel LMFBR but not better than a carbide fuel LMFBR.

Reactor Safety

General

For any ordinary condition such as failure of primary pump motor electrical
power supplies, there is no difficulty in detecting the onset of an abnormal
condition and shutting the reactor down. The available time for automatic
protective action is shorter than that on thermal reactors because of the
higher rating, but it is still ample compared with the available detection
and control rod operating times.

In LMFBRs the sodium gives the contents of the primary vessel a very high heat
capacity and decay heat of the core can be accommodated for many hours without having to have external cooling. However, if during an accident, boiling
of the sodium occurs in the operating core, this adds reactivity and may
increase the severity of the accident. The temperature rise across the core
would, however, need to increase by a factor of three before any sodium
boiling could occur.

In gas cooled fast reactors the hazard of accidental depressurisation can be
rendered small by the use of a low pressure secondary containment, and
reliable auxiliary pumping systems.

Prompt Criticality

The essential difference between fast reactors and thermal as has been
mentioned earlier is that there is no neutron moderator in fast reactors and
this means that the mean neutron lifetime is of the order of one microsec
instead of about one millisec as in thermal reactors. It also means that the
fuel is not necessarily in the most reactive configuration and that any
tendency to compress the core into a smaller space (such as, for instance, as
might occur during a core meltdown) could cause an increase in reactivity and
a tendency for the power to increase further. On thermal reactors, a core
compaction or meltdown may reduce the degree of moderation of the neutron
energy so that the core tends to become less reactive and it shuts itself
down.

In normal operation, reactor control in both thermal and fast reactors is
obtained by utilising the fact that about $\frac{1}{2}\%$ of the neutrons arise out of
fission product decay and their birth is delayed by several seconds after
neutron capture. If the excess reactivity remains less than this $\frac{1}{2}\%$, the
rate of rise of power is limited by the delay these neutrons introduce.
If the excess reactivity exceeds this threshold the reactor is said to be
prompt critical and in principle the rate of rise of power could be very
rapid. The amount of excess reactivity that would tend to cause prompt
criticality is called the dollar. In reactivity terms, one dollar is
equivalent to 0.23% excess reactivity in plutonium systems and 0.67% in
U235 systems. The delayed neutron fraction from the fission of U238 is
1.6%. In thermal reactor systems U235 and Pu co-exist in the fuel and the
dollar is of the order of 0.6% unless, of course, plutonium recycling is
introduced when its value is lower. In the fast reactor U238/Pu239 system,
the fission in U238 brings the value of the dollar to about 0.3% excess
reactivity.

If prompt criticality occurs, the fact that the prompt neutron lifetime of
fast reactors is only one thousandth of that of thermal reactors would

superficially indicate that the rate of rise of power was 1,000 times faster, and this originally led to fears that large explosions were possible.

Exhaustive calculations have shown that this is not in fact the case. As the power and temperature rise, there is an automatic negative reactivity feedback called the Doppler effect which cancels out the initial reactivity so that the temperature and power rise are matched almost solely to the rate at which reactivity can be introduced.   The mechanical energy developed in these explosions could be of the order of 1/3 full power second, but depends of course on the assumed rate of increase of reactivity.

Unfortunately, in the case of the sodium cooled fast reactor, the fact that the core is surrounded by sodium means that in an accident such as described above, some of this nearby sodium may be vaporised by the molten fuel and this sodium vapour adds to the work potential of the explosion. The degree of enhancement is of the order of two or three times the original. If spread throughout the available sodium, the rise in temperature due to this energy would be less than $10^{o}C$. Thus the sodium cooled fast reactor is provided with a strong vessel as an additional safeguard against the remote possibility of an internal explosion, in spite of the fact that the reactor operates normally at atmospheric pressure.

The chances of such an explosion occurring are very remote, possibly of the order of one per 10^7 reactor operating years. The main risk of any internal explosion occurring is only after the automatic protection has failed, and this failure condition is not normally investigated in detail in thermal reactors. It is only because of the very short prompt neutron lifetime that attention has been focused on this type of accident in fast reactors.

Plutonium Hazard

A further possibility of hazard to which people have often drawn attention is the plutonium in the core. The chance of this being released in significant quantities under accident conditions is small. The amount of plutonium in a fast reactor core is about a factor of three higher than what can be present in a water reactor (without Pu recycle) and this increase is not very significant in the order of accident that is required to be considered, as the contained fission products also offer a comparable hazard. The 20,000 year half life of Pu has been considered unacceptable by some people but some unacceptable fission products have a half life of many tens of years and the immediate energy problem is to get safely through the next 100 years or so without social collapse.

Some familiar chemicals present a toxic hazard not too different in real terms from that of Pu. The lead emitted from the exhaust of motor cars in one year in the UK is sufficient to kill the world's population if distributed "correctly" (5), yet in practice this emission appears to cause no very great harm and we tolerate it with equanimity. Similarly the mercury and cadmium present in the waste of coal fired stations would require the same amount of water to dilute it to drinking tolerance level as would the low level Pu

activity in the wastes of the nuclear fuel reprocessing plant (6). These Pu
wastes are, however, guarded very carefully, and if the Pu total level ever
becomes large, the waste can be reprocessed to obtain Pu fuel. At present
the very small amount of Pu stored in this fashion is not sufficient to
justify reprocessing. It also has been estimated that if all the waste
plutonium in a full power programme were ploughed into soil and used to grow
grain it would be responsible for 15 cancer deaths per century (8).

Summary

The present world energy situation is such that some form of nuclear power
is essential either to maintain our present level of standard of living or
to provide adequate time for transition to a lower standard of living without too rapid an economic decline and the ensuing danger of social collapse
in the next 100 years or so. The total time available before energy
resource limitations start to have an impact is not long compared with the
time taken for a large reactor building programme to be implemented.

Uranium resources and mining capacity to fuel thermal reactor systems may
not be fully adequate for this task. Whilst fusion may offer a longer term
solution, an intermediate solution is required and this can be obtained by
fast breeder reactors which increase the potential of present cheaply mined
ores by a factor of 50 and enable lower grade ores also to be mined
economically so that the supply of uranium becomes virtually unlimited. A
cycle in which thorium ores are converted to fissionable U233 is also
possible but the breeding potential may be lower than with a plutonium
breeder cycle, much more development work is necessary, and the system
offers no immediate advantage. From a fuel resource point of view there is
no short term need for a thorium cycle, and the fissile U233 is only slightly
less toxic than plutonium. In practice the presence of plutonium is not
thought to present a major increase in the total reactor hazard and this risk
in itself is small compared with risks commonly accepted in everyday life (7).
The economic and military hazards of a severe oil shortage may present a
greater total risk.

From a general safety viewpoint fast reactors can be protected against faults
as easily as thermal reactors. Gas cooled fast reactors have a disadvantage
of a high pressure coolant which may be partially lost by depressurisation
but the residual coolant pressure retained can be arranged still to give
adequate cooling.

The sodium cooled reactor has an advantage that loss of coolant can be made
incredible and there is a very high thermal capacity for absorption of decay
heat, but the liquid coolant and its vaporisation can cause fault analysis
to become complicated if boiling or fuel melting can be postulated to occur.
Sodium cooled fast reactors may also cost slightly more to build because of
the necessity for an intermediate sodium to sodium heat exchanger to ensure
full separation of the water in the boiler from the sodium in the reactor
vessel.

The shorter prompt neutron lifetime of fast reactors has caused situations
following failure of automatic protection to be analysed in very considerable
detail but this analysis has confirmed that very large explosions are not
possible.

The amount of Pu present in fast reactors is not substantially different
from that in thermal reactors of equivalent output, especially if Pu
recycling were to be used in thermal reactors. The chance of any significant quantity escaping through the agency of a reactor accident is extremely
small. The Pu wastes arising from fuel reprocessing are relatively small,
can be kept under control and, if necessary, eventually reclaimed as further
fuel. The total hazard presented by these wastes is small compared with some
more common toxic hazards that are accepted with equanimity.

Thus the risks entailed in having fast reactors are probably small compared
with the military and social disruption risks entailed in not having
adequate energy supplies. From this viewpoint fast breeder reactors can be
seen as offering a means by which the next few generations of mankind obtain
time for long term energy, population, and world resource limited solutions
to be developed.

ACKNOWLEDGEMENT

The author is grateful to the Atomic Energy Authority and to the Gas Cooled
Breeder Reactor Association for permission to reproduce details of PFR
and the GCFR design.

REFERENCES

1. Energy — from surplus to scarcity? — Proceedings of the Institute of
        Petroleum Summer Meeting, June 1973.
2. Vaughan, R. D., Uranium conservation and the role of the gas cooled
        fast breeder reactor — Journal BNES April 1975.
3. Nucleonics Week, March 6, 1975, quoting cost benefit study of LMFBR
        prepared for General Electric and Commonwealth Edison.
4. Private communication — Vaughan, R. D., Gas Cooled Breeder Reactor
        Association, Brussels.
5. Lister, B. A. J., "Nuclear Power — The Perspective of Risk" — Atom,
        May 1975.
6. Korsbech, U., Mercury from coal comparable with plutonium — Note in
        Nuclear Engineering, March 1975.
7. Rasmussen — USAEC Report WASH — 1400.
8. Cohen, B. L., Study for the Institute of Energy Analysis (Oak Ridge,
        Tenn.) as reported in Nucleonics Week, February 20, 1975.

DISCUSSION

Question: Should we invest money in uranium prospecting rather than fast
           reactor systems?
Answer: Breeder technology increases world uranium supplies by a factor
           of 50 and it is doubtful that such an increase could be obtained
           by further prospecting investment.

Question: Is there any government commitment to build a number of fast
           reactors?
Answer: The Nuclear Inspectorate is currently assessing fast reactor safety.

Question: Is the capital cost of a fast reactor higher than for a thermal
           reactor?
Answer: The capital cost will be higher for the breeder by approximately
           25% on current estimates. This is primarily due to the necessity
           of having an intermediate heat exchanger.

Question: What combination of breeder and burner reactors is required?
Answer: Owing to manufacturing problems it is easier and more efficient
           to build one reactor system. It will take a long time for indus-
           try to change over to fast reactor technology.

Question: Why does the demand for uranium ore in the U.K. decrease to zero
           in the year 2030 as shown in one of your graphs?
Answer: The breeder programme can lead to an effective increase in the
           U.K.'s fissile material inventory. Using a fast reactor programme
           the amount of uranium stockpiled in this country is equivalent in
           energy terms to the coal reserves.

Question: Must it not be borne in mind that we need H.T.R.'s for process
           heat, as breeders are only useful for electricity generation?
Answer: Although high temperature process heat can be obtained from
           H.T.R.'s, there are many material problems. This might favour
           the use of electricity from breeders for process heat.

Question: How many people would be killed by radioactive release in a fast
           reactor accident?
Answer: Between 100 and 1,000 depending on reactor location and weather
           conditions. Most of these deaths will be from cancer with an
           associated twenty year delay. These figures must be compared with
           lead emission from motor car exhausts which, from this country
           alone, has been estimated to have the potential to kill the entire
           world population.

Question: Radiation can have associated genetic effects so how can it be
           compared with lead poisoning?
Answer: Radiation levels are small and the probability of genetic effects
           is negligible. Such genetic effects are not proven.

Question: Is there not a possibility of Pu concentration in food chains?
Answer: I am not sure that such a mechanism exists.

Question: Do we not seem bedevilled by a lack of energy policy, owing to
           extreme short sightedness of our politicians?
Answer: It should be noted that in France, where a massive nuclear pro-
           gramme has been undertaken, serious manufacturing difficulties
           have resulted.

THE NUCLEAR POWER CONTROVERSY IN THE U.S.A.

R. Wilson

Physics Department, Harvard University

INTRODUCTION

On July 16, 1945, the first nuclear fission bomb exploded at Alamogordo, New Mexico; my friend, Kenneth T. Bainbridge, Project Director of the Project Trinity, has just retired as Professor of Physics at Harvard University. My colleague, Norman F. Ramsey, Higgins Professor of Physics, was at Saipan in charge of assembly of the next two bombs. As a graduate student, I worked with Hans Halban, who in 1939 with Joliot and Kowarski was the first to measure the number of neutrons (2.5) in fission and thereby prove that a chain reaction is possible. I have therefore been aware of the problems since I started as a graduate student in 1945. I do not believe that any of them have appreciably changed since then. The basic data were known, and the basic worries were present. In the intervening years, however, the issues have sharpened and the answers have become more precise.

Since the discovery of nuclear fission in 1939, scientists have been aware of the dual potentials of nuclear fission, for good and for evil. In 1945, they insisted on civilian control of the atom, through the U.S. Atomic Energy Commission, rather than leave it under military control which they distrusted.

The awful prospect of nuclear war has since been with us and will be with us till the end of civilization – and the fear is that the end may be soon.

The civilian uses of nuclear energy have steadily developed in the intervening years, and at first there was public approval. Reactors have made isotopes for medical research so that about one in four hospitalized patients have some radioactive material for diagnosis or treatment, and in the last 10 years nuclear reactors have begun to operate for electricity production.

313

The important question is – does the civilian use of nuclear fission make a
nuclear holocaust more or less likely? This is a hard question to answer. I
believe this is the question in the mind of all the nuclear critics – and it
is the only question with which I do not feel at ease. But perhaps because
it is hard to answer, and think about, critics have concentrated on other
questions which are easier to answer. I will discuss these in turn.

LOW LEVELS OF RADIATION

The first critic of note was Dr. Ernest T. Sternglass, originally a physicist
working with KeV electrons. Before nuclear power had begun to be economic
(and therefore before many power stations and their objectors) he had written
a number of papers on the dangers of radioactive fallout from nuclear weapons
tests[1]. Most scientists of my acquaintance wanted to stop these weapons
tests, and did not challenge his numbers for he appeared to be on the "right
side". It was left to Professor Rotblat in an article in Nature[2] to challenge Sternglass' methodology. In a variety of papers since, Sternglass has
tried to show that infantile mortality increases near a nuclear power station
after it begins to operate. The effects claimed are large, and would be
serious. Each of these claims, however, has been contested and I give an
example (figs. 1, 2, 3) from a paper by Hull and Shore[3]. Sternglass had
argued (fig. 1) that infantile mortality had increased 30% in Rockland and
Westchester County in New York relative to the neighbouring Nassau county,
from 1961 to 1966 in which time the Indian Point Power Station had begun to
operate. The 30% looks less impressive when the statistical errors are
assigned (fig. 2) to each point, and the curve shown in fig. 3 which gives
the full data, including other years, shows that the effect claimed by
Sternglass is, at the very least, not proven. One point we learn from this –
one eloquent critic can tie up fifty competent people in proving him wrong.

WHAT IS THE RADIATION DOSE?

The next critics were J. Gofman and A. Tamplin. John Gofman is a Professor

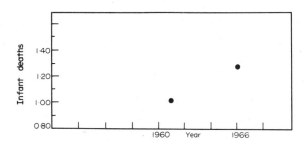

Fig. 1 - Infant mortality near Indian Point, New York, showing attempts to find an effect due to the Indian Point No. 1 nuclear power station. Plotted are normalized infant deaths in Rockland and Westchester Counties divided by normalized infant deaths in Nassau County. (Data without errors).

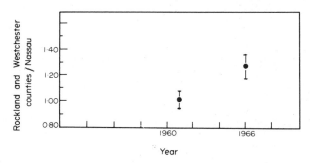

Fig. 2 - As Fig. 1. (Data with errors).

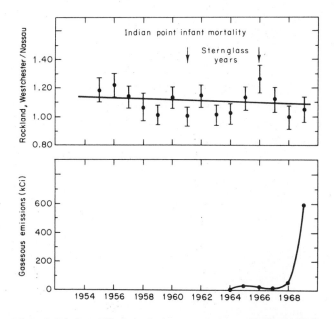

Fig. 3 - As Fig. 1 with additional data. Source Hull (1972).

of Medical Physics at the University of California at Berkeley and was in charge of a group studying effects of radiation at Lawrence Livermore Laboratories; Arthur Tamplin was an assistant.

After a study of the available information on effects of radiation, Gofman and Tamplin decided that the radiation limits, set internationally by the International Committee on Radiological Protection (ICRP), and adopted by the AEC were too high. Their arguments are set forth in a set of papers presented to a congressional hearing[4].

They estimated that if all the U.S. public were exposed to the ICRP limit of 170 millirems/year, 16,000 cancer cases would be produced. Later they raised this to 32,000 then 64,000. Since the average natural background is 100 millirems/year, this implied that 10,000 (later 40,000) of the 300,000 cancer cases in the U.S. yearly are caused by natural background. They further stated, sometimes directly and sometimes by implication, that it was the intention of the AEC to allow the dose given by nuclear power to rise to 170 mrem/year. Then the number of cancers could indeed be considered large - but not much larger than those caused by natural background, or medical x-rays.

In fact, there was no such intention. Strong regulation had already kept dosage levels low - they would have projected to an average of much less than 5 mrem/year by the year 2000 - but ineptness of the AEC and the Federal Radiation Council and public pressure have caused the releases to be reduced (at some expense) by an average of about 5 times[5]. The National Academy of Sciences examined the health effects in detail and in 1973 issued a report[6]; they found a number of 3500 cancer cases/year in the U.S.A. for 170 millirems/year, assuming (conservatively) a linear relation of cancer incidence to dose with no repair mechanism and threshold. They call Gofman and Tamplin's figure of 32,000 an "over-estimate". So also would be Pauling's figure of 100,000[7] which he used in his well-known campaign to stop above-ground weapons tests.

ICRP had recommended that, even if radiations were below the limit, they.be kept "As Low as Practicable". The AEC started a three year public hearing on what this meant, and now the radiation at nuclear power station boundaries is to be kept to 10 mrem/year[8]. These actions of NRC and the BEIR[6] have effectively silenced the opposition on this score.

ACCIDENTS

The next issue that has come up is that of safety. Reactors will not blow up
like bombs, but if the radioactive fission products were somehow to be spread
uniformly over the countryside, large areas could be uninhabitable. This was
first shown by Wigner and Smyth in 1941[9]; the numbers have not changed much
since then.

In order to release all the radioactive material including solids, the fuel
must vaporize; this is not possible in present thermal reactors; even the
gases are held firstly in the ceramic fuel, then in the fuel rods, and then
in the reactor vessel. To release all of these, the fuel must melt, the
reactor vessel must fail and the containment vessel be breached.

We have no direct experience to go on and, since we hope that an accident
such as this will never happen, we hope we never will have. Therefore, the
safety of this type of system cannot be calculated like the safety of an
automobile where 50,000 people were killed in the U.S.A. last year and we can
reliably predict that it will be between 40,000 and 60,000 next year.

It seems generally agreed that the present light water reactor in the U.S.A.
is stable against most failures; over-heating causes the moderator to expand
and shut the reactor down. But after shutdown, the large decay heat causes a
possible problem if there is an inadequate heat removal system or the coolant
has disappeared.

Figure 4 shows a graph of the decay heat as a function of time, which is
roughly the same as a graph of the radioactivity as a function of time. For
a 1000 MWe (3000 MWt) reactor it is still 240 MWt just after shut-down in a
space of about 5 cubic meters.

A possible loss of coolant accident has been extensively studied. I, for

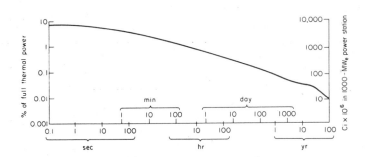

Fig. 4 - Decay power after infinite irradiation.

example, discussed some of the principal physical features at issue[10,11].
At first, attention was concentrated upon a catastropic break of a large pipe,
and whether an emergency core cooling system (ECCS) could get water to cool
the core soon enough after a pipe break. In figure 5, I show a temperature
chart illustrating the problem; the uranium itself melts at about $2700^{\circ}C$, and
radiation to the surroundings varying at T^4 should be able to cool the core.
But at much lower temperatures - $600^{\circ}C$ - the fuel rods become ductile and can
swell and block coolant passages, and at $1100^{\circ}C$ an exothermic zirconium-water
reaction liberates heat to add to the decay heat problem. This reaction oxi-
dizes the zircalloy, makes it brittle and can cause the fuel rods to collapse
and make cooling harder. Another effect is that the steam built up as the
emergency water first hits hot fuel rods might prevent further water from
entering.

There have been no full scale tests of these systems, and after one scale
test gave unexpected results four years ago, public furore, in which physicist
Henry Kendall played a prominent role, forced a re-examination of the issues.

After a two year public hearing on this problem a set of criteria was established by the AEC which were considered very conservative by most of the
objectors[12]. These will not guard against failures of the reactor vessel
itself, but a recent report[13] shows that these failures of the reactor vessel are very unlikely. I note that according to a letter of January, Sir Alan
Cottrell had not yet studied this report and the Oak Ridge work on which it
was based. All of this has been folded into a comprehensive study of reactor

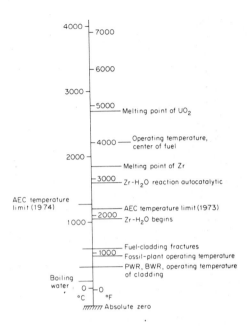

Fig. 5 - Critical temperatures in reactor operation.

safety under the direction of Norman Rasmussen[14]. This was issued as a draft report in August 1974, and various criticisms have been received from NRC, from EPA, from the Sierra Club (Kendall) and the Americal Physical Socie-ty. The valid ones will be incorporated into the final version. The report essentially supersedes all previous work and is a framework for discussion. Any argument now is about the detail of the numbers.

The report follows a fault tree approach, whereby a chain of events is fol-lowed and probabilities evaluated for each event (figures 6, 7).

The main problem with such an approach is the possibility of common mode fai-lure. However, the problem is simplified in several ways by the physics of the problem. We are only interested in fault trees which lead to a reactor core meltdown, since only these will give a radioactivity release. Moreover, these faults occur inside the containment vessel and the weather is outside, so that it is hard to envisage many ways in which a serious accident <u>inside</u> is correlated with the weather. Moreover, the <u>serious</u> weather condition,

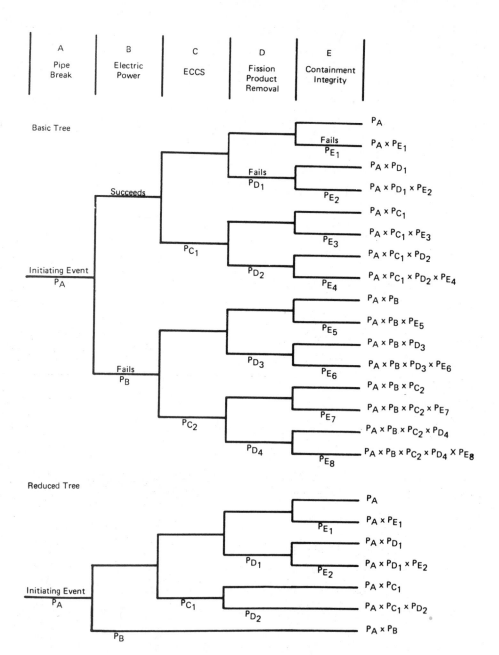

Fig. 6 - Simplified event trees for a large LOCA (accident involving
         loss of cooling).

RELEASE CATEGORY	PROBABILITY (Yr^{-1})	TIME OF RELEASE (Hr)	DURATION OF RELEASE (Hr)	WARNING TIME FOR EVACUATION (Hr)	ELEVATION OF RELEASE (Meters)	FRACTION OF CORE INVENTORY RELEASED[a]							
						Xe-Kr	Org-I	I-Br	Cs-Rb	Te	Ba-Sr	Ru[b]	La[c]
PWR 1	7×10^{-7}	1.5	0.5	1.5	25	0.8	6×10^{-3}	0.6	0.4	0.4	0.05	0.4	3×10^{-3}
PWR 2	5×10^{-6}	2.5	0.5	1.5	0	0.9	7×10^{-3}	0.7	0.5	0.3	0.06	0.02	4×10^{-3}
PWR 3	5×10^{-6}	2.0	1.0	1.5	0	0.8	6×10^{-3}	0.2	0.2	0.3	0.02	0.02	3×10^{-3}
PWR 4	5×10^{-7}	2.5	3.0	1.5	0	0.5	2×10^{-3}	0.09	0.04	0.03	5×10^{-3}	3×10^{-3}	4×10^{-4}
PWR 5	1×10^{-6}	2.5	4.0	1.5	0	0.2	2×10^{-3}	0.03	9×10^{-3}	5×10^{-3}	1×10^{-3}	6×10^{-4}	7×10^{-5}
PWR 6	1×10^{-5}	12.0	10.0	1.5	0	0.2	2×10^{-3}	8×10^{-4}	7×10^{-4}	1×10^{-3}	9×10^{-5}	7×10^{-5}	1×10^{-5}
PWR 7	6×10^{-5}	10.0	10.0	1.5	0	5×10^{-3}	2×10^{-5}	2×10^{-5}	1×10^{-5}	2×10^{-5}	1×10^{-6}	1×10^{-6}	2×10^{-7}
PWR 8	4×10^{-5}	0.5	0.5	N/A	0	2×10^{-3}	5×10^{-6}	1×10^{-4}	5×10^{-4}	1×10^{-6}	1×10^{-8}	0	0
PWR 9	4×10^{-4}	0.5	0.5	N/A	0	3×10^{-6}	7×10^{-9}	1×10^{-7}	6×10^{-7}	1×10^{-9}	1×10^{-11}	0	0
BWR 1	9×10^{-7}	3.0	2.0	2.5	25	1.0	7×10^{-3}	0.50	0.40	0.70	0.05	0.5	5×10^{-3}
BWR 2	2×10^{-6}	3.0	0.5	2.5	0	1.0	7×10^{-3}	0.60	0.30	0.10	0.04	0.07	2×10^{-3}
BWR 3	1×10^{-5}	28.0	5.0	2.5	0	1.0	7×10^{-3}	0.08	0.05	0.20	0.03	0.06	3×10^{-3}
BWR 4	3×10^{-5}	9.0	0.5	2.5	0	1.0	7×10^{-3}	0.10	0.07	0.07	9×10^{-4}	6×10^{-3}	9×10^{-4}
BWR 5	1×10^{-5}	5.0	2.0	2.5	0	0.6	3×10^{-3}	0.05	0.02	0.05	2×10^{-3}	3×10^{-3}	6×10^{-4}
BWR 6	1×10^{-4}	30.0	5.0	N/A	0	4×10^{-4}	3×10^{-8}	6×10^{-12}	4×10^{-11}	8×10^{-14}	8×10^{-16}	0	0

(a)   A discussion of the isotopes used, together with background on the isotope groups and release mechanisms, is given in ref. 14, Appendices VI and VII, respectively.

(b)   Includes Mo, Rh, Tc.

(c)   Includes Nd, Y, Ce, Pr, Pm, Np, Pu, Zr.

Fig. 7 – Summary of accidents involving core.

from the point of view of death rate in an accident is a steady night-time, non-dispersive wind blowing toward the nearest town.

WASH 1400 has produced two surprises from previous rough studies. Firstly, the probability of a partial core meltdown may be larger than previously supposed (1 in 18,000 reactor years) and is largely <u>not</u> due to the catastrophic accident so extensively discussed, but due to a lack of reliability of the residual heat removal system. This reliability can be improved with experience now that it is recognized to be important. Secondly, even a complete core meltdown will not necessarily lead to major public health consequences.

The biggest hazard in accident analysis is Iodine 131 which concentrates in the thyroid. You may remember that in the Windscale accident of 1956 about 20,000 curies of iodine went out of the stack. Milk produced in the next several weeks had to be impounded, but there were no casualties. Moreover, once persons are exposed to iodine, it is possible to remove the radioactive iodine from the system by doses of nonradioactive potassium iodide pills and thereby reduce exposure by a factor of 10.

But an American Physical Society study[15] shows that for long term hazard, the population integrated dose is greatest from Cesium 137 which can spread over many miles. Until this fact was noted, it was thought that the death rate (including cancers) from an accident would be dominated by the large doses obtained by those close to the accident. The dose from Cs^{137} will always be small. For large doses, the health hazard is well known, but for small doses we use the linear theory with no threshold. Whether or not there is a threshold seemed unimportant, but now it begins to have an important role in the analysis; we know the Cs^{137} gives doses at levels lower than natural background; but whether this will give cancers we do not know, and the APS study took the usual pessimistic view.

The effect of a nuclear accident is still quite small - comparable to that of a dam failure and much less probable. This is shown in figure 8 which compares the hazards as calculated in WASH 1400.

When we consider low doses to many people, and the chance of any one person

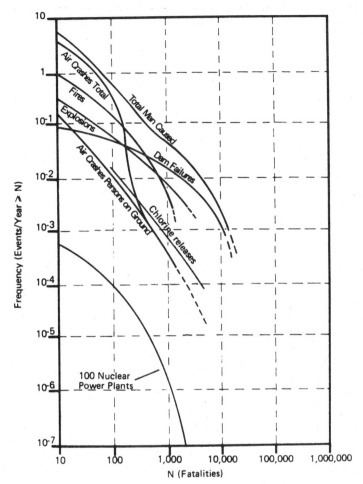

Fig. 8 – Frequency of man-caused events with fatalities greater than N.
Fatalities due to auto accidents are not shown because data
not available. Auto accidents cause about 50,000 fatalities
per year.

getting a cancer is small, it is most appropriate to calculate the total
integrated dose to the population with units in man-rems. In the worst accident (happening once every 10 million reactor years) the integrated dose was
estimated by Rasmussen (as corrected) to be about 40 million man-rems. We
could compare this to the <u>yearly</u> dose from diagnostic x-rays of about 75
millirems for each of 200 million people or 15 million man-rems. Thus, the
total cancer (and <u>genetic</u>) hazard from one nuclear accident is comparable to
the <u>yearly</u> hazard from medical diagnosis.

Another interesting point is that as a result of the recent furore, as much

money is being spent of assessing safety in the U.S., including experiments
to measure critical parameters, as on safety improvement itself. That this
is true is a credit to the original design.

WASTE DISPOSAL

The next issue is that of waste disposal. I showed the curve of radioactivity
against time for the first 100 years; but after 500 years, this curve fails to
drop any more. The transuranic elements - including plutonium - have half
lives in the thousands of years and persist.

The question arises, what do we do with these wastes? We can keep them concentrated, or we can dilute them - a choice we do not have with wastes from
coal burning for example, which spread widely over the countryside. The
intermediate case - partially dilute - is the dangerous one, because we cannot keep them out of drinking water. So we must, if we keep them concentrated,
(the present decision), know how to rapidly dilute them and spread them into
our vast oceans if accidentally the concentrate leaks.

But let us get our orders of magnitude straight. Firstly, simple arithmetic
shows that if the world provided all its energy at the 1970 level from nuclear
fission power, the wastes could be diluted in the oceans and never violate
U.S. public health service drinking water standards. If we keep the wastes
in solid form, it is hard to ingest them. We can calculate, for example, the
amount required to give a 50% chance of death from indigestion. This is one
pound of waste. Clearly, there are many more poisonous substances, such as
arsenic which we spread freely upon our food supplies. Secondly, we show in
figure 9 from Okubo and Rose how the toxicity of the wastes vary with time.
We could make the wastes into a glass and the technology is adequate to do
this. Then wastes can only get into drinking water by leaching,which is a
slow process. It is likely that the glass would be intact for several hundred years. The present plan is to remove 99.5% of the plutonium for use as
a reactor fuel, then after 500 years the waste is no more active than the
ore pitchblend (figure 10); if we remove 99.99% of the plutonium and

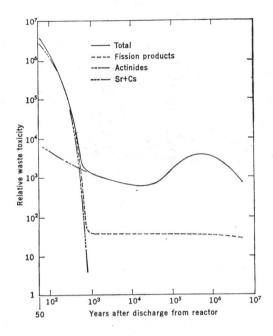

Fig. 9 - Toxicity of wastes from light water reactors, for an equilibrium fuel cycle, with 99.5% removal of uranium and plutonium. Each metric ton of fuel is assumed to deliver a total thermal energy of 33,000 megawatts x days during its operating lifetime. The turn-up at 10^6 years arises from growth of daughter products not present in the original material, which is not in decay equilibrium. (Reference 16a).

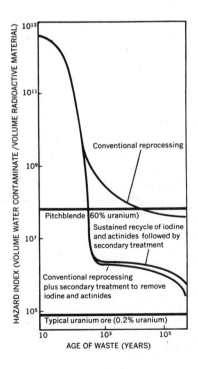

Fig. 10 - Effect of secondary treatment on the hazard index of high-level waste from a light water reactor. Conventionally reprocessed waste remains more hazardous than pitchblende for longer than one million years, whereas wastes treated by secondary processes fall within the naturally occurring range within a few hundred years, whether or not the iodine and actinides are repeatedly recycled to the power reactors. (Reference 16b).

transuranics, (which is not very expensive) the waste becomes no more active
than granite rock and can be handled similarly[16]. The extra cost is small
- about 2% of the cost of electricity. I find it hard to believe we could
not store a solid safely for 500 years nor   dispose of a waste no more
active than granite.

The disposal of wastes in salt mines - probably in New Mexico - seems a good
way. Cohen[17] calculates that even if we forget where they were and accidentally drilled into them (to try to find oil), the hazard would be negli-
gibly small. There is, in fact, too much of a tendency to feel that either
we keep the wastes permanently sequestered or they are too dangerous.

Certainly such procedures seem more satisfying than the present procedures at
Hanford of leaving wastes in liquid form to spill into the desert sands.
Probably, no harm will ever come from these spills, since calculation shows
that spilled liquid would take 1000 years to reach the Columbia river, but
why should we have to make the complex arguments necessary to prove them,
when a simple, not very expensive procedure can do better? England has successfully handled liquids without spills, but I suspect England will also
solidify wastes.

We could keep the fuel rods intact and keep them for hundreds of years. The
space needed is not anywhere near as large as is needed for waste disposal
from fossil plants. From figure 9 we can see we can store wastes for hundreds
of years before the radioactivity equals that in the reactor. Since we can
shield that, cannot we shield the wastes? At the moment we are doing this,
partially by accident. The West Valley fuel reprocessing plant is shut down
for improvement and expansion, the Morris, Illinois plant, built by G.E. with
a new process, does not work, and the Barnwell, Carolina plant is not ready.
Moreover, there is a discussion going on about how to safeguard plutonium
(see infra), plutonium recycle is temporarily banned, and meanwhile it seems
safest to keep the plutonium with the radioactive spent fuel where the probability of theft is remote.

This waste problem has received a lot of public attention; the storage method
at Hanford is indeed stupid, and the dilly-dallying of the AEC and ERDA have
not helped to establish confidence that the government knows what it is doing.

But I think the problem has many acceptable solutions.

URANIUM MINING, AND MINE-TAILINGS

Miners have died for several hundred years in the Joachimstal mines in
Czechoslovakia, and in the initial uranium mining in the U.S.A., nearly 200
uranium miners developed cancer - presumably from inhaling radon gas. In
Table I are shown some statistics on these cancers. It is clear that miners
developed more cancers than the average in the population, but the synergestic
effect between smoking and radiation is particularly interesting. As a practical matter, cancers have been reduced by only hiring miners who do not
smoke, and by much increased ventilation in the mines.

Table I
Respiratory Cancer Deaths in Uranium Miners

	Smokers	Non-smokers
Person-years	26,392	9047
Cancers observed	60	2
Cancers expected on average basis	15.5	0.5

However, we now ask what happens to the radioactive material we have unearthed.
This was the subject of recent congressional hearings[18]. Before the uranium
mining, these materials were in rocks, and did not enter the biosphere and
might have stayed there for millions of years. Now they are out and can
expose people. What are the limits?

Firstly, in Colorado builders were allowed to use these mine tailings to
build houses - and the radon gas could escape into the buildings. In some
19 buildings in Grand Junction, Colorado, the radiation level was 1000 mrem/
year as shown in Table II.

Table II

Radiation from Buildings (18)

	γ radiation (mrem/yr)	Radon-daughter concentration (working levels)
New York City (average)	–	0.002
Grand Central Station	90–550	–
Buildings in Florida phosphate areas	110–500	0.001–0.030
Buildings in Tennessee shale areas	70–190	0.001–0.038
National Capitol	30–180	–
Swedish houses		
Wooden	–	0.005
Brick	–	0.009
Concrete	–	0.018
Grand Junction		
Colorado area	120–140	0.004
2400 residences	($\cong$100)	0.01
19 residences	($\cong$1000)	0.05
Sea-level background	100	–

Fortunately, this use of mine-tailings has stopped, and federal funds have been appropriated for removal of these tailings. But Pohl[19] calculates that these mine tailings will emit radon for the whole life of the Thorium 230 present (80,000 years). Unless they are disposed of, they represent a dose commitment to future generations corresponding to tens of cancers per year per 1000 MWe power plant. On the other hand, Cohen[20] assuming that the uranium will get to the surface anyway in shorter time (compared with the half life of 4×10^9 years) shows that the net effect after 10^6 years of mining over millions of years is to reduce the dose commitment by taking uranium out of the system! All these seem crazy calculations; but they raise some important issues. Firstly, this is only important if low doses at low rates give cancers or genetic effects and we do not know whether that is so. Secondly, although nuclear power brings radioactive material up from the middle of the earth, it is only one of many of man's activities that do this; radioactivity of houses near the phosphate mines in Florida is as great as in Grand Junction, Colorado; geothermal energy is probably much worse in bringing forth devils from the deep. It is known that radon levels near geothermal fields are high. Clearly, if we worry about nuclear power activities, we should worry about these also. Thirdly, the whole question of how far into

the future we must integrate the dose commitment becomes very important. Is
it 10 years? 1000 years? 100,000 years as Pohl suggests? or 100 million as
Cohen takes in order to make a point? These are small matters.

Finally, I think the main problem with nuclear power is how to prevent insane
people - leaders of nations or individuals - from making and exploding bombs.
I can do no better in this respect than adding to my notes here an expanded
version of my comments at the European Nuclear Conference in April of this
year.

THE BREEDER REACTOR

The case for the breeder reactor seems to me to have been inadequately presented in the past. As a result of these inadequate presentations, I think
that there is great confusion in the mind of the public and of politicians
whether or not, and at what pace, to proceed with the breeder. Enrico Fermi
thought the breeder was a good idea, and for many years I suspect his word
satisfied many people. But we must now do better. In these comments I ask
questions; questions which I feel the experts must answer. I feel confident
that the answers will be forthcoming.

The purpose of the breeder reactor is to enable us to use the U^{238} and the
Th^{232} available in the earth's crust as well as the U^{235}. Whether there is
a need for immediate implementation, then, depends upon the supply and cost
of these fuels. If, for example, plentiful uranium were available at low
cost - $10 per lb. of U_3O_8 - the fuel cost would be a small part of the total
cost of a reactor system, and we could, if we wished, postpone the building
of a breeder. On the other hand, if the capital and operating cost of a
breeder are not much more than an ordinary reactor, it might well be worthwhile anyway.

At one time it was thought by many that the Liquid Metal Fast Breeder Reactor
(LMFBR) would be as cheap a reactor to produce electricity as one of the
many thermal reactor types. If this were true, immediate imp. mentation

would seem sensible, independently of the "need" as stated by a diminishing uranium supply. However, few people now have this hope, and an increased capital cost must be balanced against increased cost of uranium.

If we have a definite amount of uranium in the world, say 2 million tons, available at a cost of $10 per lb., and a definite demand curve, it is easy to see when we need a breeder and what its doubling time must be. At the European Nuclear Conference in Paris, Vaughan and Iliffe both went through such calculations[21]. If we assume a doubling time of 20 years (inverse doubling time of 0.05 yr^{-1}), we can expand supply at roughly 5% per year. They believe we need breeder reactors before the end of the century. The AEC in an environmental impact statement on the LMFBR went further than Vaughan and Iliffe in order to justify the large expenditure on Research and Development for the fast breeder in terms of the deferred benefits[22]. The draft statement was poorly done and invited the criticism it received; papers by Cochran[23], Cochran, Speth, and Tamplin for NRDC[24], Bupp and Dearian[25], and Holdren[26] have all questioned the assumptions and methodology. However, these papers themselves contain errors, and the assumptions are equally uncertain.

A paper by Stauffer, Palmer, and Wycoff[27,28] is superior to all of these and has the same aims. They calculate the present discounted monetary benefits of a breeder reactor, with various assumptions about cost, uranium supply, and growth rate. Their main contributions seem to be a correct evaluation of the discounted benefits by a new analysis of alternative uses of the money in the economy (which I assume is correct, though I have not yet verified it), a correct realization of the role taxes have in mixed public/private sectors (they are transfer payments, and if the calculation is consistently done, should not appear directly). They realize, as some others have not, the benefit to the whole energy industry of a breeder reactor economy (by keeping fuel costs down) and not merely to the one power station.

Their arguments seem impeccable, but there is still room for uncertainty and argument about the uranium supply and electricity demand forecast. They calculate for the U.S. only, and assume that there are 2.4 million tons available at $10-65/lb., and that there is little more below $150/lb. They assume the LMFBR is only 25% more expensive than a thermal reactor. Moreover, they

assume a growth rate of electricity demand of 5% per year. They calculate a
discounted present net benefit of $76 billion.

In view of the uncertainties they argue, correctly, in my view, that the
LMFBR should be considered as an insurance policy against the possibility
that their assumptions are correct: that no more low cost uranium is found,
and that the growth rate of demand remains high. Here, unfortunately, they
stop, although they have come further than anyone before them.

The $76 billion might be regarded as the benefit of the insurance policy, if
that contingency comes to pass. What premium on this policy should we be
willing to pay for this? It depends on the probability of the event coming
to pass; this is where opinions can legitimately vary. If the probability is
only 1/10, then we should only pay $7.6 billion for our insurance policy.

URANIUM SUPPLY

There are several problems with assessing the crucial role of the uranium
supply. Carbon is plentiful on the earth, and we have been searching for coal
and oil for many years. Nonetheless, it is only since recently that we have
had a methodology which has even moderate acceptance. The old method of plotting oil discovery rate versus time leads to increasing discovery of oil, but
only a moment's reflection is needed to realize that time is the wrong independent variable. It is better to follow Hendricks and to plot discovery
rate versus footage of oilwell drilled. Even here there has been room for
argument. At one time it was (effectively) assumed (by Zapp) that drilling
was random and new oil would be discovered at the same rate as before. But
M. King Hubbert, starting in 1959[29] showed that the discovery rate, plotted
versus footage of well drilled, fitted an exponential better than the random
curve. Only recently has this been generally accepted, and even in 1975 there
are dissenters who optimistically think there is more oil to be found.

The same methodology may not work so well for uranium, although there are
indications that if we try it, we have already found most of the cheap uranium.

Uranium is not one of the world's common elements, and the exploration has not been continuous, but rather a stop and go process, as firstly the military demand was urgent, then it saturated, and then the civilian demand was prematurely predicted. Prospectors have been burned twice, and they may be again. For example, Dr. Bowie, Assistant Director of the Institute of Geological Sciences, London,[30] stated his belief at the European Nuclear Conference that there could be ten times as much cheap uranium as presently known - waiting to be found. If this is true - and many people think it likely - then the need for a breeder could be postponed. Would it be a catastrophe if we did not build a breeder and found no more low cost uranium? Uranium at \$150/lb. adds only $\frac{1}{2}$ cent/Kwh to our electricity bills and mine have risen more than this in the last year. But again, we must be cautious; can it be found in time, or do we have to develop new discovery techniques? The extra uranium is probably deep below the surface, and no one knows how to find it.

ELECTRICITY GROWTH RATE

Likewise the electricity growth rate is open to question. The population growth is the U.S. is now about 2% per annum and there are many people who hope that by the end of the century this can be slowed down. Perhaps at the same time the increase with increasing standard of living will stop.

There is likely to be a major transfer from other energy sources to electricity, but this should be nearly complete by the end of the century so that a 4% electricity growth rate until the year 2000, levelling to zero after the end of the century, seems plausible. This would make the Stauffer et al present benefit drop to zero. I note that this is less than the 9% electricity growth rate of the 60's which was due, in part, to cheap energy. The important point to realize is that the breeder reactor is planned to penetrate the economy heavily only in the next century and the projections for the next century are the important ones. There could be a distinct reduction in the present advantage as calculated by Stauffer et al. I do not believe a low growth rate myself, but this seems the biggest single point at issue.

TECHNOLOGICAL PROBLEMS

The U.S. Breeder programme has left most people in a state of confusion. There
was a fine early start on liquid sodium cooled reactors. EBRI produced the
first electricity generated by nuclear power. Yet the sodium cooled reactor
in the submarine Seawolf was replaced by a light water reactor. The first
utility operated breeder, Fermi I at Laguna Beach, Michigan, has ceased to
operate. Why? We need to be told where these reactors lie on the overall
plan towards a commercially viable fast breeder reactor. What did we learn
from Fermi I? Why will the next reactors be better? [31]

The fast fuel test facility planned by Argonne National Laboratory some years
ago has been replaced by the Fast Fuel Test Facility (FFTF) in Hanford. This
will not go critical until the year 1978, will not produce electricity, and
costs too much. The Clinch River Breeder Reactor (CRBR) or Demo is a multi-
headed hydra; it is, or was, being designed by TVA, Commonwealth Edison,
Argonne National Laboratory, Westinghouse, and ERDA staff in Washington. A
recent review committee of the AEC reported that this management procedure
will not work. Perhaps we are stuck with the CRBR management, but let us get
the next one better.

What of international cooperation? There is far too little – probably because
of commercial secrecy. Too few American designers understand the French
PHENIX and the British PFR. Could the U.S. buy a Super PHENIX from France
and scrap CRBR? This might be a rapid approach to the new technology. The
French plan to buy PWR's from the U.S. Why can there not be a reciprocal
plan?

But there are uncertainties here also. PHENIX works and is a credit to the
French who are understandably pleased with themselves. But does it breed?
What changes are necessary to make it breed and can these easily be imple-
mented? Certainly it is a method to accustom us to sodium technology.

So long as all these uncertainties exist, and the past is not discussed as a
guide to the future, there will always remain the lingering feeling that the
LMFBR is proposed merely by those already committed to the technology who are

trying to keep themselves in a job.

Perhaps we need a reduced LMFBR objective. We should aim at a <u>lower</u> temperature (which should not affect breeding gain, but should improve safety margins
and reduce metallurgy problems at a slight cost in electrical output);we could
test fuel while proceeding as the French and British do.

The LMFBR seems appreciably different in design and materials from Light
Water Reactors (LWR's). The technology seems an appreciable leap and may cost
us in capital cost or reliability. For example, a heat exchanger in an LWR
can leak with no appreciable problems. If a sodium/water heat exchanger leaks
- and they have in the past - the plant shuts down. The public needs to know
that this is in order.

ALTERNATIVE BREEDERS AND CONVERTERS

Alternatives may exist, and the Gas Cooled Fast Breeder is an important consideration. The HTGR is just coming into use in the U.S. - although it may
be 20% more expensive than a PWR or BWR. The GCBR seems only a small modification in comparison with the LMFBR and may cost 10% more than the HTGR. If
R. D. Vaughan of the Gas Cooled Breeder Reactor Association is to be believed,
it promises a higher breeder gain. Why are all countries working individually
on the LMFBR? Why is there not international agreement so that one can work
on the GCBR and all share in the results? I hate to put all the world's eggs
in one basket. But again, answers are needed. What is the heat sink in case
of accident for the GCBR?

Although studies (such as Stauffer <u>et al</u>) have suggested that converter
reactors - an optimized HTGR or CANDU on a thorium cycle - only delay the
need for a breeder by a year or two, the reason for this does not come across
clearly.

Claims have been made, for example, by Canadian A. J. Mooradian[32] in a
lecture at Harvard, that a CANDU operating on a thorium cycle need only have

one fuel load of Pu^{239} or two of U^{235} and can then run on thorium (which is plentiful) indefinitely. But this probably needs short irradiation times and rapid fuel processing, so it may not be practicable. Let me assume that we use these exclusively from now on. Then uranium^{235} is needed only for the first fuel load. Can we not go on indefinitely this way? We can afford $200 per pound for uranium if it is only to pay for the expansion in electricity supply. For a 5% per year expansion, the average uranium cost for power becomes only 0.1 per kwh (although the other parts of the fuel cost must still be considered). What is wrong with this? Maybe it is not the best (most profitable) way to proceed, but is it not possible? Clearly there may be a hiatus on the maximum possible rate of uranium mining in the interim. (This would increase price but not cost, and thereby keep demand down).

Another point of confusion is the push of the industry toward plutonium re-cycle. Yet Iliffe[21], at the European Nuclear Conference, tells us that much of the advantage of the introduction of the LMFBR can be obtained by stockpiling plutonium until it is needed for a breeder. Stockpiling seems to keep our options open. But let us then stockpile plutonium as spent fuel rods - with radioactive elements included - to prevent theft. But a good economic study - looking ahead to the next century - seems indicated, rather than the simple engineering study of Iliffe.

PUBLIC OPPOSITION

All of the above would be less significant if there were not public opposition to the breeder on other grounds. I will assume, for the moment, that the thermal reactors have been - grudgingly - accepted. What new features are introduced by the fast breeder?

Firstly, it represents a continuation - and expansion - of the commitment to nuclear fission power.

Secondly, there may be increased problems with safety. Although there is no pressure vessel and piping to fail, and to this extent the LMFBR may be safer,

the higher power density and the shorter response time means that the fuel
might vaporize, whereas in a thermal reactor the worst that can happen is that
it melts. In an unusually well written report[33], R. Farmer in 1969 said
the LMFBR may be the safest of all reactor types – presumably for the first
of the above reasons. Recently, he has said it may be the most dangerous –
presumably for the second. These are not inconsistent, but they are confusing.
Maybe we should compare with hydroelectric dams. Dams are safer than coal
power stations – for there are no miners or air pollution. Yet they are more
dangerous because a single failure can wipe out 3000 people. Nonetheless, I
think that the safety issues will be resolved to general satisfaction, since
we can compare the potential accident hazards with hazards from other power
stations – dam failures, LNG tank failures, and the like.

The most important point about the public acceptance of the breeder programme
is the potential for nuclear thefts and arms proliferation. The most important feature of our nuclear age is how to prevent crazy people from making
and exploding bombs. Although this is a problem common to all reactors, most
people think it is worse for a breeder. It is probably a little worse, but
with care, we can reduce the difference. One way of doing this is by keeping
the plutonium radioactive, by keeping it with enough of its fission products
that anyone stealing the material without remote handling would die. This
would make it secure against many terrorist attacks, although the question
arises as to how soon the thief would die, since many terrorists would be
fanatic enough to die if they can complete their mission.

For Canadian reactors, the fuel is not now being reprocessed and this might
not be a bad way of dealing with radioactive wastes. Fuel from the U.S. light
water reactors is not being reprocessed either – though this is by accident,
and not by intention. If we reprocess the fuel and extract the plutonium,
there is an interval when there is pure plutonium, which must be carefully
guarded to prevent theft. This interval must clearly be kept short as possible. In EBR II fuel has been processed and refabricated, still containing
radioactive materials. It seems that this is an important design criterion
for all fast reactors.

The other important objection to the breeder in the public mind is proliferation of nuclear weapons among nations. Fast reactors are more like bombs than

thermal reactors. Computer codes for breeder reactor safety tend to be similar to computer codes for bomb design. The training of personnel for one is
similar to the training for the other. By spreading among nations the expertise of nuclear power, we spread their ability to make bombs. This is accen-
tuated by the fast breeder. This was pointed out in testimony to the U.S.
Senate Foreign Relations Committee in 1974 by Kendall, Ford, Rathjens and
Kistiakowsky[34]. It will be remembered that Kistiakowsky was science advisor
to President Eisenhower and had worked on the design of the first atomic
bombs.

One inspection system for thermal reactors is to insist that the fuel be provided and reprocessed in IAEA approved facilities - probably in the more
developed countries. But the fast fuel turn-about demanded by a fast breeder
programme almost demands that the chemical reprocessing plant be close to the
reactor - or at least in the same country. Thus safeguards against proliferation are reduced. Confidence can be established by a safeguards procedure
before extensive deployment of the fast breeder.

One way of handling this problem might be for industry to make a sincere show
of separating civilian nuclear fission power from military activities. The
recent reorganization of the U.S. AEC into the Nuclear Regulatory Commission
and the Energy Research and Development Administration leaves the promotion
of civilian nuclear power and military uses _together_ in ERDA. This leaves
nuclear power suspect. I hasten to add, of course, that bomb making must
remain under civilian control. This was a major aim of the scientists who
helped set up the original AEC. It would be a tragedy if the main effect of
the public opposition to nuclear power were to leave bomb making entirely to
the military, out of sight and mind of the public by a wall of secrecy.

Further suspicion arises when people plan to explode bombs to liberate
natural gas (project Gasbuggy, etc.) or to make heat for power (project Pacer).
If the problem with acceptance of fission power is to stop people from
exploding bombs, to deliverately explode them will not help. Nor is it clear
that the economic aspect of these proposals is anywhere close to what the
proponents claim. I believe, therefore, that the public image of the industry
would be enhanced by a repudiation of these projects.

This still does not tackle the problem of proliferation among nations. The
non-proliferation treaty is not ideal, and its worst feature is that some
important nations have not signed it. A new one should be negotiated quickly,
which France and China could sign. I think it could then have real teeth.
Signing it could be a condition for entry into the peaceful nuclear power
club; under my scheme no signatory country would sell or buy reactors or
nuclear materials from a non-signatory country. I believe this would be a
major step in public confidence.

It is clear that in addition to the nations which have exploded bombs, U.S.A.,
U.S.S.R., U.K., France, China and India, other countries must be considered
nuclear powers. In fact any country which has an isotope separation facility
or chemical processing facility can probably make a bomb within a few months.
This probably now includes W. Germany, Canada, S. Africa, Australia and
Israel. But we should try to prevent the number spreading. We could have
regional facilities under international control instead of individual country
facilities. Then any non-nuclear country would be able to use nuclear power,
but would take perhaps 2 years to become a nuclear power by building an isotope separation or chemical processing plant. This delay seems an important
safeguard.

I am shocked that W. Germany, in annoyance at being beaten in a strict commercial bid for reactor sales by the U.S., has "sweetened" a contract with
Brazil by adding an isotope separation facility and a chemical processing
plant, in which field U.S. and U.K. industry are forbidden by government policy to compete in order to prevent possible proliferation of nuclear weapons.

I am even more shocked that there has been no international outcry by the U.K.
and U.S. against what Tom Lehrer calls "our loyal ally". This could be the
opening of the dike of proliferation and the beginning of the end.

CONCLUSIONS

All of the options now open to us:(1 - increased uranium exploration as urged by Bowie, 2 - LMFBR, 3 - GCFBR, 4 - Plutonium stock-piling, 5 - converter introduction, and 6 - fusion) are expensive, need capital and have a penalty not merely for failure but also for success of one of the other options. Who would want \$30/lb. uranium if the LMFBR were an overwhelming success? Who wants any of them if fusion were as cheap and environmentally more acceptable? All must, in my view, go ahead, but these activities (e.g. uranium mining) not regulated or engaged in by governments may need national - and in some cases international - guarantee of the downside risk. Of course such a guarantee is only publicly acceptable if there is a limit to the upside gain also.

This is a large programme and one country alone cannot do it all. It is a case for the international cooperation we all need.

REFERENCES

(1) E. G., E. J. Sternglass, "Has Nuclear Testing Caused Infant Deaths?" New Scientist, July 24 (1969).

(2) J. Rotblat, Nature (1970).

(3) A. P. Hull and Shore, Nuclear News, 15 53 (1972).

(4) Environmental Effects of Producing Electric Power, Hearing before the Joint Committee on Atomic Energy, 1969-70, Part 1, Part 2, Vol. I; Part 2, Vol. II.

(5) Final report of Rulemaking Hearing Docket No. RM-50-2, Opinion of the Commissioner, Nuclear Regulatory Commission, Numerical Guides for design objectives and limiting conditions for operation to meet the criterion "as low as practicable" for radioactive material in light-water-cooled nuclear power reactor effluents, May 1975.

(6) "The Effects on Population of Exposure to Low Levels of Ionizing Radiation", National Academy of Sciences, National Research Council, Washington DC (1972).

(7) L. Pauling, Bull. Atom. Sci., September (1970).

(8) Opinion on Rule-making Hearing, RM-50-2, "As Low as Practicable", Nuclear Regulatory Commission, May 1975.

(9) E. Wigner and H. Smyth, secret report quoted in "Atomic Energy for Military Purposes", p.65. Reprinted by Princeton University Press, Princeton, NJ (1948).

(10) Richard Wilson, "The AEC and the Loss of Coolant Accident", Nature, Feb. 9 (1973), (PWR details).

(11) Richard Wilson and W. Jones, Energy, Ecology and the Environment, Chapter X, Academic Press (New York), 1974, (BWR details).

(12) Hearing of Joint Committee on Atomic Energy, U. S. Congress, January
      1974, see especially testimony of Lawson, Ybbarondo, and Rittenhouse.

(13) Report of Advisory Committee on Reactor Safeguards on Reactor Safeguards
      on Reactor Vessel Failures, January 1974.

(14) WASH 1400 (draft), U.S. Atomic Energy Commission.

(15) Study of Reactor Safety by Harold Lewis, et al., to be published in
      Review of Modern Physics, 47 Supplement 1, 1975.

(16) D. J. Rose and A. Okubo, Science 1973, J. O. Blomeke, J. P. Nichols and
      W. C. McClain, Physics Today, 26 36 (1973).

(17) B. Cohen, testimony before Joint Committee on Atomic Energy, June 1975.

(18) Use of Uranium Mine Tailings for Construction Purposes, hearings before
      the Subcommittee on Raw Materials, Joint Committee on Atomic Energy,
      U.S. Congress, October 1971. See also, Environmental Analysis of the
      Uranium Fuel Cycle, U.S. EPA-52019-73-003-B,C and D.

(19) Pohl, Cornell University preprint.

(20) B. Cohen, University of Pittsburgh, private communication.

(21) C. E. Iliffe, R. D. Vaughan, "The Sodium Cooled Fast Breeder in Relation
      to Uranium Requirements" presented to the European Nuclear Conference.

(22) Liquid Metal Fast Breeder Reactor Programme, Environmental Statement,
      U.S. AEC (1974).

(23) The Liquid Metal Fast Breeder Reactor, (book), T. Cochran.

(24) T. Cochran, Speth and A. Tamplin, Comments on AEC's Draft Environmental
      Statement, included in reference 22.

(25) I. C. Bupp and J. P. Dearian, Nuclear News (1974).

(26) J. Holdren, Uranium Availability and the Breeder Decision, EQL Cal Tech
      No. 8, 1974.

(27) The LMFBR, Assessment of Economic Incentives, T. R. Stauffer,
      H. L. Wycoff and R. S. Palmer (1975). See also T. R. Stauffer's testi-
      mony to U.S. Congress, May/June 1975 (Humphrey Committee, Udall Commi-
      ttee, JCAE).

(28) See A. Manne and O. Yu, Nuclear News (1974).

(29) e.g. M. K. Hubbert in Resources and Man, Freeman, San Francisco (1969).

(30) Dr. Bowie, Panel on Uranium Resources, European Nuclear Conference,
      April 1975.

(31) A paper by E. L. Alexanderson, "Contributions of the Fermi Project to
      Fast Breeder Reactor Technology", British Nuclear Energy Society Con-
      ference on Fast Reactor Power Stations. (Page 13 is incomplete in this).

(32) Lecture to Economics 2590, Harvard University, A. J. Mooradian (Director,
      Chalk River Laboratories AECL).

(33) R. Farmer, An Assessment of LMFBR Safety, UKAEA, Risley, 1969.

(34) Testimony on possible sale of reactors to Egypt; U.S. Foreign Relations
      Committee, August 1974.

DISCUSSION

Question: Is it true that 'smoking' uranium workers had 10 times more cancer
            deaths than non-smokers?
Answer: Yes, in Table 1, I showed that there was a 4-fold increase in the
            natural rate for smokers because tobacco destroys the cell's
            natural repair mechanism.

Question: What happens in a core melt-down such as that suffered by EBR1?
Answer: The core just melted and stayed in the 'pot' - a bigger reactor
            might have been more violent. A breeder is unlikely to explode
            because of the strong negative reactivity due to Doppler
            broadening.

Question: How long must one store waste before glassification?
Answer: At least 3 months in cooling ponds on the reactor site.

Question: In the fault-tree analysis of WASH 1400, can you be sure that the
            probabilities are independent?
Answer: Some are and this was taken into account.

Question: How effective is fault-tree analysis in other situations?
Answer: I don't know although I have been told it has been successful.

(This discussion was curtailed by shortage of time, but further points were
 raised in the panel discussion "Prospects of Nuclear Power").

FUSION POWER

K. V. Roberts

UKAEA, Culham

Abstract

The three potential major new energy resources after the turn of the century, when the world price of uranium used in thermal fission reactors is expected to escalate, are the fast breeder reactor, controlled thermonuclear fusion, and solar power. Since historically it has taken some 60 years to bring in a new energy source, and since none of the three has been fully demonstrated as yet, it is important that work on all three lines should be actively pursued together - we do not have very much time. A large international nuclear energy industry is expected to develop during the next few decades, based on a limited number of multinational companies, and it is suggested that in order to avoid balance of payments problems due to excessive imports many countries will find it necessary to build up their own nuclear technology to the point at which such companies are attracted to set up local plant. The world is likely to grow more technological, not less.

Fusion has reached a more advanced stage than solar power although less so than the fast breeder which is already in operation, and a significant advantage over the fast breeder is that it could in due course remove some of the hazards that such a world-wide fission industry might entail, namely plutonium proliferation, radioactive waste fission products, and the possibility of an explosive energy release in a reactor accident. However, the scientific and technological problems of fusion are severe and feasibility has not yet been demonstrated. One of the characteristic features of a fusion reactor compared to most other engineering devices is that for physical reasons it must be intrinsically large, even at the prototype stage - this means that feasibility can only be demonstrated as part of a carefully-planned national or international project. Several alternative magnetic confinement geometries are being investigated at the present time in addition to laser fusion.

The views expressed in this article are those of the author.

1. ENERGY STRATEGY

At a time when the world is so obviously and increasingly short of power it
is necessary that all potential sources of energy should be actively explored.
This is particularly true because the lead time needed to bring a major new
energy source into full commercial use is very long indeed, usually about
60 years as Figure 1 illustrates. We clearly cannot afford to wait until one

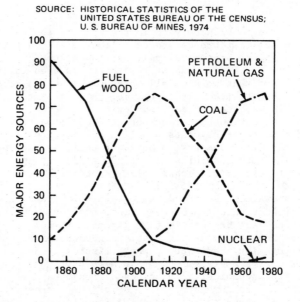

SOURCE: HISTORICAL STATISTICS OF THE
UNITED STATES BUREAU OF THE CENSUS;
U. S. BUREAU OF MINES, 1974

Figure 1. U.S. Energy Consumption Patterns [1]

form of energy runs out before investigating the next, and several options
need to be brought to the stage of commercial feasibility in parallel with
one another in order to provide adequate insurance. This point has been well
brought out in a recent U.S. ERDA report [1] :

  'Finally, two additional major resources exist, both of which can
  represent essentially inexhaustible sources of energy if the
  technology to use them can be perfected. These are solar energy
  and fusion energy. In both cases the potential is substantial,
  although significant problems remain to be solved in their
  development. These sources, together with nuclear breeding,
  represent the major candidates for meeting energy needs of the
  future. Even if these technologies should prove successful in

satisfying the technical, economic, institutional and environmental requirements for implementation, their major energy supply contributions will occur in the twenty-first century.

There are other potential sources of energy production, many of them limited in one way or another. A full list of sources is shown in Table 1.

Coal	Oil
Crops	Oil Shale and Tar Sands
Fertile Nuclear	Sunlight
Fissile Nuclear	Tides
Geothermal	Waste Heate
Hydroelectric	Waste Materials
Natural Gas	Water (fusion & hydrogen)
Ocean Heat	Windpower

TABLE 1. Fuels and Energy Sources

This transition to new energy sources must be made more swiftly than ever before.

The historical perspective of Figure 1 shows that in the past it has taken some sixty years from the point at which a transition to a new energy resource was first discernible until that resource, in turn, reached its peak use and began to decline relative to other sources. Domestic supplies of oil and gas appear to have reached that sixty-year peak. Their relative shares in the U.S. energy market are expected to decrease with time. It is essential, therefore, to plan now for the transition from oil and gas to new sources to supply the next energy cycle. The Nation cannot afford to wait another 60 years to complete the next transition. Only an aggressive program of technological development can expedite this process. It is urgent to begin now.

To accomplish this transition a framework of national energy technology goals has been established.'

In fact the energy difficulties that now face us are more concerned with economic problems such as the balance of payments, and with sociological, environmental, political and planning questions, than with any real immediate shortage of fossil fuel, since world coal reserves have been estimated as high as 7.5×10^{12} tons compared with a total usage of all forms of energy (in 1968) of only 6×10^{9} tons of coal equivalent/annum. Even at the greatly enhanced rate of energy consumption that some people have predicted for the end of the century and beyond, there would still be enough coal to last for more than 200 years if it could just be extracted from the ground and rationed out to those who needed it. Figure 2 shows some U.S. estimates of recoverable domestic energy resources. Oil will disappear much more rapidly than coal,

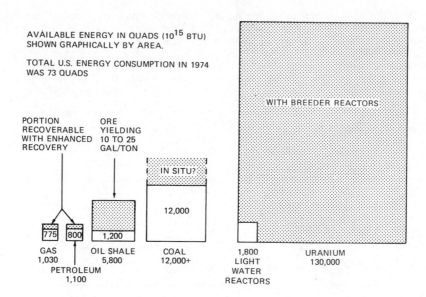

AVAILABLE ENERGY IN QUADS (10^{15} BTU)
SHOWN GRAPHICALLY BY AREA.

TOTAL U.S. ENERGY CONSUMPTION IN 1974
WAS 73 QUADS

PORTION RECOVERABLE WITH ENHANCED RECOVERY

ORE YIELDING 10 TO 25 GAL/TON

WITH BREEDER REACTORS

IN SITU?

12,000

775 800 1,200

GAS 1,030

OIL SHALE 5,800

COAL 12,000+

1,800 LIGHT WATER REACTORS

URANIUM 130,000

PETROLEUM 1,100

Figure 2.   Available Energy from U.S. Recoverable
            Domestic Energy Resources [1]

but suitable synthetic fuel for road, marine and air transport can be made
from coal at a cost which compares favourably with current crude oil prices.

One of the difficulties in simply following the fossil fuel option, however,
is that the locations where oil and coal are found in quantity do not happen
to coincide with the great concentrations of world population. The economic
system does not allow for world resources such as fuel, food and raw materials merely to be transferred from the places where they occur in Nature to
the places where they are required for use without some corresponding transfer of resources of equivalent value back to the originating country - in
other words they must be paid for in real money. Energy consumption is
already so great that the world balance of payments is seriously distorted
unless most countries can, in effect, be responsible for providing their
energy supply from their own indigenous resources. Two further disadvantages
of coal are that existing methods of mining are dangerous and unpleasant, and
that the production and burning of fossil fuels on an ever-increasing scale
may seriously damage the environment.

Nuclear fission will go a long way towards solving the economic problems if
the fast breeder reactor can be successfully introduced, mainly because the
fuel is much more concentrated - 1 ton of uranium burnt in a fast breeder
reactor is equivalent to more than 2 million tons of coal. Balance of payments difficulties associated with the supply of raw fuel should therefore
become unimportant in due course. Although the capital costs of nuclear
power stations, fuel processing plant and other facilities are relatively
high, these can in principle be met within the local economy once a domestic
nuclear industry has been established. There is enough uranium and thorium
in the world to supply the planned breeder reactors for many millennia as

Figure 2 indicates (although the energy reserves in U^{235} alone are only comparable with those of gas and oil). Small mobile units such as cars, trains, most aircraft and many ships cannot carry their own reactors because of the capital cost and the weight and bulk of the shielding that would be required, and especially for reasons of safety in case of accidents such as ordinary crashes and collisions, but these can be powered by synthetic chemical fuel which is simply built up from its elements using nuclear energy - for example carbon obtained from limestone and hydrogen from water.

The recent rise in oil prices means that fission power has suddenly become particularly attractive to those regions of the world where the local supplies of fossil fuel are inadequate to meet the energy needs of the population, such as Europe, Japan and many of the developing countries. A large and widespread programme of nuclear power station construction is likely to be embarked upon in order to reduce the current balance of payments problems. The practical scale of this programme may be seen by noting that <u>with nuclear power, fuel costs tend to be replaced by capital costs. Therefore the scale of annual capital expenditure on nuclear power stations is likely to correspond to the present annual expenditure on oil and coal supplies, multiplied by a further factor to allow for the increase in world energy consumption</u>. Enormous multinational nuclear energy companies can therefore be expected to develop on a scale larger than the existing oil companies. The complexities of nuclear power stations will make it advisable for these companies to concentrate on a limited number of standard designs built from mass-produced modular components, as has already happened in the computer industry. Because time is short and manpower limited it will become essential to establish a safety case for a reactor <u>type</u> (as with aircraft) rather than for each individual installation separately. A fully-developed nuclear economy will require several new reactors to be brought into operation and an equal number of old reactors decommissioned every day.

Experience with the computer industry suggests that these multinational companies will be mainly U.S.-based and that it will eventually be just as difficult for other countries to compete in the production of complete reactor systems as it is for them to compete with IBM in the production of computers at the present time. However, this does not mean that other countries can afford to 'go slow' on energy R & D and expect to be able to buy the power stations or even the technological know-how to make them once the groundwork has been carried out in the U.S. To avoid a new series of balance-of-payments difficulties due to fuel-processing costs and the import of expensive capital equipment and trained manpower, <u>each country that relies extensively on nuclear power will find it necessary to develop its own industrial potential and scientific and technological expertise to the point at which it is able to take an appropriate share in this multinational activity</u>. This argument applies to the UK, the EEC as a whole, to Japan, to the developing countries and also to the OPEC countries once their present supplies of oil are no longer available, and it justifies an individual country embarking on a nuclear energy programme provided that this is carefully planned to take advantage of the trading situation that is expected to develop, e.g. by the production of specialized components for international types of reactor, the undertaking of civil engineering contracts and so on. In this sense the present popular trend away from technology may be seen as a short-term aberration : all three of the energy resources with greatest future potential (namely solar energy, fusion and nuclear breeding) involve low fuel costs, high capital costs and complex technology, and it is difficult to see how a country which does not maintain its technological expertise at a high level

will be able to pay for its energy supplies. Of course this does not mean
that it can afford to neglect 'low technology' products which are also
economically important.

Fission power is already well established technologically, and so far it has
proved in practice to be much cleaner and safer than virtually every other
form of industrial process known. It may be several years before a satisfactory commercial fast breeder reactor is available, but there seems little
doubt that this goal can be achieved in due course even though the plutonium
doubling time (which at present is measured in decades) gives some cause for
concern. There are many important environmental and safety questions to be
resolved but these are virtually certain to have adequate <u>technological</u> and
<u>economic</u> solutions and the real difficulties are much more likely to be
political and practical. Before the world commits itself to relying on
fission for an indefinite period, detailed and careful thought should clearly
be given to the dangers that might arise and how they can best be avoided.

A full-scale fission economy will require thousands of nuclear power stations
to be distributed throughout the world together with fuel processing plants,
radioactive waste storage and large stocks of plutonium. Economies of scale
will ensure that individual aspects of this overall activity tend to become
localized in a limited number of places, and this in turn will mean that
plutonium, enriched uranium, spent fuel rods, waste products etc. are continually being transported from point to point and from country to country.
Some of the equipment and material is statistically certain from time to time
to find itself in areas where minor conventional wars, revolutions and various forms of terrorist activity are taking place and it could be accidentally
or deliberately damaged as a result, with the possibility of a consequent
release of radioactive material. The dangers of illicit use of plutonium for
military or terrorist purposes have by now been well publicized and some
international safeguards have been formulated although these are probably not
yet adequate. The expected rapid expansion of the world fission economy will
presumably mean that some of the staff are insufficiently trained and might
therefore make mistakes. Some of the ships carrying radioactive material or
which are themselves powered by nuclear reactors (now expected to be economically attractive compared to oil-burning vessels [2]) must surely be involved
in ordinary accidents or collisions at sea just as some oil tankers are, and
precautions must be taken against any consequent release of radioactivity
into the environment.

All the nuclear hazards would disappear if solar energy were already available, and many of them would be removed or reduced with controlled thermonuclear fusion. Unfortunately they are not with us yet, and we do not in
fact know whether either is economically feasible at all. Even the scientific
feasibility of a self-sustaining controlled fusion reaction has not been
established so far, although this is expected with some confidence to occur
before 1985 and may happen sooner if the promising new technique of laser
fusion [3] is found to work. Once the goal of scientific feasibility has
been reached the necessary technology has still to be developed, and in view
of the 60-year time-scale mentioned earlier it is likely to be several
decades before fusion power stations can be in operation on a substantial
scale unless a crash programme of extraordinary magnitude can be mounted.
Even so, on current estimates this stage would not be reached before the year
2010 at the earliest. Fission is expected to be in large-scale operation
well before this.

What then should be the world energy strategy? In an ideal world we might try to take advantage of our existing oil and coal resources to delay the introduction of a full-scale nuclear economy until it became quite clear that this could be done safely. Economic forces seem certain to make this course impracticable, and in any case it might well be undesirable on environmental and sociological grounds to burn much more coal than at present. Since it therefore seems inevitable that by the end of the century more than half our electrical power will have to be generated from thermal fission, which is the only major new energy source that could be effective on such a timescale, the international fission reactor industry needs to be built up rapidly to the point at which it is able to take on the job and the time taken to design, approve, construct and commission the power stations should be reduced by mass production and other techniques. Continued public press-ure for adequate safeguards is essential, but unreasonable opposition to fission energy in the absence of any constructive and quantitatively valid alternative is likely to be counterproductive, since the hardship to develop-ing countries, universal loss of morale and sources of potential conflict caused by a world shortage of energy could well prove worse than the envis-aged dangers of the fission reactors themselves.

It is suggested that three other courses should be pursued in parallel with the build-up of fission energy :

> Firstly, steps should be taken by international agreement to analyse the potential hazards of nuclear fission more realis-tically than at present and to reduce them as much as possible. For example, the power stations and other nuclear plant might be 'hardened' against damage by conventional explosives by burying them underground, and such a precaution should also help to reduce the dangers of radioactive waste dissemination from any reactor accidents that may occur. More effective pro-tection should also be sought for fissile and waste material in storage and transit.

> Secondly, because the stakes are so high it would be prudent as a double insurance policy to accelerate to the maximum extent both the controlled thermonuclear fusion programme and an investigation into the large-scale exploitation of solar energy in order to provide a back-up to the fast breeder in case of unforeseen difficulties. This is the policy currently followed by the U.S. An important advantage is that it minimizes the period during which the world remains at risk if the hazards of nuclear fission do in fact turn out to be too great to be accep-ted for more than a few decades. There appears to be no prospect of avoiding these hazards altogether, and indeed we have already lived with them successfully at a rather lower and more con-trolled level for 30 years, but assuming that there is a certain low but finite risk each year we may not wish to prolong this situation indefinitely.

> Lastly, the need for an ever-increasing number of developing countries to start relying extensively on fission power over the next decade or so in order to secure their energy supplies might be considerably reduced by a deliberately rapid exploitation of alternative and simpler forms of energy, e.g. small-scale domestic solar energy devices, which would be much easier to build locally than nuclear power stations and would therefore

minimize balance-of-payments problems as well as the hazards
of proliferating fissile material. Local production of non-
nuclear components of fission power stations might also be
encouraged for a similar reason.

2. FUSION REACTOR SIZE AND TIMESCALE

Fusion technology differs from previously-known technologies in the scale,
cost and complexity needed for a minimum-sized working thermonuclear reactor.
Familiar devices such as dynamos and steam engines which were introduced in
the past could be built and made to work empirically and quickly on a small
scale by individual inventors and then steadily increased in size and optimized, initially by small private companies. Even the supercritical fission
chain reaction was first tried out on a small scale at negligible power
level as part of the Manhattan project after the principles had been established by university research in a number of countries. In the case of CTR
it appears that a reactor cannot run at all unless it has a size as large or
larger than that of most existing power stations. Although this is not
necessarily a disadvantage for a commercial reactor it does make the technology difficult to develop, since one needs considerable knowledge and
confidence to design the first prototype reactor from basic principles and
be sure that it will operate successfully, and yet funding is naturally hard
to come by from government (and virtually impossible from private industry)
if the outcome is uncertain. It is therefore important to be clear which are
the physical parameters that determine the minimum reactor size, and so far
as possible to establish an R&D strategy based on a series of experimental
devices of increasing scale on which the theory can be checked step-by-step
and the scaling laws and other empirical data can be accumulated.

Apart from their lower cost, small-scale devices can be built more quickly
and are easier to operate, but some of the information that is relevant to
the working conditions in a reactor can only be obtained from large-scale
experimental equipment and this is why increasingly larger Tokamaks and
other devices are being designed and built at the present time. Such devices
are likely to take 3 years for the design and approval stages, 5 years to
build, and perhaps 2 years to obtain significant results, i.e. 10 years
altogether, after which they may then be modified for a further stage of
operation. This long timescale constitutes a major problem. According to
Marshall [4], "controlled thermonuclear power looks to be the most difficult
technological enterprise yet tackled by mankind". There are many physical
and technological aspects to be investigated and optimizations to be
performed, and because of the urgency of the energy problem it would introduce too much delay if one had to wait 10 years for the results of each stage
before embarking on the next; a considerable degree of parallel planning is
necessary even though this may lead to some wrong turnings. Thus several
options such as Tokamaks, Reversed Field Pinches, Stellarators, High-β
Stellarators, mirror machines and laser fusion are being investigated
together, and plans are being made to study the technological problems of
fusion before it is known whether or not the physics problem of plasma containment can be solved.

The minimum reactor size depends on a number of physical parameters to be
discussed in §6.

3. THERMONUCLEAR ENERGY SOURCES

It is clear from the shape of the well-known binding energy curve [5] that a vast amount of nuclear energy is potentially available in ordinary matter, amounting to 8.4 MeV/atom or 8.1×10^{11} joules/gm in the case of protium (ordinary hydrogen, H). Since the most tightly bound nucleons are those of nuclei with atomic weights in the range $50 < A < 60$, with the maximum occurring at Fe^{56}, energy can in principle be released either by breaking down the heavier nuclei or by combining together those that are lighter. The helium nucleus He^4 is particularly tightly bound, so that rearrangement of other light nuclei such as H, D, Li^6 to form He^4 is able to release a considerable fraction of the total available energy.

To exploit this source of energy it is however necessary to find a self-sustaining reaction process. This appears to be more difficult than in ordinary chemistry since nuclei are constrained from approaching one another by their positive electric charge (Coulomb barrier), and they are protected against spontaneous break-up by the resistance of their surfaces to changes of shape, although some naturally occurring isotopes are radio-active on a very long timescale. In the fission process this shape barrier is overcome by temporarily supplying the binding energy of an additional neutron which is able to penetrate the nucleus because it has zero charge; more neutrons are then released by the fission event itself so that the neutron population can multiply and the chain reaction builds up. In nuclear fusion the charge nuclei must be made to approach one another with a speed high enough to enable their mutual Coulomb barrier to be penetrated with an appreciable probability so that the reaction can take place.

Charged nuclei can readily and efficiently be accelerated to adequate energies by means of suitably-designed electric fields, but unfortunately it does not seem to be possible to exploit this idea to produce a self-sustaining fusion process in any straightforward way, since fast charged particles moving through ordinary cold matter lose most of their kinetic energy to electrons before a nuclear reaction has time to occur with adequate probability. The simplest way of avoiding the difficulty is to heat both the nuclei and the electrons to the same high temperature, and then to hold the gas together and keep it hot for long enough to enable the fusion process to take place with adequate energy yield. This method is referred to as thermonuclear fusion because of the high temperatures used. These are of the order $10^7 - 10^9$ °K, corresponding to particle energies in the range 1 - 100 keV, and under these conditions most of the atoms of the gas (other than high-Z impurities) are completely stripped into their constituent nuclei and electrons so that we are dealing with a fully ionized plasma.

Thermonuclear fusion processes may be distinguished firstly by the confinement method used to hold the plasma together (gravitational, inertial or magnetic, to be discussed in §4), and secondly by the reactions that take place.

Fusion is the energy source for normal stars, the important overall reaction in stars such as the sun being the burning of protium to form helium:

$$4H + 2e^- \rightarrow He^4 + 2\nu + 26.7 \text{ MeV} \tag{1}$$

where e^- represents an electron and ν a neutrino. This does not take place directly, but as a consequence of two main alternative chains of simpler processes known as the proton-proton chain and the CNO (carbon-nitrogen-oxygen) cycle [5]. Although an enormous amount of energy is locked up in the protium of the oceans, the reaction (1) is extremely slow (taking place in the sun during times of the order 10^{10} years) and - perhaps fortunately - no way is seen at present by which we might release it on earth. The overall reactions envisaged for the CTR project are therefore mainly the two much faster reactions:

$$3D \rightarrow He^4 + p + n + 21.6 \text{ MeV} \tag{2}$$

$$D + Li^6 \rightarrow 2He^4 + 22.4 \text{ MeV} \tag{3}$$

Additional energy is released when the neutron in reaction (2) is captured in some suitable absorber, e.g. sodium, followed by radioactive decay:

$$n + Na^{23} \rightarrow Mg^{24} + e^- + \bar{\nu} + 12.5 \text{ MeV} \tag{4}$$

In both reactions (1) and (4) part of the energy is irretrievably lost in the form of neutrinos (ν) or antineutrinos ($\bar{\nu}$).

The reaction rate and therefore feasibility of thermonuclear reactions depends to a large extent on accidental details of the light nuclei involved. Reaction (2) involves 4 distinct thermonuclear processes:

$$\begin{aligned}
D + D &\rightarrow He^3 + n + 3.3 \text{ MeV} \\
D + D &\rightarrow T + p + 4.0 \text{ MeV} \\
D + T &\rightarrow He^4 + n + 17.6 \text{ MeV} \\
D + He^3 &\rightarrow He^4 + p + 18.3 \text{ MeV}
\end{aligned} \tag{5}$$

of which the first two (D-D reactions) are the slowest. Although deuterium is only a minor constituent of naturally-occurring hydrogen with an atomic abundance ratio of $1.5 \times 10^{-2}\%$, the energy content of ordinary water from reactions (2) + (4) amounts to 1.82×10^7 joules/gm so that the energy content of the world's oceans is 2.5×10^{31} joules. This would be enough to supply our total energy requirements for billions of years.

The D-D reaction is comparatively slow and it is not yet clear that it can be exploited for CTR, at least using magnetic confinement, due to competing forms of energy loss. Present attention is therefore concentrated on (3) which can proceed most rapidly as a 2- or 3-stage reaction:

$$\begin{aligned}
D + T &\rightarrow He^4 + n + 17.6 \text{ MeV} \\
n + Li^6 &\rightarrow He^4 + T + 4.8 \text{ MeV} \\
n + Li^7 &\rightarrow He^4 + T + n - 2.5 \text{ MeV}
\end{aligned} \tag{6}$$

of which only the first stage is thermonuclear. The D-T reaction is the fastest thermonuclear process known and produces 4/5 of its energy in the form of 14.1 MeV neutrons which simply escape from the plasma region. These must be slowed down and captured in a surrounding lithium blanket in order

to convert their kinetic and absorption energy into heat and to regenerate
the tritium (T). Tritium is radioactive with a half-life of 12.26 years and
is thus not naturally available, but the world lithium supply appears to be
sufficient to last for many hundred years and during this time it is hoped
that the D-D scheme will become practicable. A useful multiplication can be
achieved both from the third reaction (6) and also from (n,2n) reactions in
various other materials in order to make good any losses of neutrons or
tritium and to increase the total supply of tritium fuel. It appears that a
tritium doubling time of a few months could be achieved, which is much
faster than for fast fission breeder reactors.

In addition to the main nuclear reactions which have been listed, there are
many (n,γ), (n,2n), (n,p) and (n,α) reactions which occur due to the interaction between the neutrons escaping from the plasma and the material of the
surrounding blanket structure. The target nuclei involved include all the
isotopes of the main blanket materials, together with those of minor alloy
constituents, impurities, transmutation products and so on. It is worth
emphasizing that reaction (3) involves one neutron / 20 MeV of released energy
compared to one neutron/60 MeV of released energy in fission, and also that
the fusion neutron spectrum is much harder than the fission spectrum so that
additional reactions can occur. Radiation damage and induced radioactivity
of the reactor structure therefore represent very significant problems to
CTR. There are however no fission products or long-lived actinides, the
waste products (p,He^4) are stable, and the only volatile radioactive isotope
is likely to be tritium itself which is a weak β-emitter with a relatively
short physical and biological half-life. Fusion is therefore potentially
cleaner and safer than fission [6].

4. PLASMA CONFINEMENT

A fission reactor is constructed from conventional materials which may be
solids, liquids or gases, and it can operate at temperatures which are
familiar in standard engineering practice because the neutrons carry no
electrical charge and are therefore able to enter the nucleus at any velocity since they are not repelled by it. The fact that they do not interact
electrically with either nuclei or electrons also means that a high-energy
neutron energy distribution with an 'effective temperature' of order
10^9 - 10^{10} °K can if necessary (as in a fast reactor) readily be maintained,
even in a region of space which is already occupied by ordinary matter at a
much lower temperature. There is thus no difficulty in testing a fission
reactor at negligible temperature and power level, and in this respect it is
even simpler than an ordinary furnace or chemical plant which has to run hot.
In practice an operating fission reactor must run at a high enough temperature to achieve adequate thermodynamic efficiency, but this has little to
do with the nuclear reaction itself.

A fusion reactor employs a hot plasma whose interaction with ordinary matter
is extremely strong, due to the electric charges on its constituent electrons
and nuclei, so that the two forms of matter must be kept physically separate
from one another until the plasma has had time to react or it will immediately
be cooled. This is the origin of the so-called 'confinement problem' which
has been the main focus of attention of the CTR project so far. The required
reaction time τ_r is inversely proportional to the particle density n, and
according to the Lawson criterion [6,7] a value

$$\Lambda \equiv n \, \tau_r \geq 3 \times 10^{14} \ sec/cm^3 \tag{7}$$

is necessary for an efficient D-T reaction in which more useful energy is produced than is needed to heat the plasma in the first place.

How are hot plasmas to be contained? Unless the plasma is restrained by some long-range field of force its constituent nuclei will simply fly apart with their random thermal velocity V_{th} of order 10^8 cm/sec and the lighter electrons will follow them in such a way as to maintain overall electrical neutrality. If the dimension of the plasma is R this gives an <u>inertial confinement time</u> of order

$$\tau_c \simeq R/V_{th} \ secs \tag{8}$$

which for a plasma of radius 3 metres and density 10^{14} atoms/cm^3 is about 10^6 times too short.

There appear to be three principal confinement methods by which a thermonuclear reaction can be made to work - gravitational, inertial and magnetic (Figure 3). Direct confinement of a hot plasma by an external electric field seems to be ruled out, since any plasma with a realistic particle density

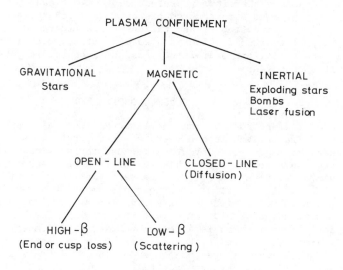

<u>Figure 3</u>.   <u>Plasma Confinement Methods</u>

necessarily has to be almost neutral in order to avoid enormous electrostatic energies, and then the electric forces on the oppositely charged nuclei and electrons just cancel one another out. However internal electric fields do often play an important role in holding the two particle species together.

The simplest method, gravitational confinement, is exploited in a star such as the sun. Here the plasma is held together for an indefinite period in

empty space by its own gravitational attraction, so enabling even very slow
thermonuclear reactions taking 10^{10} years to be utilized. This method has
attractive thermostatic properties since if for any reason the thermonuclear
reaction rate decreases the star automatically contracts, so releasing
gravitational energy which raises the temperature and causes the reaction
rate to increase again. However it is quite impracticable under terrestial
conditions since there is no way to obtain the strong gravitational fields
required.

The next simplest method is called inertial confinement and is used in
thermonuclear explosions. According to equations (7) and (8), if the product
ρR of density ρ (proportional to n) and radius R can be made sufficiently large, the inertial confinement time τ_c will exceed the reaction
time τ_r and the plasma will have enough time to react before it succeeds
in blowing itself apart. A simple calculation brings out the fact that

$$M \sim \frac{1}{\rho^2} \qquad (9)$$

where M is the mass of material which can react efficiently. At normal
solid densities the explosive yield of an efficient thermonuclear device is
around 1 megaton TNT equivalent, which is obviously difficult to use for
industrial purposes although perhaps not inconceivable if all other forms of
energy available to us were ultimately to run out. If the density can
however be increased by adiabatic compression by (say) a factor 10^4 before
the reaction is allowed to start, then the explosive energy yield is reduced
by a factor 10^8 to about 10 Kgm TNT which can probably be controlled quite
readily since it is only enough to boil a few gallons of water. Although
the potential blast damage from such an explosion might still appear to be
rather high, in fact this is reduced by a further factor of more than 10^3
because the masses of expanding reaction products involved are tiny, and
each explosion would probably have no more effect than that of a large fire-
work inside a rugged metal container. These ideas are the basis of the new
concept of laser fusion, which will however not be discussed further in this
article because an excellent account has recently appeared [3].

The third and most complex method of confinement employs a magnetic field to
contain the hot plasma. Because the plasma is a good electrical conductor,
it can in principle be trapped within a suitably shaped magnetic 'bottle'
for long enough to enable it to react. The charged nuclei and electrons of
the plasma no longer travel directly outwards in straight lines, but move in
helical orbits along the direction of the magnetic field, with some departure
from these ideal orbits due to magnetic inhomogeneities, electric fields,
instabilities and turbulence of various kinds as well as collisions between
the particles themselves. The particle orbits must then not be allowed to
intersect the walls of the container until the reaction is sufficiently complete. At density $n \simeq 10^{14}/cm^3$, the required confinement time is of order
3 seconds, during which the ions travel some 3×10^8 cm and the electrons
about 60 times as far. Excellent particle containment is therefore required.

5. CLASSIFICATION OF MAGNETIC CONFINEMENT SYSTEMS

The various magnetic confinement systems that have been devised so far can
be classified in several alternative ways as indicated in Figure 3 and the
lower part of Figure 4.   To a good approximation each particle behaves like

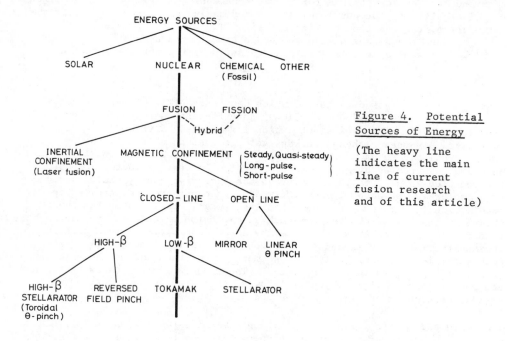

Figure 4. Potential
Sources of Energy

(The heavy line
indicates the main
line of current
fusion research
and of this article)

a little magnet (current loop) with a constant magnetic moment

$$\mu = \tfrac{1}{2} m V_{\perp}^{2} / B \tag{9}$$

where m is the particle mass $V_{\perp}$ is the velocity transverse to the magnetic
field, and B is the field strength. It thus sees an effective potential
energy

$$W = \mu B = \tfrac{1}{2} m V_{\perp}^{2} \tag{10}$$

which can cause it to be reflected from regions of strong magnetic field
('mirrors')if the parallel velocity $V_{\parallel}$ is sufficiently small. This makes
it possible to distinguish two main geometric classes of magnetic field configuration, open-line and closed-line (Figure 5).

In open-line magnetic mirror devices a particle whose velocity ratio $V_{\parallel}/V_{\perp}$
is sufficiently small will travel backwards and forwards between the magnetic
mirrors at either end until this ratio is reduced by deflections of its
direction caused by scattering, either due to classical Coulomb collisions
with other particles or to electric field fluctuations resulting from plasma

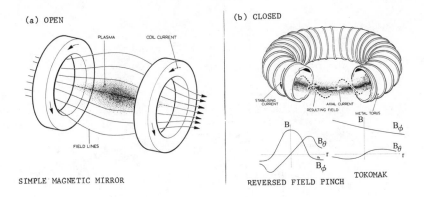

(a) OPEN

SIMPLE MAGNETIC MIRROR

(b) CLOSED

REVERSED FIELD PINCH   TOKOMAK

Figure 5. Open and Closed Field-line Configurations

turbulence, after which it is immediately lost. The velocity-space distribution function is therefore anisotropic, having an empty 'loss-cone' into
which particles are gradually scattered by velocity-space diffusion.
Magnetic mirror devices have many advantages as potential thermonuclear
reactors but unfortunately they have one critical flaw, namely that the ratio
between the D-T thermonuclear reaction rate and the classical Coulomb loss
rate, which is independent of particle density n but a function of temperature T , is marginally not high enough at any temperature to adequately
meet the Lawson criterion (3), so that a straightforward mirror reactor would
either not be self-sustaining or would have an unacceptably high circulating
power density. Many laboratories have therefore ceased to work in this area
although a strong effort continues at the Lawrence Livermore Laboratory,
based on the concept (Figure 6) of a steady-state device in which beams of
fast neutral atoms are continuously injected into the reaction region, and
the directed kinetic energy of the escaping plasma is converted back into
high-voltage DC by an efficient direct converter which would probably need
to be very large (Figure 7).

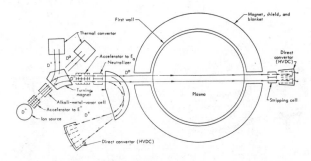

Figure 6. Schematic diagram of a neutral-beam
injection system for a magnetic-mirror reactor

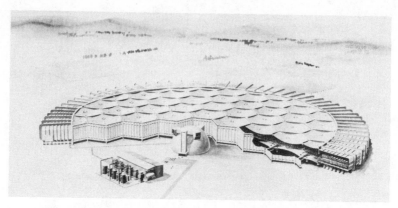

Figure 7.   Overall view of the LLL D-T magnetic-
mirror fusion power plant.   The direct-conversion
equipment is shown behind the spherical reactor shield.

In topologically toroidal, closed-line devices the field lines in the containment region lie to good accuracy on a set of nested 'magnetic surfaces'.
During any limited period of time a charged particle is confined to one of
these surfaces, except for periodic deviations due to its finite Larmor
radius, but it gradually diffuses from one surface to another due to classical Coulomb collisions or to turbulent fluctuations.   The velocity-space
distribution function is almost isotropic and the important transport
coefficients (diffusion, heat conduction, electrical resistivity) relate to
real physical space.   Mirror effects do however play an important role
because the magnetic field strength B will normally vary along each closed
field line, so that particles with small $V_{\parallel}/V_{\perp}$ can be trapped in local magnetic mirrors while others circulate continually around the torus.   The
Tokamak device has this geometry and is currently the most favoured choice
for a reactor.

In another class of high-density open-line device, illustrated in Figure 8,

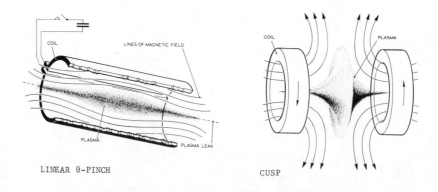

LINEAR θ-PINCH   CUSP

Figure 8.   High Density Open-Line Devices

the classical collision rate is large and the velocity-space distribution
therefore nearly isotropic, so that the plasma flows out of the 'holes' in
hydrodynamic fashion with a velocity of order V_{th} .   However the rate of
escape can be reduced either by making the device very long (linear θ-pinch)
or the holes very narrow (cusp).   It is not clear whether either type of
device can be made into a practical working reactor: for example a linear
θ-pinch working at a density 10^{16} atoms/cc and a reaction time of order
0.1 sec would need to be several kilometres long - a shorter length could be
achieved by working at higher density but this is limited by the transverse
magnetic field strength that material walls are able to support.

Equilibrium

So far we have discussed only the containment of single particles.   In any
magnetic containment device with reactor potential the timescale τ_c is
long compared to the time taken for acoustic or magnetohydrodynamic waves to
cross the system, so that an accurate balance

$$\nabla \underline{p} \;\; = \;\; \underline{j} \times \underline{B} \tag{11}$$

must be achieved between the plasma pressure tensor $\underline{p}$, (isotropic pressure
p in the case of toroidal geometry) and the force exerted by the magnetic
field, where $\underline{j}$ = Curl $\underline{B}/4\pi$ is the current.   Essentially the thermal pressure
of the plasma has to be transmitted by the magnetic field, to the supporting
mechanical structure, across an intervening region which serves as a good
thermal insulator.   The thermonuclear power density at given temperature T
is proportional to p^2 and is constrained by

- A.   The maximum ratio β of plasma to magnetic pressure that
  the field geometry can support.

- B.   The maximum strength of practicable mechanical structures.

- C.   The maximum field strength allowed by superconductors,
  (when superconducting coils are used to generate the
  confining magnetic field).

It is economically advantageous to obtain a high power density from a reactor
of given size, and therefore methods are sought by which A, B & C may be
maximized.

Straight open-line devices such as mirrors and linear θ-pinches allow β to
be high, of order 1, but with toroidal closed-line devices the plasma ring
must be supported against a tendency to increase its volume by outward
expansion.   Usually this is achieved by means of a poloidal field B_p ,
which may either be produced by a toroidal current I flowing within the
plasma itself as in the Tokamak or Reversed Field Pinch (Figure 5b), in
which case the configuration can retain its azimuthal symmetry, or else by
shaping the magnetic surfaces as in the stellarator (Figure 9a).   An alternative geometry is the bumpy torus which resembles a ring of magnetic
mirrors (Figure 9b).   If the β-value in a toroidal closed-line device is
too high the magnetic surfaces are pushed against the outer wall and the
equilibrium breaks down.   The theoretical upper limit for β in a Tokamak is
relatively low, of order 5%, and attempts to maximise this value lead to the
use of small aspect ratios R/a (major radius/minor radius) and non-circular
cross-sections.   The Reversed Field Pinch can have a considerably higher
β-value and this may represent a substantial engineering advantage if the

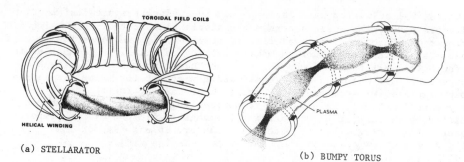

(a) STELLARATOR

(b) BUMPY TORUS

Figure 9.   Asymmetric Configurations

more complex field configuration shown in Figure 5 can be established and
maintained.   Experimental evidence from Zeta and other devices indicates
that the necessary reversal of the toroidal B_z field at the outside of the
plasma occurs automatically and there is good theoretical support for this
[8].   The equilibrium in high-β asymmetric devices is not fully understood
as yet.

Stability

An acceptable equilibrium must be essentially stable, that is it must
correspond to a ball resting in a trough rather than at the top of a hill.
The range of possible instabilities of a reactor configuration is very wide,
with timescales extending from fractions of a microsecond up to seconds, and
space-scales extending from fractions of a millimetre up to 1 metre.   Both
purely-growing modes and unstable waves may occur.   When an instability
starts it will continue to build up for a time with a well-defined eigen-
function and complex frequency, $\omega = \omega_R + i\,\omega_I$ (linear regime), after which
(if it does not disrupt the equilibrium altogether) it may either be limited
by non-linear effects and lead to a new stable equilibrium or a stable large-
amplitude oscillation, or it may interact with other modes, perhaps eventually
leading to a steady convection process or to turbulence.   The linear regime
has been studied theoretically for many instabilities and the results are
well confirmed by experiment, but much less work has been done on the non-
linear states to which they lead.   It is generally believed that most real
plasmas are likely to be at least weakly turbulent, like the gravitationally
contained atmosphere or ocean, or a gently boiling kettle or a star with a
convectively unstable zone.   Such turbulence enhances the electric field
fluctuations in the plasma above their classical value, leading to increased
transport coefficients (toroidal devices) or velocity-space diffusion (mirror).
The important problem is therefore to be able to predict and to limit the
magnitude of this enhancement since it affects the feasibility and the minimum size (§6) of a reactor.

Instabilities may be classified into several types including:

    (a)   Ideal MHD (gross distortion in which the plasma and
         field move together).

    (b)   Resistive MHD (change of field topology, or escape of
         plasma across the field).

    (c)   Microinstabilities (finescale, velocity space).

    (d)   Thermal.

A major MHD instability will cause the whole configuration to distort rapidly
and is unacceptable. Less severe MHD instabilities may only cause a mild
turbulence or 'boiling' of the plasma in regions of limited radial extent, as
in a star, although this concept has not been worked out fully as yet and in
general the non-linear consequence of most MHD instabilities is not properly
understood: the calculations are difficult analytically because they are
non-linear and difficult computationally because they are 3-dimensional.
Two powerful methods for removing or controlling many instabilities are
firstly the <u>magnetic well</u> in which the magnetic field B (or its average along
a field line) increases outwards, and secondly the use of a <u>sheared magnetic
field</u> in which the angle of the field lines changes with radius.

The importance of the magnetic well was first realized in the case of the
simple mirror (Figure 5a), which is unstable against 'flute instabilities'
in which flux tubes occupied by plasma change places with those outside
which have larger volume. This interchange process allows the plasma to
expand, so releasing thermal energy which drives the instability. Changing
the configuration to that of Figure 10 by the introduction of extra current-
carrying rods produces a minimum B or equivalently a maximum volume $\int d\ell/B$
at the centre and so stabilizes the plasma as many experiments have confirmed.
Later it was realized that an average magnetic well also occurs in a Tokamak
due to its toroidal geometry.

One of the effects of shear is to
cause any radial distortion of the
plasma to bend the field lines, so
increasing the magnetic energy and
inhibiting the distortion. An
important necessary stability condition [9] due to Suydam

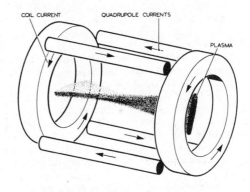

$$- \frac{dp}{dr} < \frac{r}{4} \ \frac{B_\varphi^2}{8\pi} \left(\frac{d}{dr} \log \mu \right)^2 \ , \ \mu = B_\theta/rB_\theta$$

$$(12)$$

illustrates the fact that many MHD
instabilities can be stabilized by
the shear μ'/μ if the pressure
gradient is not too high.   The
Reversed Field Pinch (Figure 5) has
a high shear in the outer regions
of the plasma which is good for

<u>Figure 10</u>. Magnetic Well

stability, although the shear is very weak near the centre so that the inner
core of the plasma may remain weakly turbulent.

The overall MHD stability of the Tokamak is largely determined by the <u>safety
factor</u>

$$q(r) = \frac{r}{R} \frac{B_\varphi}{B_\theta} \qquad (13)$$

which must exceed unity everywhere and must increase outwards. For helical
magnetic field lines the MHD instability modes are helical displacements
following the pitch of the lines so that these are not too distorted and thus
exert a stabilizing influence. When q > 1, the plasma cannot be unstable to
a mode which displaces it into a helix of circular cross-section since the

motion cannot be accommodated, and so on for trefoil distortions which are stable for q > 3, etc.   In present experiments q is chosen to be ⩾ 1 at the centre and ⩾ 3 at the plasma edge r = A. This limits the current since I = a B_θ :

$$I = a \cdot \left(\frac{a}{R} \frac{B_\varphi}{q(a)} \right) \qquad . \qquad (14)$$

Microinstabilities are caused by departures from the velocity distribution function from isotropy due to the loss cone (mirror machine), electric currents, and density or temperature gradients etc.

Thermal instabilities are caused by the shape of the function f(T) that determines the net heating rate :

$$\frac{dT}{dt} = f(T) \qquad . \qquad (15)$$

Thermal equilibrium requires f(T) = 0 and will be

$$\underline{stable} \quad if \quad df/dT < 0$$
$$\underline{unstable} \; if \quad df/dT > 0 \qquad . \qquad (16)$$

Stability is not necessarily easy to achieve since in many situations the heating rates increase with temperature and the loss rates decrease so that some form of feedback control may be required. One example occurs in toroidal reactors, operating at fixed magnetic field and β and hence fixed pressure p = 2nkT. A rise in temperature forces the density n to decrease in proportion, so that since the reaction rate/unit volume has the form $n^2 r(T)$ the best choice of operating temperature lies at $r(T) \sim T^2$, i.e. at about 15 keV. This will not normally satisfy the stability condition (16) since at this point the heating rate is rising rapidly with temperature. A second example occurs in present-day Tokamaks, in which the ohmic heating rate is proportional to the conductivity $\sigma \sim T^{\frac{3}{2}}$. The current therefore tends to concentrate in the hot region at the centre, thus heating it further and causing q(0)/q(a) to decrease. This either leads to instability at the centre or forces the current I to be reduced.

Many other subtle effects are likely to be discovered and brought under control as the physics of reactor plasmas becomes better understood.

6.   PHYSICAL FACTORS DETERMINING THE SIZE

Critical Minor Radius of a Toroidal Reactor

An early task in the case of fission systems was to determine the critical mass or radius of a fissile assembly, e.g. of a bare sphere of Pu^{239} or U^{235}, a sphere surrounded by a moderator, a graphite + U^{238} pile, and so on. A similar concept appears to be valid for a toroidal thermonuclear reactor, but the concept itself is not yet fully established and the numerical value of the 'critical radius' is not accurately known.

The critical radius of a fissile assembly is determined by the fact that a neutron must just be able to reproduce itself before it escapes from the system, so keeping the chain reaction going at a constant level. In the diffusion approximation the containment time τ is related to the radius a and the diffusion coefficient D by

$$\tau \sim a^2/D \qquad (17)$$

so that (other factors being equal) the critical radius a_c is proportional to $\sqrt{D}$. In the case of a thermonuclear reactor working at zero net power output, an element of plasma thermal energy ϵ must be able to generate an extra amount of useful energy $\epsilon(1-\eta)$ before it escapes, where η is the efficiency with which the escaping energy can be recovered. This corresponds to the Lawson criterion (7). Once the operating density n is prescribed the condition

$$\tau_E = \tau_r = \Lambda/n \qquad (18)$$

leads to a critical plasma minor radius a_c if the energy containment time τ_E is an increasing function of a, which is expected to be true for a toroidal device in which much of the energy is lost by the radial diffusion of particles or heat across the magnetic field, although not for a mirror machine in which energy is mainly lost by the escape of particles from the ends of each individual tube of force.

Maximum operating density

An upper limit to the particle density n is determined by the maximum plasma pressure $p = 2nkT$ that can in practice be supported by the magnetic field and ultimately by the external engineering structure

$$n_{max} \sim \frac{\beta_{max} B_{ext}^2}{T} \qquad (19)$$

where T is the operating temperature, and β_{max} is the maximum ratio of plasma pressure to the magnetic field pressure at the external coils. B_{ext} may have to be reduced if the main part of the field is provided by superconducting windings, but the principal factor limiting the density is β_{max}. Thus the Los Alamos-ANL reference θ-pinch reactor (RTPR) operates in pulsed mode with $n = 2.5 \times 10^{16}$ cm^{-3}, $B = 110$ KG, $\tau = 0.07$ sec, $a_W = 50$ cm while for the University of Wisconsin UWMAK Tokamak conceptual reactor design $n = 8 \times 10^{13}$, $B = 87$ KG, $\tau_E = 4$ sec, $a_W = 500$ cm with a cycle time of 5790 sec [7], a_W being here the radius of the first wall.

Broadly speaking, therefore, the higher β_{max} in RTPR compared to a Tokamak enables n to be $\sim 10^2$ larger, τ correspondingly $\sim 10^2$ smaller, and hence a to be ~ 10 smaller if a law of the form (17) applies.

Containment time scaling

The dependence

$$\tau_E = f(a,n,B,T) \qquad (20)$$

of energy containment time on radius, density, magnetic field and temperature is not yet fully known for any containment geometry. In particular it is not known whether a scaling law of the form (17) applies, and if so, what is the numerical magnitude of the diffusion coefficient D. This is economically important since for a sequence of reactor designs with fixed plasma geometry, density n , temperature T , field B and fixed total thickness Δ of blanket, magnetic field coils, structural supports, shield etc. one may expect that

$$\text{Capital cost} \sim \text{volume} \sim a\,((\Delta+a)^2-a^2) \sim a\,\Delta^2\,(1+\tfrac{2a}{\Delta}) \tag{21}$$

$$\text{Power output} \sim a^3 \tag{22}$$

so that for $a \gtrsim \Delta/2$

$$\text{Minimum cost} \sim a^2 \tag{23}$$

and for a law of the form (17)

$$\text{Minimum cost} \sim D \tag{24}$$

Classical scaling

Although the mechanism for the particle and heat loss that is observed in present-day experiments is not fully understood, a number of classical, neoclassical and pseudoclassical theories all lead to a law of the form (17) with a coefficient of the type

$$D = \Gamma_c\,T^{-\frac{1}{2}}\,n/B^2 \tag{25}$$

where Γ_c is a constant, so that the Lawson criterion now appears in the form

$$I = B_\theta a > \text{constant} \sim (\Gamma_c\Lambda)^{\frac{1}{2}} \tag{26}$$

in the case of a pinch or a Tokamak of given magnetic configuration where I is the toroidal current flowing in the plasma. A similar criterion determines the containment of the α-particles that are responsible for maintaining the plasma temperature. Thus it is believed at present that a self-sustained reaction can be maintained in a Tokamak provided that I is sufficiently large. It is anticipated that scientific feasibility studies can be carried out with current $I \approx 5$ MA but it should be emphasized that

$$\text{cost} \sim I^2 \sim a^2 \sim \Gamma_c\Lambda \tag{27}$$

and that at present both Γ_c and Λ are uncertain within factors of order unity.

Bohm scaling turbulence

An alternative law which fits some experiments is (17) with the 'poloidal Bohm' diffusion coefficient

$$D = \Gamma_B \cdot \frac{1}{16}\,\frac{KT}{e\,B_\theta} \tag{28}$$

with $\Gamma_B \le 0.01$. Scaling from existing Tokamak devices then requires a larger critical radius than is given by (25).

Finally, it is not obvious without a detailed understanding of the physical mechanisms responsible for diffusion that a law of the form (17) must necessarily apply at all: for example straightforward hydrodynamic turbulence might be expected to lead to a scaling law

$$\tau \sim a\,T^{-\frac{1}{2}} \tag{29}$$

It is therefore necessary to carry out a series of experiments with increasing minor radius a and parameters B,T,n approaching the reactor regime. An example is the proposed European JET experiment with a = 1.25 - 2.10 m (non circular), B ≃ 30 KG, I = 3 MA.

Beam injection

Equation (26) suggests that for an experimental prototype there is considerable advantage in reducing the required value of Λ as much as possible in order to decrease the cost. One way of doing this is to maintain the temperature at a thermonuclear level by supplying auxiliary heating instead of relying entirely on α-particle energy. This should enable Γ and Λ to be measured under 'sub-critical' conditions which can nevertheless be extrapolated to those in a working reactor.

A related possibility is to increase the fusion reaction rate that can be achieved by the maximum working pressure by maintaining controlled non-Maxwellian ion energy distributions and $T_e < T_i$. In the TCT (two-component torus) concept a beam of high-energy deuterons is fed into a tritium target plasma. This allows the electron gas to be cooler than the usual thermonuclear temperature ∼ 10 keV, and so to exert a lower pressure, while some enhancement of the fusion reaction rate is obtained from the injected fast deuterons before they slow down. A further improvement can be obtained by injecting beams of both tritons and deuterons.

Mirror reactors

The working of a mirror reactor is not governed by spatial diffusion so that the radius a can in principle be smaller; the total size is also considerably reduced by the geometrical fact that the plasma does not have to be bent into a toroidal ring. If a mirror reactor can be made to work the prototype can therefore be smaller and cheaper than a toroidal reactor and run at lower power output. Unfortunately there is a fundamental upper limit to the parameter Λ which is governed by the scattering of ions into the loss-cone due to classical Coulomb collisions, and this limit may not be high enough to enable a mirror reactor to be a net power producer although the situation is still uncertain at the present time. Any enhanced scattering due to plasma turbulence would however reduce Λ still further.

Mirror reactors have many engineering advantages, however, since many of their components (injector, direct converter etc.) can be developed and tested separately, and they can operate as steady-state devices without the problems of fuel injection and waste product removal encountered in toroidal systems.

Laser fusion

In case of a laser fusion system the Lawson criterion for ρR determines in practice a minimum laser implosion energy needed to initiate a self-propagating thermonuclear burn. The exact value of this energy is not yet known, since it depends on the physics of the implosion and of the subsequent triggering process, but it may lie in the range 10-100 kJ. Lasers suitable for experiments at this energy level are already being built, and provided that a self-propagating burn can be demonstrated during the next few years the R&D path towards a working reactor appears to be relatively well mapped out. However it is by no means certain as yet that such a reactor would be economic. The main questions are first, whether the laser can be made efficient enough to achieve a net power output, and second, whether the cost of the components including that of pellet production and the maintenance of the optical system can be made less than the value of the electricity produced.

Neutronics

Reactor size is also determined by a number of other considerations. The most obvious of these is the overall thickness Δ of the blanket needed to absorb the neutrons to produce heat and regenerate tritium, to protect the magnet against damage and deposition of heat due to neutrons and X-rays, and finally to serve as a biological shield. The necessary thickness determined by nuclear cross-sections is of order 2 metres as illustrated in Figure 11.

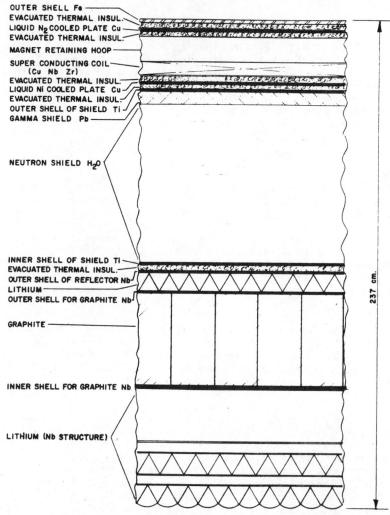

Figure 11.   Cross-section through the Reflector-
Shield-Superconducting Coil Region

The minimum blanket volume is $4/3 \pi \Delta^3$ for a zero-radius plasma, but when allowance is made for a minimum-radius plasma of toroidal shape together with a divertor, this volume is greatly increased as the conceptual design of Figure 12 indicates, and it is evidently sensitive to $\Gamma\Lambda$ if this is too large. Again mirror and laser fusion reactors have an advantage because of their simpler geometry.

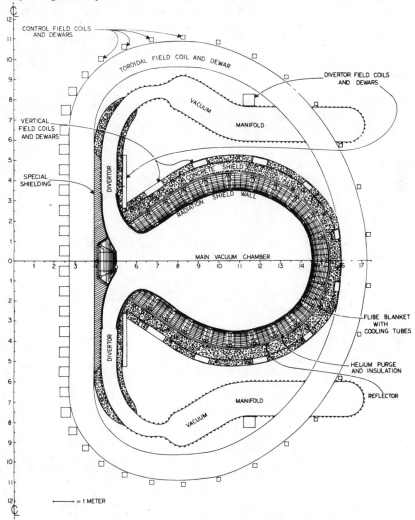

Figure 12. Radial cross-section of the PPPL Tokamak
reactor divertor and toroidal magnet coil

The capital cost of the reactor will depend on its mass or volume, other things being equal, and it is important to exploit this fully by operating it at a high power level. The thermonuclear power output is proportional to n^2 and therefore according to (19) to $\beta_{max}^2 B_{max}^4$. This gives considerable incentive to optimize β_{max} (by choosing the best shape, or by using pinch rather than Tokamak geometry), and if possible to increase B_{max} by using improved superconductors and stronger materials. However, the total neutron dose and

X-ray dose received by the structure are proportional to the total energy
produced, and there may also be an element of cost arising from the need to
replace damaged components. This particularly applies to the first wall
which is subjected to bombardment by ions, electrons, neutral atoms and
X-rays in addition to neutrons and X-rays. According to present ideas a
thermonuclear reactor should be built in modular form so a damaged section
can be removed and a new section substituted without undue interruption to
the working of the system, the original section subsequently being repaired
off-line, and automatic handling is likely to be needed because of the
induced radioactivity of the structure and the presence of tritium. The
development of materials that are less subject to radiation damage would
clearly be advantageous.

The power output envisaged for the UWMAK conceptual Tokamak reactor [7] is
4690 MW(th),1454 MW(e), and though large this is not out of line with
projections of future power station sizes.

7. OVERALL REACTOR STRUCTURE

The overall structure of a Tokamak reactor is indicated schematically in
Figures 13 and 14. For a discussion of other conceptual reactor types
together with a bibliography the reader is referred to ref [7].

At the beginning of the cycle the
plasma must first be raised to its
ignition temperature ~ 5 keV by a
combination of ohmic heating due to
the main toroidal current I,
together with some auxiliary heating
source. There are several possibilities including the use of RF
power, laser beams, turbulent heating,
relativistic-electron beams, adiabatic
compression and injection of neutral-
atom beams, the two latter being
currently the most favoured. It must
then be maintained at the chosen
operating temperature by self-heating
due to α-particles produced by the
thermonuclear reactions, which may
require some ford of feed-back
stabilization to avoid thermal
instabilities mentioned in §5. Heat
is lost during this period by
bremsstrahlung, impurity radiation,
synchrotron radiation, ion and electron thermal conduction, plasma
diffusion, and charge-exchange with
neutral atoms emitted by the first
wall.

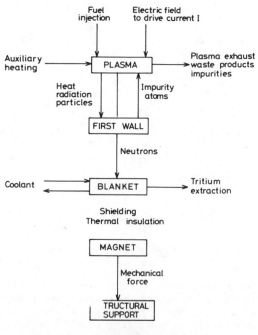

Figure 13.   Main Components of a
Tokamak Reactor

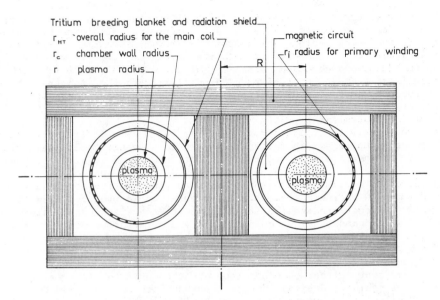

Figure 14.   Schematic description of a Tokamak-type reactor

The Tokamak cycle is not expected to last indefinitely because the current
I has to be excited by transformer action using an iron core, but the
expected burn time of 5400 sec in UWMAK is long compared to the energy confinement time of 4 sec. The total cycle time is 5790 sec. If the current
can be maintained by neutral injection (Ohkawa current) or by plasma diffusion (bootstrap current) then it may be possible to operate a Tokamak as a
steady state device.

Because of the length of the cycle it is necessary to provide for the removal
of waste products and impurities and the addition of fresh fuel and these
constitute major problems at the present time. It is envisaged that plasma
will be continuously extracted by deforming the magnetic field lines at the
outside and several types of divertor have been devised, an azimuthally-
symmetric 'poloidal divertor' being illustrated in Figures 12 and 15.
Fresh fuel might be provided by firing in pellets of D-T ice, by inward
diffusion of neutrals from a 'gas blanket', or by the injection of neutral
beams.

The interaction between the plasma and the first wall is critical for both
components. The first wall is bombarded by 14.1 MeV neutrons, lower-energy
neutrons, γ-rays, X-rays, synchrotron radiation, plasma ions and electrons,
and fast neutral atoms and absorbs some 10% of the total energy produced in
the reactor. The success with which it can be protected from damage arising
from this bombardment and from effects such as fatigue due to thermal
cycling sets a limit to the power/unit wall area that the reactor can produce and is therefore economically important. A current working estimate
for the wall loading is 1-2MW/m^2. In any case it is expected that the first
wall will have to be replaced more than once during the life of the reactor,
so that the whole structure has to be made modular, demountable, and readily
and quickly maintained in a radioactive environment.

The impurity atoms emitted from the first wall can move into the plasma,
become ionized in a series of stages, and diffuse rapidly into the plasma
where they cause energy loss by radiation.) They represent a serious problem in a quasi-steady-state device , but attempts are being made to reduce
their effect by measures such as the use of a low-Z wall, minimizing the
interaction of high-energy ions with the wall, divertor action, reduction of
inward impurity diffusion by controlling the temperature and density profiles,
preferential ejection (e.g. by resonant RF), or the natural stirring of the
plasma due to turbulence.

The blanket is intended to multiply the neutrons by the (n,2n) and (n,n'T)
reactions, to moderate them, to regenerate tritium from lithium, to generate
heat, and to act as a neutron and γ-ray shield) It is subjected to an
intense flux of 14.1 MeV and lower energy neutrons and therefore suffers
radiation damage including creep, embrittlement, void production due to
helium and hydrogen produced by (n,α) and (n,p) reactions, and nuclear transmutation (which changes alloy compositions and structures for example). It
also becomes radioactive, depending on the major or minor isotopic constituents present. The limiting neutron fluence that a given component can
tolerate before it has to be replaced is an important factor in determining
the economic cost of thermonuclear power. If some of the more exotic blanket elements could be used again once they had been rendered radioactive
then logically they ought to be counted as 'fuel', and their availability
might dominate that of the lithium (and especially deuterium) supplies, but
with 'third millenium technology' it should be possible to re-fabricate components from radioactive material and to re-assemble them by remote control
as well as to dis-assemble them from the reactor, or alternatively to remove
the radioactive isotopes using laser isotope separation.

Blanket cooling may be by liquid lithium, fused lithium salt or helium gas.)
Two potential disadvantages of liquid lithium are firstly the extra energy
needed to pump a metallic conductor across the strong magnetic field, and
secondly the tritium release that could occur in the event of a lithium fire.

Tritium extraction at the rate of 0.7 kg/day (UWMAK) is relatively straightforward and does not require a large plant, but since tritium gas diffuses
readily through hot solid materials in the same way as hydrogen the containment problems are quite severe, partly because it is a valuable fuel which
must not lose itself in inaccessible regions of the reactor or become mixed
with ordinary water, and partly because it is radioactive and so must not be
allowed to escape into the environment. Analysis indicates   that these
conditions can be met.

A Tokamak reactor is expected to operate with a superconducting magnet in
order to avoid excessive heat dissipation due to the electric current in the
toroidal field winding. This must be thermally insulated and shielded
against neutrons and γ-rays) as indicated in Figure 15, (and the outward pressure of the B_θ-field must be balanced by an appropriate mechanical structure)
The limiting strength of this structure, together with the maximum field that
the superconductor can provide, determines the maximum plasma density and
hence the thermonuclear power density as indicated in §6.

Since the main part of the mechanical structure is likely to be outside the
superconducting coil and therefore in a region of low radiation flux, radiation damage considerations are unlikely to be important and the main economic

criterion is that of strength/unit cost. If the inner parts of the blanket are supported from the outside, those sections that are subjected to an increasing flux carry progressively less weight and therefore need not be so strong.

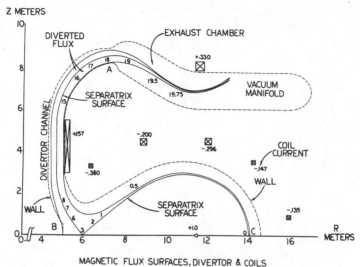

MAGNETIC FLUX SURFACES, DIVERTOR & COILS

Figure 15. Magnetic flux surfaces, divertor and coils

8. SUMMARY AND CONCLUDING REMARKS

Three of the main tasks facing us over the next few decades are to avoid nuclear war, to control world population growth, and to secure a permanent energy supply. The provision of adequate food and raw materials is also urgent, but given an abundant supply of energy and good organization this problem can probably be solved by the use of artificial fertilizers, farm machinery, irrigation, synthetic food production, recycling etc.

World coal supplies are sufficient in principle to last some 200 years, but economic problems associated with the balance of payments appear to prevent us from relying entirely on coal, and in any case nuclear fission power is both cheaper and cleaner than power from fossil fuel although considerably more complex. A large international fission power industry is expected to develop between now and the end of the century, together with a growing industry for the production of synthetic liquid and gaseous fuel for road, marine and air transport. Many countries will find it necessary to develop their technological and management potential to the point at which they can play an active role in these international industries in order to avoid serious balance of payments problems comparable with those which at present arise from the import of oil. Thus the world is likely to become increasingly dominated by technology, notwithstanding a current trend of opinion in the opposite direction.

As the supply of U^{235} from cheap uranium ore begins to diminish and the price to rise it will become economically attractive to introduce reactors which use plutonium fuel.   Initially these are likely to be thermal reactors burning the plutonium that such reactors already produce, then fast breeder reactors with an initial charge of plutonium from thermal reactors, and finally fast breeders providing their own plutonium.

Although there appears to be no alternative to the use of thermal, U^{235}-fuelled fission reactors for the next 3-4 decades, in the longer term there is a potential choice between the fast breeder reactor, fusion power, and solar power. Each course has its disadvantages; the fast breeder because of possible hazards associated with safety and with the proliferation of plutonium, and fusion and solar power because they have not yet been shown to be scientifically or technologically feasible. The importance and urgency of securing an abundant and safe world energy supply after the year 2000, and the long timescale needed to achieve it, are such that all three courses need to be actively pursued now.

REFERENCES

[1]   'A National Plan for Energy Research, Development and Demonstration: Creating Energy Choices for the Future', Vols.1 & 2, U.S. Energy Research and Development Administration Report ERDA-48 (Washington D.C. 1975).

[2]   Kinsey, R.P. and Llewelyn, G.I.W., 'Nuclear Merchant Ship Propulsion - The Present Status in the U.K.', International Conference on Nuclear Merchant Ship Propulsion, Atomic Industrial Form Inc., New York, May 1975. Reprinted in Atom No.227 (September 1975), published by UKAEA.

[3]   Brueckner, K.A. and Jorna, S., 'Laser-Driven Fusion', Rev.Mod.Phys. 46, 325 (1974).

[4]   Marshall, W., talk given at a New Scientist conference 'Towards a Self-Sufficient Britain', London (March 1975). Reprinted in Atom No.223 (May 1975), published by UKAEA.

[5]   Clayton, D.D., 'Principles of Stellar Evolution and Nucleosynthesis', Mc Graw Hill,(New York, 1968).

[6]   Kammash, T., 'Fusion Reactor Physics ; Principles and Technology', Ann Arbor Science Publishers Inc. (Ann Arbor, Mich. 1975).

[7]   Ribe, F., 'Fusion Reactor Systems', Rev.Mod.Phys.47, 7,(1975). 'Fusion Reactor Design Problems', Proc.Culham IAEA Workshop, Nuclear Fusion, Special Supplement (1974).

[8]   Taylor, J.B., Phys.Rev.Lett. 33, 1139 (1974).

[9]   Suydam, B.R., Proc.2nd International Conference on the Peaceful Uses of Atomic Energy, Vol.31, p.157,(IAEA Vienna, 1958).

DISCUSSION

Question: Fission technique is taking a long time to develop. How long will it take before the first fusion reactors appear?

Answer: Feasibility may be demonstrated within the next decade, while a prototype may be built at the end of this century.

Question: How is heat removed from the device?

Answer: Lithium, lithium salt and helium may be used as coolants, although electro-magnetic losses may be associated with lithium.

Question: How long does it take to recharge a tokamak after an hour's run?

Answer: It takes a few minutes to recharge, but a more serious problem is that of thermal shock.

Question: Would you comment on laser fusion?

Answer: To attain Lawson's criterion at normal density, large mass is required. By using lasers, material can be condensed by a factor of 10^4, so that the mass required to reach Lawson's criterion is reduced by 10^8. Currently, a large amount of work on the design of D-T pellets is being undertaken in the U.S.A.

Question: What are these pellets made of?

Answer: They are glass pellets filled with D-T gas and, for an economically available reactor, must be made at the cost of a penny each.

Question: What is the breeding ratio of tritium from lithium?

Answer: This is not well known at this moment but experiments are carried on at various centres. In Britain an experiment is being carried out at Birmingham University.

Question: When would we run out of lithium?

Answer: In about two hundred years.

Question: Will the JET-project satisfy Lawson's criterion?

Answer: The aim of the experiments is to measure the diffusion coefficients in order to determine the minimum size of a fusion reactor.

Question: Would the money for fusion research be better spent on solar energy?

Answer: All energy alternatives must be investigated, and the total amount of money spent on energy research in U.K. is small in comparison with certain aircraft development projects, for example.

Question: How does European fusion research compare with U.S.A.?

Answer: European fusion research is performed under the auspices of EURATOM, centred in Brussels. The total budget is similar to that of U.S.A., but it is possibly less well directed.

Question: How safe is a fusion reactor?

Answer: Unlike plutonium, tritium cannot be made into bombs, so it may be politically acceptable. There is no possibility of a nuclear explosion although there is some possibility of fire.

Question: Is it proposed that rare materials be used in fusion reactors?

Answer: Yes, niobium for example.

PANEL DISCUSSION :
THE PROSPECTS FOR NUCLEAR POWER

Panel: Dr. D.B.R. Kenning, Lincoln College, Oxford, (Chairman).
        Mr. W.B. Dale, UKAEA, Risley.
        Prof. D.C. Leslie, Queen Mary College, London.
        Mr. D. Merrick, National Coal Board, London.
        Dr. K.V. Roberts, UKAEA, Culham.
        Prof. R. Wilson, Harvard University.

Kenning: I will remind you of Professor Wilson's comment that the voiced criticisms of nuclear power are very often concerned with what seems to be the relatively minor aspects of particular systems when, in fact, the critics really have in their minds much more frightening possibilities, things like acts of war, terrorism, or perhaps genetic damage to populations. I would like you to start the discussion on these major topics followed by the rather more specific points associated with particular reactors.

Question: How many nuclear plants are there actually operating in the United States?

Answer: Approximately 50 and they are almost all light water reactors; pressurised water reactors made by three manufacturers Westinghouse, Babcock and Wilcox, Combustion Engineering, and the boiling water reactors only by General Electric. There is one other type coming up, a high temperature reactor which might be regarded as a development from the British gas cooled reactors. There is a small one working and a big one which has had teething troubles during the last three years.

Question: How many incidents have there been?

Answer: It depends what you mean by incident. There are two ways you can define it. There is the strict definition of the Nuclear Regulatory Commission defined as abnormal operating experience which can include anything from one of the panel meters tripping at 2% over normal. One of the reasons why these have to be reported is that by such means one obtains statistics on the reliability of the instruments which we put in the safety analysis. Each one of these reactors has 15 or 20 incidents every year. Nothing has come really close to being a major accident. There was one accident which shut a power plant down. That happened several months ago at Browns Ferry in Alabama where someone with a candle shut down a 2000 MW power plant. It was rather an absurd accident, but it is an example of what happens in accidents. Both reactors had their control cables pass through a common spreader room which was pressurised so that radioactive gases do not enter. An electrician replaced a small cable and did something which is very common in the electrical trade but which is not allowed in the nuclear trade. He was testing to see whether he had sealed up the hole by looking for airflow with a candle flame. The cable material is supposed to be fireproof but it wasn't and, of course, the lower

pressure of the reactor pulled the flames along the cable spreader room out of
reach. The redundant control circuits were supposed to be 6 feet apart but
they caught fire.   Thus, the reactor was shut down. It was nowhere near a
severe accident because of three other manual controls, but it was enough to
shut the reactor down for at least three months and the Nuclear Regulatory
staff will insist on inspections for another month or two. Substitute fuel
will cost $50 million.

Question: Do your figures for incidents cover reactors only or also the processing side of the industry?

Answer: Reactors only. The processing side of the industry has had no incidents. I am leaving out non-nuclear incidents, as these happen in other plants
also, though slightly less often in nuclear plants because of their better
attention to safety. Statistically, there are slightly fewer automobile accidents among the reactor staff.

Question: I would like to ask you about the candle accident. Has Rasmussen
accounted for this?

Answer: It is in accordance with the numbers that Rasmussen was using. The
particular accident with the candle was not anticipated, but the number of
occasions that two systems – two redundant systems – were simultaneously inoperative was included; the amount of time they were inoperative is in agreement
with Rasmussen's calculations. Rasmussen never said they should be inoperative
for any specific reasons and I do not think it would be appropriate to start
calculating this.

Question: On the question of the reactor vessel failure, can you prove the
rate is $1/10^6$ years?

Answer: That's a very tricky one, but I do know I could justify $1/10^5$ moderately well on experience and it is justified in reports on the experience of all
the pressure vessels which have been used in all industries.

Comment: These probability projections do involve the collection of a fantastic
amount of plant equipment data and it is going to take many years to accumulate
this information. People at UKAEA Risley have been collecting now for eight
to ten years and I think they have the best collection. They have been collecting from all over the world and I think this information is available to people
all over the world as well.

Comment: I do not think it is given away. It is exchanged certainly for good
valid information from other sources.

Question: When the nuclear industry starts to expand, will the quality control
and the attention to detail be as good as it is at present?

Answer: This will be a problem not only in the States but all over the world,
particularly in the developing countries where there may be a considerable
training problem.

Question: The number of nuclear power plants projected for the States is 200
or 210 by 1990. If that is the case, have the manpower studies been done to
indicate where the number of nuclear engineers, the trained technicians and

personnel that are going to man these plants are going to come from?

Answer: We have done some preliminary calculations on what would be required for the UK nuclear programme going forward to the year 2050 and there would be less employees than are actually employed by ICI and this is in the design, the construction and the operation of the power stations and including the full reprocessing of the fuel.

Question: On what time scale do we need to begin?

Answer: It is happening now. The people will simply be drawn away from other areas of industry where perhaps they will be less well paid. The other parts of industry will make do without them.

Comment: We are talking about the prospects of nuclear power and we have assumed so far that the major difficulty with nuclear power is its safety, whether you consider this in the context of accident or ultimately in the misuse of fissile material. I am by no means convinced that the economic case on behalf of nuclear power is well made. It certainly is different in different countries at different times, but I would like to comment on the consequences of the Browns Ferry fire. That fire knocked out 15% of the operating capacity of the Tennessee Valley Authority in half an hour and the consequences in the local grid system must have been fairly traumatic. Certainly the cost was and is very substantial indeed. We have also seen evidence, including that from ERDA, the successor of AEC, and from the Atomic Industry Forum, that the capacity factor of the operating stations in the US is, at the moment, under 60% and is hovering, in the case of some stations, below 50%. This means that if you invest in a station, you must assume that you are going to have to build two for the yield of one. This compares with the 80% which is usually assumed as the operating capacity factor for mature stations. Some of you will, undoubtedly, be familiar with the work done by David Coney, who is a very well informed nuclear critic based in Chicago; he has pointed out that the problems which are beginning to emerge with stations that have been operating up to six and seven years are not the "learning curve" problems in which the stations are presenting the usual problems of just getting the hardware operating properly. They are problems associated with active maintenance, with crud accumulation, or problems that you might call "premature senility". We know that there are a number of nuclear stations that have had to be shut down after comparatively short lengths of time for exactly this reason. Now it seems that we have a very wide variety of resource allocation options available to us and it does seem that we are short-changing them by devoting so much of our money, manpower and materials to this one narrowly-focussed option. I know that in the rest of the Summer School attention is being paid to these other options, but I would like to remind people that the economic case for nuclear power is not proved; in my opinion and before I see my share of the gross national product in the UK devoted exclusively to development of nuclear energy, I would like to see a lot more convincing salesmanship on its behalf.

Question: What is the availability of fossil plant in the USA on the same scale?

Answer: About 65 - 70%.

Comment: The difficulty is that there is this alleged economy of scale when you must build a 1000 MW plant instead of ten 100 MW plants. We have the economy of scale so far beyond what is optimum for purposes of maintenance, construction lead time, etc. For purposes of financing, a discount rate of over 10%

might suggest we construct more plants of smaller size. This argument applies to both nuclear and fossil plants.

Comment: The British nuclear plant capacity factors appear to be 90, 69, 76, 94, 14 (an experimental plant), 70, 72, 90, 91, 66, 77, 80 and 84%.

Comment: I would like to comment briefly on economies of scale. Our decision to build SGHWR meant we would not adopt 1320 MW but 660 MW for just the reasons mentioned. The Canadian experience is that it took seven years beyond the programme date to get the prototype CANDU at Douglas Point working acceptably. Since then, the operation of Pickering really has been quite brilliant, marred only by some fabrication errors in some rolled joints which is a one-off thing that will be put right. I think this is prolonged childhood and not premature senility.

Question: I have the distinct impression that the amount of money we will have to pour into fusion research over the next fifty or sixty years would produce a much better effect if it was pumped into, say, solar energy where we are not going to use quite so many resources of the earth. Sunlight is not going to drop in intensity over the next two or three centuries.

Answer: If solar energy can supply energy economically on the scale we require, it is preferable. At the moment, fusion looks more likely than solar energy but you have to put money into both as well as into fission. The scale is small compared with the GNP of the world. Far more is being spent on drinking and smoking than on energy research.

Comment: It is important to point out that in the UK, work in the Open University has established with reasonable certainty that within the present economic structure of the UK, the end use energy requirements are about 55% for low temperature heat and only about 5% for uses which are very unambiguously electrical for example, smelters, electronics, etc. We have a bad mismatch between the supply and demand, in terms of thermodynamic quality of energy. It would be well to decide whether we intend to improve the match.

Comment: Regarding the comparative costs of coal-fired and nuclear power stations, we do not know what it costs to build a nuclear station in the UK. Not many have been completed since the increases in fossil fuels prices. For chemical engineering plants in general, prices have generally increased in sympathy with the trends of energy prices. It still requires to be demonstrated that nuclear power is significantly better on costs than fossil fuels, even for base load operation, and uranium prices are uncertain in the future. The cost of generating electricity from a nuclear station is not particularly sensitive to this, but we cannot be sure of the multiplier factor to use over a 25 year life.

Comment: There is no recent British experience but there is a fair amount of American experience. The inflation has been greatest of all in field work and the extra cost of nuclear plant compared with fossil plant is mostly in factory fabrication which has not escalated as much as field work. The cost gap is now reckoned in the States to be only about 25% and I do not see why it should be any greater in this country.

Question: What system will keep the waste safe in the year 2400 AD when the breeder reactors have run out of fuel and there is still 19,600 years to half life?

Answer: The fission products are kept until their activity has fallen when the plutonium can be recovered and used as fuel.

Question: Is there any previous occasion in history when such technological commitments were made on behalf of generations yet to come?

Answer: Every generation in the past has always made commitments for the future either explicitly or implicitly but we have been unusual in this generation in even thinking about what commitments we are making. The amount of radioactivity we have in the natural environment is so great and it is so much around us that it is very hard for me to conceive that we cannot find a good method for handling the waste. Whether we are doing it or not is another question.

Question: I'd like to ask about the so-called "nuclear dustbin" to burn waste products.

Answer: With 14 MeV neutrons you can, in principle, burn up undesirable products by means of (n, 2n) reactions and so on. It does not look too feasible at present.

Comment: The resources in the UK and in the US are being disproportionately distributed between the specifically nuclear energy supply option and other alternative options. In the UK, for example, there is something like £92 million allocated specifically for government funded energy research and development in 1973/74. £68.5 million of this is specifically allocated for fission research, including £33 million for the fast reactor on which we have been working on for twenty-five years with no full-power operating prototype yet. We might possibly share the wealth more evenly between nuclear and other forms of energy research. For example, the government should allocate some money to build an operating prototype fluidised bed combuster to remove SO_2.

Question: What happens to the waste material?

Answer: It can be regenerated and recycled. Limestone or dolomite is converted into a sulphate and can then be regenerated to recover the sulphur as elemental sulphur for use in chemical processes.

Question: Who is going to make decisions on the nuclear programme? We need a great deal more publicity so that people are given the facts we have heard at this School.

Answer: It takes a great effort to become competent to pronounce on this issue, though participatory democracy is appropriate for political issues. The technical issues are probably best left to the people who are prepared to attend to them.

Comment: I do not think this is a question of technical issue because we have just been talking about safety. When we talk about $1/10^6$ as being an acceptable risk, that is not only a technical point.

Answer: I agree it is not a technical figure but it was decided in comparison with other acceptable risks.

Comment: I have taken part in a public enquiry and a question that was put to me by local councillors was not, "What is the risk and how do you calculate it?".

but "Does the Nuclear Inspectorate know what it is doing and is it to be trus-
ted?". That is the question that ordinary people ask.

Comment: Comparing the risks of fossil fuel and nuclear plant is very difficult because in fossil plant there is a high risk of a series of comparatively
small occurrences, while with nuclear plant there is a small risk of a catastrophe. Perhaps the public is less inclined to take even a small risk if it
is going to have an affect on a large number of people and successive effects
over generations. It is not a straight forward comparison.

Question: Regarding nuclear safety, nuclear bombs and not nuclear reactors
are the more frightening, and we have spent no time so far discussing the connection between civil nuclear proliferation and the availability of fissile
material for malevolent purposes. There seems to be no way out of this and it
seems that governments can see no solution to this either. We mentioned the
West German-Brazilian deal earlier in the day and this is a travesty of international responsibility. Canada now seems to be bent on making a similar deal
with Argentina and possibly also with South Korea — two countries of uncertain
stability. There is no rationale for this export programme except by a direct
analogy with the armaments trade which, in order to be able to support the
industry on the scale required for domestic needs, you have to reduce unit
costs by building more of whatever it is you are building, whether missiles,
planes or reactors so you can sell overseas. The scale of the international
nuclear industry has reached this level and I would be interested in any panel
comments on how you might possibly deal with this.

Comment: Over the last year the Fabian Society has been looking at Government
Science policy and is interested in trying to democratise scientific decision-
making. Our best hope is people like "Friends of the Earth", i.e. pressure
groups. The technical problems involved mean that ordinary people cannot get
involved so they can't make a decision. Their best hope is to get pressure
groups on both sides to really thrash out the arguments.

Comment: I would expect the "Friends of the Earth" to point out the consequences of not 'going nuclear" to be prepared to accept the consequences of not
doing so and to be prepared to persuade people to accept the lower standard of
life that would follow.

Comment: Of course the "Friends of the Earth" make a great point of saying
"Don't do this, do that instead". They never say "don't" without saying "do".
Accepting a lower standard of life is difficult and people will fight to maintain their standard of life. If the alternative is that people would have to
do without energy, they would have to accept a lower standard of life, they
would have to make their clothes last longer, they would have to do without
automobiles. However, we may not have to do without energy. There are many
different sources of energy which have not received enough attention.

Question: Regarding proliferation of plutonium, what can be done? Public opinion in any of the developing countries would surely be in favour of nuclear
power. Even the equivalent of "Friends of the Earth" in most of the developing
countries of the world would be in favour of nuclear power and would not be too
concerned with the consequences. If they are not to have nuclear power, from
what source will they derive their power in the short term — perhaps ten or
twenty years? Fusion is, in principle, a solution but not on this time scale.
Nuclear power will proliferate; nobody knows how to prevent this proliferation,
and we are going to be living in a very dangerous situation.

Answer: We can insert a time delay. Argentina has a graphite reactor and it is rumoured that a graphite reactor was bought because fuel loading was faster and plutonium 239 free of Pu^{240} could be obtained for possible bomb manufacture. That was some years ago. However, the possession of reprocessing plants and isotope separators is crucial if a mad dictator should emerge. If a country has those, within about six or eight weeks enough Pu could be obtained for a bomb. However, if a reprocessing plant and an isotope plant has to be built, it would take two years and that delay is enough for the world to come to its senses. So the crucial thing is to arrange for regional facilities under IAEA control so that a Hitler could not build a bomb. The Russian and Chinese governments have learned the dangers of this,for example; there is certain stability in this, so a two-year delay may be quite important.

Comment: Regarding energy supply for the third world, this is very important and in those third world countries in which there is a "Friends of the Earth Affiliated", the latter are not enthusiastic about importing reactors. Also, in the present state of technology, the only way to distribute the output from a reactor is through an electricity grid and most of the countries which are possible customers for reactor systems do not have an electricity grid, nor do they have the electricity demand which is required for keeping a grid stable. What they do have is low latitude almost without exception, and to do some research that will be a real value to the third world countries and low latitude areas, we should concentrate on solar energy systems because they will bring about the natural resource redistribution on this planet.

Comment: In order to establish good data on which elitists or others may make choices, can the public have any faith in conclusions reached about alternative sources of energy in the UK when most people doing the studies are employed by the atomic energy organisations? Even ETSU is, presumably, composed of people who are out of work in atomic energy.

Blair: Take that man's name!

SECTION 4

ALTERNATIVE ENERGY SOURCES

GEOTHERMAL ENERGY

E. R. Oxburgh

Geology and Mineralogy Department, University of Oxford

INTRODUCTION

The use of geothermal energy involves the extraction of heat from rocks in the outer part of the Earth. It is relatively unusual for the rocks to be sufficiently hot at shallow depth for this to be economically attractive. Virtually all the areas of present geothermal interest are concentrated along the margins of the major tectonic plates which form the surface of the Earth. Heat is conventionally extracted by the forced or natural circulation of water through permeable hot rock. Steam generated in this way is used in turbines for electricity generation. The many areas where impermeable hot rock is at shallow depth are not at present exploitable, but if current research into the generation of fissure systems by explosive or other means is successful, they may become so. The number of regions of geothermal interest would also be significantly increased by the development of commercially attractive vapour turbine cycles which were able to use low enthalpy waters for power generation. Low enthalpy waters ($\sim$ 100 cal/gm) are at present used for agricultural or domestic heating. If geothermal energy is used in Britain it will probably be in these latter ways.

Geothermal energy has been used to generate electricity at Larderello in northern Italy almost continuously since 1904 by using the energy of hot natural flows of water or steam. By 1972 Italy had installed geothermal power stations with a total capacity of 390 Mw at a capital cost of about $100/Kw. Total production costs are estimated at 5-9 US mills/Kwhour. In addition, both New Zealand and the United States have major geothermal power stations in operation and a dozen or more other countries are either actively exploring the possibility of such installations or have small schemes already working.

385

In this paper we review the main features of geothermal power and discuss its possible future development and limitations. We take the use of geothermal power to mean the "economic extraction of energy either for driving machines or direct heating, from naturally hot rocks in the upper part of the Earth's crust".

THE ENERGY SUPPLY

The Earth is believed to be close to a state of thermal equilibrium where the energy which is received at the surface by solar radiation is lost again at night, and the much smaller amount of energy which is generated by the decay of unstable isotopes of Uranium, Thorium and Potassium distributed within the Earth is balanced by the small continuous heat flux from the Earth's interior to the oceans and atmosphere.

Heat generation within the Earth is approximately 2700 Gw, roughly an order of magnitude greater than the energy associated with the tides but about four orders less than that received by the Earth from the sun. The mean surface heat flux from the Earth is about 1.5 Heat Flow Units (1 HFU = 10^{-6} cal cm^{-2} sec^{-1}).

Temperature distributions within the Earth depend on:
1) the mean surface temperature (which is controlled by the ocean/atmosphere system).
2) the abundance and distribution of heat producing elements within the Earth.
3) the thermal properties of the Earth's interior and their lateral and radial variation.
4) any movements of fluid or solid rock materials occurring at rates of more than a few millimetres per year.

Of these four factors the first two are of less importance from the point of view of geothermal energy. Mean surface temperatures range between $0°$ and $30°C$, and this variation has a small effect on the useable enthalpy of any flows of hot water. Although radiogenic heat production in rocks may vary by three orders of magnitude, there is much less variation from place to place

in the integrated heat production with depth. | The observed range in depth-
integrated heat production for the upper (and most variable) part of the
crust is equivalent to about 1.5 HFU.

The latter two factors, however, are of great importance and show a wide
range of variation.| Their importance is clear from the relationship

$$\beta = \frac{q}{k}$$

where, for a steady state, β is the thermal gradient, q is the heat flux and
k is the thermal conductivity. The first requirement of any potential geothermal source region is that β be large i.e. that high rock temperatures
occur at shallow depth. Beta will be large if either q is large, or k is
small, or both.

By comparison with most everyday materials, rocks are poor conductors of heat;
values of conductivity may vary from 2×10^{-3} to 10^{-2} cal cm^{-1} sec^{-1} $^{\circ}$C^{-1}.
Rocks are also very slow to respond to any temperature change to which they
are exposed, i.e. they have a low thermal diffusivity, $K = \frac{k}{\rho C_p}$, where ρ and
C_p are density and specific heat respectively.

Diffusivity values are of the order of 10^{-2} cm^2 sec^{-1} and are so low that the
effects of diurnal surface temperature variation are damped out at depths of
less than a metre and effects of annual variation become insignificant at
depths of about 3m. The effects of major and longer term climatic variations
such as ice ages may be recognisable at depths over 1500m or more. However,
these effects change temperatures by only a degree or so at that depth. Thus
fluctuations in surface temperatures have little effect at depth, but the
presence of a thick surface layer of low conductivity rock such as clay
(Jurassic clay from the south-east English midlands has a mean conductivity
of about 2.5×10^{-3} cal cm^{-1} sec^{-1} $^{\circ}$C^{-1}) has the effect of a surface insulating blanket which can maintain unusually high temperatures within and
below it, although the heat flow itself may be of average value.

We turn now to variations in the Earth's surface heat flux, q. As indicated
above, the heat flux ultimately derives from the Earth's radioactivity, and
in consequence, regions with higher radioactivity will tend to have a somewhat higher heat flux. Very high heat fluxes, however, are related in a more

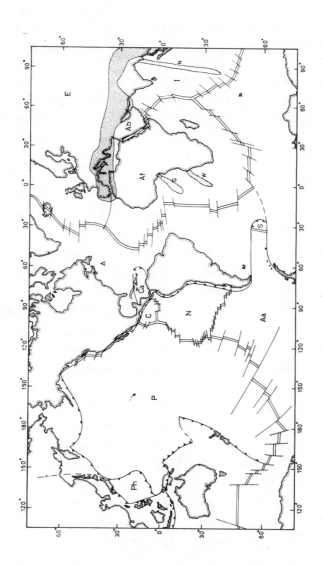

Fig. 1. - The main tectonic plates of the Earth's surface: Ph, Philippine plate; P, Pacific plate; Aa, Antarctic plate; Nasca plate; A, American plate; E, Eurasian plate; Af, African plate, I, Indian plate; Ab, Arabian plate. Toothed lines indicate zones of plate convergence, double lines zones of plate separation and single lines, plate margins along which plates slide past each other. Stippled area Alpine-Himalayan belt where two plates have recently locked.

indirect way to radioactivity.

It is now known that the surface of the Earth comprises a small number (a
dozen or so) of relatively rigid tectonic "plates"; these have a thickness
of the order of 100 km and lateral dimensions of the order of thousands of
kilometres (fig. 1). These places are continuously in motion with respect
to each other and form part of a large-scale, solid-state convective flow
system which involves at least the outer 700 km of the Earth. This flow
occurs because the interior of the Earth is unable to lose heat by conduction
as rapidly as it is generated by radioactivity and consequential convective
instabilities develop.

A variety of processes occurs through interaction between plates along their
margins, which lead to partial melting in and below the crust (i.e. at depths
from 15 to 200 km). The liquids produced in this way are about 10% less
dense than the material melting to produce them, and in consequence these
penetrate the surrounding rocks and rise rapidly (at rates of cm/yr to
cm/day) towards the surface. Those which reach the surface give rise to volcanic activity, others may come to rest in the middle or upper part of the
crust (i.e. depths less than 20 km) where they crystallise to form the intrusive igneous bodies familiar to geologists. Depending on their depth or
origin and the exact composition of the material undergoing partial fusion,
liquids (magmas) will arrive in the upper part of the crust at temperatures
between 800° and $1200^\circ C$.

The cooling and crystallization of igneous bodies which quite commonly have
characteristic dimensions of several kilometres or more, will give a very
high local heat flux for thousands to hundreds of thousands of years depending on local circumstances. Although on an Earth time scale of 4.5×10^9
years these are very short lived phenomena, an area of present igneous
activity and high heat flow can be expected to remain so on all time scales
relevant to our present energy problem. Active igneous areas commonly have
surface heat flux values of 10 HFU and occasionally 100 HFU or more.

To summarize, very high heat flux values and thus the majority of active
geothermal areas, tend to concentrate around the margins of the major litho-
spheric plates (compare figs. 1 and 2). High thermal gradients can also be

associated with abnormally low values of thermal conductivity; strata with
these characteristics may be found on any part of a continent. Table 2
gives the characteristics of many of the geothermal fields shown in fig. 2.

TABLE 1

	lower	average	upper
q (HFU)	0.8	1.5	3.0 (non volcanic) ∿100 (volcanic)
k cal cm^{-2} sec^{-1} ^{o}C^{-1}	2×10^{-3}	6×10^{-3}	12×10^{-3}
β ^{o}C/km	8	20	60 (non volcanic) ∿300 (volcanic)

Note these values are simple intended to give a general idea of the normal
range of geothermal parameters. In volcanic regions, in particular, both q
and β can vary considerably, and the upper values given are somewhat notional.

Values of Geothermal Parameters

TABLE 2

Field	Reservoir Temp., ^{o}C	Reservoir Fluid	Enthalpy, cal/g	Average well depth, meters	Fluid salinity, ppm	Mass flow per well, kg/hr	Non-condensable gasses, %
Larderello	245	Steam	690	1,000	<1,000	23,000	5
The Geysers	245	Steam	670	2,500	<1,000	70,000	1
Matsukawa	230	Mostly steam	550	1,100	<1,000	50,000	<1
Otake	200+	Water	∿400	500	∿4,000	100,000	<1
Wairakei	270	Water	280	1,000	12,000	-	<1
Broadlands	280	Water	400+	1,300	-	150,000	∿6
Pauzhetsk	200	Water	195	600	3,000	60,000	-
Cerro Prieto	300+	Water	265	1,500	∿15,000	230,000	∿1
Niland	300+	Brine	240	1,300	260,000	∿200,000	<1
Ahuachapan	230	Water	235	1,000	10,000	320,000	∿1
Hveragerdi	260	Water	220	800	∿1,000	250,000	∿1
Reykjanes	280	Brine	275	1,750	∿40,000	∿400,000	∿1
Namafjall	280	Water	260	900	∿4,000	400,000	6

Characteristics of Selected Geothermal Fields, from Koenig in Kruger & Otte, 1973.

HEAT EXTRACTION

So far we have considered only the distribution of hot, near-surface rock.
For such rock to provide a useful energy supply, however, the heat must be
extracted by some form of heat-exchanger. In many geothermal areas natural
heat-exchangers exist in the form of large scale, sub-surface water circulation systems, which commonly give rise to hot springs or geyser activity at
the surface.

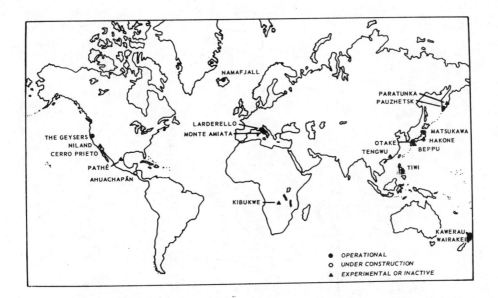

Fig. 2. – Geothermal electric power stations from Koenig in Kruger & Otte,
          1973.

It is at once evident that if heat is to be extracted either by natural or
artificial means the hot zone must be permeable, and at present this further
requirement prevents the exploitation of a number of otherwise promising
areas.) We return to this problem later, but for the present we consider the
characteristics of permeable regions in more detail.

Fig. 3 shows an idealized system. A large body of slowly crystallizing
magma lies at a depth of six kilometres or so below the surface. This has
penetrated rocks which are for the most part impermeable. In the region
immediately above the magma body the impermeable (crystalline) rocks are
overlain by a localized pocket of permeable strata. These strata are bounded
by and cut by steep faults (fracture zones along which some relative motion
has occurred – shown as heavy black lines on the diagram).

The upper levels of the permeable strata are less permeable than the lower.
In nature this could occur by chance through original variation in rock
properties. More commonly, however, it will be the result of the precipitation of dissolved solids from ascending hydrothermal fluids. Water in the

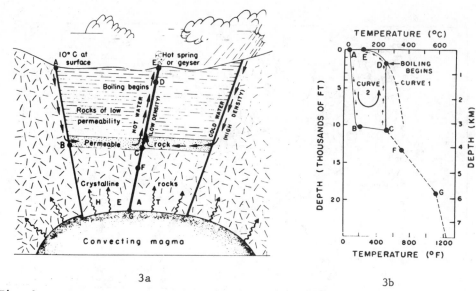

3a   3b

Fig. 3a. - Schematic representation of a geothermal system.
Fig. 3b. - Schematic temperature distribution expected in a); curve 1 is the
boiling curve; see text for discussion. From White in Kruger & Otte, 1973.

temperature range 150° - $300^{\circ}C$ readily dissolves silica and a number of other
rock constituents. Hydrothermal waters tend to be unusually rich in SiO_2, Cl,
B, Na, K, Li, Rb, Cs and As; occasionally they may contain significant amounts
of H_2SO_4. Therefore, the effect of hydrothermal circulation is often to
increase permeability by solution in the deeper, high temperature zone, and
to reduce permeability by precipitation in the cooler, lower pressure, near-
surface region. Thus many hydrothermal areas have a convenient self-sealing
property which means that after circulation has gone on for some time, surface flows may be largely or entirely restricted to places where holes have
been drilled for the purpose.

Fig. 3a shows the convective circulation of ground water; it circulates downwards from the surface along the marginal fault zones, moves laterally and is
heated passing through the permeable layer and finally rises to give a hot
spring or geyser at E. The permeable layer is heated by conduction of heat
from the magma chamber. The variation of temperature around the system is
shown in fig. 3b. Note that the gradients which are governed by convection
(AB, CD and G downwards) are much shallower than the conductive gradient (CG).

We now examine the shallow depth behaviour of the ascending flow by reference
to fig. 4. The heavy curve shows the variation of the boiling point with
depth of pure water under hydrostatic pressure. In nature the curve will be
modified through the effect of salts in solution and in many cases the pressures will not be hydrostatic. As the heated water rises, it undergoes
relatively little temperature change (CD, fig. 3b). Provided that its "base
temperature" (C, fig. 3b) is significantly above 100°C, at some depth its
temperature path will intersect the boiling curve (e.g. A, fig. 4) and the
water will begin to boil.

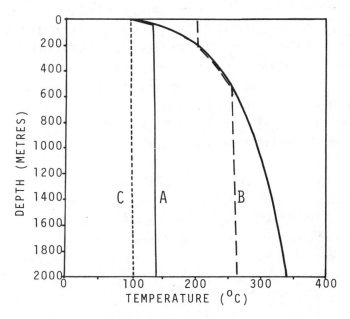

Fig. 4. - The boiling curve (heavy); see text for explanation of A, B and C;
after White in Kruger & Otte, 1973.

As it continues to rise its temperature drops, following the boiling curve,
with the generation of more steam. It now becomes clear that natural hydrothermal systems fall into three groups:-

1) Dry steam systems: these have a relatively high base temperature and a
   deep intersection of the boiling curve; all water is converted to steam
   which is super-heated when it reaches the surface (e.g. fig. 4 curve B).
   Both Larderello and the Geysers field in California are of this kind.

2) Wet steam systems: the boiling curve is intersected at shallower depth

but not all the water has been converted to steam by the time it reaches the surface, where there are commonly violent geyser eruptions of water-steam mixtures (e.g. fig. 4 curve A).

3) Hot water systems: in these the boiling curve is not intersected, or at most only in the upper few tens of metres. Hot springs are the normal surface manifestation (e.g. fig. 4 curve C).

In nature, however, the situation is rarely as simple as represented here. In cases where the near surface layers have been sealed by precipitation of silica, pressures may be significantly higher than hydrostatic at shallow depth. As discussed below, drilling into such systems may allow the con-trolled flashing of the hot water. With highly permeable surface layers, on the other hand, boiling may occur at lower temperatures and/or greater depths than indicated by the boiling curve. Because the hydrostatic assumption is no longer valid, once boiling has begun in a system which is in communication with the atmosphere, the pressure is controlled not by the weight of a column of water, but by the weight of a column of mixed water and steam.

Further complications arise in connection with the balance between flow of hot water out of the system, and the recharging of the system from ground-water. When the latter is inadequate to compensate for the surface losses, the nature of the system will gradually change. As the volume of water available for heating decreases, fluid pressures will tend to fall, boiling will occur at greater depths and the system may evolve from a wet-steam to a dry-steam type.

EXPLOITATION OF GEOTHERMAL AREAS

In general, exploitation requires the drilling of a number of holes at care-fully selected sites in a geothermal field which has previously been delin-eated and evaluated by geological and geophysical methods. Steam, hot water or both are then piped as short a distance as possible to a power station.

Economic viability depends largely upon the number, diameter and depth of holes which have to be drilled and the useable enthalpy and ancillary charac-teristics of the geothermal fluid. Drilling is an expensive business (fig. 5), and much effort has been devoted to securing the hole profile which optimises the fluid discharge/unit drilling cost ratio. At the Geysers field holes are typically about 2 km deep with a surface diameter of about 51 cm reducing by degrees to about 22 cm for the lower third of the depth.

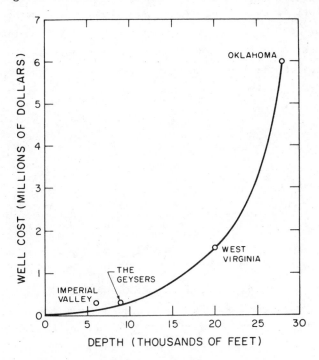

Fig. 5. – Drilling costs increase rapidly with the depth of the well (from Anderson in Kruger & Otte, 1973).

At the power station the treatment of the geothermal fluid largely depends upon its character. In general, the higher its temperature the better. In the commercially attractive, high-temperature (T > 200°C) dry steam systems, such as Larderello or the Geysers, the steam may be expanded directly through a steam turbine, to generate electricity. Fields of this kind are rather rare but produce over 70% of the world's geothermal electricity output at present.

In wet steam, sealed systems suitably located and pressure regulated wells may make it possible to control the depth at which flashing occurs; the steam may be separated from the water at the surface and again used to drive a turbine.

In both wet and dry systems there are various operational difficulties, although in the latter they are rather less. Dry steam at high pressures tends to have high velocities and to carry with it particulate material which can coat pipes and turbine blades; coarse fragments can damage the installation. Wet steam and the water associated with it can be very highly corrosive because of the dissolved impurities, and it is sometimes necessary to put it through a heat exchanger rather than into a turbine directly. In addition, cooling of the fluid leads to precipitation of the impurities, which in time leads to serious problems of furring of the pipes and other parts of the installation.

In recent years considerable attention has been devoted to plants using a heat exchanger and a vapour turbine cycle. The heat exchanger accepts liquid water from the well and uses it to heat a low boiling point fluid such as propane or freon, which in turn is expanded through a turbine and then condensed and reused. The schematic layout of such a scheme is shown in fig. 6.

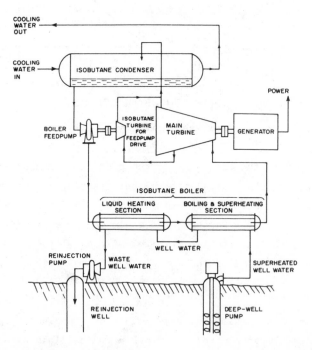

Fig. 6. – Schematic representation of a vapour-turbine cycle; from Anderson in Kruger & Otte, 1973.

The economic viability of such systems is as yet unproven although a small freon-based system has been in operation in the USSR for several years. Their advantage is that they may permit the profitable exploitation of much lower enthalpy geothermal fluids than is at present possible.

Although the major interest in geothermal power must stem from its potential for generating electricity, there are also a number of schemes in operation by which geothermal water too cool for electricity generation by present methods is used directly for town or agricultural heating purposes. Such waters which tend to be cooler than $150^{\circ}C$ have relatively low impurity levels and are generally not seriously corrosive. Systems of this kind are in operation in Iceland, Oregon USA, Hungary, the Soviet Union and a number of other places. Where hot waters are available close to urban areas, they often offer a very inexpensive form of heating. In all probability, it is in this way that geothermal energy will be used in Britain. Some comparitive cost information is given in Table 3.

TABLE 3

Geothermal field	Geothermal production	Local average, other fuel
Electricity, U.S. mills/kwh		
Namafjall, Iceland	2.5 - 3.5	-
Larderello, Italy	4.8 - 6.0	∿7.5
Matsukawa, Japan	4.6	∿6.0
Cerro Prieto, Mexico	4.1 - 4.9	∿8.0
Pauzhetsk, U.S.S.R.	7.2	∿10.0
The Geysers, United States	5.0	7.0
Space heating, U.S.$/Gcal energy		
Reykjavik, Iceland	4.0	6.7
Szeged, Hungary	3.0	11.0
Regrigeration, U.S.$/Gcal energy		
Rotorua, New Zealand	0.12	2.40
Drying diatomite, U.S.$/ton		
Namafjall, Iceland	∿2	∿12

Selected Comparative Cost Data for Geothermal Energy, from Koenig in Kruger & Otte, 1973.

RESERVES

As with any other naturally occurring commodity, reserves may be estimated in two ways: we may estimate the absolute abundance of that commodity without regard to its concentration, or alternatively we may regard as reserves only

those occurrences which, under prevailing economic conditions and with
present technology of extraction, appear capable of profitable exploitation.

To view the Earth's thermal reserves in the first way is meaningless because
although the Earth's total heat content is enormous ($\sim 10^{35}$cal) nearly all of
it is inaccessible to us and likely to remain so. On the other hand, to
apply the latter criterion at present is almost equally meaningless because
of the relatively small amount of effort which, until recently, has been
devoted to developing the technology of heat extraction from the Earth. It
is most unlikely that there remain many high temperature dry steam geothermal
areas to be discovered; their surface manifestations are so obvious that in
populated areas (where they are likely to be of use) they would have almost
certainly been discovered. Fig. 7 shows the major undeveloped geothermal
fields of the world which by present criteria (i.e. high temperature and
permeability) have potential for power generation. If this represents the
limit of the Earth's commercially exploitable geothermal resources, geothermal energy is unlikely to increase its share of the energy market (at present
about 1% in the United States).

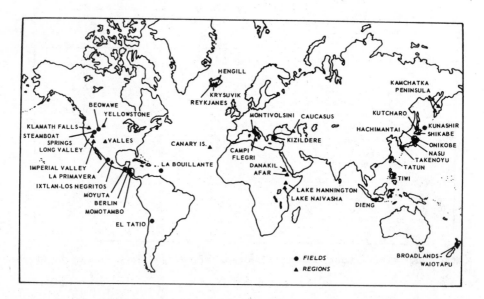

Fig. 7. - Major undeveloped geothermal fields or regions under active exploration; from Koenig in Kruger & Otte, 1973.

Although cautious, such an assessment of the future is perhaps unduly pessimistic. Technological advances in three areas could together or separately
considerably improve the prospects of geothermal power, although its economic attraction will as always remain dependent upon a comparison with other
energy sources available at the time.

The first possible area of advance is that of generating fissure systems in
otherwise impermeable rocks and thus allowing heat to be extracted from
extensive areas of high geothermal gradient which cannot be exploited at
present because water cannot circulate through them. A number of different
techniques are being used. Hydraulic fracturing by generating very high
pump pressures at restricted depth intervals in holes has been used for a
number of years in the oil industry to increase rock permeability. By itself
this method appears inadequate to generate fissure systems sufficiently
extensive for geothermal purposes. There is, however, now the possibility of
causing fractures generated in this way to undergo self-propagation as thermal contraction cracks as heat is extracted by circulating fluid. Alternatively, the fracture system may be initiated by chemical or nuclear explosions down the hole. In some systems of this kind it might be necessary to
force the circulation of the fluid by surface pumping, particularly in the
early stages. If techniques of the kind outlined above became fully developed and were routinely and relatively cheaply employed, the present geothermal resources (as defined at the end of the previous paragraph but one)
would be increased by a factor of between one and two orders of magnitude.

The second area of advance has already been mentioned above. At present,
enthalpies of > 200 cal/gm are sought for power generation and > 100 cal/gm
for direct heating. The lowering of one or both of these requirements by
the use of vapour turbine cycles or other methods would enormously increase
the volume of resources. The amount of increase would be related to the
degree of lowering of the requirement, but halving the power generation
requirement should again increase reserves by one to two orders of magnitude.

The third and, perhaps, least likely area for advance lies in drilling technology. Advances here are least likely only because much more time and
effort has been devoted to drilling methods over the years than to the other
problems. If, however, new and cheaper methods were developed (e.g. drilling

by rock melting is at present being tested), the depth range over which it
was profitable to extract heat from rocks could be considerably increased.
Most geothermal wells in use at present are less than 2 km deep. If this
range could be increased by a factor of two or three without large cost
increases, and if fluid circulation could be maintained at those depths,
perhaps half the area of the continents would be underlain by rocks capable
of yielding water with enthalpy of 100 cal/gm.

In summary, it is at present not reasonable to count on a major contribution
to the world's energy budget from geothermal sources. Advances in one of
several fields of technology could, however, abruptly change the situation.

PRACTICAL DIFFICULTIES

There are various practical difficulties and disadvantages associated with
the use of geothermal power. Some of these have been touched upon in the
previous sections and will only be recapitulated here.

1) The geothermal fluid is often highly chemically corrosive, or physically
    abrasive as a result of the entrained solid matter it carries. This may
    entail special plant design problems and unusually short operational
    lives for both the holes and the installations they serve.

2) Because the useful rate of heat extraction from a geothermal field is in
    nearly all cases much higher than the rate of conduction into the field
    from the underlying rocks, the mean temperature of a field is likely to
    fall during exploitation. The rate and significance of such temperature
    reductions will depend on local conditions (Larderello has operated
    successfully for 70 years and at an increasing power output). In some
    low rainfall areas there may also be a problem of fluid depletion.
    Ideally, as much as possible of the geothermal fluid should be reinjected
    into the field. However, this may involve the heavy capital costs of
    large condensation installations. Occasionally, the salinity of the
    fluid available for reinjection may be so high (as a result of concentra-
    tion by boiling) that is is unsuitable for reinjection into the ground.
    In that case, it must either be diluted before reinjection, purified, or

disposed of elsewhere. Occasionally the impurities can be precipitated and used, but this has not generally proved commercially attractive.

3) Transmission: geothermal power has to be used where it is found. In Iceland it has proved feasible to pipe hot water 20 km in insulated pipes but much shorter distances are preferred.

4) Environmental problems: these are somewhat variable and are usually not great. Perhaps the most serious is the disposal of warm high salinity water where it cannot be reinjected or purified. Dry steam plants tend to be very noisy, and there is release of small amounts of methane, hydrogen, nitrogen, ammonia and hydrogen sulphide; of these the latter presents the main problem. At the Geysers field it is present in the vented steam at the level of about 225 ppm; this is equivalent to about one quarter the rate of release of a coal fired power station (using coal with 1% sulphur) and the same power output.

Changes in porosity and fluid pressure in geothermal reservoirs could give rise to small amounts of seismicity especially if the rate of inflow of water is much less than the rate of withdrawal. It is thought that this kind of problem can be monitored and controlled fairly easily.

CONCLUSIONS

At present geothermal energy makes a very small, but locally important, contribution to world energy requirements. This situation will not change unless important technological advances are made. Environmentally it is probably the least objectionable form of power generation available at present with the exception of hydro-electric methods.

Although geothermal exploration of Britain is at a fairly early stage, all the signs indicate that the only resources available are likely to be low enthalpy, relatively pure waters. These may locally offer the possibility of very inexpensive agricultural or even domestic heating. The extent of such resources in Britain is not yet known and there is little in the way of domestic technology available for their exploitation. If any of the major

technological advances mentioned earlier occurred, the position in Britain
would be dramatically changed.

SELECTED REFERENCES

Kruger, P. & C. Otte (Eds.) 1973. Geothermal Energy – Resources,Production,
Stimulation. Stanford University Press, Stanford.

Leardini, T. 1974. Geothermal Power. Phil. Trans. Royal Society London,
A. 276, 507–26.

Open University, 1973. The Earth's Physical Resources & Energy Resources.

Proceedings of the United Nations Conference on New Sources of Energy: Rome,
August 21st–31st, 1961, 1964. Volumes 2 and 3 Geothermal Energy, New York,
United Nations.

Proceedings of the United Nations Symposium on the Development and Utilization
of Geothermal Resources, Pisa,Italy, September 22nd–October 1st, 1970, 1971.
Geothermics, Special Issue 2, 2 volumes.

DISCUSSION

Question: What would be the costs of drilling to extract geothermal heat?
Answer: It would be about one million US dollars for a 4.5 km bore-hole
            based on costs in 1972. Present day costs are undoubtedly greater
            but it does not seem more economical to drill several small dia-
            meter holes instead of a single large hole.

Question: Is a systematic geothermal survey being made of U.K.?
Answer: A map showing the geothermal character of the U.K. does not exist
            yet, but work is active in this respect and a map should be avai-
            lable in a few years. Unfortunately, the cost of drilling explora-
            tory bore holes 300 m deep is high and so progress on mapping has
            been slow. Holes drilled for other purposes have been used but
            often the requirements for these conflict with those necessary to
            make geothermal energy use worthwhile. This is particularly so
            for oil drilling which, for obvious reasons, is not carried out
            near the geological faults at which the magma approaches the sur-
            face. The magma is, of course, at such a high temperature that
            effective use for it cannot be found in our present technology.
            Should there be a likelihood of magma rising up the bore hole,
            evacuation of the drilling area might be necessary. There have
            been some geothermal surveys in the Paris basin which consists of
            fairly thin alternate layers of permeable and non-permeable mater-
            ial. Low enthalpy sources probably exist there which would be
            ideal for direct domestic space heating with water at about 50 –
            60°C but the generation of electricity is extremely doubtful.

When the U.K. geothermal map is complete, we should then be able to predict the possibility of district heating schemes with water temperatures up to about 80°C. An alternative procedure using a heat pump could be envisaged in which the geothermal energy is used to create a heat reservoir for the pump of, say, 40°C; however, solar energy absorbed within the first 3 m or so of the earth's crust could be used for the same purpose, provided that diurnal temperature variations could be accepted or attentuated.

Question: Will there be any seismic effects from heat extraction from deep holes?

Answer: Seismic effects might occur especially if equilibrium is not maintained between hot water extraction and cold water input to the permeable layers. These effects would probably be small. Underground nuclear explosions to increase permeability could be dangerous but little is known about this at present.

Question: Why is East Africa considered a potential site as it lies in the middle of a plate?

Answer: Because the edges of geological plates are the ideal locations for geothermal energy use, the interest shown in East Africa seems at first sight to be unlikely. It is now thought that, in time, a fault similar to that under the Red Sea will develop in a southerly direction from the Mediterranean Sea through East Africa causing an upwelling of magma.

Question: Would there be any advantage in locating heat exchangers underground?

Answer: This would be impractical because of the difficulty of removing furring from the exchanger surfaces. Although noxious gases would be suppressed, the heat transfer efficiency would undoubtedly be low.

Question: Can the water be used after heat has been extracted?
Answer: Common salt is obtained in California together with fresh water from condensation.

Question: What contribution could geothermal energy make to our power requirements?

Answer: Prospecting methods are still being developed and radon and micropulsation techniques still need to be made more reliable. When geothermal sources can be assessed with fair accuracy, it is reasonable to expect that, on the basis of present technology, some 1500 MW of geothermal power could be used to generate electricity by the year 2000. The widespread use of hydrofracture could increase this power by a factor of 3 or 4. It we could use low grade heat, this factor could be increased to 10 - 100, however, this might not be feasible because we cannot presently utilize the vast quantities of low grade heat rejected by power stations.

THE ATMOSPHERE AND THE OCEANS
AS ENERGY SOURCES

D. T. Swift-Hook

Marchwood Engineering Laboratories, CEGB

INTRODUCTION

Power from "natural sources" has always had great appeal. Barrels of oil,
lumps of coal, even uranium, come from nature but the possibilities of almost
limitless power from the atmosphere and the oceans seem to have special attraction. The windmill provided an early way of developing motive power (see Fig.1)
and, more recently, a project was completed to obtain power from the rise and
fall of the tides in the Rance Estuary but neither of these systems has been
considered to be economically viable in recent years. The massive increases
in fuel prices over the last year or two have, however, made any scheme not
requiring fuel appear to be more attractive and to be worth re-investigation.
In considering the atmosphere and the oceans as energy sources, the four main
contenders are wind power, wave power, tidal power and power from ocean thermal
gradients.

WIND POWER

Available Power

Winds are caused by pressure differences between various regions of the atmosphere and, throughout the British Isles, the biggest differences are caused by
the differential heating of land and sea, i.e. wind power comes ultimately
from solar energy. Winds are greatest around the coast and the wind contours
or isovents have a tendency to follow the coastline, as shown in Figure 2. It
is well known in principle how to capture the energy in the wind. Working
windmills are still to be found around the countryside, not all of them having

Figure 1.

An old Norfolk windmill. The
blades were covered with canvas
and the miller himself had to
turn the mill to face the wind.

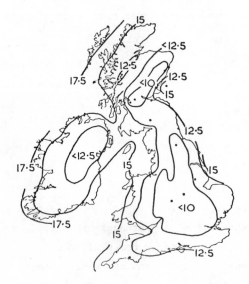

Figure 2.

Winds around the British Isles
are caused by differential
heating of land and sea and so
the isovents follow the coast
line.

been relegated to the status of "Ancient Monuments" or to the archives of
industrial archaeology.

The kinetic energy per unit volume is $\frac{1}{2}\rho v^2$; over a swept area A this corresponds to an energy of $\frac{1}{2}\rho A v^2$ per unit distance in the direction of the wind.
The wind's energy is convected along at velocity v and so the total energy
flux per second – the power – is $\frac{1}{2}\rho A v^3$. Not all of this power is available
in practice since, to extract it all, the air would have to be brought completely to rest and a device that did so would accumulate static air around it
which would prevent its operation. Only 16/27 of the total amount is theoretically available (even ignoring drag and frictional losses) or just over 59%.
Although some special designs can reach up to 80% of theoretical efficiency,
most modern designs of windmill do not often achieve efficiencies better than
about one third (older designs were much worse); in a fairly strong wind of
10 m/s, around Beaufort Force 5 in nautical terms, the maximum power is
0.3 kW/m^2. With sophisticated aerodynamic design of the windmill, drawing
wind from an area larger than the swept area, it might be possible to improve
this available output, but any such increase would almost certainly lead to
increased costs and there is, therefore, little prospect of major increases
in the power available per unit area.

This variation of power as the cube of the wind speed means that wind generators have a very non-linear performance and produce their greatest amount of
power when the wind is blowing as hard as can be handled. Above this rated
wind speed, a generator is governed so that it does not exceed the rated power.
When the wind blows especially strongly the windmill has to be shut down for
safety reasons. There is also a minimum wind speed needed to justify operation
– the "cut-in" speed – where system losses equal extracted wind power.

Wind data is usually given in terms of average wind speed, $\bar{v}$, and the average
power available will be greater than that given by inserting this average value
into the formula because the average of a cube is greater than cube of an
average. The difference depends very much upon the distribution, i.e. on how
much the wind speed varies. In practice it is found that the average power
obtainable is likely to be between 1.0 and 1.8 times that corresponding to $\bar{v}$,
the lower figure applying to windy sites with average speeds of more than
10 m/s and the higher figure applying to average sites with mean wind speeds
of 5 m/s or so.

The ratio of peak rated wind speed to average wind speed is an important parameter which governs the overall performance of a windmill system; it determines
the balance between the size of the windmill itself and the rating of the generator which it drives. For a given generator capacity or rated power, a low
peak/mean ratio calls for a large wind rotor to obtain full power at low wind
speeds, while a larger ratio calls for a smaller rotor giving full power only
at high wind speeds. A large-rotor machine is clearly more expensive but gives
greater average output and these two factors must be optimised. The economic
optimum is usually when rated wind speed is $1\frac{1}{2}$ to 2 times the average wind
speed. Because of the cube law, the average output of the generator is then
considerably less than the rated power, since the load factor is only 20% to
40%. Such low load factors make it difficult to compare wind driven generators with conventional generators of the same rating but much higher load
factors. The inconsistent nature of the wind makes it essential in most cases
to provide some form of storage or an alternative power supply. Modern experience with windmills producing up to 100 kW is that more than half the overall
cost of the energy supplied by a windmill-battery combination can be due to
the battery.

Available Sites

The British Isles are set in one of the windiest regions on Earth. The winds
are strongest around the west coast of Ireland, Scotland and Wales (7.8 m/s
average wind speed), fairly strong around the other coasts (5.6 to 6.7 m/s
average wind speed) and rather lighter over most inland regions (around
4.4 m/s average wind speeds). Golding (1955) published the results of a survey of potential wind-power sites in Britain which was conducted by the
Electrical Research Association. He suggested that the economic generation
of electrical power becomes feasible where the average wind speed is more than
8.9 m/s and he found 39 sites where such conditions were met.

Windmill Design

An amazing variety of windmill designs has been built over the years (Gimpel,
1958). In the early days, control was achieved by the miller adjusting the

cloth which covered the blades and he had to turn the whole mill to face into
the wind by hand; subsequent techniques used slats which would open automatically at high wind speeds to let air through the blades. Modern designs of
windmill take advantage of aircraft technology by using blades with aerofoil
cross sections to reduce rotational drag and to enhance the "lift" forces
which drive the sail and from which power is extracted. These aerofoils can
be feathered to run close to stalling in order to limit the torque produced
by any sudden gusts or increase in wind velocity. Unfortunately, experience
with modern machines shows that many of the technical possibilities such as
universal joints, shock absorbers, centrifugal governors and blade pitch control all lead to high costs and to loss of availability. Simplicity appears
to be essential for both cheapness and reliability. Vadot (1957), who also
reviewed a large number of designs, concluded that only the horizontal axis
propeller type of windmill, with two or three blades, is suitable for the
generation of electrical power. Certainly this is the only type which has
been built in large numbers or in large sizes.

The largest wind driven generator ever built was the 1.25 MW Smith-Putnam
machine (Putnam, 1948) which had two variable pitch blades, with an overall
diameter of 53 m (Figure 3). They were mounted downwind from the tower since
experiments showed that pressure fluctuations due to the blades passing the
tower were more serious than those due to normal levels of turbulence so that
better performance would be achieved with the sails to the leeward. The steel
lattice tower was over 30 m high and the drive was transmitted through a gearbox and a hydraulic slipping coupling to a synchronised generator. A hinge at
the root of the blades allowed them to lean away from the wind, rotating to
sweep out the surface of a cone rather than a disc. With this hinge and the
"coning" action, the bending moment at the blade root was appreciably reduced.
This machine underwent trials for several years in the early 1940's in the U.S.A.
but one of the blades was destroyed by strong winds shortly after it entered
commercial service in 1945 and that put an end to the project.

In the past, vertical axis machines have always produced very low efficiencies
but an improved design has been produced recently by the National Research
Council of Canada (South and Rangi, 1974) and built by the Sandia Corporation,
which uses a catenary or troposkein shape of blade to eliminate bending stresses
(Figure 4). The blades of this small experimental device have a symmetrical

Figure 3.

The largest windmill ever built:
the Smith-Putnam 1.25 MW machine
built in 1941.   The tower was
30m high and the blades 53m
across.

Figure 4.

The Sandia Corporation built
this experimental, vertical axis
windmill designed by the Canadian
NRC.   The blades bow out to
about 4.5m diameter.

aerofoil section and bow out to a diameter of 4 m or so to generate 8 kW in a
strong (13 m/s) wind, i.e. 30% of theoretical efficiency.

The Prospects for Wind Power

The economic viability of windmills is better in situations where conventional
transmission costs are extremely high (e.g. because of inaccessibility and
only a small load exists), or where continuous availability of supply is not
essential so that only a limited amount of storage or standby power need be
provided.

Even if economic problems are disregarded, the total amount of wind power
available is not large. A barrier of windmills 50 m high around the entire
coastline of the British Isles, which would include some of the windiest
places on earth, would only provide an available power capacity of, say, 5 GW.
That would represent only a small part of the country's power requirements
(less than one tenth of the present installed capacity) and such enormous
installations situated in areas of high amenity value would certainly be
unacceptable.

Wind power is therefore not likely to make a _major_ contribution to this
country's overall power requirement but it may find individual, small (about
1 MW), local applications. It is interesting to note that the Dutch are
considering a national research programme on windmills.

TIDAL POWER

Tides Around the Earth

When considering power from the oceans, attention is usually focused first on
the tides. They were virtually unknown to ancient sea-faring civilisations
of the Mediterranean while around British coasts they always dominated the
nautical scene - the very word "tide" is almost synonymous with "sea". Both

these extremes are brought about by accidents of geography; normal tidal ranges around the world, although quite marked, are also quite modest.

Figure 5 shows schematically the way that the waters of the oceans are distributed around the earth to produce the twice-daily tides. Such a "di-pole" distribution of water is driven by the gradient of the gravitational force field rather than by the field itself, i.e. it is proportional to M/R^3 where M and R are the mass and distance of the heavenly body producing the field. Since the mass of the sun is nearly 27,000,000 times that of the moon while its mean distance from the earth is about 390 times that of the moon, the gravitational force gradient due to the sun is in the ratio of $27,000,000/(390)^3$ to that due to the moon, i.e. about 5/11; tides on earth can therefore be said to be driven about 30% by the sun and 70% by the moon. Thus, the rise and fall of the tides nearly twice a day are associated with the rotation of the earth every 24 hours in relation to both sun and moon.

In most parts of the world, tidal ranges are only a metre or so and therefore not of very great interest for power generation. In enclosed inland seas such as the Mediterranean the tidal range is so small as to have escaped the notice of the ancients. Around the British Isles, however, the various coastal complexities produce a remarkable richness of tidal patterns due to resonant effects produced by the local geography. There are particularly large peak amplitudes in the Rance Estuary, the River Severn and across Morecambe Bay. The very complex tidal patterns which are found around the shores of Great Britain can be analysed into simple Fourier components related to the sun, the moon and these local geographical resonances. Figure 6 shows the tidal pattern around the British Isles and it can be seen that the most advantageous site is in the Rance Estuary in Brittany where tides as high as 10 metres occur. A pilot scheme has been in operation there for many years (Editorial, 1967) and, although it has run reasonably satisfactorily, there have been problems, e.g. with silting; no other schemes have been constructed anywhere else around the world. It can be seen from Figure 6 that the next most favourable site around the British Isles would be the Severn Estuary and after that Morecambe Bay.

Tidal Barrages

To extract energy from tides, the water must be trapped at high tide behind a

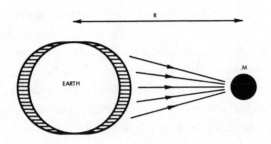

Figure 5.

Tides on opposite sides of the earth are driven by the gradient of the diverging gravitational field.

TIDES - FT.

Figure 6.

Tides around the British Isles.

barrage or dam and then made to drive water turbines as it returns to the sea
at low tide. The energy available is proportional to the square of the tidal
amplitude and so it tends to be concentrated around the regions of highest
tide level. The total exploitable energy around the British Isles is about
30 TWh per year and about two thirds of this is concentrated in the Severn
Estuary which could provide up to 10% of the country's present power consumption.

Because of the square law variation with tidal height and the variations in
tidal pattern, the available energy varies widely from day to day. It may
therefore be advantageous to build multiple barrages or dams as, for example,
in Figure 7, so that the different basins can each store water by pumped
storage and smooth the power output, as shown in Figures 8 and 9. Such an
improvement involves engineering complications which lead to higher capital
costs and some reduction in the net energy output. Although the simpler
scheme using a single basin is cheaper, it only provides an intermittent output which follows the phases of the sun and the moon rather than the load
curve of the national power system. When the source of power is not constant,
alternative plant must be built as a standby and there is then no saving in
capital cost, although there is a saving in fuel.

Engineering Problems

Those regions which look most promising for tidal power generation tend to be
the very ones which are difficult from the point of view of barrage construction. Estuaries frequently carry large silt burdens, have shifting mud banks
and carry loose material in the estuary bed which can be moved around by any
change in the tidal flow pattern. These factors lead to silting difficulties,
blocking of shipping channels, etc.

Power stations on the estuaries might be affected by the change in their cooling water flows that a barrage would cause. Water tank and laboratory studies
of the effects that a barrage might have on silting, tidal flow and heat dissipation can give some indication of the problems, but experience at the Rance
shows that such studies are not always reliable.

Dams are best constructed on bed-rock, but estuaries usually have a

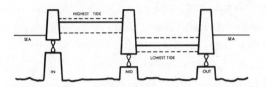

Figure 7.

A two basin tidal scheme produces
continuous power. The upper basin
fills during high tide, the lower
basin empties during low tide and
there is always sufficient head
between the two basins to drive
the turbines between them.

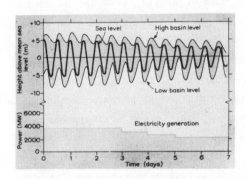

Figure 8.

A two basin scheme with unequal
areas (586 and 222 square
kilometers). 6GW capacity in the
middle turbo-generators, 2.5 GW
reversible turbo-pump/generator
in, and 2.5 GW turbo-pump out.
Firm power is 4 GW during
springtides and 2 GW during
neaps over a 7 day cycle.

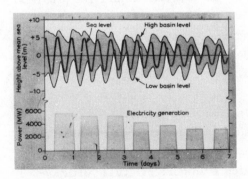

Figure 9.

Generation over only 16 hours a
day gives 5.5 MW of firm power at
springs and 3 GW at neap tides
although the total generation is
substantially less than with
continuous generation (Denton et al,
1975).

bed-covering of unconsolidated sediments and loose deposits. What characteristics these have as far as the design of foundations is concerned are not very
certain and this is one feature which makes the determination of precise cost
estimates particularly difficult. If necessary, all the dams could be formed
from concrete caissons but it would be about 30% cheaper to make a major part
of the dam an embankment with outer zones of rock and a central core of sand
or other loose material dredged from the estuary bed.

A barrage across the River Severn would present much more severe engineering
difficulties than that across the Rance Estuary because the total tidal flow,
water depth, etc. are much greater and the whole construction would be on a
much larger scale. In particular, the cofferdam method of construction of the
Rance would probably not be appropriate in the Severn.

The biggest problem when building a dam is that of closing it. As the constriction becomes narrower, flow velocities become greater, the maximum velocity
depending upon the total volume of the flow, i.e. on the area enclosed and the
height of the tide. Fortunately, the Dutch have a well established method
using cellular caissons which are fabricated on shore, towed into position
with the plant partly installed and allowed to sink on to prepared foundations.
Power house construction of this sort might be possible for a Severn barrage
although specialist skills would be needed.

The Rance scheme has operated successfully since 1966 with a number of hori-
zontal-shaft, bulb-type units for generation and pumping duties which have
performed very reliably and such machines would be suitable elsewhere; they
can now be built with diameters up to 11 m. In addition to silting problems
special measures are needed at the Rance to combat corrosion and marine fouling, but no doubt the solutions which have been found successful there could
be applied to other schemes. Various developments which are taking place in
the hydro-electrical field may also be relevant to future tidal power schemes.

Where wind power is difficult to contemplate in very large units, the reverse
is true of tidal power. Tidal schemes cannot start with a small installation
and grow over a period of time as more and more capacity is installed. The
uncertainties must therefore be investigated even more carefully with such a
scheme than with other systems. The long construction period will also represent a grave disadvantage. The interest payable during construction could

easily double the cost of such a scheme over the period of 10 to 15 years that
it might take to build.

It is not appropriate here to undertake a detailed economic assessment of
tidal power, but there is a great deal of interest throughout the country.
Very careful and detailed consideration is being given to individual schemes
by various interested parties, including the CEGB (Denton et al, 1975). The
engineering uncertainties make it difficult to estimate the costs very accurately
but, on the basis of optimistic estimates, such a barrage assessed solely as
a method of electricity generation could not produce electricity more cheaply
than by other means. However, a tidal scheme could make a significant contribution to the country's energy needs (say, 10% of present CEGB demand) so it
may well receive further attention.

THERMAL SEA POWER

Another way that is often proposed for extracting power from the oceans would
use the temperature difference between surface water and colder currents
beneath. This method can only be seriously considered in tropical regions
such as the Caribbean where temperature differences of up to $20^{\circ}C$ are available with theoretical Carnot efficiencies of more than 6% and practical efficiencies of 3% or so. Thermodynamic efficiency is not important from the
point of view of energy cost - the warm water is there for the taking - but
it is vital in determining the size and cost of the plant required to produce
a given amount of power. The scale and cost of plant depends principally upon
the amount of heat, the cost per kilowatt of electricity is inversely proportional to efficiency.

In British waters, temperatures are low and temperature differences very small,
particularly in winter when power is most wanted. Nevertheless, for completeness we will briefly consider the technical possibilities of thermal sea power.

A previous attempt to obtain "Power from Tropical Seas" (Claude, 1930) used a
low-pressure steam cycle. Many of the problems which Claude found insuperable,

such as corrosion by salt and de-aeration, have now been overcome in the development of "Controlled Flash Evaporation" for sea water (Roe & Othmer, 1971), but
alternatively another working fluid can be considered. Propane has suitable
pressure-temperature characteristics but, with any change of working fluid,
the heat exchanger is a problem; with such small temperature differences, very
large surface areas and mass flows are called for (more than twenty times those
for a conventional power station) which are very costly. Ingenious designs
have been put forward (Anderson & Anderson, 1966) which use the water pressure
at various depths to provide a pressure balance across thin-plate heat exchange
surfaces and these could solve some of the economic problems.

Various schemes of this sort are being considered for tropical regions
(Walters, 1971; Zener, 1973) and a possible scheme is shown in Figure 10. It
remains to be seen whether the engineering problems can be overcome but there
is no prospect of application around the British Isles.

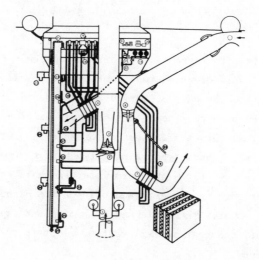

Figure 10.

A proposed propane cycle for thermal sea power would use a thin plate
heat exchanger as shown in the inset.

WAVE POWER

Wave Characteristics

The major problem with wind power is that it is so diffuse. If means could
be found of collecting the energy in the wind over a large surface area and
concentrating it into a relatively small volume, the prospects would be considerably improved and that is what the oceans do. Oceanographic surveys
(see Figures 11 & 12) show that typical wave heights throughout the year are
2 or 3 metres. The energy per square of surface, stored in the continuous
circular motions (amplitude a) under gravity (acceleration g) of the individual particles of sea water (density ρ) and in their variation of potential
energy above and below the average water level, is $\frac{1}{2}\rho g a^2$ corresponding to
more than $10 kJ/m^2$. In the Atlantic the wave period T is between 7 s and 14 s
for most of the time, with an average of around 9 s. The speed with which
the waves propagate, $gT/2\pi$, is therefore around 14 m/s and the speed with which
the energy is transported is half of this. Multiplying the stored energy by
its velocity of propagation gives a power flow of more than 70 kW across every
metre of wave front. This is a very considerable amount of power, particularly
bearing in mind that it is an average throughout the year. Along a coastline
of more than 1500 km of the British Isles, the mean power available is more
than 100 GW, i.e. more than double the present demand for power in the U.K.

One important characteristic of this wave energy is that, although it is variable, it has a seasonal peak in the winter, closely matching the demand pattern
for electric power, see Figure 13; in this respect it has a great advantage
over, say, solar power. Nevertheless, there are periods (about 1% of the
time) during which there is a serious drop in power level and they would have
to be accommodated in any practical power system. The lack of firm capacity
would require the provision of adequate storage or standby plant. To instal
standby plant such as gas turbines would add to the capital costs (although
fuel demand would be very low because of the low load factor) but this would
not be necessary until wave power took an appreciable part of the total load.

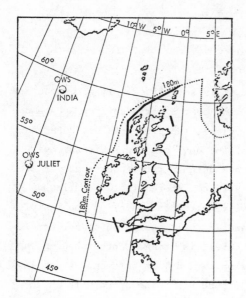

Figure 11.

Favourable areas for UK wave
power schemes.

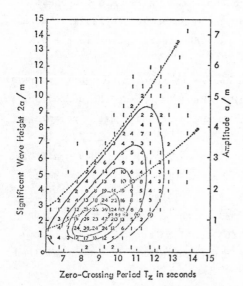

Figure 12.

Wave heights and periods are
widely scattered but the waves
are in a 2 to 1 frequency range
99% of the time.   Data from
weather ship "India" (Draper,
1967).

Wave Power Conversion Devices

The patent literature is full of devices for extracting energy from waves -
floats, ramps, flaps, converging channels have all been suggested. Small generators driven from air trapped by the rising and falling water in the chamber
of a buoy are in use around the world. It is sometimes suggested that a float
bobbing up and down and a vertical plate flapping backwards and forwards operate on different water motions. Such thinking is somewhat misguided since
each particle of water moves with essentially constant speed in a circle and
it is only possible to reduce the horizontal and vertical motions together.
Similarly it is not possible to extract potential energy without also extracting kinetic energy since they are equal when averaged over a wave length.

Although very large amounts of power are available in the waves, it is important to consider how much power can be extracted. A few years ago only a few
percent efficiency had been achieved. Recently, however, several devices have
been studied which have very high efficiencies. Once more it must be emphasised that when the source of energy is free, efficiency is only important in
as far as it affects the size and hence cost of the device which is required
to provide a given power output. Where the scale is determined by the input
power that must be handled, the cost per kW generated is inversely proportional
to efficiency.

The highest efficiencies to date have been obtained with the device shown in
Figure 14 invented by Salter (1974) at the University of Edinburgh. It simulates the action of a vertical flapping plate at the front where energy is
absorbed but rotates with a circular rear section, so that the water behind
the plate is not disturbed and none of the power is transmitted onwards. Wave
tank tests have shown (Swift-Hook et al, 1975) that conversion efficiencies of
more than 90% can be achieved with this device. Such high efficiencies depend
upon resonance effects but quite reasonable efficiencies of more than 50% have
in fact been achieved over a very broad range of frequencies (2 to 1) corresponding to the range found under realistic sea conditions as recorded in
Figure 12. Because of its asymmetrical shape, the response of this device
depends upon its loading and so counterbalancing is necessary. Adjustments
to the inertia and effective mass of the system change the resonant frequency,
as shown in Figure 19, but very large bandwidths are still available.

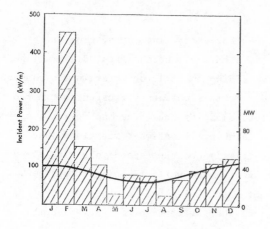

Figure 13.

Annual variation of wave power
measured at ocean weather ship
"India" compared with a typical
CEGB peak load curve.

Figure 14.

Rocking float which has extracted 50% of the wave power in the
Laboratory over a 2 to 1 range of frequencies (Swift-Hook et al, 1975).
The front simulates a vertical flapping plate but the rear is
cylindrical so that the water beyond the device is not disturbed as
it rotates.

Substantial turbulent losses were found at shorter wavelengths but it is anti-cipated that they would not be serious on a larger scale.

A conceptual full-scale device of this type is shown in Figure 15; it is near-ly 20 m deep, about 20 m across, 300 m long and would generate 10 MW with a displacement of about 100,000 tonnes (a cubic metre of sea water weighs just over a tonne). The scale and the level of technology are commensurate with those required for the construction and operation of supertankers. If the costs are comparable with tanker costs (per tonne), the overall cost of wave power might be no more than a few hundred pounds per kW, i.e. no more than the cost of nuclear power.

Other devices could have much smaller displacements and hence, possibly, much lower specific costs. An example is the wave contouring raft system invented by Sir Christopher Cockerell of Hovercraft fame (see Figure 16). Efficiencies approaching 50% have been measured over a wide range of frequencies (Woolley & Platts, 1975). This system looks very promising and might be improved from the point of view of loading and inertia if the rafts were below the surface and covered with water.

Another system, invented by Masuda (1974) of the Japanese Marine Science and Technology Centre, uses a bell-shaped chamber filled with air which is pumped through an air-turbine by the rising and falling motion of the water. Several hundred devices of this sort are in continuous use around the world each pro-viding about 70 W for marine buoy lights. Such devices are inherently quite small and omni-directional but a number of them could be combined together to provide a large floating structure, for example, a ring buoy which would be omni-directional or an arrow-shaped triangular system for mounting near the shore, Figure 17.

Directionality will be important since ocean waves do not come from the same direction all the time (see Figure 18). Our measurements have shown that Salter's device is not very highly directional but nevertheless it would need to be steered, or allowed to take up its own direction across the wave front. Other devices are more omni-directional.

Significant programmes of research are in hand at the Marchwood Engineering

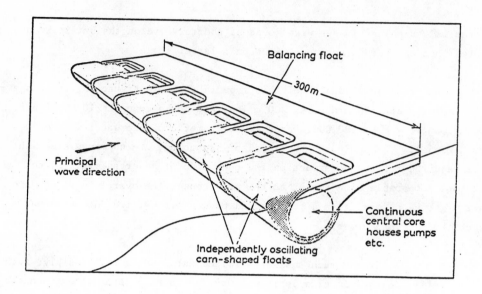

Figure 15.

A possible engineering realisation of a 10 MW wave power generator.

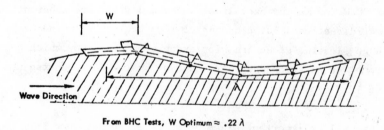

Figure 16.

Wave contouring pontoons suggested by Sir Christopher Cockerell.
Tests at British Hovercraft Corporation have shown that optimum
efficiencies can approach 50% when the wavelength spans 4 or 5
of the hinged rafts.

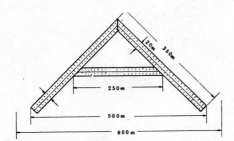

Figure 17.

Masuda system of floating buoys.

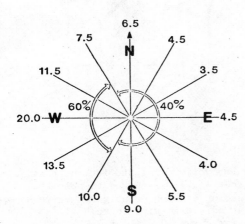

Figure 18.

Directional distribution of
waves in the Atlantic. The
figures are percentages in a
30º sector. NPL data.

Laboratory at East Kilbride, the Mechanical Engineering Department of
Edinburgh University and the National Hydraulics Laboratory at Wallingford.
There are many other groups interested in theoretical studies and already a
fair measure of agreement is found between theory and experiment on the laboratory scale (see Figure 19). No doubt many of the large computer programmes
used by those involved in North Sea technology and the design of super-tankers
will be relevant for full-scale devices.

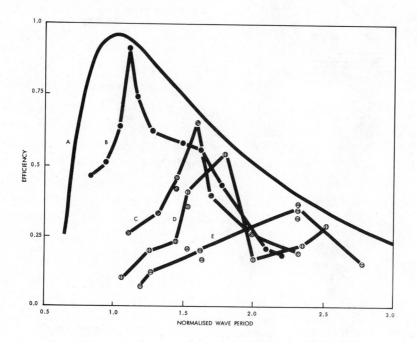

Figure 19.

Experimental performance of Salter device for various weights (or inertias, I)
relative to the minimum added weight (or inertia, I_o) required to hold the
device level. The theoretically optimum loading at each point gives curve A.
Experiments with a 0.1 m diameter device at Edinburgh give curve B (I/I_o = 0.44).
Experiments with a 0.5 m device at Hythe give curves C (I/I_o = 0.92),
D (I/I_o = 1.42) and E (I/I_o = 1.92).

Overall System Considerations

It must be emphasised that the device itself is only one component, albeit the most important one, in the total system if wave power is to make a significant contribution to the country's total power resources; Figure 20 shows the many other aspects that will need to be covered. No single problem appears to be insuperable but each of them may be significant; for example, if the power is to be brought ashore electrically, an under-sea cable will be required. There is a considerable amount of experience on submarine power cable throughout the world including the 160 MW cross-channel link and so the technological problems and the costs (perhaps £50/kW for each 100 km) should not be too severe provided the devices do not need to be too far off-shore, say, no more than 100 km. Alternatively the power could be used on board floating factories around the oceans of the world e.g. extracting uranium from sea water.

Some form of storage will be essential on a second-to-second and minute-to-minute basis to smooth the fluctuations of individual waves and wave packets but storage from one day to the next will certainly not be economic; that is why provision must be made for adequate standby capacity.

It is interesting that very high efficiencies have been achieved in the laboratory with a wide range of different devices and that crude initial estimates look promising. In many other fields (for example, solar cells) extensive research and development would be required to improve efficiencies and reduce costs very substantially before a system could be considered seriously. For quite a small investment in research, wave power could be tried out on a small scale and developed very rapidly through a series of stages to a prototype, after which replication would bring all the potential benefits of mass production. If successful and if required, a wave power programme would make a significant contribution to the country's energy resources within a relatively short time and with existing technology.

ENVOI

The true reason for the interest in alternative power sources lies in the desire for an energy insurance policy. The U.K. and the world have plenty of coal but suppose the miners refuse to dig it? Who could blame them? There

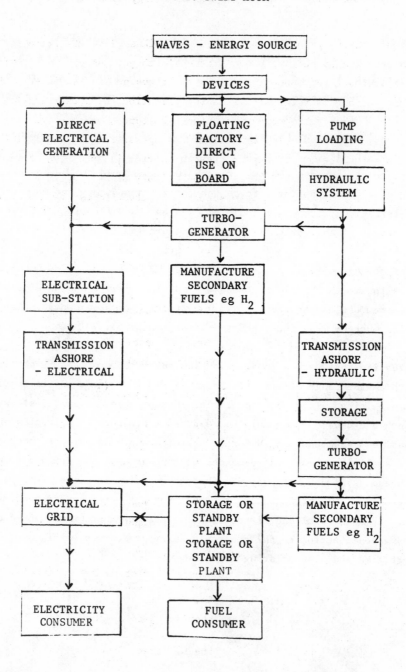

Figure 20.

Possible systems for exploiting wave power. Each element represents an essential link in the chain from sea waves to consumer.

is plenty of oil in the ground but suppose the Sheikhs raise its price yet
further? There are many nuclear power stations with good safety records and
many more are planned, but suppose there is a serious nuclear accident? The
power industry has a good record for reducing pollution and preserving the
environment but suppose those conservationists who oppose advances in technology are successful? In any of these situations, power from "natural" sources
would have a strong appeal. If any of these scenarios is possible, diversifi-
cation of our energy supplies is a prudent aim simply as an insurance. Wave
power is one possibility that has been selected by the CEGB and by the country
as being worthy of further detailed investigation.

REFERENCES

Anderson, J.H. and Anderson, J.H. Jr., (1966). "Thermal Power from Seawater".
Mech. Engg.88, 41.

Claude, G. (1930). "Power from Tropical Seas". Mech. Engg. 52. 1039.

Denton, J.D. et al. (1975). "The Potential of Natural Energy Sources". CEGB
Research 2, 28.

Draper, L and Squire, E.M. (1967). "Waves at Ocean Weather Ship India".
Trans. Roy. Inst. Nav. Arch, 109, 85.

Editorial (1967). "Rance Tidal Power Scheme". Water Power, January 1967, p.7.

Gimpel, G. (1958). "The Windmill Today". Engineering, 185, 686.

Golding, E.W. (1955). "Electrical Energy from the Wind". Proc. I.E.E., 102, 677.

Masuda, Y. (1974). Private Communication.

Putnam, P.C. (1948). "Power from the Wind". Van Nostrand-Reinhold, New York.

Roe, R. C. and Othmer, D.F. (1971). "Controlled Flash Evaporation". Mech.
Engg., 93, 27.

Salter, S.H. (1974). "Wave Power". Nature, 249, 720.

South, A. and Rangi, R.S. (1974). "Wind Tunnel Investigation of a 14ft. Diameter Vertical Axis Windmill". Reports LTR-LA-74, LTR-LA-105. National
Aeronautical Establishment, Ottawa, Canada.

Swift-Hook, D.T. et al., (1975). "Characteristics of a Rocking Wave Power
Device". Nature, 254, 504.

Vadot, L. (1957). "A Synoptic Study of the Different Types of Windmills, La
Houille Blanche, 13, 189.

Walters, S. (1971). "Power in the Year 2001. Part 2 - Thermal Sea Power".
Mech. Engg., 93, 21.

Woolley, M. and Platts, J. (1975). "Energy on the Crest of a Wave". New
Scientist, 66, 241.

Zener C. (1973). "Solar Sea Power". Physics Today, 26, No. 1., p.48.

DISCUSSION

Question: Is there an economy of scale for wind-power devices?
Answer: I do not feel this to be a fruitful line of approach owing to the
enormous complexity of very large wind-mills in the megawatt range.
Some technical problems could be alleviated by adopting recent
developments on vertical-axis machines but even then it would be
preferable to build up a large power capacity from a series of
smaller units, transmission and linking techniques being well
known and not particularly costly. Again, on the grounds of cost
and complexity, the ducted windmill, i.e. one where air is captured and accelerated in a venturi, must be discounted. The best
applications for wind power are most likely to be where (a) power
requirement is small, say up to 10 kw; (b) storage is unnecessary
or of low capacity; (c) new transmission installations are costly;
and (d) sites are relatively inaccessible. For the average wind
speeds in the United Kingdom, which are generally quite high compared with other areas of comparable habitation, cost estimates
carried out at CEGB Marchwood indicate that competition from wind
power in the sectors presently covered by fossil and nuclear fuels
is extremely unlikely in the foreseeable future.

Question: How can thermal pollution, environmental and ecological problems
be predicted and avoided in the Severn tidal estuary scheme?
Answer: These will need careful examination. By their very nature, tidal
schemes have to be large and costly. Considerable expenditure on
scale experiments is justified in order to identify the fluid
mechanics and thermal dissipation in the large water flows likely
to occur. Although, for example, the Severn tidal scheme may not
prove economically viable as a producer of energy for general use,
there could be other overriding reasons for completing it, e.g.
improving harbour facilities or enabling an airport to be built on
the reservoir walls. Neighbouring holiday resorts may have far
different views!

Question: What about the problems of storage and storm damage associated
with wave power?
Answer: For short periods, say a few hours, this can be done using wave
devices in a back-to-back fashion. The cost increases by a factor
of five or so if storage over some days is required but, fortunately, those areas in which energy could be extracted from ocean
waves suffer from few calm periods. Rough seas are, of course, a
problem but better predictions of abnormally high and steep waves
should be possible. Damage could be avoided by incorporating a
mechanism for submerging the raft to a depth where wave amplitudes
are tolerable. The technology necessary is not far removed from
that used in the design of supertankers and oil rigs.

Question: How do you see future wave power developments?
Answer: In the long term, the jointed raft device would probably be the
best although so far the oscillating cam has received the most
attention. Should research on wave energy continue, it is conceivable that within two years a test barge could be in the sea
and a prototype device in the megawatt range working some two
years after that.

SOLAR ENERGY

B. J. Brinkworth

Mechanical Engineering Department, University College, Cardiff

Though we have come to rely heavily upon oil for our primary energy, it is
clear that we can do so only for a few more decades (1).   Other orthodox
sources may be more enduring, but are not without serious disadvantages.
Coal is plentiful, though there is concern about despoilation in winning it
and pollution in burning it.   Nuclear power has been developed with
remarkable timeliness, but is not universally welcomed;   construction of the
plant is energy-intensive and there is concern about the disposal of its
long-lived active wastes.   Even a shortage of fissionable fuel is already
envisaged.   A fusion reactor, expected to defer a general energy shortage
more-or-less indefinitely, is proving elusive.

Most of the discussion about future energy supplies is concerned with
capital sources, yet in the long term we must learn to live within our energy
income.   This and the adjacent presentations of the course are about some
potential sources of income.   It will be sensible to enquire whether we are
now at the point at which some of the resources allocated to the continued
development of capital sources could be more profitably applied to these.
Some of the unorthodox sources may involve long R & D lead-times, so that
work should be started now;   others might be capable of contributing to our
supplies even in the short term and we might begin to phase these in quite
soon.

In assessing a possible energy source, we need first to be satisfied that it
is of sufficient magnitude to contribute significantly to meeting our needs
and that the technology for its exploitation exists or can be foreseen with
confidence.   Lastly, we need to assess the likely economics of its
exploitation in relation to other contending sources at the time of its
introduction.   This presentation is about the possibilities of the direct use
of solar energy.   There is no question of its sufficiency in general terms;
that falling on a region 100 km square in a favourable location exceeds the
entire world energy demand at the present time.   We shall be looking at the

nature and characteristics of this source and of known means of exploiting it, to provide a basis for discussion of its merits in comparison with other sources.

Certain indirect uses of solar energy, for example, via rainfall, winds, waves raised by winds, and ocean thermal gradients are not considered here, as they will feature in other presentations. Even so, it will be possible to cover only the main features of this large and multidisciplinary field (2).

SOLAR RADIATION

At the top of the atmosphere, the energy flux of the solar radiation is about 1.35 kW/m^2, with a spectral distribution resembling that of a black-body radiator at about 5 800°K. A variety of scattering and absorbing processes in the atmosphere reduce the maximum flux to about 1 kW/m^2 at the surface. Most of the u.v. radiation is removed by absorption processes involved in the oxygen-ozone balance above 25 km and there are substantial fissures in the energy spectrum in the i.r. due to H_2O and CO_2 absorption (3). There is some scatter by atmospheric constituents and pollutants in clear conditions, but the main causes of scattering are the water droplets in clouds. As these are mostly much larger than the wavelengths, the scattering is predominantly forward and much of the scattered radiation reaches the ground, even in cloudy environments such as that of Britain. It is convenient to distinguish the two modes of reception of the radiation at the surface: the direct and the diffuse. In the UK more than half of the radiation reaching the surface is in the diffuse mode, and this is important, because this part cannot be brought to a focus by a reflector.

Because of the inclination of the earth's axis to the plane of its orbit, the maximum altitude of the sun and the length of the day vary substantially with latitude and season. In summer these are compensatory, so that on a clear day the total energy falling on a horizontal surface has a similar value (around 30 MJ/m^2) at all latitudes up to the polar circle. In winter the effects of day length and solar altitude combine to produce a large dependence on latitude. At the latitude of Britain, a clear day in winter yields about one-tenth of the mid-summer value. The effects of cloud cover are smaller than might be imagined, the fall in the direct component being partly balanced by a rise in the diffuse component. In Britain, the average daily total is about 16 MJ/m^2 per day in summer and about 2 MJ/m^2 per day in winter (Fig.1).

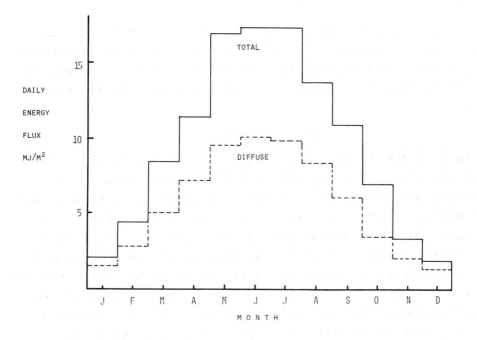

Fig 1.   Monthly mean daily energy flux for southern UK

The total energy received by a horizontal surface in a year is about 3 500 MJ/m^2, whilst the corresponding value for Australia, which enjoys quite the opposite reputation for weather, is about 6 500 MJ/m^2, less than twice as much.   Most of the world's inhabitants live between the 5 000 MJ/m^2 contours.

COLLECTION OF SOLAR ENERGY

It seems to have been recognised only in recent years that penetration of solar radiation through windows and absorption on the surfaces of buildings represent significant factors in their thermal behaviour (4).   As insulation standards rise, these will assume greater importance and they are now becoming major preoccupations in the architectural world.   It is convenient to distinguish between this passive collection, in which the collecting surfaces form part of a structure, having an architectural function, and active collection, which is largely an engineering problem.

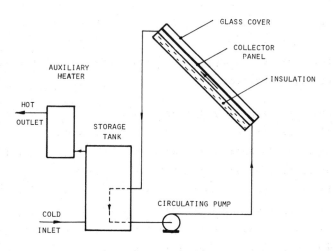

Fig 2.   Flat plate collector and water-heating system

The simplest active collector is the flat-plate type, shown in Fig.2.
Radiation passes through a transparent cover plate and is absorbed on the
collector panel.   This in turn communicates heat to a fluid flowing in
passages in contact with the panel and it is by this fluid that energy is
extracted from the system for use elsewhere.   The efficiency of conversion
of solar energy into heat depends on the losses to the surroundings and hence
on the temperature at which the collector is allowed to operate.   The cover
plate reduces convection losses and also acts as a radiation rectifier, as it
is transparent to the incoming radiation but opaque to the i.r. radiation
emitted by the collector.   Further reduction is possible by treating the
absorber surface with radiatively selective coatings.   These give the
surface a high absorptivity for solar radiation, but low emissivity at the
operating temperature, made possible by the wide difference in wavelength in
the two cases.   These measures enable a useful fraction of the incident
energy to be furnished as heat in the collector fluid at moderate temperatures;   say 50% at 55°C for British summer conditions.

It will be evident that the efficiency falls as the operating temperature

rises and that there is, for any given conditions, a maximum attainable
temperature at which the losses equal the input.   For a good collector
operated in Britain, this approaches 100°C in summer, but only 40°C or less
in winter.   Higher temperatures can be obtained only by using a concentrating collector which focuses the energy falling on a large area onto a
much smaller one.   Since the area from which losses occur is small, the
efficiency is greatly increased.

Many types of concentrating collector are possible.   The most effective is
the paraboloidal reflector, which can produce focal temperatures of over
3 500°C.   Its main disadvantage is that it must be tracked round to follow
the apparent motion of the sun.   Sometimes the paraboloid is stationary
and the radiation is directed into it by one or more moving flat reflectors
or heliostats, each appropriately driven, as in the 1 MW solar furnace at
Odeillo.   Heliostats may also be used to reflect the radiation directly
onto the heated object (5).   A less elaborate collector is the two-
dimensional parabolic trough, which is mounted with its focal axis permanently aligned E-W, as shown in Fig.3.   A suitably sized collecting tube on

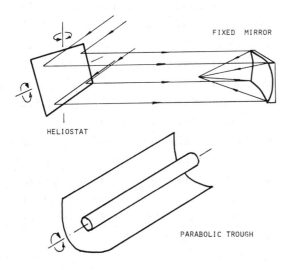

FIXED MIRROR

HELIOSTAT

PARABOLIC TROUGH

Fig 3.   Concentrating collectors

the axis then receives a useful fraction of the energy falling on the trough over the central part of the day.   A small movement in altitude ($\pm 23\frac{1}{2}^{\circ}$) is required during the year to allow for the variation in the sun's declination; this, or movement of the focal tube, may be used to extend the period of useful collection during the day.

Apart from the cost of construction, the main disadvantage of concentrators, particularly from the point of view of operation in Britain, is their inability to accept the diffuse component of radiation, so that the collection efficiency is limited here to a maximum of about 40% on an annual basis.

When collection on a large scale is envisaged, it is expected to be brought about by the replication of collectors of the types described over a large area of ground.   Probably the cheapest large collector would be a body of water, perhaps a specially constructed solar pond or even the free ocean. Methods have been described for reducing the losses from ponds by suppressing buoyant circulation, so that quite useful temperatures can be obtained (6).

THERMAL USES

Use of solar energy for heating purposes is already well established and can be expected to spread to higher latitudes as fuel costs rise and techniques improve.   A traditional dwelling house in temperate latitudes requires in mid-winter about 350 MJ per day for space heating and about 50 MJ per day for domestic water heating.   Since the usable roof area might be about 50 m^2, it is seen that even with high collection efficiency, the space-heating demand cannot be met at this time of year, though it may be in autumn and spring.   There is sufficient area to collect the energy needed for domestic water heating, but it cannot be obtained at the required temperature in winter.   Thus, for both uses, solar heating must be seen as a means of fuel saving rather than a complete replacement for traditional heating methods.

A typical hot-water heater was illustrated in Fig.2 and a space-heating installation in Fig.4.   Both involve a measure of energy storage to even out disparities in the supply and demand and to control the temperature of operation.   Storage can be accomplished in two main ways: in the form of

sensible heat or latent heat.   Work is in progress on the analysis and
design of thermal storage units for minimum volume.   Although this involves
some difficult heat and mass transfer problems, the development of suitable
storage units is emerging as a key element in the improvement of performance
of solar heating systems.

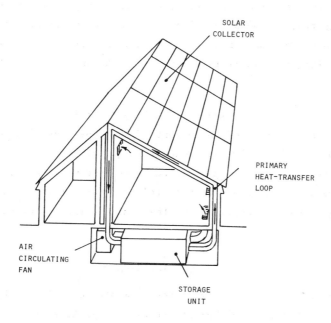

Fig 4.   Solar space-heating system

Two other thermal systems worth emphasising are the solar still and solar
drier.   In the simplest kind of still, shown in Fig.5, water from a saline
or polluted source is admitted to a blackened tray where it is heated by
radiation passing through a transparent cover.   Vapour condensing on the
cooler cover drips into a collecting trough.   Simple stills of this kind
have been built on a large scale, up to a hectare or so in area, and yield
the equivalent of 3-5 mm of rainfall per day.   More elaborate multi-effect
stills have been investigated (7).   Solar driers have been used to
accelerate the drying of numerous crops and products and can greatly reduce
wastage through bacterial action.   Some studies of design principles for
solar driers have been reported (8).

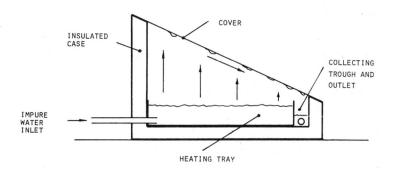

Fig 5.   Single-effect solar still

THERMODYNAMIC CONVERSION

Solar-driven heat engines were first operated at least a century ago, but
have not been economically competitive during the era of cheap fuel.   If
the heat source is a flat collector, the low output temperature severely
limits the thermodynamic efficiency of the engine, whatever its nature, so
that the overall energy-conversion efficiency is very poor - typically about
3% (2).   However, there is little technical difficulty in making plant of
this kind and few places in the world where it could not be produced locally.
Although the capital cost would be several times that of alternative systems
such as the diesel engine, the costs of power from a solar engine may be
competitive now in places with high fuel costs.   More promising economics
have been claimed for low-temperature engines driven by ocean thermal
gradients(9).

Higher thermodynamic efficiency calls for higher maximum temperatures in the
engine cycle, with the attendant difficulties of using concentrating
collectors.

Proposals have been worked out in some detail for large-scale thermal power
plants to be built in desert areas.   In some designs the solar radiation is
directed onto a central tower-mounted boiler (5).   Others, consisting of
elements such as those shown in Fig.6, would probably employ parabolic
trough collectors, with energy transfer by liquid metals or heat pipes to
storage units operating at about 300°C (10,11).   The cost of power from
plants of this kind has been claimed to be intermediate between that from
conventional thermal plant and nuclear plant.   Because the collecting
areas involved are comparable with those employed in agriculture , proponents
of these plants have termed them 'solar farms'.   It has been suggested also
that if they were built in desert areas, it might be possible to cultivate
the land in the shade of the collectors and farm this in the conventional
sense.

Another important thermodynamic use may be in refrigeration and air-
conditioning, since the demand for cooling is greatest at the time of
maximum solar energy availability. For typical operating temperatures, the
CP* of the vapour-compression refrigerator is about 5, but a mechanical input

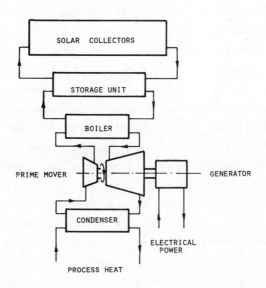

Fig 6.   Elements of solar power plant
* Coefficient of Performance.

to the compressor is required.   It is not very practicable to provide this
by a solar-powered engine because of the complications mentioned above.   On
the other hand, the absorption refrigerator, which requires mainly a thermal
energy input, has a CP of around 0.5 when the separator is operated at the
low temperatures available from flat plate collectors.   Although a solar-
powered refrigerator was first operated at least a century ago, and there
has been a continuing effort in the field since then, a cost-effective
solution is still proving elusive (12).   A similar situation applies to
heat pumps.   It will be advantageous to use such a device for heating only
if the energy delivered to the warm space is greater than that released in
burning the fuel used to drive the prime mover.   In practice this means
that the heat pump CP needs to be at least 3, and much greater values would
be required to cover the equipment capital costs.   Machines operating from
ambient sources usually have lower values of CP than this (13).   The CP can
be increased by raising the source temperature, and consideration is being
given to the practicability of doing this by solar heating.

CONVERSION TO ELECTRICITY

Thermionic and thermoelectric generators require the maintenance of a temperature difference for their operation and hence are subject to thermodynamic limitations.   When the cathode temperature of a thermionic
generator is sufficiently high to give a good current density (around
1 500 $^{\circ}$K) the conversion efficiency may be 20% or more (2) but this
requires a concentrating collector and arrangements to limit space-charge
effects.   Further limitations apply to the thermoelectric generator, which
utilises the Seebeck effect arising from the diffusion of charge carriers
down a temperature gradient.   The highest conversion efficiencies are
obtained in very degenerate semiconductors and quite thorough studies have
been made of the most promising materials for use at various temperatures.
Maximum conversion efficiency is obtained when a cascade of thermoelements
is used, with each operating at its optimum temperature.   A two-stage
device has been reported, having a conversion efficiency of 13.5% with hot
and cold junction temperatures of 1 300°K and 300°K respectively (14).
Devices operated with flat collectors, though having an efficiency of less

than 1%, continue to be of interest because of their simplicity and ability to make use of the diffuse component of the solar radiation.

In photovoltaic cells, electron-hole pairs generated by the absorption of radiation are separated by an electric field and the electrons can then be led through an external circuit to do work before returning to be recombined with the holes.   The electrons have to be given enough energy to cross a Schottky barrier or an internal energy gap onto a conduction band.   There is now no intermediate heating stage and hence no thermodynamic limitations such as those afflicting other types of converters.   Nevertheless, high conversion efficiency is not easily achieved.   Photons having energies less than the barrier height or band gap cannot free an electron at all, whereas any energy in excess of this contributes nothing to the external work and is given up in heating the ion lattice.   In consequence, there is an optimum energy gap for radiation of any given spectrum and in the case of solar radiation, it is around 1.2 eV, corresponding with a wavelength of about 1.1 μm.   About 25% of the solar energy has longer wavelengths than this, and the shorter wavelength radiation cannot be fully converted, so that the maximum possible conversion efficiency by a single-stage photoelectric device is found to be about 45%.

The degree to which the theoretical maximum efficiency can be approached depends upon the rapidity of separation of electron-hole pairs before they can recombine and on losses such as those due to Joule heating in the photo-cell material.   There has been long experience with silicon homojunction cells, which have been highly developed for use in spacecraft and more recently GaAs/GaAlAs heterojunction cells have been showing great promise (15).   Both types have conversion efficiencies up to about 15%.   This high efficiency has been obtainable only at great cost.   Suitable properties are obtained in materials in single-crystal form, suitably doped to obtain p - n junction formation.   The level of doping required is quite low, of the order of 1 ppm.   Thus the basic material must first be cleared of impurities to extraordinarily low levels, perhaps of the order of 0.01 ppm. The expense of this, the slow crystal growth, and the subsequent treatment produces capital costs of the order of £10-100 per W of installed capacity. Thus, to be at all competitive with other sources, their cost would need to be brought down by a factor of at least 100.

Much work is in hand to achieve this, because of the great convenience of
direct conversion to electricity.   Methods are being studied of the growth
of continuous ribbons of silicon which retain the single-crystal form (16)
and of obtaining good conversion efficiency in thin-film polycrystalline
cells, deposited epitaxially from the vapour phase, which would be suitable
for large scale manufacture.   The most thoroughly-studied device of this
kind is the CdS/Cu_2S heterojunction cell, reported to have efficiencies up
to 8% (17).

PHOTOCHEMISTRY AND PHOTOBIOLOGY

Photochemical processes with high yields are already widely used for the
manufacture of solvents, insecticides, pharmaceutical chemicals, vitamins
and others. ˙In general, the photon energies in solar radiation are too
low for these processes and u.v. sources are employed in practice.   A
process needs to be found which operates within the solar spectrum and leads
to the formation of stable species from which energy may be recovered at will.
The most desirable would be one leading to the photolysis of water into
hydrogen and oxygen, yielding a storable and transportable fuel from a cheap
and abundant raw material.   This is not possible in sunlight in a single
step, but the example of photography holds out the promise that sensitisation
may be possible.   Sensitisers are apparently able to promote the retention
of absorbed energy long enough for more than one photon to take part in the
process.   To have long-term value, the sensitiser should be a catalyst,
being regenerated in the process also.   If such a catalyst is found, it
would perhaps mediate a photochemical reduction

$$2H_2O + 4M + radiation \rightarrow 4H^+ + 4M^- + O_2,$$

followed by an ordinary chemical oxidation

$$4H^+ + 4M^- \rightarrow 4M + 2H_2.$$

It is probable that a practicable solution, if one is ever developed, will
operate within a photoelectrochemical cell in which recombination is
prevented by the removal of electrons into an external circuit.   If photons
are absorbed on an electrode, having a band gap enabling a useful fraction
of the solar energy to be acquired by electrons, hydrolysis may occur at the
surface:

$$2H_2O + radiation \rightarrow 4e^- + 4H^+ + O_2.$$

Hydrogen is released at the dark electrode by electrons returning via the external circuit:

$$4e^- + 4H^+ \rightarrow 2H_2.$$

This and similar processes have been demonstrated but with very low efficiency (18).

Meanwhile, natural photosynthesis offers an enviable example of the possibilities in this area.   Energy can be recovered from plant materials by direct combustion or by pyrolysis or fermentation and other microbiological actions leading to combustible liquids such as alcohol or gases such as methane.

Vigorous plants growing under tropical conditions produce material at a rate which would permit about 3% of the incident solar energy to be recovered as heat on an annual basis.   It remains to be seen whether the breeding of plants specifically for maximum production of combustible material could improve on this significantly.   Another possible source would be the lower plants, particularly the unicellular algae, grown in open tanks and nourished on waste products such as sewage (19).   Although these organisms have a better conversion efficiency than the higher plants, the overall efficiency based on the area of the system exposed to radiation is likely to be low.

There is a major engineering element in the consideration of photobiological energy conversion systems.   The economic possibilities are heavily dependent upon such factors as the energy used within the system itself, for harvesting, or felling, transporting, drying and processing a very low-grade fuel;   the efficiency of energy conversion into work;   and the environmental consequences of discharging the products of combustion or digestion.   It has been interesting to note the recent comment (20) that much experience in these matters exists in the Irish Republic through the use of peat as a fuel for thermal power stations.

POINTS FOR DISCUSSION

In discussing the relative merits of solar energy and other sources, it will be important to keep in mind the difficulties arising from the nature of the energy supply, in particular its low intensity and its intermittency.   Many

of the devices whose principles have been described have been thoroughly
studied and most operated in prototype form, so that their feasibility is
not in doubt.   Much R & D will be required before they could be put into
service on a large scale, but no insuperable technical problems can be
foreseen which would rule this out in principle.   The problem is to see if
it can be done economically.

The low intensity of the solar radiation flux means that very large collection areas have to be envisaged if significant amounts of power are to be
obtained from it.   No available system has a conversion efficiency greater
than about 20% (21) so that an output equivalent to that of a conventional
power station (about 5×10^{10} MJ per year), even in a favourable climate,
would demand an aggregate collector area of about 4×10^7 m^2, or about 6 km
square.   The construction of solar farms on this scale would be very
demanding of materials at a time when concern is developing over the supply
of some of those involved, and a very long service life would have to be
assured to provide a reasonable balance on the energy account.

The uninhabited regions of the earth are amply large enough to accommodate
collectors capable of furnishing all the world's energy needs (2).   In most
cases, however, they are far from the main centres of consumption so that a
large investment in transmission equipment would have to be added to the
capital cost.   Centralised power generation and reticulation to consumers is
appropriate to the more highly-developed countries, whose economies are based
on industrialisation and urban civilisation.   Many of these also have high
population densities and can least spare the necessary space for large-scale
collectors and storage systems.   If we were to envisage providing the equivalent of Britain's annual electricity supply by solar energy, for instance,
we would have to allocate a collection area for this exceeding that of
Greater London (22).

These considerations have led many to the view that the most appropriate
scale of solar energy utilisation should be small and local rather than large
and central.   On a domestic scale, there are already some millions of solar
water heaters in use throughout the world and this seems certain to grow.
Even in Britain a heater with a few square metres of collection area can
provide half of a household's annual needs for domestic hot water.   For
every 1% of dwellings fitted with these, the annual energy saving would be

the equivalent of at least ½ million tons of oil.

A number of dwellings with solar-augmented space heating and air conditioning, distributed throughout the world, are building up a good case for this
also.   Löf and Tybout concluded in 1973 that a further doubling of gas
prices would make solar heating and cooling competitive in almost all parts
of the United States (23).   Moderate-scale systems of various kinds for
schools, hospitals, hotels, factories and farms are now under review.
Through the introduction of these, solar energy could make a substantial
contribution to the energy requirements of developed countries without the
drawbacks that large-scale centralised collection systems would entail.
It should be noted that about a third of our primary energy usage takes place
in domestic and commercial buildings and that most of this is for heating
(24).

The implications may be even more important for the developing countries (25).
Often well endowed with solar energy, they have been badly hit by rising
fuel costs and desperately need power for the provision of work and amenity
among scattered populations with poor communications.   Further studies are
needed of the extent to which this could be assisted by the wider introduction of solar driers, stills, irrigation pumps, refrigerators, air
conditioners and power plants used locally.   More than 80% of the world's
population lives between the 40° latitudes, providing opportunities for
responses from the industrially-developed nations extending across a wide
spectrum from technical aid programmes on the one hand to new export
business on the other.

The mounting R & D activity in the solar energy field throughout the world,
though largely aimed at the needs of the countries concerned, is to be
welcomed for the addition that it must make in due course to the common
stock of knowledge.   Funding for US work under the programme of Research
Allied to National Needs has now reached $50 million per year, proceeding
along lines recommended by the National Science Foundation in 1972 (26).
This envisaged a commitment of $3 500 million, with the contention that
solar energy could provide 35% of the nation's heating and cooling requirements for buildings, 40% of its fuels and 20% of its electricity.   A
comparable programme has been started in Japan (Operation Sunshine), with a
budget for solar energy work rising to 103 000 million yen (about £100
million) by 1990.   An Academy of Sciences report recommends a similar

programme for Australia (27).

It is noteworthy that together these programmes represent an investment in solar energy technology of a scale hitherto seen only in the largest national enterprises, such as the Apollo moon-landing project.   In view of the manifest need to husband the world's resources and to preserve a healthy environment this will seem a wise and timely investment.

It is uncomfortable to reflect that, although significant contributions to the state of the art have been made in Britain over many years, official funding here has so far been negligible.

REFERENCES

1. King Hubbert, M., World Energy resources, Proc. 10th Commonwealth
      Min. & Met. Congr., Canada, 1974.

2. Brinkworth, B.J., Solar Energy for Man, Compton (UK) 1972,
      Wiley/Halsted (US) 1973.

3. Robinson, N. (ed), Solar Radiation, Elsevier, 1966

4. Davies, M.G., Thermal comfort in a solar heated school,
      Steam & Htg. Engr., 42, 1973, 44-47.

5. Hildebrand, A.F., and vant-Hull, L.L., A tower-top focus solar
      energy collectors, Mech.Engng., Sept. 1974, 23-27.

6. Rabl, A., and Nielsen, C.E., Solar ponds for space heating,
      Sol.En., 17, 1974, 1-12.

7. Howe, E.D. and Tleimat, B.W., Twenty years' work on solar distillation
      at the University of California, Sol.En., 16, 1975, 97-105.

8. Selcuk, M.K., Ersay, O., and Akyurt, M., Development, theoretical
      analysis and performance evaluation of shelf-type solar driers,
      Sol.En., 16, 1974, 81-88.

9. Anderson, J.H. and Anderson J.H. jnr., Thermal power from sea water,
      Mech.Engng., 89, 1966, 41-46.

10. Meinel, A.R. and Meinel, M.P., Physics looks at solar energy,
      Physics Today, Feb. 1972, 44-50.

11. Jordan, R.C., Eckert, E.R.G., Ramsey, J.W., Schmidt, R.N., Sparrow, E.M.,
      and Wehner, G.K., Terrestrial solar-thermal power systems,
      Proc. UNESCO Congress 'The Sun in the Service of Mankind,
      Paris 1973, paper E137.

12. Swartman, R.K., Vinh Ha and Newton, A.J., Review of solar-powered
      refrigeration, ASME, 1974, paper 73-WA-SOL-6.

13. Heap, R.D., Domestic heating in Britain using heat pumps,
      Proc.Symp. 'Re-assessing Energy Sources', NEL, January, 1974.

14. Rosi, F.D., Thermoelectricity and thermoelectric power generation,
      Solid-state Electr. 11, 1968, 833-868.

15. Davis, R. and Knight, J.R., Operation of GaAs solar cells at high
    solar flux density, Sol.En., _17_, 1975, 145.

16. Currin, C.G., Ling, K.S., Ralph, E.L., Smith W.A. and Stirn, R.J.,
    Feasibility of low-cost silicon solar cells, Proc. 9th IEEE
    Photovolt. Spec. Conf., Johns Hopkins Univ., 1972.

17. Coste, G., Fremy, J. and Nguyen, D.T., Technological improvements
    on CdS solar cells. Solar Cells, Gordon & Breach, 1971, 187-200.

18. Paleocrassas, S.N., Photocatalytic hydrogen production, Sol.En.,
    _16_, 1974, 45-51.

19. Noguchi, T., Recent developments in solar research and applications
    in Japan, Sol.En., _15_, 1973, 179.

20. Lalor, E., Solar Energy for Ireland. Rep. to National Science Council,
    Dublin, 1975.

21. Brinkworth, B.J., Prospects for solar power, Chart. Mech. Engr.,
    _20_, 1973, 86-90.

22. Brinkworth, B.J., Making the best use of solar energy, J.IEE.,
    (Elect. & Power), _20_, 1974, 356-359.

23. Löf, G.O.G. and Tybout, R.A., The design and cost of optimised
    systems for residential heating and cooling by solar energy,
    Sol.En., _16_, 1974, 9-18.

24. Rogers, G.F.C., Energy conservation - choice or necessity?
    Ch. Mech. Engr., _22_, 1975, 65-69.

25. - Solar Energy in Developing Countries: Perspectives and Prospects,
    Nat.Acad.Sci., Washington, 1972.

26. - An Assessment of Solar Energy as a National Energy Resource,
    NSF/NASA Sol.En. Panel Rept., Univ. Maryland, 1972.

27. - Solar Energy Research in Australia
    Aust. Acad. Sci. Rept. No.17, 1973.

DISCUSSION

Question: What collector coatings can be used to improve performance?
Answer: Several different coatings of the polished plate collector are
            possible. These produce selective absorption of short wave solar
            radiation with a minimum of long wave emissivity. These coatings
            could take the form of (a) particles of copper oxide about 1 µm
            diameter, (b) a semiconductor such as silica, or (c) layered coat-
            ings of calcium oxide and magnesium oxide utilizing interference
            effects, but which are significantly directional in selectivity.
            Unfortunately, all coatings are expensive and can become detached
            before their economic life of, say, 20 to 25 years has been reached.
            Laboratory studies indicate that the ratio of absorption to emiss-
            ivity is about 15 which shows the effectiveness of coating the
            collector.

Question: What information exists about the direct and diffuse solar
            components?
Answer: Considerable data are available on the direct solar radiation dis-
            tribution in the U.K. as records were started by the Meteorological

Office some 25 years ago. There are many areas for which more data
are needed because reliance is placed at present on extrapolation
curves. Most measurements are of total receipts. Where receipts
are separated, the partition is usually between direct and diffuse
radiation including circumsolar radiation in the diffuse part.
Recently, about a dozen stations have started separate measurements of the circumsolar radiation, which is as useful as direct
radiation for the purposes of optical concentration. Bearing in
mind that the availability of wind energy and solar radiation
energy appear to be greatest in the winter and summer respectively
in the U.K., it is conceivable that a combined system could be
devised with each working to its best advantage. The idea is not
new and no detailed economic studies have been made but it is
unlikely to be viable on a large scale. Nevertheless, the case
for substantial financial investment in solar energy research and
development seems as strong as that for fusion research. After
all, solar technology is straightforward, well-explored and
thoroughly demonstrated and is held back only by economic disadvantages.

Question: What contribution can solar energy make domestically?
Answer: A collection area of 5 m^2 on a house roof in the U.K. could supply
about 40 per cent of the annual domestic water heating load.
Although more difficult to estimate, perhaps, a contribution of
1/3 to 1/2 of the space heating load could be contemplated. Since
less of the traditional roofing materials would be needed, the
cost of installing a domestic solar heating system might be no
more than £1,000 for the average house in the U.K. Solar-assisted
heat pumps are very attractive as the results of crude estimates
indicate that even a rise of 20°C or so of the reservoir temperature above that of the surrounding air could increase the C.O.P.
(coefficient of performance) of the pump to 5 or 6. A rather
awkward limitation for domestic use could result from the deposition of dust on the collector panels but undoubtedly washing techniques would be developed to overcome this.

Question: What factors must be considered when designing more efficient and
economic solar conversion systems?
Answer: There are indications, e.g., those put forward by workers in South
Africa, that increasing the efficiency of solar converters adds
significantly to capital cost. Other important factors in the
system must be taken into account such as consumer demand, pumps,
storage, amount of glass area available, and the effect of latitude on the incident energy. The scale of the installations is
also important. One intriguing example of large scale use in
desert areas is the solar farm where desalination of water leads
to irrigation and thus fertilization of the areas resulting in
the widespread growth of crops. Although we may wish to disso-
ciate solar energy from pollution effects, it must be remembered
that the ultimate dumping of thermal energy may occur after transmission of the energy to regions quite remote from the solar generator. This may lead to local thermal effects. Solar ponds are
another large-scale possibility for the utilization of solar
energy, but, apart from some quite impressive work in Israel,
little experience of them has been acquired generally.

ENERGY FROM WASTES

W. Sabel

Science Department, Oxford Polytechnic

At a simple and very broad level it may be argued that in the context of
energy production, waste is any source from which it is possible to obtain
energy - the energy is there and merely needs to be released. With such a
view, we may consider all sources of energy, whether normally used as such
or not and ask ourselves the following questions:

  i) Is it possible to increase the amount of useful energy obtainable
  from conventional fuels?

  ii) Is it possible to extract from sources not normally used as fuels,
  energy in a usable form? eg wind, tides, sun or geothermal
  sources.

  iii) Is it possible to improve the efficiency of utilisation of the
  liberated energy - that is, can the wastage of available energy
  be reduced?

  iv) Is it possible to increase the utilisation of useful materials
  for which the production requires energy consumption?

It is immediately obvious to anyone with any idea at all of science and
technology that the answer to all of these questions is in the affirmative.
Obviously the recovery of useful energy from conventional fuels can be
increased, we can set up wind or tide driven electricity generators, there
is no special problem in reducing fuel losses in machinery by such methods
as improved combustion, better thermal insulation, or more efficient bearings;
there is always the possibility of minimising or further reducing the loss
of useful products for which energy has been consumed. For example, in
recent times, with the increased cost of energy there has been considerable
interest in the possibility of utilising much of the straw resulting from
grain cultivation and for which traditionally there has been little use, and
the very problem of disposal by burning off after harvest has presented

difficulties and dangers.

From all of this it follows that problems of energy production from waste can be dealt with in a variety of ways, but there is one feature that they have in common - the main constraints are not technical. This is a reasonably safe generalisation in virtually all matters of technological development which may be defined for purposes of the present discussion as the application of science to achieve conditions and effects within a social system. In the vast majority of cases, sufficient science is already available, so there is at hand the possibility of achieving the desired effects. All that is required in addition to these resources is the willingness to use them and to pay the price. In this context the price has to be measured in terms broader than those of money values. We have to take account of social and economic implications, and allow for the fact that these may be beneficial in some respects but disadvantageous in others. In all cases a compromise is necessary.

Obviously there are still some gaps in our scientific skills and knowledge and there is some substance in the commonly advanced reason for not utilising some procedures and processes, namely, that the technology is not sufficiently developed, but even this statement conceals one of the points that needs to be made much more explicit. In discussing the scientific means at our disposal, and our willingness to use them for any purpose such as energy production, there is underlying all judgements an element of evaluation - or worthwhileness - which means, essentially an appraisal of cost. Many of the debates about technical feasibility are in fact dominated by cost considerations. Indeed a judgement on cost and value is embodied in the very word waste, which forms the theme of this discussion and it is useful to remind ourselves that waste resulting from one activity is often the raw material of another. This applies as much to energy as to production of materials such as metals, and indeed it is especially important now, with the cost of primary fuel sources so high and the incentives to conserve it correspondingly large. In these circumstances, interest in energy conservation has increased enormously, and this in itself, whether manifested as better insulation of equipment and buildings, or as improved combustion efficiencies can be seen as one aspect of obtaining energy from waste. It has now been dignified with a

new title - Total Energy. We also have the new approach of Energy Audit -
a simple concept applied to account for all the energy inputs and outputs
to an installation such as a factory and so have the means of tracing
sources of unnecessary loss.

Many of the cures for such losses - leaking steam valves or insufficient
insulation - involve the barest minimum of technology and merely require
the application of very modest managerial skills under the motivation of
economics. Further examples of this are provided by the location of manufacturing plant in such a way that heat normally wasted from one unit can
be utilised by an adjacent one. Thus one can visualise an effective
arrangement whereby a minor user of heat, such as a brewery, draws its
supplies from that discarded by its neighbour engaged in aluminium smelting.
Systems of district heating based on heat otherwise lost in a power station
are examples of a similar kind.

In all of these, there is no particular technological problem and the main
determinant in their application is economics.

References to cost can be misunderstood; costs embrace more than monetary
values even though some are difficult or even impossible to quantify. For
example, one must include consideration of the effect of the activity in
question on certain amenities or the impact on the environment generally.
All of these cost elements can for present purposes be regarded as 'social'
values. Thus we find ourselves considering social values as elements of
the overall cost-benefit analysis which provides the basis for decision on
such technological projects as the production of energy from wastes. It
is not sufficient to consider only the cost of resources for producing the
energy balanced against the corresponding monetary value of the energy
produced, or of the alternative fuels saved or indeed of the cost of other
methods that might have been used for disposal of the waste; this is all
too simple because in many cases there are significant social costs and
benefits, easily enough exemplified by reference to a hypothetical rubbish
disposal system depending on pyrolysis to produce useful energy products,
but accompanied by a pall of black smoke and a foul smell. Do the local
residents then welcome the energy savings involved? Perhaps they would
be more appreciative of the energy saved if the pyrolysis plant were in

the next city out of sight and smell. This illustrates the additional parameter which must always enter into these considerations - the "distance
factor". Those local residents away from the sight and smell of the tip
are much more likely to applaud the local authority for its enterprise in
producing cheap energy from waste than will those unfortunate enough to
have it outside their sitting room windows.

Although some of the earlier semantic discussion in this paper about the
nature of waste can have its value in establishing lines of thought in
considering such a major topic as producing energy from waste, we need
also to examine more specific cases.

For normal purposes of dealing with disposal problems we can consider two
sources of waste - domestic and industrial. Each type has some particular
characteristics; industrial waste may have some constituents not normally
found in domestic materials and could include some noxious or toxic compounds requiring special disposal treatment. It may also contain material
such as metals which have great importance in the context of energy
conservation and production. Indeed the metal content of such industrial
wastes can very often be of sufficient value to justify significant expenditure in removing it before final disposal of the residues. Many industrial operations use raw materials in which the production of large amounts
of energy was involved. Steel is a good example of this and it is clearly
advantageous in terms of energy economy to minimise wastage of this raw
material which costs so much energy to produce. It is not just a matter
of recovering the steel as scrap, but rather of ensuring that the
quantities of discarded material are minimised and that it is not diluted
with other substances such as paper or wood so as to make difficult or
costly its separation or its preparation for re-use. One of the advantages
in dealing with industrial waste is that it is relatively easy compared
with domestic refuse, to separate the former into appropriate types;
there is the further advantage that it is usually much easier and more
economic to collect and treat these industrial wastes occuring in large
quantities in relatively few sites. One of the main cost elements in
dealing with domestic refuse is that of collecting it from a very large
number of sources and then carrying out any necessary sorting and processing. One must not overlook the possibility that the energy consumed in

collecting and processing the waste can more than offset the value of the
energy finally recovered. There is always the possibility that using a
rigorous energy audit many of the benefits presumed to arise from policies
of producing energy from waste are more apparent than real, but even this
glimpse of the obvious conceals further problems, because costs and benefits
are not absolute values, but depend very much on the criteria. We cannot
ignore the fact that we are not yet living in a homogeneous social and
economic system, but in a large number of sub units. It is the criteria
applied by these - whether national or even local village - that determine
choices and decisions about what is good or bad in a technological proposal such as waste disposal.

All of these considerations raise still bigger problems, inevitable in a
free society. Essentially, it is a matter of whose interest is to be
served. Who pays and who receives the benefit? The national interest
might very well be best served by reducing energy consumption for the
production of aluminium from its ore, or making glass from soda ash, and
adopting an effective system for re-cycling scrap material or re-using
glass bottles, but it does not follow that individuals or companies or
local communities can derive sufficient economic benefit by their own
activities in this direction. In a similar vein, it would appear to be
highly desirable in order to minimise energy consumption to derive the
maximum possible life from items in general use. A motor car represents
a substantial energy consumption merely to produce it from the iron ore.
Having thus converted the ore to steel sheet, and that into a motor
vehicle, it would seem logical to make it last as long as possible by
reducing loss from rust and corrosion. This is not particularly difficult
technically but what would be the effect of such a policy on the motor
manufacturing industry and the national economy? What are the social and
political implications of policies of this kind? This raises immensely
difficult problems but in principle, it seems to be a matter of adapting
a society, developed over many generations into a particular pattern of
priorities and values, to deal now with a new situation in which very
suddenly one of the main cost elements - that of energy - has changed
dramatically. The problem has been made immensely more complex because it
involves more than a quadrupling in the price of oil. It seems reasonable
to argue that the social and economic impact has been greatly increased

because of the fact that oil, the main energy source, is found in relatively few locations, many of which are outside the main stream of the world industrial development that has taken place so far. Thus the consequences of the rapid increase in oil prices is a very considerable perturbation in the hitherto established pattern of international economics based on the wide circulation of the world money supply. It is therefore scarcely surprising that the subject of economics concerned with the study of the use of scarce resources and never a very precise tool for dealing with real problems even when these occurred in situations made familiar by long experience, should prove to be peculiarly unsuited to deal with this monumental change in the long established patterns of resource availability and distribution.

In discussing energy supplies it may be too easy to concentrate on fuels for machinery and in so doing overlook human fuels - that is, food.

Perhaps the main distinction between human and mechanical fuels is that the former depends solely in the direct sense on solar energy while the latter are mainly concerned with energy reserves contained in fossil fuels. The distinction is not complete however because agriculture depends very largely on the use of fertilisers, the production of which requires substantial amounts of fossil fuels - oil or natural gas. The fact remains that food crops are direct users of solar energy and thus provide the means whereby it can be stored for relatively short periods and used to supply the energy requirements of people and animals.

It is also necessary to remind ourselves that metabolism of foodstuffs by the animal kingdom is no more than a special case of the combustion of organic materials with liberation of useful energy, more commonly observed when these same materials, such as cellulose are burned in steam raising equipment for example.

Prevailing economic conditions in Britain and the global difficulties of providing adequate nutritional levels for a human population now approaching two thousand million and increasing rapidly, focuses attention on the need for the proper use of food, and introduces the twin problems of achieving the maximum efficiency in using it for human nutrition and of minimising the consumption of fuel in its production.

Social custom plays a large part in all this. In dealing with the problem
of Britain's food supply and the increasing economic pressures that can
constrain our ability to afford to import it in times of increasing world
demand, people like Mellanby and Pirie have commented on the alleged
inefficiency of animal husbandry as a source of human nutrition. It is
claimed that the use of vegetable material to produce animals to provide
human food is seriously extravagant of energy and that a far larger human
population could be supplied by persuading people to rely on vegetable
material as a source of protein. Pirie has made an extensive study of
the possibilities of using leaf protein for this purpose.

It is unlikely, however, that meat will, in the near future, be totally
removed from man's preferred diet, and therefore it remains necessary to
consider the serious energy loss represented by animal excretion. If this
could be economically returned to the soil as a fertiliser or used in some
other way, it would make a significant improvement in the energy balance
of meat production and result in a direct saving in the oil or gas used
in the manufacture of synthetic fertilisers. But the problem is not simple
and we have the present anomaly that in the main cattle rearing areas
depending on battery type feeding in particular, disposal of the excreta
from the animals is a serious problem of pollution of water-ways etc.
while the same material could be very effectively used as a crop fertiliser in other regions. It has even been suggested that pipe-lines or a
tanker system of transportation might be justifiable but it seems unlikely
that in terms of energy balance the cost of handling and distribution
would be offset by the value of the energy saved. This is another illustration of the serious conflict between the view of energy utilisation
seen in global terms and that of smaller units operating within a particular local situation. The economic use of wastes from animal rearing
as a fertiliser presents problems, but the possibility of collecting
locally and using it as a substrate for methane production as a fuel or
chemical source is possibly a more attractive proposition in many cases.

Fermentation provides other interesting possibilities. For example,
bagasse the fibrous waste material left after expressing the sap from

sugar cane has traditionally been used as a fuel for producing the process
steam in the sugar refinery but is now being widely investigated as a
substrate for producing protein by fermentation.

Other fermentation procedures are well known for producing such materials
as ethanol, butanol, acetone etc. These methods, developed earlier in the
present century have been used commercially, but failed to compete economically with production based on oil when the latter was in the $1 - $2
per barrel price range. Now that these costs of oil have increased five-
fold the economics of fermentation, which can be seen as a means of utilising current solar energy absorbed by growing vegetable matter, should
be re-examined; there seems to be a strong probability that present costs
are such that there will be many cases where this fermentation procedure
can be justified economically. As an example of this, it is known that
in India, I.C.I. is operating a plant for the production of polyethylene
using ethylene made from ethanol produced by fermentation of plant
materials.

The method could be used either for the primary purpose of producing
chemicals and would then depend upon such materials as carbohydrates as
substrate, or it could be applied as part of a waste disposal operation
to deal with municipal domestic wastes or for effluents from industrial
operations such as food processing or animal husbandry. In all of these
cases it is clear that adequate technology is already available and the
only major constraint is economics or willingness to pay.

In the case of fermentation of carbohydrates there is the additional
problem that the same substrate can generally be used as food. As a
generalisation 'if you can ferment it you can eat it' and the political
and social implications of this possibility of competition between nutrition
and industrial production of chemicals and fuels must not be under-estimated.

The problem could be particularly acute in the Third World. There are
many regions in the tropics such as Africa or South America where there
is an enormous amount of vegetable material at present going to waste
which if properly harvested and fermented could be the source of considerable amounts of useful products such as fuels and solvents. It might be

argued that theoretically the same ground could be cultivated for purposes
of food production, and indeed there must be cases where this should be
done, but there will be other regions where the growth and collection of
suitable vegetable material for fermentation for chemicals production
would provide an enormously important labour-intensive industry that would
have substantial benefits in terms of economics and social well-being
for the local community. In some cases such operations would be fully
justified. In this type of consideration it seems most important to
distinguish between global and local situations. There is a risk, I would
suggest, that discussion of these problems at global level can be misleading because it tends to obscure the effect of special situations
prevailing locally. In spite of the problem of world food shortages and
the clear need to eliminate them wherever possible it is still reasonable
to argue that in certain particular regions, in South America for example,
the production of local vegetation for fermentation purposes does not in
any way act to the detriment of the world food supply but does on the
other hand generate local wealth and alleviate poverty and suffering
of all kinds.

DISCUSSION

Question: How can one place a numerical measure, in terms of money, energy
            or whatever, on the various social factors thrust upon our daily
            lives?
Answer: Many believe this is not possible, such problems were amply dem-
            onstrated by the many misleading conclusions derived from the
            various analyses for siting of the third international airport in
            the south of England. Nevertheless, I suggest that explicit
            quantification of social factors, however unreliable, is prefer-
            able to implicit assumptions, such as of unchanging social empha-
            sis, since the explicit can be examined and modified if necessary
            whereas the implicit factors must first be found.

Question: Are social costs dominant in waste disposal decisions?
Answer: I think that cost-benefit analyses should be performed separately
            at economic, social and environmental levels and then be drawn
            together by the decision makers who must independently assess the
            level of importance of these issues. The ideas expounded in my
            paper are an attempt to offer some sort of guide line in this
            respect.

Question: Why do we not recycle more materials than we do?
Answer: The utilization of waste increases its financial value and, unless
            this can be properly assessed, an almost fanatical response from
            the public can result, as illustrated by the present flood of
            paper for recycling. The answer to such a problem is unclear and
            one must do one's best to avoid overloading existing facilities
            for recycling. Even so, there are numerous openings for recycling
            of waste and a particularly interesting example concerns the in-
            corporation of straw in paper-making. Transport of straw could
            be made economical by ensuring high-density bales but does the
            industry need this raw material? One can think of other examples
            such as the recovery of wood waste or excess paper for use as fuel.
            There is also the possible conflict of requirements, e.g. power
            stations get the cut of the oil barrel for which the refineries
            have no application. Small local authorities reject schemes
            using wastes for fuel, but adopt schemes designed to improve both
            public amenities and the environment. Thus, pulverised waste is
            used to fill in excavations for subsequent use for agriculture or
            leisure facilities

Question: Can you suggest who should be involved in assessing the costs and
            benefits involved in a particular decision?
Answer: In all questions of social cost there is a strong feeling that
            individuals should have the right to express an opinion based on
            all available evidence. The view has often been expressed that
            too much decision making is taken by central government. Economic
            evidence seems to show, however, that most people choose solutions
            to problems on the basis of financial cost in any case. Perhaps
            the answer lies in a fundamental change of present social attitudes?

SECTION 5

TRANSMISSION AND STORAGE
OF ENERGY

LARGE-SCALE ELECTRICAL POWER GENERATION AND STORAGE

J. K. Wright

Research Department, GEGB

batt-vehicles
p-546

INTRODUCTION

This lecture is in two parts. The first is concerned with large central
power stations for electricity generation. It describes the present methods
and possible improvements. It does not cover topics such as the generation
of electricity from tides, wind, waves, solar or geothermal sources which,
it is assumed, are covered in other lectures. The second part of the lecture
is concerned with possible ways of storing electrical energy on a large scale.[†]

Fig. 1 - Thorpe-Marsh Power Station, I.P. turbine line.

[†] Reprinted from CEGB Research Volume 2, by permission of the Central Electricity Generating Board.

Today's modern fossil-fired electricity generating station is a logical
development of the first public power station equipped with turbines, which
was commissioned at Forth Banks in Newcastle in 1890. It consists essentially of a large combustion chamber in which fuel is burnt; the hot combustion
products are first used to boil water in tubes which form the walls of the
combustion chamber and are then passed to a superheat section in which the
temperature of the generated steam is raised to 535-565°C. This steam is
expanded through a steam turbine which, in turn, is used to drive an electricity generator, Figure 1.

Although the basic principles have remained unchanged over the years there
have been great improvements in detail. Power station efficiencies are now
in the region of 35% and their capital cost has steadily reduced in real
terms as the size of the station and steam pressures have increased. A
single large modern power station will now supply about 4% of the UK electricity and it is only possible to use this effectively because of the existence
of a suitable high voltage grid to distribute it. By connecting the power
stations together with the grid it is possible to obtain high overall
reliability and it is possible to meet a winter peak demand by installing
only 20% extra capacity over the mean anticipated winter load. The great
bulk of the world's electricity is produced in this way. Table 1 gives a
breakdown of the primary source of energy used for electricity generation.

TABLE 1. Sources of primary energy for electricity generation. Figures are
percentages of electricity produced from various sources.

Sources	World	UK
Fossil	75	89
Nuclear	3	10
Hydro-electric	22	1
Tidal	0.01	-
Geothermal	0.08	-

Plant Mix

It is important to observe that no one type of electricity generating plant
is ideal for the system - a mixture is required. The basic reason for this
is that the electricity demand to be supplied is not a steady one but varies
between night and day and between summer and winter. The typical base load

plant must have a low running cost (in terms of p/kWh) but, because it runs for a high proportion of the year its capital cost can be relatively high. This role is ideal for nuclear plant. At the other end of the spectrum the plant which is run only at times of peak demand must have a low capital cost but, since the number of hours run per year is low, the efficiency and running cost are less important. Gas turbines or old steam generating plant of low efficiency are often used in this role. Between the two there is scope for mid-merit plant. In some countries there is scope for special plant of this type, but in the UK it seems likely that the large quantity of modern coal and oil-fired plant will fill this role. It is for this reason that combined cycle plant, fluidised bed boilers and MHD generation is not being actively pursued by the CEGB, whilst there is interest in these developments by some overseas electrical utilities.

The Steam Cycle
The requirements of a high efficiency of conversion of heat to mechanical work (or electricity) are:
(a) A high top temperature and low bottom temperature for the cycle.
(b) All the heat should be put in at the top temperature.
(c) There should be no irreversibilities.

A Carnot cycle is shown on a temperature, entropy plot in FIGURE 2.

AB represents the heat input at T_1 and is $T_1 (S_B - S_A)$

CD represents the reject heat and is $T_2 (S_C - S_D)$

FIGURE 2.

The work done is the difference between the two and is equal to the area of the rectangle A B C D.

The efficiency in this case is $\dfrac{T_1 - T_2}{T_1}$

In a steam cycle it is not possible to put in all the heat at a high temperature. Such a cycle is shown in FIGURE 3.

A.E.C.—P

AB represents water heating

BC represents boiling at constant
temperature

CD represents heating the steam
produced (superheating)

DE represents expansion through a
turbine stage

EF represents reheating the steam

FG represents expansion through
the rest of the turbine

GA represents condensation of the
expanded steam

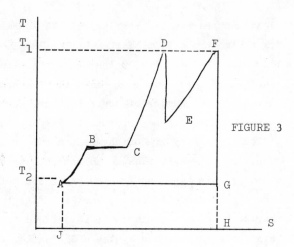

FIGURE 3

This diagram is not   rectangular and the efficiency $(= \dfrac{\text{area ABCDEFG}}{\text{area JABCDEFGH}})$ is
less than that of a Carnot cycle at the same top temperature.

In order to improve the efficiency it is desirable:

(a) To raise the boiling temperature (point B) as high as possible - this
means high pressures.

(b) Superheat to as high a temperature as possible (point D). Upper temperatures of about 565°C. are often used. In fossil fired power stations the
temperature is determined by the properties of a corrosive deposit on the
fire-side.

(c) Use one or more stages of reheat to improve the shape.

(d) Condense at as low a temperature as possible. The heat is often rejected
at only about 10° above ambient - the pressure of the steam before
condensation being, of course, well below atmospheric. In large modern
power stations this leads to very large low pressure turbine stages
(FIGURE 1).

In addition to the above features it is also possible to obtain some improvement in efficiency by bleeding steam from the low pressure turbine stages and
using it to heat the water before it enters the boiler. From this it will be
clear that the modern steam power station is highly optimised and it is
unlikely that many radical improvements to the steam cycle itself will
evolve.

Dual Cycles

In a fossil fired power station combustion temperatures are often about
1600°C so there is scope for utilising some of the heat at a high temperature

in some device before rejecting to the steam cycle. If all the heat of combustion is passed to the topping device and a fraction η_T is converted to electricity and all the remainder $(1 - \eta_T)$ passes to the steam cycle where a fraction η_S is converted into electricity, it is easy to see by considering the reject heats that the overall efficiency η is given by:

$$1 - \eta = (1 - \eta_T)(1 - \eta_S)$$
$$\text{or} \quad \eta = \eta_S + (1 - \eta_S)\eta_T$$

In practice all the heat does not pass through both topper and steam cycles and the improvement in efficiency by going to a dual cycle tends to be rather less than presented in this equation.

Gas Turbine Combined Cycle

The basic principle is shown in FIGURE 4. In the simplest arrangement air is compressed, a clean fuel added and burnt in a combustion chamber before being expanded in a gas turbine. The exhaust gases are passed to a boiler and used to raise steam. It is possible to design to extract different proportions of energy in the steam and gas turbine by varying the input pressure to the gas turbine. Overall efficiencies of up to about 45% can be realistically obtained.

In practice the plant burning a "dirty" fuel, such as coal or residual oil, tends to be more complex. For example, the CEGB has, in the past, considered gasifying coal to provide a clean fuel which is fed to the gas turbine. The residual char, together with the exhaust from the gas turbine, can be fed to the boiler and used to raise steam. Not all the energy in the fuel passes through the gas turbine part of the cycle and so the overall efficiency is reduced below the above theoretical value.

MHD Generation

This is another possible dual cycle system. The principle of MHD generation is very simple. Electrically conducting gases are passed along a duct at high speed normal to a strong magnetic field. The induced currents are collected on electrodes and passed to the electrical load (FIGURE 5). MHD is potentially an attractive topping device since the electricity is generated in the hot combustion products themselves and it is possible to cool the walls of the duct. In this way the advantages of high temperature conversion

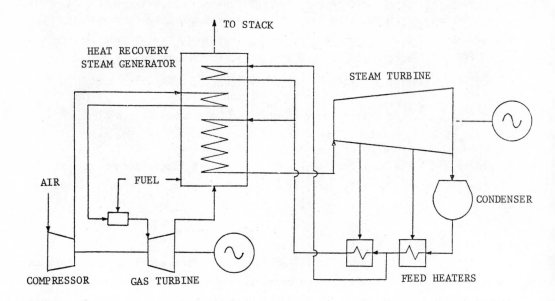

Fig. 4 - Basic concept of combined cycle gas-steam plant (unpressurized
         steam generator).

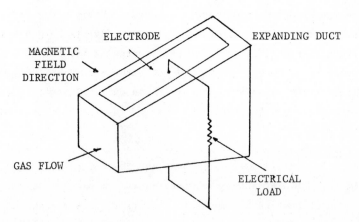

Fig. 5 - Schematic of an MHD generator.

can be obtained without extreme materials conditions - at least in principle.

However, the practical realisation of an MHD power station will require the solution of formidable technical problems. The principal features are:

(a) An air preheater - needed to raise the combustion temperatures to around $2500^{\circ}C$. to raise the electrical conductivity of the gases.

(b) A pressurised combustion chamber with low heat losses.

(c) Seed injection facilities - a low ionisation potential seed material (probably potassium) is required to give sufficient electrical conductivity.

(d) A water cooled but electrically insulating MHD duct with long life electrodes.

(e) A large superconducting magnet (required for economic reasons).

(f) Seed recovery apparatus - required for both economic and environmental reasons.

There was a great deal of international interest in MHD generation during the early 1960s. This tended to decrease, except in Russia where a large pilot plant has been operating for several years. There has recently been a renewed interest in various countries - particularly the USA and Japan.

Thermionic Generation

An alternative possibility which has been considered is to use thermionic generators as toppers. These work on the principle that, when a cathode is heated electrons "boil off" and will cross to a nearby anode, thereby creating electric current. The concept would be for the devices to be mounted outside the boiler tubes. Heat rejected at the anode would be used to raise steam in the normal way. The concept gives rise to a large number of low voltage (about 1 V) DC units which are difficult to handle practically. There is less interest in this method than in MHD generation.

Freon and Potassium Vapour Turbines

A form of dual cycle which has been considered is to use a turbine driven by freon in place of the low pressure stage of the steam turbine. Thermodynamically the large LP stages can be avoided in this way but only at the expense of a difficult steam to freon heat exchanger.

Similarly, it has been suggested that potassium vapour would be preferable to steam at the high temperature end of the cycle.

These multiple turbine proposals are interesting but have not found general commercial use.

Nuclear Gas Turbines

One interesting possibility is to use the gases which are used to cool the fuel elements in a high temperature gas cooled reactor to drive a gas turbine. The complexities of the steam cycle are avoided but at the expense of having turbines which must be extremely reliable, have long periods between maintenance and cause no safety hazards to the reactor. Reject temperatures are relatively high and could be used for process steam or district heating, to drive an auxiliary low temperature heat engine - possibly using freon as a working fluid - or exhausted to a dry cooling tower which, because of the high temperatures, could be relatively cheap (see below).

Fuel Cells

In theory a fuel cell should enable electricity to be generated extremely efficiently by a process which is the reverse of electrolysis. In practice it is necessary to work with a clean fuel (such as hydrogen produced from available fossil fuels by reforming) and either work at very high temperatures or use expensive catalysts to obtain the high current densities required for an economic plant. Although the total plant can be built up from small modules, development costs are not cheap. For example, Pratt and Whitney (USA) are engaged on a $60m programme aimed at developing an acid electrolyte hydrogen oxygen cell which would be capable of generating electricity in city centres at overall efficiencies and capital costs comparable with those for large fossil fired plant.

The Reject Heat

With limited cycle efficiencies it is inevitable that there should be large quantities of reject heat. Since this is normally at only 10-15°C above ambient it gives rise to large quantities of tepid water which must be rejected or preferably usefully used.

Reject to Sea or Rivers

A modern directly cooled 2000 MW power station requires 1200 million gallons per day of cooling water, which is far greater than the dry weather flow of any British river. Direct cooling for large power stations, therefore, is

limited to coastal or estuarine sites and a careful study is needed at each
location to ensure that there is adequate separation between the cold inlet
and warm outlet water. With increasing power station sizes the dispersal of
the hot water plume becomes a potential problem. Analytical techniques have
recently been developed by the CEGB to enable this problem to be studied at
any particular site. Our findings show that coastal sites up to at least
10 GW are feasible provided there is a moderate residual drift current in the
sea in addition to the normal tidal currents.

Cooling Towers

With limited river cooling water available it has become customary to use the
latent heat of evaporation of water using cooling towers and thereby reduce
water requirements. The large cooling tower bulk and the long plumes could
be considered to be environmentally disadvantageous. The plume can be eliminated by using "dry" cooling towers in which the heat is removed by heating
the air. But these are expensive and very bulky (FIGURE 6). The bulk can be
reduced (and the plume problem slightly improved) by using a fan to blow air
through a normal wet tower. Such a tower has been developed by the CEGB and
is being built at Ince B.

Finally, we are working on a hybrid cooling tower design which could well
contribute significantly to both the bulk and plume problem. By heating the
moist air in the cooling tower it is possible to keep the moisture below
saturation as the plume subsequently cools, thus eliminating the plume visi-
bility problem.

District Heating

With increased fuel charges relative to capital costs, district heating is
now more interesting than hitherto, particularly for redeveloped city centres.
District heating can either be from simple fossil fuel fired boilers or the
heat can be supplied from power stations. In the latter case it is necessary
to supply the heat at a temperature considerably greater than the normal
reject temperature of the power station and this results in a loss of electrical output. For every three or four power stations converted to district
heating one additional power station must be built to compensate for the
reduced electricity generated. If the power station is several miles from
the city the heat transmission costs become significant. The economics,

Three conventional natural draught towers

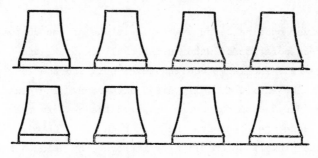

Eight dry towers (of Rugeley A type)

One assisted draught tower (Ince B)

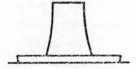

One hybrid wet/dry assisted draught tower

Fig. 6 - Arrangement of cooling towers 1,000 MW power station.

therefore, are quite complicated and each case must be examined on its merits.

Other Uses

The CEGB are conducting research into other ways of making use of the power station waste heat. The studies include possible applications to horticulture (particularly tomato growing) and fish farming.

Storing Electrical Energy on a Large Scale [*]

The CEGB system already uses energy storage on a small scale, to help with the problems of maintaining 'spinning reserve' - the arrangement whereby some generating plant is normally kept running at part-load so that it can increase its output very rapidly on demand. Pumped-water storage schemes of the sort we have at Ffestiniog and are building at Dinorwic provide us with an alternative means of generating additional power very rapidly. They have clear economic advantages over conventional spinning reserve, with its inefficient use of equipment and high running costs.

However, there is no need to have more spinning reserve than the amount of power at risk through the loss of a single transmission line or generating set. Dinorwic and Ffestiniog, providing potentially some 2000 megawatts of stored capacity, will suffice for some time to come to cover this particular need.

Another important advantage of storage is that it can help to smooth out the daily variation of power to be supplied from main generating plant and can thus lead to improved reliability and reduced maintenance costs. This point will become significant when the proportion of nuclear plant on the system has grown to the point where some of it would have to be shut down at night. Reliability is specially important for nuclear stations, since any loss of scheduled output must be made good by conventional stations with much higher operating costs.

At the moment, the CEGB's nuclear capacity is around 4000 megawatts. As the FIGURE 7 shows, this is well below the minimum demand, so the need for this type of storage does not arise. But by the 1990's, the available nuclear

[*] This section has been written jointly with G. C. Gardner, A. B. Hart and R. D. Moffit, and is reprinted from CEGB Research Vol. 2, by permission of the CEGB.

capacity will begin to exceed the minimum demand at night. This will extend
to more days in the year as the proportion of nuclear capacity grows.

The Economics of Storage

With storage, a surplus of nuclear energy available at night could be used in
the daytime in place of relatively expensive power from fossil-fuelled stations
The resulting fuel-cost savings, as shown in the FIGURE 8, increase quite
rapidly through the late 1990's, until nuclear plant accounts for about 70 per
cent of the total on the system. The overall savings depend appreciably on
the size of the nuclear programme, on the efficiency of the storage device
(efficiency being the ratio of the energy recovered to the energy put in) and
on the costs of fossil fuels. Further savings would accrue from using the
storage system in winter to provide additional peak capacity, thereby
reducing the need for other peaking plant such as gas-turbines.

The FIGURE 9 indicates the total present-worth benefits of storage, assuming
an interest rate of 10 per cent, the smaller of the two nuclear programmes,
and fuel prices that remain static in real terms. They are plotted as a
function of the annual capital charges for the particular storage device
installed. The important thing to note is that the sums of money involved
can be very large. The obvious conclusion is that storage warrants a substantial research investment in the hope of having a suitable system available
in good time.

A large-scale energy-storage system for smoothing the daily load cycle would
be required to charge up in about six hours and to be capable of releasing
its energy over 12 hours. The capital-cost target we believe such a system
must meet is about £200 per kilowatt of output power (including interest
charges during construction), which is equivalent to £17 per kilowatt hour of
energy discharged. This figure is based on a 30 year lifespan and involves
reasonable assumptions about the amount of load cycling nuclear plant would
otherwise be forced to do later in the century.

It takes account of the probable cost differential between nuclear energy and
energy from fossil fuel at that time and includes the saving of peak capacity
that storage would allow. An installation with a shorter life, such as a
battery, would need to meet lower capital-cost targets if it were to achieve
the same annual capital charges.

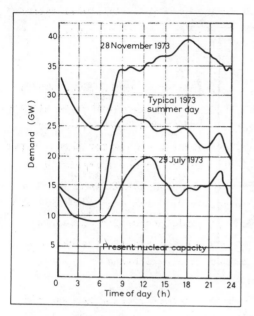

.Fig. 7 - The level of demand for electricity in the U.K. varies widely through the day, but always shows a characteristic trough in the small hours. Large-scale storage plant could occupy spare generating capacity during this period, releasing the energy at peak-demand times later in the day. During the 1990's there will begin to be spare nuclear capacity available on summer nights.

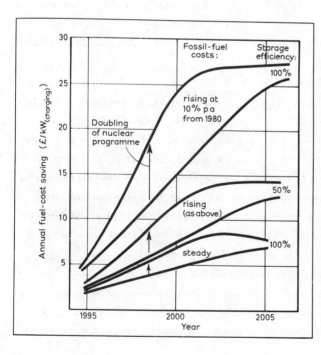

Fig. 8 - The potential fuel-cost savings made possible with storage rise steadily until nuclear plant accounts for about 70% of the total generating capacity. The savings depend on the efficiency of the storage device and on the rate of escalation of fossil-fuel prices, as well as on the rate of installation of nuclear plant. The lower curve in each pair assumes that nuclear capacity will reach 115 GW by 2005. The upper curves assume a nuclear programme double this size.

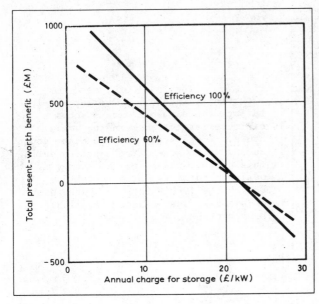

Fig. 9 - The present-worth benefits of storage can be very large. This graph is based on the smaller of the two nuclear programmes used for Fig. 8 and on fossil-fuel prices that remain steady in real terms. Benefits (discounted to 1980) are plotted against annual storage costs per kilowatt of discharging power.

Targets for storage[*]

	£/kw	£/kWh
Capital costs	220	17
Annual charges	22	1.8

[*] The targets quoted relate to a 12 hour discharging phase and capital costs are based on an assumed 30 year life. For less durable equipment, such as batteries, the annual charge is a more appropriate measure.

Pumped-water Storage

The only large storage systems already in operation are pumped-water systems, such as Ffestiniog. Britain's pumped-storage capacity at present totals about 1000 megawatts, but will increase to about 2500 megawatts with the completion of the Dinorwic project in Snowdonia in the early 1980's. There are two reasons why the CEGB cannot simply rely on expanding pumped-storage capacity to meet the needs of a predominantly nuclear-fuelled system. One is that the costs of a big scheme such as Dinorwic, althouth attractive on the basis of a 'spinning-reserve' cycle with generation for 5.2 hours per day, will not necessarily be so for the high-capacity schemes that would be needed in the

1990's. The other, more serious objection is that the number of suitable sites available to the CEGB is small, particularly for the sort of scheme that uses a lake at the top of a mountain and a lake at the bottom, with water pumped between the two.

One alternative proposal is simply to dig a big hole in the ground to act as a lower reservoir and to use a lake, sea or estuary for the upper reservoir. The cost of excavation would clearly be a significant item, although it might be partly offset by choosing a site close to a load centre and so saving on the lengthy transmission lines associated with natural-reservoir schemes. It might also prove possible to excavate material of intrinsic value. However, on closer examination, under-ground pumped storage of water looks rather less promising. For a given volume of under-ground reservoir, it would be better to store compressed air (see below) since to yield the same power output as a compressed-air scheme, the reservoir would have to be very much deeper and could encounter appreciable geothermal temperatures. Bearing in mind that the electrical generating plant would have to be operated at the lower reservoir level, we cannot rate the underground pumped storage of water very highly.

Compressed-air Storage

No large-scale compressed-air storage scheme is yet in operation, although there have been many proposals over the past 25 years. All have sought to improve on the basic inefficiency of the peak-load gas-turbines, which typically requires two thirds of its work output to drive its own air compressor, leaving only one third for the generation of electricity. With a relatively straightforward rearrangement of the shafts, the compressor can be driven independently by a motor running on off-peak electricity, the compressed air being cooled and stored in large under-ground caverns. During peak-demand periods, the air can be released, used to burn a distillate fuel oil and expanded through the turbine. The whole of the turbine power is now available for electricity generation. The same electrical machine can, of course, double as compressor-motor and generator.

Such a scheme effectively up-rates the gas-turbine by a factor of about three and promotes more effective use of the oil fuel. But it does not strictly store energy. On the contrary, it discards most of the electrical energy

absorbed, because it is necessary to cool the very hot compressed air before
storing if the cost of creating suitable subterranean reservoirs is to be
kept reasonably low. The Marchwood Laboratory has proposed a way of avoiding
this drawback by retaining the heat of compression separate from the air in a
regenerative heat store. The scheme would then involve true energy storage.

Superconducting storage

The idea of storing electricity in a very large superconducting magent is, at
first sight, a very attractive one. Such a storage system would have negligible losses; it could feed straight into the electricity system and, in principle, could be built on a very big scale.

The obvious problem is that the cost would be very high. For a given magnetic-
field strength, the cost of the superconducting windings is proportional to
the surface area of the coil, while the stored energy is proportional to the
volume. Hence, the cost per unit of stored energy is inversely proportional
to the linear dimensions of the coil (or the cube root of the storage capacity) and, to be economic, the storage system must be made up from large units.

Several big superconducting magnets have already been built for research
apparatus. For example, the CERN bubble-chamber magnet, which stores 0.22
megawatt hours - very little compared with the storage we shall have to concern ourselves with - costs around £8000 per kilowatt hour of storage capacity.

For a projected 2.5 gigawatt (electrical) fusion-power device, the superconducting magnet used for containing the plasma would have a typical stored
energy of 50 megawatt hours. The generally acceptable cost target to make
such a magnet economically attractive for a fusion device is £1000 per kilowatt hour stored. This reduction relative to the cost for the difference in
storage capacity according to our simple scaling rule.

The particular size at which a storage magnet would break even and meet our
proposed target of £17 per kilowatt hour is difficult to define. It depends
on how we anticipate possible improvements in performance and reductions in
the cost of superconducting materials. At the Los Alamos Scientific Laboratory and at the University of Wisconsin the view is being put forward that,
apart from the superconducting materials, it is necessary to economise on the

mechanical structure holding the storage magnet together, which at present
accounts for at least half the cost. What is proposed is that the entire
structure should be built underground in a cave, so that the natural rigidity
of the rock would serve this purpose.

Fig. 10 - This artist's impression shows a tentative scheme being studied
at the Los Alamos Scientific Laboratory for combining superconducting magnet
energy storage with a 1000 km, 5000 MVA, dc power-transmission line. Since
the transmission line requires much more refrigeration during cool-down than
during steady operation, the idea is to use the surplus regrigeration capacity
to cool the storage units, which would be located every 20 km or so along the
line at the regrigerator sites. About 100 such units, each storing 300 MWh,
would serve to 'peak-shave' the system fed by the line.

However, even the most optimistic assessments do not envisage superconducting
storage breaking even with other possibilities below capacities of 10 gigawatt
hours (or 1000 megawatts for 10 hours). On our own scale of priorities -
although we have arrived at no final conclusions as yet - we would put superconducting storage magnets fairly close to the bottom.

Flywheel Storage
Flywheels have been used for many years to maintain steady rotational speeds
in machines, and for other short-term energy-storage purposes.

Pulses of high power for the magnets of a nuclear-particle accelerator, for
example, may be drawn from a flywheel store in which the energy has been
built up over a longer period. The simplest sort of flywheel is a circular
rim - like a hoop - rotating about a central shaft. The energy is stored in
the mass of the rim, as kinetic energy. The principal stress is a circumferential or hoop stress, caused by the centrifugal forces tending to tear
the rim apart. The maximum storage capacity is achieved at the point of
mechanical failure of the flywheel. For safety's sake, of course, the stored
energy must be kept well below this limit.

An elementary theoretical analysis of the flywheel shows that the energy
storable, per unit mass, is proportional to the allowable hoop stress divided
by the material density. Contrary to most people's intuitive feeling, the
maximum storage capacity, for a given load on the bearings, is achieved with
a fly-wheel made, not of heavy metal, but of a material which combines low
density with high tensile strength. Suitable materials are the recently
developed fibre-reinforced plastics and it is only since their introduction
that it has become possible to envisage flywheels capable of storing energies
as great as 10 megawatt hours for recovery over a period of a few hours.
Carbon fibres, which spring immediately to mind as a reinforcing material,
are expensive. The current price is about £15 per kilogram and even on the
fairly optimistic assumption that this will fall by a factor of three, the
cost of the flywheel alone comes to about £70 per kilowatt hour stored. It
seems unlikely that the price could fall by the further factor of five or more
that we should need to make a carbon-fibre flywheel worth considering for a
storage system.

Professor Richard Post of the University of California, who has recently been
responsible for reviving interest in flywheels, argues that glass fibres,
rather than carbon, offer a brighter prospect. Glass fibres are intrinsically
as strong as carbon fibres, but, at first sight, they have two disadvantages.
Firstly, they tend to be more elastic than the normal bonding materials so
that the bonding material crazes under high strains. Secondly, the glass
suffers from stress corrosion on exposure to moisture (which may gain access
to the surface as the bonding crazes) and this greatly reduces the ultimate
tensile strength of the fibres.

Post offers solutions to these difficulties. He suggests that the composite should be made in an integrated factory so that the glass can be prepared, bonded and sealed under dry conditions. To reduce stress corrosion, which proceeds at a very temperature-dependent rate, he suggests operating the flywheel at a deep-freeze temperature, and finally he proposes an effort to develop bonding materials of much greater elasticity.

At this stage, there is not much experimental evidence to support Post's conjecture that such measures could yield glass-fibre structures as strong as carbon-fibre structures for a fraction of the cost. On the basis of presently-established materials' properties and costs, carbon-fibre composites have a clear advantage. But even on optimistic cost projections, they are too expensive to permit the cost target of £17 per kilowatt hour to be achieved. Using glass fibres, it would be necessary to increase their effective strength by a factor of at least 20 above current values for the system to be of interest. Even then there would be significant safety problems, since the rupture of a 10 megawatt flywheel at full speed could release energy equivalent to about one ton of high explosive and substantial containment would need to be provided. Flywheels, like superconducting-magnet stores, do not rank very high in our list of priorities for research at the moment.

Battery Storage

Electric storage batteries would be well suited as a replacement for spinning reserve. They could also cope with the problems of smoothing the daily load cycle. They can be switched off instantly while charging and brought instantly up to full load (and, briefly, taken well beyond their rated level of discharge). They could be relied upon to function automatically and with minimum maintenance. Because their optimum module is small, say up to 0.5 megawatts, they could be sited close to the consumer and so help to moderate the load on the grid.

For the CEGB, the question that arises is whether it would be justifiable to embark upon what experience has already taught would be a long and expensive research programme to develop a new high-energy-density battery. There already exist more than a dozen battery systems that might be considered for large-scale storage, not excluding the time-honoured lead-acid battery, whose development potential is by no means exhausted yet. The performance and

estimated costs for some of these batteries are given in the Fig. 12. Batteries have a relatively short life, so a better basis for comparison is the
annual cost per unit of output power available. Our target here is £22 per
kilowatt per year.

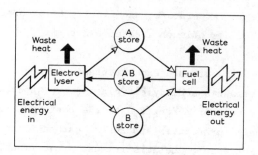

Fig. 11 - The essential functions of a storage battery are: (i) the charging
(or electrolyser) function, in which the chemical AB is electrochemically
decomposed to A and B; (ii) the storage function, in which A and B are held
apart; (iii) the discharge (or fuel-cell) function, in which A and B are
reunited, with the simultaneous generation of electricity. In conventional
batteries, the electrolyser and fuel-cell functions are combined within the
same cell. This is a convenient but not a necessary arrangement.

A major incentive for battery makers and others already exists in the form of
electrically powered vehicles of greater endurance and speed. Success in this
field might go some way towards defraying the cost of developing a bulk-
storage battery and would incidentally also reduce the need for one by encouraging the use of off-peak electric power for transport.

The nickel-cadmium, nickel-iron and lead-acid types of battery are the only
ones currently commercially available. Lead-acid is the cheapest at about
£40 per kilowatt per year, this price relating to mass-produced 'starting
lighting and ignition' batteries with a life of 600 to 800 cycles. If their
low manufacturing cost could be combined with the much longer cycle life of
large traction batteries, lead-acid costs would come down to £21 per kilowatt.
Costs in or below this region are predicted for the more advanced combinations,
such as sodium-sulphur, lithium-sulphur and lithium-chlorine. A great deal
of work is being done by British Rail and by the Electricity Council Research
Centre at Capenhurst on the sodium-sulphur battery for transport applications.
Lithium-sulphur is being actively developed at various laboratories in the
U.S.A. for the same purpose.

	Hydrogen-oxygen	Lead-acid	Nickel-iron	Nickel-cadmium	Sodium-sulphur	Lithium-sulphur	Lithium-chlorine
State of development	Commercial electrolysers Fuel cells for military and space use	Commercial 1-2MW installations and s.l.i. vehicle batteries	Commercial traction batteries	Range of commercial batteries	Laboratory and proto-type traction batteries	Laboratory	Laboratory
Electrolyte	Various: aqueous KOH to solid ZrO_2/CaO	Aqueous 30% H_2SO_4	Aqueous 25% KOH	Aqueous 25% KOH	Solid β-Al_2O_3	Molten lithium halide eutectic	Molten alkali halide eutectic
Approximate operating temperature (K)	300-1250	300	300	300	600	630	830
Open-circuit voltage (V)	1·2	2·0	1·4	1·3	2·1	2·2	3·6
Approximate efficiency	0·46	0·8	0·6	0·65	0·6	>0·8	>0·8
Estimated annual cost (£/kW output)	46	40	50	80	26	12-25	17

Fig. 12 — The table lists characteristics of some of the rechargeable battery systems that might be considered for large-scale storage. The cost is calculated for each cell type on the basis of a module of power output 0.5 MW and storage capacity 6 MWh. It includes the cost of associated electrical equipment, cooling, and storage of liquid or gaseous reactants, but does not include building or site costs.

Conceivably, these batteries could prove cheaper still if they were developed specifically for large-scale electricity storage. The difference is that transport batteries must be optimised for providing a large amount of power for a minimum weight, whereas the CEGB wants a large amount of power for a minimum cost. But it is by no means clear at the moment that a major research investment in developing a battery for the CEGB's particular needs would be worthwhile.

Electrolysis Storage Systems

An electrolysis storage system may be thought of as a battery in which the electrolysis and fuel-cell functions are physically separated. Hydrogen produced by electrolysis could be subsequently burned in fuel cells to regenerate electricity. Electrolysis systems are already available in quite large sizes and of proven reliability and long life. There are also prospects of substantial cost reductions; but even so, it does not look as though electrolysis will become economic for storage on a daily basis. There remains, of course the possibility that hydrogen and oxygen could be stored for much longer

482   J. K. Wright

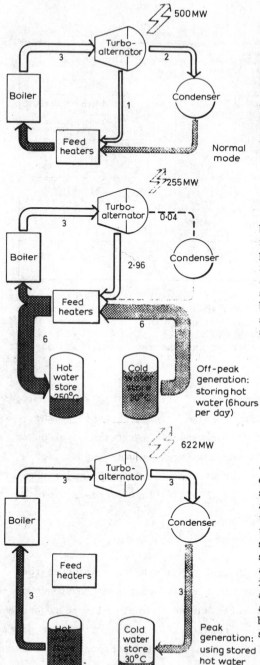

a) In a normal 500 MW generating set,
about one third of the steam flowing
through the turbine is bled off and
passed to the feed heaters, where it
is used to raise the temperature of
water returning to the boiler from
the condenser. This improves the
thermo-dynamic efficiency of the
boiler

b) At night, when electrical demand
is low, the boiler is operated at full
power, but additional steam is bled
from the turbine to the feed heaters
and used to heat extra water for
storing. Electrical output is reduced
by about 50% for the extreme case
shown here, where practically all the
steam flow is bled off.

c) During the day, at periods of high
electrical demand, the supply of bled
steam to the feed heaters is cut off
and hot water is drawn from the store
to feed the boiler. The boiler operating conditions thus remain at a constant level but the removal of the
steam bleed increases the turbo-
alternator's output by about 25% and
its rated capacity must be increased
accordingly. The diagram again shows
an extreme case, in which all of the
boiler feed water is supplied by the
store.

Fig. 13 - These diagrams show one of a number of possible ways in which
stored hot water can be used to boost the output of a conventional steam
cycle. Numbers indicate the relative energy flows.

periods to help cope with seasonal variations in demand.

Heat Storage

Heat is transported by various media through a thermal power station - by water, steam, carbon dioxide or molten sodium. If one of these, together with its heat, could be stored economically, there would be a double advantage. Most importantly, all plant up to and including that used to transfer the heat could be run under constant conditions, independently of electrical demand, since the stored heat could be used to satisfy the fluctuations. Second, the energy would be readily available so that load fluctuations could be met swiftly. In practice, the heat capacity per unit volume of gaseous media is too low to be economically attractive for this purpose, leaving water and sodium as the possible storage media.

In a modern steam cycle where superheated steam is expanded through a turbo-alternator, about 30 per cent is bled off half-way through the turbine and used to pre-heat water returning to the boiler. The rest continues through the turbine and is condensed in the normal way. The obvious thing to do to store some energy is to increase the steam flow to the feed heaters at the expense of the flow to the condenser and to use the additional heating capacity of the feed heaters to provide additional hot water, which can be stored, (see FIGURE 13).

The important point is that the same amount of water would be flowing through the boiler as before, so that the steam flow through the first parts of the turbine would be unchanged. The only change as far as the turbine was concerned, would be the steam flow to the condenser. The next day, when one wanted to produce some additional power, one could use the hot-water store to feed the boiler, cutting off the feed heaters altogether and passing the whole of the steam flow to the condenser. In this very simple way, a storage device could be incorporated into power stations, and on present estimates the cost appears to be within our targets.

There are some snags, of course. The total system would become more complicated and this would inevitably reduce the all-important factor of reliability. We should have introduced another possible source of leakage into the primary circuit and this could have safety implications if the source of heat were a boiling water reactor. Nevertheless, the possibility is extremely interesting and we feel it could be worth pursuing further.

The other possibility, hot-sodium storage, is necessarily restricted to use in conjunction with the sodium-cooled fast reactor. The scheme would simply involve extra sodium flowing through the boiler circuit and being temporarily intercepted by a reservoir. The efficiency of energy recovery would be high, but there are clearly more costs involved than there are for hot-water storage schemes. At current sodium prices, the economics of sodium storage do not look attractive.

In summary, it is clear that in 20 years time there could be a case for installing substantially more storage plant on the CEGB system. It is not yet clear which of the options we should pursue. Those schemes which present the most interesting technical challenges, such as flywheels of low-density fibre-composite materials or superconducting electro-magnetic stores, seem to be the least attractive economically. More attractive are the electro-chemical methods and the storage of compressed air, although the latter is of little interest unless the heat of compression can be usefully conserved as well. Cheapest and simplest of all could well be the storage of additional hot water in the boiler circuit of a thermal power station. But there is certainly a lot more work to be done before we can make our final choice.

Further Reading

Electrical Energy Storage by G. C. Gardner, A. B. Hart, R. D. Moffitt and J. K. Wright (CEGB report; RD/L/Ra906, 1975).

Electrochemical Batteries for Bulk Energy Storage by A. B. Hart and A. H. Webb (CEGB report: RD/L/R1902, 1975).

Compressed Air Storage by I. Glendenning (CEGB report: R/M/N783, 1975).

Energy Storage and its role in Electric Power Systems by J. L. Haydock, (World Energy Conference Paper 6.1-21, 1974).

Pumped Storage is the Cheapest Way of Meeting Peak Demands by F. P. Jenkin (in Electrical Review, 195, 196, 1974).

Flywheels by R. F. Post and S. F. Post (in Scientific American, 229, (6), 17, 1973).

Thermal Energy Storage in Rock Chambers by P. H. Margen (in Proceedings of the Fourth International Conference on the Peaceful Uses of Atomic Energy, 4, 177, 1972).

Proceedings of the Ninth World Energy Conference, Detroit, 1974. Especially Division 4.

Energy in the 1980's. Royal Society, 1974.

Heywood & Womack. Open Cycle MHD Power Generation. Pergamon 1969.

Analysis of Engineering Cycles, Haywood, February, 1975 Second Edition. Pergamon Press (in paperback form).

Engineering Thermodynamics, G. F. C. Rogers and Y. R. Mayhew, Longmans, 1962.

Gas Turbine Theory, H. Cohen and G. F. C. Rogers, Longmans, 1962.

Steam Turbines and their Cycles, J. K. Salisbury, Wiley, 1950.

Thermodynamic Cycles and Processes, R. D. A. Hoyle and P. H. Clarke, Introductory Eng. Series, Longman, 1973.

Thermodynamic Cycles of Nuclear Power Stations, D. D. Kalafati, IPST, 1963.

Energy R & D, Problems and Perspectives, O.E.C.D. Publications, Paris, 1975.

DISCUSSION

Question: What efficiency is reached with the prototype MHD power station in the USSR?

Answer: I understand that a power level of 4 MW (10% to 15% efficiency) has been obtained in continuous runs of up to 100 hours. The maximum power generated was about $12\frac{1}{2}$ MW for $\frac{1}{2}$-hour, with an enhanced gas flow. These figures should be compared with the design output of 25 MW. However, the plant represents a major step forward in MHD technology.

Question: What load factors are being achieved with the present nuclear generating plants in the UK?

Answer: The Magnox stations average between 70% and 80%.

Question: What are the projections for the future capacity of nuclear plants in the UK and how will the unit size change?

Answer: By the time the first batch of SGHWRs are commissioned in the early 1980's it is expected that the UK will have a nuclear capacity of around 14 GW. The early SGHWRs are likely to have an output of 660 MW. The choice and quantity of the future programme will be reviewed by the Government in 1977 and 1978.

Question: Is the CEGB considering dual fluid cycles?

Answer: In the past the CEGB have given considerable thought to this topic. In particular, a detailed theoretical and experimental study was carried out of the possibility of using freon in place of steam for the low pressure stages of a turbine. A freon turbine would be smaller and cheaper than a conventional one but this would be offset by the cost and losses associated with the water/freon heat exchanger. Moreover, there is much to be said for keeping generating plants as simple as possible to obtain high availability. As a result of these investigations sufficient information was obtained to make it clear that the topic should not be pursued further at this stage.

Question: What scope is there for battery storage of electric power at the
            generating station?
Answer: A variety of possible forms of electrical energy storage have been
            considered, including batteries. The most attractive possibilities
            seem to us to be storing hot water bled from a turbine which is
            subsequently used for boiler feedwater and compressed air storage
            combined with heat store. The CEGB development effort is likely
            to be most heavily concentrated on these most promising methods.
            However, batteries run a close third and we will watch with interest
            the development work being carried out for transport applications
            which, if successful, could readily be adapted to power station use.

Question: How does the economy of scale in electric power plants affect the
            prospects for using rejected heat in towns or in industry?
Answer: Large plants (as opposed to a large number of small ones) have been
            chosen because they are cheaper, easier to site and have less en-
            vironmental impact. The heat output from a large 2000 MW station
            would meet the needs of a large city.

Question: What are the prospects for use of power stations rejected heat?
Answer: It is mainly an economic problem. The main heat pipe is not very
            expensive, approximately 1p to 2p per GJ per km of pipe length and
            such pipes might well be economic in lengths of 20 km. However,
            the heat distribution system near the point of use is very costly.
            All CEGB regions and the area electricity boards in the UK have
            been advised to investigate possible district heating loads. Some
            district heating and horticultural uses will be found but is should
            be noted that the 3000 acres of glasshouses in the UK could all be
            supplied with the rejected heat of a single 2000 MW generating
            station.

Question: What are the prospects for flywheel energy storage in central
            power stations and has a cost estimate been made?
Answer: We have examined the possible materials and, unless there are
            radical improvements in the strength of glass fibre structures,
            they do not appear to be as attractive as other storage alternatives.

ELECTRIC POWER TRANSMISSION

W. T. Norris

Central Electricity Research Laboratories, CEGB

My topic is electric power transmission, most known to the public by the transmission towers, those finely shaped tall sculptures in metal tracery which are
stood in noble and purposive array adding so handsomely to parts of our countryside with graceful catenaries of line suspended between them and leading the
eye and mind to far horizons and distant cities.

I take as a beginning for this discussion that electricity is a desirable
commodity, that it is best made in electricity generating stations, and is
best conveyed thence to consumers by transmission lines and a distribution
network.

I will describe some features of the transmission system we have in Britain,
and then go on to talk about some of the new ideas that are being discussed
which might serve in the future to cheapen an admittedly already cheap part
of the electricity supply arrangements, and to make it an even more agreeable
part of our civilisation.

There is little need or benefit to rehearse the debate of the relative merits
of alternating current and direct current systems, or single phase or 3-phase
arrangements, save perhaps to point out that a.c. allows transformation of
electric power from one tension or voltage level to another. The higher the
power to be transmitted the higher the most economic voltage to use, the
added electrical insulation being balanced by the saving in conductor material
and power lost by resistive dissipation.

Thus in principle one carries high power from a power station over transmission
lines to the area of consumption feeding so called "bulk supply" points whence
a number of lower voltage lines disperse the power to other substations for
transformation to still lower voltages and yet further division. Several
voltage levels are used:- 415V, 3.3kV, 11kV, 33kV, a limited amount of 66kV,
132kV, 275kV and 400kV in this country. The quoted voltage is r.m.s. and
between phases; from phase to ground is $1/\sqrt{3}$ of the quoted voltage. There are
corresponding, but often slightly different voltages in other countries. Abroad
there are voltages going up to 800kV and the Russians will shortly be building
lines at 1100kV. It is customary to refer to voltages from 132kV and upwards
as EHV or 'transmission' voltages, lower voltages as 'distribution' voltages.
The term UHV usually refers to 1000kV and higher, but often to voltages
over 400kV.

In Britain whilst the transmission system serves to bring power to the great
conurbations from power stations situated on rivers and the coast, where
cooling for the turbine condensers is readily available, or from power stations
near coal fields, it, the transmission system, also acts as a pool of electrical
power allowing economy by saving construction of power stations to satisfy a
merely regional demand, to allow the use of stations of higher thermal
efficiency to satisfy the national need, to provide a buffer against rapid local
fluctuations in demand and to provide local power if local stations should fail.
It was to especially reap the advantages of this pooling of resources that the
original 132kV grid was constructed in the late 1920s to be superseded in the
1960s by the 400kV grid in which there are now some 4000 km of double circuit

Fig. 1   400kV twin conductor transmission line majestically
            leading over the horizon.

Fig. 2   The British 400kV network; note the strong connections
            to the conurbations.

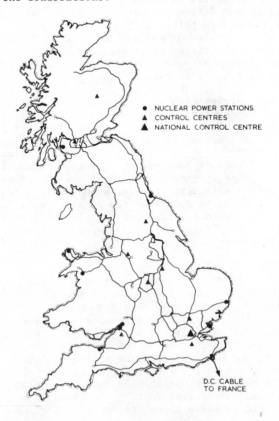

line.

Let us look at some of the elements that are used as building blocks. I am
afraid we will have to make this survey at something approaching a gallop, but
I will slow to a canter when I turn to new ideas that might be useful.

Consider first the most prominent feature - the transmission line itself.

In its rudimentary form it is a row of poles holding up wires. But we must
first ensure that the lines are insulated from ground so they are suspended
by strings of porcelain insulators sufficiently far from the towers to prevent flashover directly from line to tower. The surface of the insulator is
extended by the shape of the porcelain sheds. This inhibits discharges
propagating across them when they become, as they do become, covered with
damp electrically conducting dirt. We know a good deal about how insulators
such as these work, but still have a great deal to learn.

Metal protrusions from conductor and tower called arcing horns provide the
shortest airgap from line to tower and, if extra tensions occur across the
insulator this gap breaks down and the arc does no damage to the insulator.
The arc may extinguish as the voltage comes to zero, but it may not, in which
case a circuit breaker will have to operate.

The conductors themselves are stranded. The central strands are of steel for
strength and the outer strands of aluminium for electrical conductivity. Two
factors determine the size and number of conductors in a bundle. Firstly the
required current rating which itself depends on the cooling and the maximum
temperature allowable without excessive sag or creep of the conductors, or
without loss of the corrosion preventing grease from the interstices of the
stranding. The second requirement is that the electric fields at the surface
of the conductor, and at fittings, are low enough so that there is no
bothersome radio or T.V. interference or excessive audible noise from corona
discharges.

The fittings must all be tested for such interference effects and to ensure
that they have a long life and do not harm the conductor mechanically.

Wind is an important matter and the wind loading of the conductor and so of
the tower is a major design factor. More irritating is the occasional habit
of lines to vibrate and sometimes to gallop - a remarkable occurrence in
which whole spans solemnly wave in the wind: typical wavelengths are a span
long and amplitudes are measured in tens of feet; conductors clash, flashovers
occur and circuits can be unusable for extended periods. There is a world
wide search for palliatives and many ingenious damping arrangements have been
devised, but few are well proved. Occurrences of galloping are mercifully few
for the peace of mind of the system operator, but they are therefore
infuriatingly elusive to researchers seeking to know more of them, but the
formation of thin coatings of ice which changes the cross section of the
conductor and the concurrent appearance of wind from an unpropitious
direction seem to be associated with most displays of the phenomenon in
Britain.

Lines must have adequate clearance from ground and objects on the ground so
that flashovers to such objects do not occur and so that electrostatic currents induced in grounded objects are small.

The last major consideration I must mention is the line route itself. This
excites much interest at public enquiries and the careful selection of routes
to avoid visual intrusion is important. Satisfying taste in landscape can be
expensive.

In substations the main elements are the busbars, the transformers and the
switchgear.

Busbars are aluminium and supported on rigid post insulators, again made of
porcelain. Substation layout is complex and incorporates provision for
rerouting power in different ways if a line or piece of equipment is out of
service for maintenance or because of a fault. The complexity is eminently
apparent from a mere glance at a substation.

Transformers are amongst the cheapest and most efficient of transmission system
plant. The traditional design of a laminated iron core and copper coils has
proved eminently serviceable and reliable over the years. Many attempts at
improvements by using different materials, such as superconductors, have
foundered early on and one must look for improvements in the design of
transformers only in a carefully thought out refinement of existing designs,
and perhaps the judicious occasional use of new insulating materials.

Switchgear is designed for emergency conditions. If all that was required of
the switches on the transmission system was to turn off and switch on loads,
then they could be made very much smaller than they are. The problem is that
flashovers occur across the insulating parts of transmission lines and cables
from time to time with the result that very substantial fault currents flow.
The inductance of the system limits the level of fault current which flows
(resistance being negligible), but influences the voltage induced across the
switch as current is interrupted.

Almost all breakers used for transmission are based on the gas blast principle,
that is electrodes are pulled apart and a blast of gas is directed between
them so as to remove as much as possible of the heat generated by the arc. As
one loop of the alternating current falls to zero, the blast is sufficient not
only to remove the heat that is generated, but also to ensure that the arc
channel has lost so much of its electrical conductivity that as the current
attempts to rise to negative values through the current zero, then it finds
the gas so poorly conducting that it cannot do so. For voltages over 132kV
it is customary to have a number of interrupters in series to provide a
single circuit breaker. Most existing circuit breakers use air as the blast
gas, but a number of designs are now available in which sulphur hexafluoride
is used. The advantage of sulphur hexafluoride is two fold. First of all it
tends to form negative ions, which are less mobile than electrons (i.e. the
gas when hot is a poor conductor of electricity), but perhaps most importantly
it has a high thermal conductivity in the temperature range where the electrical
conductivity is falling most rapidly, so that a conducting column near current
zero is more readily cooled than with other commonly used bases. There are
numerous other circuit elements we have not time to pause to look at. For
example surge diverters to limit the overvoltages on the system; current and
voltage transformers for measurement purposes, reactors to limit short circuit
currents and provide reactive load, and in some systems, but not to a great
extent on British systems, non-linear reactors to provide voltage stabilisation
during fault and overload conditions. I have not either mentioned isolators,
switches to work at zero current to allow safe isolation of circuit elements so
that when these are being worked on, no overvoltages on the system attack them.

Fig. 3   A 400kV transmission tower: 2 circuits, 4 conductors per
          phase bundle; double insulator strings: arcing horns provided
          to space flashover arcs from insulators; Stockbridge dampers
          on lines near to tower to suppress aeolian vibrations

Fig. 4   A 400kV outdoor substation where power is collected and
          routed to other parts of the system.

Fig. 5a   A 400kV airblast circuit breaker made by G.E.C. Ltd. Fault
           rating is 35 GVA. There are four interrupter heads, two on
           each post. In the photograph the breaker is not yet connected
           up.

Fig. 5b   A 1000MVA autotransformer connecting the 400kV and 275kV grids.
           The isolators either side of the transformer are open. The
           transformer is made by Parsons Peebles Ltd.

The whole of this system requires very careful control so that it is run
economically and with the most economic use of the various power stations
available to provide electricity, and so that it is run stably and is able
to withstand the sudden loss of a transmission connection without losing
synchronism.

There are several ways of looking at the basic nature of power transmission.
On a rather imprecise, but in some ways rather satisfying mechanistic view,
one could regard the conductors as containing a long railway train of
electrons. Power is transmitted by shaking the electrons at one end of the
train so that a vibration passes down the train to be recovered at the other
end. Considering typical current densities in copper conductors, and the
mobility of electrons in copper, it turns out that the average amplitude of
vibration of electrons for 50Hz transmission system is a few tenths of a
millimetre from peak to peak.

Another more rigorously defendable way of looking at the transmission of
electric power is to consider the Poynting vector. You will recollect that
the Poynting vector is the cross product of the electric field and the
magnetic field and that its integral across a surface is a measure of the
power flowing across the surface. If we consider two parallel conductors
both the electric field and the magnetic field lie in the plane perpendicular
to the conductors so that the Poynting vector lies parallel to the conductors
themselves and power is indeed transmitted along the lines - but in the space
between them. The importance of both electric fields and magnetic fields in
determining power transmission is well illustrated by the consideration of
the Poynting vector.

Overhead lines have a rather poor distribution of electric fields and
magnetic fields because not only is the surface electric field limited
($\sim$1-2 MV m^{-1}) but the conductors being small both electric field and magnetic
field fall off fairly rapidly as one moves away from the conductor, and in
order to get the energy transfer it is necessary to have fairly substantial
spaces between the various phase conductors.

In a cable the electric field and the magnetic field are controlled rather
more carefully and both have a fairly high and uniform value over all the
area between the go and return conductors (electric fields $\sim$5-10MV m^{-1})

Much attention has been given over the past 50 years to achieving high electric
fields in cables without there being excessive danger of breakdown. However,
over the last decade or so there have been significant developments from
recognising that the magnetic field is important. Put another way, although
you can double the power carrying capacity of the cable by increasing the
working voltage, it can be done equally effectively by increasing the current.
One of the disadvantages of increasing the current is that the heat
dissipation rises substantially. For a cable buried in the ground the heat
loss from the cable is a moderately complicated process, but one factor that
is extremely important is the conductivity of soil laying between the cable
and the ground surface. More recent high power cable development, however,
has shortcircuited, as it were, this difficulty by providing more direct cooling
of the cables. In its most elementary form one can simply lay pipes with
cooling water flowing through them alongside the cables. The benefit to be
derived from this is substantial, resulting roughly speaking in a doubling of
the power capacity of a particular cable. Of course one has to do this
exercise with some circumspection. It is essential that the cooling pipes

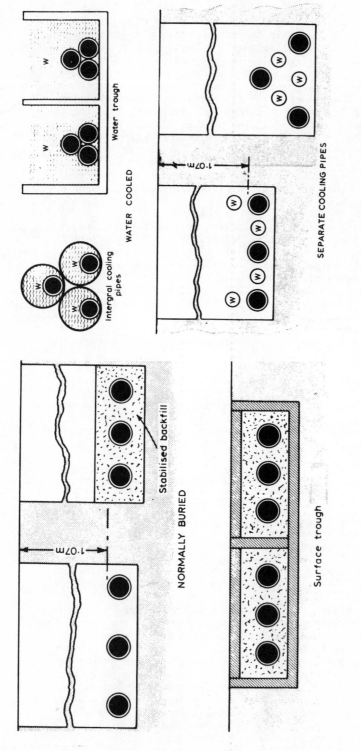

Fig. 6b

Fig. 6a

Methods of burying cables. Normal burial and the use of improved cooling by using (a) stabilised backfill (i.e. a backfill or weak mortar with good and reliable thermal conductivity) and (b) a surface trough are shown.

Even better cooling and so higher current and power ratings can be achieved by putting the cable in a pipe with water (e.g. the Severn Cable tunnel), using troughs with water flow controlled by weirs (Fawley-Southampton tunnel) or using separate non-metallic cooling pipes laid alongside cables.

are not metallic otherwise eddy current pipes will be induced in them, and
there will be additional losses, and the benefit of the cable cooling will be
lost. However, polythene piping and piping reinforced by fibres are now
readily available and can work at fairly high pressures so that this difficulty is obviated.

Laying cooling pipes alongside cables provides a very simple installation,
but by putting the cable inside the cooling pipe even higher powers can be
transmitted, although with considerable complication in the installation
procedure. Another development is to pass a coolant, say oil, or even water,
down the very centre of the central conductor so that the heat is removed as
soon as it is generated: this makes for even more complication, but development
is in hand.

A significant feature of EHV cables as they are presently used is that
they have a paper and oil insulation. The paper being wrapped round in tape
form to make many layers which are then impregnated with the oil. This has a
fairly high thermal resistance, but, worst of all, the paper having a substantial
dielectric constant makes the cable of high capacity. This capacity has
various unfortunate effects on the system of which the one that I ought to
point out here is that for cable lengths of much more than 10 or 20 miles the
capacitive current when there is no load being taken through the cable is so
high as to easily reach the current loading at the excited end of the cable.
The effect is that no power can be safely transmitted down the cable without
exceeding acceptable operating temperatures. This is one of the reasons why
direct current is used for long distance submarine cables. However, the
difficulty can be overcome to a degree by the use of insulating systems with a
low dielectric constant, as we shall see in a moment or two.

One of the difficulties arising from the use of alternating current is that
reactive power flows are extremely significant. Reactive or wattless power
is when current is being drawn by inductive or capacitive loads. No net real
power is transferred, the Poynting vector is imaginary, but currents can be
large. The inductance of transformers and overhead lines and the capacitance
of cable is of considerable importance. The phase differences along lines and
through transmission plant can lead to instabilities in the generators which
can cause them to pole slip, having lost synchronism.

Roughly speaking the alternating frequency of the system is regulated by
control of the power input. As more load is put on the system ultimately this
means that there is a braking force on the turbines and a tendency for the
frequency to fall. It is precisely to keep this frequency constant that the
governors are used on the turbine generators. Most of the voltage variation
in the system arises because of reactance and therefore voltage control is
achieved by careful balancing of the various reactive elements, such as series
reactors, shunt reactors, transformers, etc.

Instrumentation and control systems are essential to whole schemes. The
so-called protection systems are logic control systems used to activate circuit
breakers. The careful design of the controls is essential if faults are not to
lead to excessive loss of connection to or synchronism of generators which
could in the worst case lead to a collapse of the system. One proceeds very
conservatively.

Having taken a short gallop past the system, so to speak, as it is, let us turn
to ideas for the future.

A.E.C.—Q

Indubitably as the demand for electricity rises the power to be handled by the transmission system will rise. Some time we must expect the use of higher voltage lines in this country, but at the moment the 400kV network that we have is extremely strong and considerable work is being done to seek out better ways of using the system so as to delay the necessity for installing extra lines at 400kV or newer lines at higher voltages.

I will start off with what you will doubtless consider to be a very adventurous proposal. I have embellished, but only to a small extent ideas that were propounded by M. Maurice Magnien at his Presidential address in 1970 to the S. F. E. One could conceive of a fusion reactor in which one induces a laser action as a means of extracting the energy. M. Magnien himself had thought of other ways of generating laser light. One then directs the light thus produced down a tube to its destination. Power densities greater than 10 MW cm^{-2} - far higher than can be achieved even in a super-conducting cable - are then transportable. It would of course have to be a tube that one would use since line of sight could not be contemplated because of the likely damage to objects that got in the way. At the arrival at the end of its route the light would be focussed so as to produce a highly conducting gaseous plasma which one could then use to drive a magneto hydrodynamic generator of high efficiency - with a little waste heat, possibly, for district heating.

Such aqueducts of light are doubtless a long way off. Another proposal made more recently by some American scientists was to use the kinetic energy of electrons instead of an electromagnetic wave. One could imagine accelerating an electron beam to a power of several kilovolts, then directing it along an evacuated tube. It would not be necessary for this tube to be straight, since bends could be negotiated by the application of an appropriate magnetic field transverse to the tube. The electrons could be kept secure on their path in the middle of the tube by the use of quadrupole magnets fitted at suitable intervals. At the far end of its trip the electron beam would be reconverted to alternating current by the use of the inverse of a linear accelerator.

One has the problem then of returning the current, for otherwise large electro-static differences will be set up which will staunch the flow of electrons. The current could be returned along the tube which contains the high energy beam of electrons. As a delegate said at the C.I.G.R.E. meeting where this was discussed, "Here is another idea to put on the table". I expect it will stay there for quite a long while.

Another of the rather more adventurous notions, but one which has been explored more extensively is to use microwaves. All that is needed for the transmission connection is a hollow metal pipe. The transmission losses will be very high, unless the pipe is very accurately machined and laid to a high degree of precision. There has been much work done in various places on how to reduce such losses. One idea is to coat the inside of the pipe with a thin plastic coating on which is laid a thin layer of metal. But attacks on the reduction of losses in the transmission line itself are largely pointless for one would have to convert back to 50 cycle power at the receiving end and generate microwaves at the sending end. Existing devices for radar have efficiencies approaching 90%. This equipment is fairly well understood and there is little likelihood that we shall ever do very much better. No device yet exists for converting microwaves back to 50Hz and it is unlikely to be much more efficient than a generator of microwaves. The overall losses in the two converter stations at either end are far too high to be acceptable even if the transmission ducts could be constructed for nothing. The conversion losses make the whole scheme economically untenable from the outset. Only the perverse pursue such ideas.

Let me now turn to some more promising directions in which advances are being made. Firstly consider overhead lines. Extrapolation of existing designs and design criteria to higher voltages can readily be made. For the higher voltages the towers are certainly large, but set against a distant skyline may not seem excessively out of scale with the landscape. This particular question of visual impact has occupied quite a lot of attention in recent years. There have been various attempts, particularly in America, to beautify lines, one of which abandoned the lattice structure for more bulky structural elements. The design costs a great deal more than conventional structures. The solid members that are used are not necessarily more attractive in any given landscape situation. There is a strong element of artistic value judgement here which both engineers and the general public will need more time to consider and reconcile with the call for electric power in home and factory.

There have been lots of ideas on what would be suitable alternative structures for UHV lines of 1,000kV (Paris, 1970). They all have a degree of elegance, and although each of us have his own particular preference, I am sure the Gallup poll would show there is little to choose between them, and that the engineer might be best advised to select that which is cheapest for his particular application.

One way of overcoming the problem - if such a problem exists - of visual intrusion, is to attempt to make the lines invisible. One way to do this is to try to make them smaller and Prof. Reggiani produced a design of line for 1,100kV operation which was about one fifth the height of conventional designs. The line is slung at a very low height with the three phases arranged in a horizontal configuration and very close together. A problem of bringing the lines so close together is that it is necessary to reduce the electric field at the surface of the conductor so as to ensure that there is no noise from corona discharges and so that the breakdown voltage is high, is that one must substantially increase the diameter of the conductors (say 0.4m dia. at 10 m. spacing). He envisaged an aluminium tube with joints every so often supported in the middle either with a steel tube or with a central steel cable. The electric field induced at ground level would be extremely high, and so it would be necessary to have the right of way purchased by the electricity company (as is common practice in America) and the public excluded. Of course such a line would have to cross roads and railways at not infrequent intervals, and special provision would have to be made at such places by raising the conductor to a much larger height above the ground. Professor Reggiani calculated that such a design might cost twice as much as a conventional UHV line and much of the cost would arise because of the large costs of the frequent supports that would be necessary. Personally I am not sure that the requirement to take complete possession of the right of way would be acceptable in this country, but we would certainly have a line which would be visible from only relatively nearby.

One new direction has been to look at the use of plastic insulators. Plastic itself is not strong enough, but if one were to use, as has been the case, a fibreglass core to provide the strength and a suitably chosen plastic for the sheds, then advances can be, and have been, made. The product tends to be rather more expensive than conventional porcelain and the surface cannot be made as durable as porcelain, but indications are that it can be made durable enough.

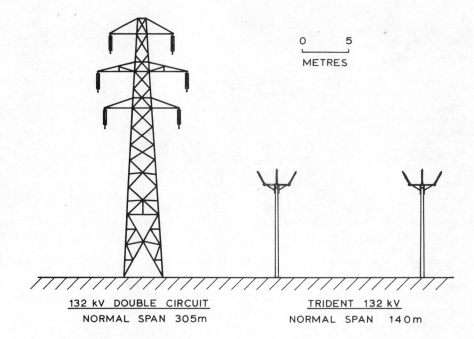

0   5
METRES

132 kV DOUBLE CIRCUIT   TRIDENT 132 kV
NORMAL SPAN 305m   NORMAL SPAN   140m

Fig. 7   132kV line designs. On the left a conventional metal tower
            double circuit line; on the right two single wood pole lines
            with "trident" rigid glass reinforced plastic insulators
            giving lower profile, but a wider corridor.

Fig. 8   Transmission line designs. (a) Tower for an 1100kV single
            circuit line; (b) Tower for conventional 400kV line as used
            on the British grid; (c) a proposed low level 1000kV single
            circuit line – the fence to prevent public access is not shown;
            (d) side view of the low level line; supports are placed two to
            three times more often than for a high level 1100kV line. The
            low level line would cost about twice as much as a high level
            line.

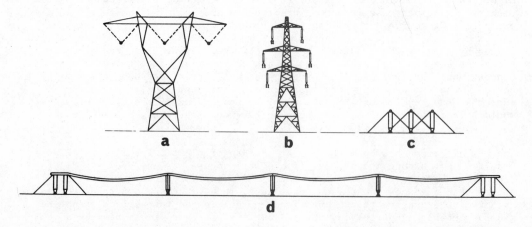

One of the advantages of a reinforced glass core plastic insulator is that it can be made with considerable cantilever strength and there is great promise for application in this direction at 132kV, again attacking the problem of making lines less visible. One notes especially the use of rigid insulators mounted on poles in the so-called Trident construction 132kV line with the insulators rising at a slight angle to the vertical from a metal cross arm. One uses here a "fully insulated" philosophy rather than an overhead earth wire. Wood poles are used for support rather than steel towers. The result is a structure that is considerably lower and smaller than existing 132kV lines, but has supports rather more frequently. The cost of such lines and their performance is not altogether certain and widespread application is not to be expected for a little while yet, until a little more experience has been obtained.

Plastic insulators also have the advantage of being lightweight and allowing a fairly intricate shed design to be used to get a long creepage path, and there is indication that they may be applicable at UHV voltages or for lines being installed in regions where their lightweight is an advantage.

Another way to make transmission lines invisible is bury them. This is fearfully expensive - the costs of cable as installed are something between fifteen and twenty times the cost of overhead lines and have been so for some time. And cables are not always altogether invisible, if the cable is buried in shallow troughs to get good cooling. But under propitious circumstances the final result might well be a gas main for all one can tell from above ground.

However, for the readily foreseeable future most cable installations will be made in circumstances where they are technically the only viable alternative such as in urban areas where it is impossible to squeeze in a transmission line or to cross waterways too broad for a single span of overhead line.

Nonetheless cables will be used and attempts are being made to evolve cheaper ways of undergrounding. This is part of a continued effort typified in the development of intensely cooled cables of which I have spoken - an excercise aimed at, as Mr. Cherry suggests, containing the cost of undergrounding. If we achieve cost reductions of a few percents we shall do well, a few tens of percents will be excellent progress, a halving will be revolutionary - but one can be hopeful.

A first step is to use a better dielectric than oil and paper - and one looks especially for a lower capacity cable - i.e. with an electrical insulator with a lower dielectric constant so as to reduce the reactive power generated; and also for an electrical insulation that is a good thermal conductor.

One method being pioneered by B.I.C.C. is to use a laminated tape with paper on the outside of the sandwich and plastic film between. Another is to go to a lapped polythene tape and impregnate with pressurized sulphur hexafluoride. The basic structure of tapes wound round a central conductor is as for conventional EHV cable.

Both these cable types are being actively explored and both promise the twin advantages of lower dielectric capacity and better heat transfer. In both cases there are problems in predictably making a reliable insulating system and whilst great advances have been made we are not yet confident that all the crucial elements are sufficiently well identified to provide a sound basis for

a properly controlled manufacturing process to be assured. The work is
promising, but the development not yet ended.

An ideal solution would be to use a solid extruded polythene insulation: such
cables are common enough at 11kV and even 33kV. They have been installed at
132kV and even trial lengths at 275kV, but there are uncertainties still and
the designs for 132kV and 275kV are of uncertain economic benefit. There are
technical problems too.

A problem close to being understood is the occasional appearance in the
polythene after some time of operation of small worm holes penetrating deep
into the plastic. These so-called "trees" are believed to occur by some
complex electrochemical action and do not necessarily lead to dielectric
breakdown; but at present no engineer would knowingly accept them in his
cable. The second problem is that it is known that to achieve a high
electric field in the insulation the material must have no holes and, to
transliterate an expression in a French paper, be exempt from impurities.
This is quite an engineering problem and expensive to solve.

But the promise of a cable with good physical properties and, hopefully no
need for auxiliary equipment to maintain oil or gas pressure inside it is
attractive.

Another possibility is to simply have a conductor suspended in the middle of
a pipe and fill the pipe with high pressure gas - sulphur hexafluoride for
preference. This technique is already used to a moderate extent in the
so-called metal clad substations which are very compact indeed. Short
lengths of transmission link have been used in a semi trial situation, but
the electric fields have been modest and the designs too expensive to
contemplate for lengths longer than a few hundred metres.

Nonetheless a gas-spacer cable, as it is sometimes called, has superb
physical characteristics in all ways except size. It is of low capacitance
(the dielectric constant of most gases is close to unity) and thermal
convection currents in the gas promote good heat transfer. The end
terminations are fairly easy to design too.

One problem is the design of supports to carry the central conductor. Dust
is a bugbear for supports and cleanliness of the gas is essential. But
research and design work is going on world wide looking at methods of making
the gas a reliable insulator, searching for the best choice of conductor,
gas, gas pressure and pipe construction, and exploring methods of installation -
a factor which is likely to be crucial. American opinion is almost uniformly
highly optimistic; European views on the promise vary from mildly optimistic
to downright scornful.

Superconductors are also offering their services. They are a little reticent -
or may be aristocratic - declaring that they are ready to serve when only
really high powers are to be carried - say 5 GVA or 10 GVA - roughly two to
four times the top summer loading of a 400kV circuit as used on the British
grid, or the power of one of the largest British power stations. This is
because cooling them to liquid helium temperatures is expensive and only if
the cryogenic envelope contains really valuable material will the containment be readily paid for.

A cable core will probably look very much like a conventional cable core.

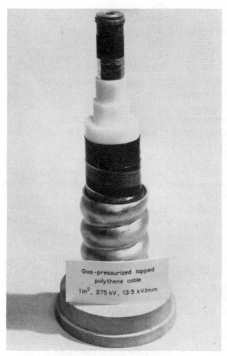

Fig. 9

A section of a proposed cable using lapped
polythene tapes impregnated with SF_6 to
form the main insulation. Note the metal
coated and carbon loaded tapes both
adjacent to the central copper conductor
and surrounding the polythene; these
conducting elements are to provide inti-
mate contact with the dielectric and
prevent damaging discharges in the gas
spaces. A corrugated aluminium sheath
protects the cable from mechanical
damage and contains the gas.

Fig. 10

An early design of
spacer for a gas
insulated cable. Gas
convection within the
large spaces provides
good heat transfer.

The electrical insulation is of lapped tapes - polythene - which at helium
temperatures has a low enough dielectric loss to be acceptable. The
superconductor itself is made of copper strips covered with a thin layer of
niobium. The niobium carries the supercurrent and the copper is there to
provide electrothermal stabilisation and carry some overload fault current.
More complex conductor structures are possible using two layers of superconductor, the lower one to carry fault currents.

The strips are wound helically. This arrangement is designed so that on
cool down all the thermal contraction is taken up. Other designs using
tubes have problems in this respect.

Three such cores are formed into one cable. Each core has a go and return
conductor to reduce magnetic fields in the other metallic parts of the cable,
for such magnetic fields would induce currents and so ohmic heat dissipation
which is not acceptable at all. Cores are contained in tubes carrying the
helium coolant and the assembly enclosed in the so-called cryogenic
envelope - a vacuum assembly with radiation shields and intermediate
temperature coolant.

All this work is still in the laboratory stage of development. Moves are
being made in the U.S.A. - and I believe in the U.S.S.R. - to build a trial
installation: this will show up the feasibility of installing this
relatively complex assembly of tubes in the field, of inserting conductors
and making vacuum tight all that should be vacuum tight. This will
certainly be a testing occasion; we may then see superconductors leap to the
front or lay down their aristocratic position for, say, more imperial
pretensions. It will be some years before the situation is absolutely clear,
but if superconducting cables are economic only at 10 - 20 GVA then they
are unlikely to find application even on a system as large as that in the
U.K. with its present total load of about 40 GW.

If superconductors do in the final analysis show cost advantages, they will
bring technical advantages too (which have, of course, to be counted in the
financial reckoning). Firstly they provide a very narrow power route which
would be of use in metropolitan installation conditions and secondly the
refrigeration stations can be situated quite a long way apart (say 20km)
compared with the 1km or $\frac{3}{4}$km required for water cooled conventional cables.
They are, like the other cable types also of low capacitance.

I have dealt at some length with lines and cables partly because I know more
about them and partly because they are more in the public eye.

There are advances in other directions that are being made. Switchgear using
SF_6 as the blast gas is already in use and is likely to be found more
frequently. Substations with the conductors (and other gear) enclosed in
metal and the containment filled with SF_6 are much more compact than air-
insulated substations and are being built and operated. They are likely to
become common for new installations.

Using vacuum as an arc interrupting medium is also a possibility. Simply pull
contacts apart in an evacuated vessel (or 'bottle' as the technical jargon
has it). Whilst current flows the conduction across the electrode space is
essentially through metal vapour which mostly recondenses as the arc current
falls to zero - thus leaving no good conducting medium between the parted
electrodes. These devices are fairly common at lower voltages and are used

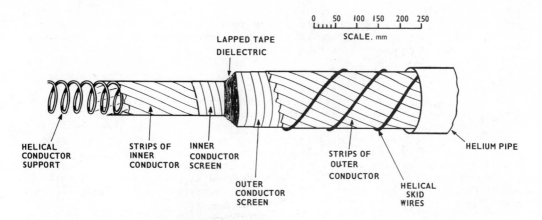

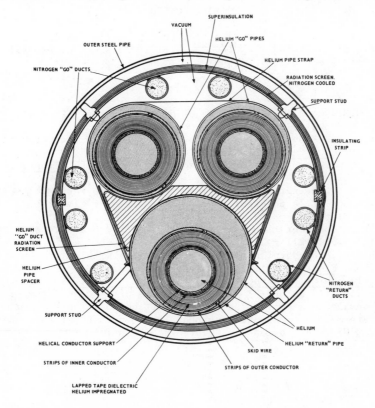

Fig. 11a   Single phase core for an experimental a.c. superconducting
            cable. The niobium copper conductors in strip form
            are laid helically to assist in absorbing the thermal
            contraction on cooling down. The insulation is lapped
            polythene tape impregnated with helium.

Fig. 11b   Section of a design for a 4 GVA superconducting cable.

extensively by British Rail. For high powers electrode melting by constricted arcs and vacuum voltage breakdown problems need to be studied more since present bottles are not good enough – too many interrupters would be required in series and the actuating mechanism would be gross.

Examples of progress in almost every aspect of transmission engineering can be found; use of optical links, optoelectronic instrumentation and high speed digital control for system protection; refinement of transformer design by a better understanding of what happens in them; better use of overhead lines from better understanding of meteorological effects on conductor cooling; more reliable ways of joining conductors so as to avoid overheated joints by, for example, the use of cold pressure welding of aluminium – a technique most promising for earthing strips in substations – there is as much conductor at earth potential in a substation as there is at high voltage; D.C./A.C. converter designs are advancing for submarine and very long distance power links.

The market for electric power could not at present be described as buoyant. I think it will become so again. We have in Britain a good 400kV transmission system which will stand the strain of rapid increases in demand. And the armoury is being well stocked with techniques and designs to strengthen the 400kV system when it is necessary, and to build new supply lines in a way that will add to the splendour of civilisation and life in Britain.

ACKNOWLEDGEMENT

Whilst I am anxious to acknowledge that this paper is based on my experience whilst working for the C.E.G.B. it does not necessarily reflect the policy of my employers.

REFERENCES & BIBLIOGRAPHY

1. Cherry, D.M., Containing the cost of undergrounding. Proc. I.E.E. 122 (3), 293–300 (1975)
2. Gardner, G.E., Lloyd, O., and Urwin, R.J., Heavy-current arcs in Switchgear, CEGB Research, No.1, (Dec. 1974) pp 20 – 28.
3. Kind, D., Gas Insulated Energy Transmission Systems. Trans. S.A. I.E.E., 65(12), (Dec. 1974)
4. Magnien, M., L'avenir de l'electrotechnique des courants forts – L'industrial l'utilisateur et le novateur. Revue Generale de l'Electricite, 79(3) (March 1970) pp 201 – 209.
5. Paris, L., Considerations sur les lignes a tres haute tension de l'avenir, RGE 79(3) (1970) 211 – 224
6. Reggiani, F., Study of non-conventional lines with extremely reduced dimensions, Electra, No. 27, (March 1973) pp 25 – 34.
7. The Philosophical Transactions of the Royal Society, No. 1248, Vol. 275A pp 33 – 253, (August 1973) gives the papers presented at a meeting on recent advances in heavy electrical plant and contains authoritative papers on transformers, switchgear, overhead lines and substations, underground cables, D.C. transmission and power system control.

DISCUSSION

Question: If superconducting cables are ever introduced in power systems,
            will they allow fewer transformers and circuit breakers to be
            used as a result of lower transmission voltages?

Answer: It is doubtful whether there would be any such saving and at best
            it could only be of marginal importance.

Question: Are there any problems in joining superconductors to circuits
            operating at normal temperatures?

Answer: There are no great difficulties in doing this but there is an
            additional heat inflow at each junction probably equivalent to
            that from about 1km of cable.

Question: What are the future transmission needs of the UK?

Answer: The reduction in demand around 1974 is likely to revert to the
            longer term trend of rising consumption. With 30% extension of
            the transmission network it would probably be adequate for four
            times the present system peak demand, i.e. 200 GW. Incidentally,
            the transmission system capital cost accounts for 5% to 20% of
            electricity cost to the consumer and the losses in transmission
            for 1% to 3%.

Question: Are there any proposals for a UHV transmission system in the UK?

Answer   We may need a UHV system operating at 1 MV or above about 2000 AD,
            but it may be considerably later before it is needed.

Question: Is there any justification for alarm at the possible biological
            effects of high electric fields that will exist under the UHV
            lines that are proposed?

Answer: There is no evidence to suggest that harm is done to man or
            animals by subjecting them to such fields. A study on the medi-
            cal history of operatives engaged on live line working in the US
            has shown no indications of ill effects.

Question: Is it cheaper to transport coal or to transmit electricity?

Answer: This depends very much on local circumstances but it has been
            established that it is quite economic to transport coal a distance
            of 1000 miles in the US in some cases.

ENERGY STORAGE

I. E. Smith

School of Mechanical Engineering, Cranfield Institute of Technology

Introduction

It is difficult to discuss the storage of energy as a single topic because it
seldom, if ever, occurs as an isolated process. Energy storage is usually
accompanied by other changes, or governed by criteria that are dependent on
the system as a whole including, of course, the eventual application of the
stored energy.

Storage may, for example, be associated with concentration as in the case of
solar energy storage that occurs in living organisms, or by dispersion as
happens when a massive body receives energy from a concentrated source.
Frequently storage is achieved by conversion to another energy form such as
sensible to latent heat or radiation to thermal energy, and the energy form
in store may or may not be the one that is ultimately required. The storage
or transmission of energy for the purpose of providing transport is dominated
by constraints that do not arise in other applications. Indeed, transmission
itself may result in the accumulation of energy, and examples which spring to
mind are the quantity of gas that is contained within a pipeline, or the mass
of coal in transit at any given time. The effect of this energy accumulation
may reflect significantly both on the dynamics of the transmission system
and the amount of storage necessary at the terminals.

Thus the factors which govern the selection of storage systems are many, and
it is simply not possible to discuss the subject comprehensively under a
single title.

In order to simplify matters and provide a rational foundation for discussion
it is useful to establish boundaries according to the quantity and grade of
energy, the duration for which storage is required, and the use to which the
energy will finally be put, and within such a framework consider the means
and mechanisms that are available or that may become available in the future.

Quality of Storage

QUALITY, or grade, is the entropy which is a measure of the orderliness of
the energy form. When thinking in terms of thermal energy we may regard the
quality as being proportional to the absolute temperature, for it is obvious
that a parcel of energy at 1000°K is worth more than a similar parcel at
500°K.

However, energy is never utilised down to the absolute zero of temperature, but rather at a finite temperature depending on the particular application. It follows that energy is only useful if it is available at or above this temperature.   In the above example, were the energy required at 500°K then the first parcel would be worth twice as much as the second.   There exists, therefore, an exchangeability between quantity and quality, and the rate of exchange depends upon the final application of the energy.

Whilst this concept is not a difficult one if purely thermal energy is considered, the situation becomes less clear when one considers the energy that is stored in, say, a hydrocarbon fuel.   Chemical energy is generally calculated in terms of a heat equivalent with respect to a reference temperature, which is usually 273°K or 298°K.   However, if the energy is required for steam raising at 600°K, it is obvious that less energy will be available than that calculated from the standard heat of combustion.   It is important to remember that in the utilisation of energy the real reference point is the temperature at which it is required.

The use to which the energy is to be put thus defines the minimum grade at which it may be stored after allowance has been made for any degradation that takes place during the retrieval process.   It follows that processes which require high grade energy also need a high grade of storage, such as high temperature thermal, chemical, electrical, potential or kinetic systems. These high grade systems are clearly costly and complex in terms of capital and operation in comparison with lower grade thermal storage systems.

Their adoption is sometimes called for, when low grade energy is required, on grounds of size and convenience.   A familiar instance is the domestic 'night-store' heater where the energy is stored at a high temperature yet is utilised at a much lower temperature.

It will be appreciated that the optimum form of energy storage in a given situation is a complex compromise involving the output quality in association with economic and convenience considerations.

Storage Duration
The length of time for which energy is required to be stored has an important bearing on the manner of storage in that the losses which take place during this period determine the efficiency of the system.   To take a rather extreme example, the windage loss that arises at the surface of a flywheel during one revolution of a reciprocating engine represents a negligible loss of energy as compared to that which the flywheel stores.   However, were the flywheel called upon to store energy for a period of hours or days then the loss through air friction would become very significant.

In many situations where energy is consumed by humans the storage time is dictated by their di-urnal cycle of waking and sleeping.   Even in the field of transportation where a portable energy store is required the time element is of similar magnitude and is measurable in terms of hours rather than minutes or days.

Hence discussion will be restricted to the storage of energy for periods of about half a day, a process that is known in the electrical industry as 'peak lopping'.

Storage Scale

The scale on which storage is required is important for it influences the optimum manner very considerably. Storage methods that merit consideration on one scale may be quite inappropriate on another. An illustration is that of pumped water energy storage which is economical only on the large scale where natural reservoirs are available at different altitudes.

This largest scale of storage, such as is required by the public utilities, has been dealt with elsewhere and only medium to small scale storage will be considered here. Energy storage schemes such as are applicable to smallish factories and domestic establishments are appropriate subjects for consideration, as is the storage of energy for transportation purposes.

If these areas appear unduly restrictive, it must be remembered that the consumption of energy domestically accounts for over 25% of the total U.K. energy expenditure, and private cars contribute a further 10%. If the chemical industry is also included, which is one of the larger consumers of medium and low grade heat, the total becomes 55%.

The factors by which the utility of the store may be judged are (a) the energy density ($J.m^{-3}$) and the specific energy content ($J.Kg^{-1}$), (b) the quality of the stored energy, (c) the storage efficiency, i.e. the fraction of the input energy that is usefully recovered, (d) the storage and retrieval rate, (e) the energy involved in fabricating the store, which may be regarded as energy capital, and finally (f) the capital and operating costs of the system.

METHODS OF ENERGY STORAGE

Chemical Energy

Pre-eminent in terms of volumetric and gravimetric merit is, of course, fuel itself. Hydrocarbon fuels have an energy density of around 40 $MJ.Kg^{-1}$ and when burnt in oxygen or air release this energy at a high temperature. It is, in a sense, unfair to compare fuels with other forms of energy storage since they are generally regarded as a primary energy form, their synthesis having taken place naturally over millions of years. Only for rather special purposes are fuel deliberately synthesised, such as in rocket propulsion, but in these cases the fuel may be legitimately regarded as an energy store.

The use of fuel-stored energy is not without certain disadvantages. In order to make the energy available the fuel has to be burnt with about fourteen times its mass of air, and furthermore, the waste products frequently contain appreciable amounts of atmospheric pollutants. It is for this reason especially in the field of transport, efforts are being made to develop non-polluting forms of energy storage.

A particular fuel around which has aroused interest is hydrogen, partly on account of the fact that it can be produced directly from electrical energy and water, but also for other reasons. Clean burning and efficient as a fuel, it is appealing to the technologist and environmentalist alike.

The main drawback to hydrogen is, in fact, its poor storage qualities. Liquid hydrogen which boils at 21°K has a density only one fourteenth that of water, and thus is volumetrically inefficient. The combination of this with the need for a thermally insulated container which adds to the bulk places liquid hydrogen in an impossible position, even neglecting the problems associated with the handling of a cryogenic liquid.

The only prospect of economically storing hydrogen seems, at present, to be in the form of chemical storage. Hydrogen will combine loosely with a number of metallic elements forming hydrides. Once formed a relatively modest increase in temperature or reduction in pressure will cause them to de-compose in a reversible manner, liberating gaseous hydrogen. Many of the most suitable metals are in the rare-earth series (for example lanthanum) and are therefore expensive, but of the more common elements magnesium shows promise. The hydride of this metal, which contains two hydrogen atoms, has a theoretical storage density based on the hydrogen alone that is 50% greater than liquid hydrogen. In terms of the contained energy magnesium hydride has seven times the mass of its hydrocarbon energy equivalent, but it only occupies four times the volume.

The main problems associated with this form of energy store are those of storage and retrieval, and research is currently directed at investigating these problems in other metals and alloys.

An interestingly energetic reaction that has possibilities both for storage and transmission is that between methane and water which forms an equilibrium with hydrogen and carbon monoxide according to the equation:

$$CH_4 + H_2O + 6 \text{ MJ.kg}^{-1} \rightleftharpoons 3 H_2 + CO$$

At temperatures above 800°C the equilibrium lies well to the right hand side and thus energy is absorbed forming hydrogen and carbon monoxide. If the reverse reaction is then frozen by a rapid cooling, the gases may be stored or transmitted with little tendency to revert to methane and steam. However, when the energy release is required the reverse reaction may be initiated by the application of an activating energy pulse or the presence of a suitable catalyst. Following this the steam may be condensed and rejected as water, and the methane re-cycled to the source of energy. This form of energy transfer has the advantage over combustion reactions that there are no noxious products rejected and there is no involvement with high temperatures. However the heat absorption occurs at a high enough temperature to be of interest in the field of nuclear reactors, and the heat release is at a temperature that is useful for steam raising.

Sensible Heat Storage
The simplest form of heat store is an inert body whose temperature is raised when heat is absorbed and lowered when heat is withdrawn.

For heat to flow into and out of such a store necessitates a temperature gradient, and hence on transfer a certain degradation of the energy takes place. This degradation is further enhanced by the temperature change that takes place when the element is heated and cooled, and from a practical point of view this change in temperature with time may be a disadvantage. One way in which this can be minimised, and a more isothermal manner of operation achieved, is to store the energy at a much higher temperature than that at which it is required, but with the obvious penalty of degradation.

Of the criteria which govern the selection of a storage material clearly the specific heat at the operating temperature is important. If volumetric con-siderations have to be taken into account then the product of specific heat and density are the significant parameter. Table 1 lists the specific heat

TABLE 1.   SENSIBLE HEAT CAPACITY OF SELECTED MATERIALS

Material	Specific Heat $J.Kg^{-1}K^{-1}$	Volumetric Capacity $J.m^{-3}K^{-1}$ x 10^{-6}	Max Temperature °C	Thermal Diffusivity m^2s^{-1}
Water	4180	4.18	100	liquid
Mineral Oil	2717	2.36	250	liquid
Diphenyl/Diphenyl Oxide	2400	1.92	400	liquid
Sodium	960	0.91	880	liquid
Aluminium	896	2.63	660	8.4×10^{-2}
Iron	501	3.93	1000+	1.7×10^{-2}
Magnetite	752	3.85	1000+	5×10^{-7}
Concrete	1128	2.53	1000+	7×10^{-4}
Stone	878	2.41	1000+	7.5×10^{-7}
Brick	830	1.87	1000+	3×10^{-4}

and volumetric capacity for a number of common materials. Also listed is the
maximum operating temperature, for this influences the total amount of heat
that can be stored.   Although water has the highest capacity, the low maximum temperature at atmospheric pressure is a disadvantage in many situations.
On the other hand the cheapness and particularly the mobility of water are
attractive features, the latter favouring high heat transfer rates.   Other
fluids listed in the table are metallic sodium (M.Pt. 97.5°C), which has a
rather poor thermal capacity, and the so-called 'thermal fluids' used in the
chemical industry for process heating, none of which are without hazard at
their operating temperatures.

The remaining materials are solids and all possess a much higher temperature
capability. However, in considering the use of a solid as a heat store it
must be remembered that the energy has to be transmitted through the body of
the material by a process of conduction. This sets a limit on the rate at
which heat can be stored or withdrawn for any given transfer area and depth
of material.   The controlling parameter is the thermal diffusivity and for
comparative purposes this is listed in the final column of the table.   It
will be observed that the two solid metals, iron and aluminium, have thermal
diffusivities several orders of magnitude greater than the other materials,
permitting a rapid exchange, but it must be remembered that these metals are
inherently expensive to produce in terms of energy consumption, and hence
their 'pay back' period is much longer than lower grade materials such as
stone, cement or brick.

In spite of the low thermal diffusivity of the structural materials listed,
provided a sufficient area can be presented for heat transfer and the heat
flux restricted to fairly modest values, they can and do form a useful heat
storage medium in buildings and other structures.

So far as the general subject of sensible heat storage is concerned it seems unlikely that any new materials will come to light having significantly superior properties and yet be sufficiently inexpensive to be of interest in this application.

Latent Heat Storage

The melting of solids is accompanied by an absorption of heat, which is released when solidification takes place, and this latent heat can be used to store energy. Pure materials melt at a sharply defined temperature, and thus a latent heat store can operate at a constant temperature.

TABLE 2. LATENT HEAT CAPACITY OF SELECTED MATERIALS

Material	M. Pt. °C	Heat of Fusion $J.Kg^{-1} \times 10^{-5}$	Volumetric Heat of Fusion $J.m^{-3} \times 10^{-8}$	Temp. change in mass Equiv. Water
Calcium Chloride Hexahydrate	29-39	1.74	2.84	42
Sodium Carbonate Decahydrate	32-36	2.67	3.85	64
Sodium Sulphate Decahydrate	32	2.41	3.55	58
Calcium Nitrate Tetrahydrate	41	2.09	3.82	50
Hypophosphoric Acid	55	2.14	3.22	51
Sodium	98	1.15	1.09	28
Lithium	180	6.28	3.32	150
Lithium Nitrate	250	3.67	8.76	88
Ferric Chloride	304	2.65	7.68	63
Sodium Hydroxide	322	2.09	4.45	69
Potassium Magnesium Chloride	487	3.16	6.8	76
Aluminium	660	4.02	10.5	96
Sodium Chloride	800	4.86	10.3	116

The latent heat and volumetric latent heat of a number of selected materials are listed in Table 2 together with the melting point. The last column, which has been included for comparison purposes shows the temperature change that would take place in an equivalent mass of water corresponding to the latent heat.

It must be stressed that these materials also have useful specific heats, and that their full potential may only be realised by utilising their sensible heat capacity as well as their latent heat.

The same fundamental problem associated with heat storage in solids arises with the heat abstraction process in latent heat systems. Inevitably the heat transfer surface will face solidified material, and thermal diffusivity will limit the rate at which heat can be removed. This does not arise during the storage part of the cycle for the heat input will result in melting of the material adjacent to the surface, and convection-aided heat transfer will take place.

The situation may further be complicated for the hydrated salts where loss of water occurs during heat addition. Under these circumstances three phases may exist, the unmelted solid in equilibrium with the anhydrous solid and an aqueous solution. For re-crystallisation to take place molecular diffusion of water into the anhydrous salt has to occur, and this is usually a slower process even than thermal diffusion. Furthermore, if physical separation of the two phases occurs owing to density differences, mechanical mixing may be required if the exothermic process is to take place at all.

Heat of Solution and Reverse Osmosis
Two further physical-chemical properties of salts and solutions can be utilised as a means of storing thermal energy.

The first is the fact that nearly all salts have either a positive or negative heat of solution. Furthermore, all salts possess a positive temperature coefficient of solubility, i.e. they dissolve to an increasing extent as the temperature is raised. It follows that if a mixture of a salt which absorbs heat on solution and its saturated solution is heated, the solution of more salt will absorb energy, and the apparent specific heat of the mixture will be greater than that of its components alone. Conversely if a saturated solution is cooled, crystallisation will take place with the evolution of heat, for the process is a reversible one. Apart from enhancing the heat absorption capabilities, the presence of the salt will also raise the boiling point of the solvent, as a result of which energy will be storable at a rather higher temperature than is possible in the solvent alone.

Relatively little work has been done on this form of storage system, and heats of solution in concentrated solutions are not well known (heats of solution are normally specified for a high degree of dilution). However, the product of the heat of solution and the coefficient of solubility should provide a useful guide in the search for likely materials. Table 3 lists a number of possible salts and tabulates the effective specific heat due to solution heat at 20°C and 100°C with water as the solvent.

Values are listed at two different temperatures since the apparent specific heat alters with salt concentration. Many saturated solutions contain very much more salt than solvent, as a result of which the salt constitutes the main mass and volume of the mixture. An example, perhaps the most extreme, is that of ammonium nitrate for a saturated solution at 100°C contains only 10% water.

Similar problems concerning the diffusion of heat into the mixture occur as have been noted previously, and an additional limitation may arise on account of the finite rate at which re-crystallisation takes place. Indeed, many salts exhibit the phenomenon of super-saturation, and require the temperature to be lowered to below the theoretical value before re-crystallisation will take place at all.

The second possibility of storing energy which utilises the heat of solution depends on osmosis.

Osmosis occurs between solutions separated from their solvent by means of a semi-permeable membrane, when the solvent diffuses preferentially into the

TABLE 3. SOLUTION HEAT CAPACITIES OF COMPOUNDS

Material	Heat of Solution $MJ.mol^{-1}$	Solubility change $mol.Kg^{-1}K^{-1}x\ 10^3$	Solution Heat Capacity $KJ.Kg^{-1}K^{-1}$	
			20°C	100°C
Ammonium Nitrate	27.09	1.31	12.1	3.5
Potassium Nitrate	36.13	0.27	7.4	2.8
Calcium Nitrate Hexahydrate	33.45	0.18	2.0	1.3
Sodium Phosphate	54.43	0.075	3.7	2.0

solution.   If this diffusion is resisted, osmotic pressures are set up across the membrane that may be of the order of tens of atmospheres. However, if a pressure in excess of the osmotic pressure is impressed, the solution will lose solvent and become more concentrated. Thus for a salt having a positive heat of solution, cooling will take place or, under isothermal conditions, heat will be absorbed.   If the pressure is removed, then the solvent will pass back into the solution with the liberation of heat.

An advantage of this mechanism of heat storage over those previously described is the absence of any solid phase with its attendant limitations of thermal diffusion.

KINETIC ENERGY STORAGE

Whilst the kinetic energy of any rapidly moving mass represents a store of energy that may be recovered at will, a linear motion of the body is usually inconvenient. If, however, the motion is rotational and the energy is contained in the form of a flywheel, a more useful energy store is possible. Flywheel energy storage has received attention for application in two fields: (a) the large scale storage of energy for electricity production where capacities in the order of tens of megawatt-hours are required and (b) smaller scale storage for transportation purposes in cars and lorries where a few tens of kilowatt hours would suffice.

The amount of energy stored by a flywheel depends on the mass of the wheel and the square of its rotational speed. The limit to the amount of energy that can be stored is, of course, determined by the stresses that arise from the centrifugal force. At first sight, therefore, one might imagine that a wheel of a high density metal was called for, rotating at as high a speed as was consistent with safety.

In fact the criterion that governs the energy that can be stored per unit mass is the strength/density ratio of the material of which the wheel is made and from this it follows that for a given strength the lightest possible material should be chosen.

To explain this apparent anomoly, consider two flywheels of identical size and strength, one having a density twice that of the other. As they are accelerated the denser one will reach its stress limit first, and contain a certain amount of energy in store. The same stress level will be reached in the other wheel at a speed $\sqrt{2}$ times the speed of the first, when the total energy contained will be just the same. Hence the specific energy capacity $(J.Kg^{-1})$ is twice as great for the lighter wheel as it is for the heavy one.

The theoretical energy store capability is given by the relationship: $E_{max} = \frac{1}{2} \sigma/\rho$ where σ is the maximum stress and ρ is the density. In practice, of course, even if a flywheel could be constructed to have a uniform stress level at all points, the theoretical value would have to be modified by an appropriate safety factor.

TABLE 4. MATERIALS OF CONSTRUCTION FOR FLYWHEELS

Material	Density Kg.m^{-3}	Tensile Strength N.m^{-2} x 10^{-10}	Energy Storage MJ.Kg^{-1}
Aluminium alloys	2600	0.41	0.076
Maraging Steel	8000	0.28	0.17
E-Glass	2500	0.33	0.68
Carbon Fibre	1520	0.28	0.77
S-Glass	2480	0.48	0.95
Kevlar	1480	0.36	1.26
Fused Silica	2160	1.50	3.13

Table 4 lists some possible materials from which an energy storing flywheel might be constructed and tabulates the density, tensile strength and storage capability. The fact that emerges is that composite materials appear far more promising candidates than the isotropic materials. An additional important factor in their favour is that if the safety margin should be exceeded and the wheel burst, these composite materials shred into dust and fine particles which can do relatively little harm, in contrast to the metals that fracture into large fragments.

A design problem arises in that the stresses in a normal flywheel are bi-directional, the circumferential stress reaching a maximum at the hub and the radial stress peaking at the half-radius. Since most composite materials develop their strength uniaxially the problem is how to design a wheel having unidirectional stresses only. One suggestion has been to construct the wheel as a series of concentric annuli with an elastic bonding material between successive elements in order to eliminate radial stresses. An obvious drawback of such an arrangement is that only one annulus (the outermost) would be fully stressed, and therefore the remainder of the wheel would be less efficient than it might be. A solution that has been suggested is progressively to increase the density of the material as the centre is approached say by the addition of metallic powders to the composite. In this way the strength of the material would be affected but little, and a uniform stress level throughout could be achieved.

From the figures shown in table 4 and making a few simple assumptions, it is
interesting to consider the mass of a flywheel necessary to power an automobile, and to compare its energy efficiency with conventional methods of
propulsion.

Suppose a journey length equivalent to the consumption of 20 Kg. of fuel
(i.e. 5 gallons) is specified before the system has to be recharged with
energy. The total energy released by this amount of fuel is roughly 800 MJ,
of which about 15% is transmitted, under favourable circumstances, as
mechanical power to the wheels. The nett mechanical energy requirement is
thus 120 MJ. With an electrically driven flywheel storage system the overall
in/out efficiency might well be as high as 90%, particularly if regenerative
braking were employed, and the flywheel would thus be called upon to store
133 MJ. If an efficient flywheel were to be designed and manufactured out
of Kevlar having a capacity of 1.26 MJ.Kg^{-1}, the total mass of the material
would be only 105 Kg. Even applying a suitable safety factor and making due
allowance for the case and mountings, the total weight of the system would
be by no means outrageous.

However, in terms of the overall energy efficiency the flywheel driven
vehicle shows a considerable advantage over the one driven by an internal
combustion engine. Even if one assumes that the electricity is generated at
an efficiency of 30% the primary energy requirement is only about one half
that of the petrol driven vehicle.

Peripheral speeds for such energy storage are necessarily high and lie in
the range 1500 - 2000 m.s^{-1}. This velocity corresponds to rotational
speeds of the order of 10^5 revolutions per minute for rotors of a realistic
size. It follows that evacuated containment and magnetic suspensions would
be necessary in order to reduce the frictional losses to acceptable values,
and of course contra-rotating systems to contain gyroscopic forces.

It may also be noted in passing that a vehicle having similar requirements
but powered by lead-acid accumulators would require approximately 1500 Kg
of cells, i.e. fifteen times the equivalent flywheel mass. Furthermore, were
a rapid re-charging cycle adopted for such a power unit the storage efficiency would fall to about 50%, and the system would show little if any energy
advantage over the conventional petrol driven vehicle.

ACCESS AND LOSSES

Whatever form of storage is adopted access is called for in order to transfer
energy either into or out of the store. Furthermore there will be a natural
tendency for the store to lose energy, either via the access points or other
channels, and such losses must be guarded against if the store is to function
efficiently.

It is probably true to say that the only really 'watertight' form of energy
store is one involving chemical energy, say as a fuel, since this energy can
only be released through the application of an activating energy which occurs
in the ignition process. Only when this key has been inserted does the
stored energy become available. In all other forms of storage the energy is
immediately available, and it follows that leakages will occur wherever the
opportunity exists.

With thermal energy storage losses can only be prevented by effective thermal insulation, and the requirement in terms of efficiency is dictated by the period for which the store has to operate and the penalties associated with energy loss.

It is possible, of course, to design thermal insulation systems that approach perfection in this respect, the highest standards having been reached in the field of cryogenics where the problem is that of preventing heat gains rather than losses. However, the attainment of such standards requires materials and techniques that are both expensive and bulky, which, for many applications, cannot be justified on economic grounds.

The two factors, cost and bulk, are those that generally determine the optimum insulation, for insulants are usually of such a low density that their mass hardly enters the picture. In many instances it is the cost alone that must be considered, as, for example, in houses and buildings.

No matter how effective is the insulation in reducing the loss, there remain the points of access and these may provide the most serious leakage path.

In many instances nothing more elaborate is required than an on/off switch which can be provided by a valve if fluid is used to transfer heat to and from the store. In other situations it may be desirable that the store should be accessible whenever energy is available, an example of which is a solar energy store.

Recently a number of 'thermal rectifiers' have been described which permit the flow of heat in a preferential direction. Conductive devices that rely on the distortion that takes place owing to temperature gradients in metals having dissimilar coefficients of expansion are one example, and convective rectification based on the fact that heated fluids rise with respect to cooler fluids also appears possible. Although these devices are still very much in their infancy, if developed their application might radically alter the design and use of energy stores.

It is factors such as this and many others relating to the system as a whole that will determine whether or not any particular form of energy store is likely to be successful, for, as was mentioned in the opening paragraph, storage is by no means a process that may be considered in isolation.

Bibliography and Further Reading

Hottel, H.C. and Howard, J.B., New Energy Technology - Some Facts and Assessments, M.I.T. Press (1971).
Brinkworth, J., Solar Energy for Man, Compton Press (1972).
Post, R.F. and Post, S.F., Flywheels, Scientific American, December (1973).
Clauser, H.R., Advanced Composite Materials, ibid. July (1973).
Haydock, J.L., Consulting Engineer, Storage of Energy, November (1974).
Berg, C.A., Science, July (1973).
Energy Prospects to 1985, Vols.1 & 2, O.E.C.D., Paris (1974).
Energy R & D Problems and Perspectives, O.E.C.D., Paris (1975).
O'Callaghan, P.W. et al, Developments in Thermal Rectification, Sixth I.Mech.E. Thermo and Fluid Mechanics Conference, Durham, April (1976)(to be published).

DISCUSSION

Question: Why have existing flywheel systems, as in buses, not been
            continued?

Answer: They have been inefficient but people have been looking into the
            future with composite materials and new designs of flywheels. At
            the moment it is not possible to design them with efficiencies
            over their whole mass comparable with the efficiency at the rim.

Question: What about the gyroscope effect when cornering?

Answer: To overcome this, one uses two contra-rotating gyroscopes and lets
            the bearings take the strain.

Question: What about the safety aspects of a gyroscope system in the event
            of a catastrophic accident?

Answer: There is a hazard, but one must bear in mind that cars carry
            around 10 gallons of highly inflammable gasoline, which is also
            potentially very dangerous.

Question: What about the extraction of heat from the latent heat storage
            system?

Answer: This is mainly limited by thermal diffusivity for extraction of
            heat, as in this case there is a solid adjacent to the heat source,
            whereas when putting heat in to the system, there is a liquid next
            to the heat source.

THE HYDROGEN ECONOMY

J. K. Dawson

Energy Technology Support Unit, Harwell

INTRODUCTION

To the user of ever more costly carbonaceous or hydrocarbon fuels, which in
some cases are also inconvenient to use and are polluting towards the
environment, the prospect of a new fuel which could be considered inexhaustible and therefore relative stable in price, and which would be virtually
non-polluting in use, may appear very attractive. Hydrogen has been
suggested as a substitute fuel with these attractive attributes by many
authors over the past few years: indeed the concept which we now call the
hydrogen economy occurred to Jules Verne over 100 years ago (he saw hydrogen
as a substitute for coal).

However, hydrogen is not a primary fuel. Energy must be expended to produce
it from water. It is therefore akin to electricity, i.e. it is to be
regarded as a convenient energy storage or transmission medium and as a
potential secondary fuel. It has indeed already been used as a fuel in
commercial quantities for space vehicle propulsion in the USA and for road
vehicles in Germany during the Second World War. Why then, if the concept
has been around for so long and there is some pioneering experience available,
are we not already moving rapidly into the hydrogen economy? Table 1 will
give some clues as to the reasons for lack of progress, but above all there
is the question of cost - we simply do not yet have a method of producing
hydrogen from water which is economic in terms of the value of hydrogen in
comparison with its competitor fuels.

Let me say at this stage that I have no strong prejudices for or against the
widespread use of hydrogen. I hope to present you with some facts and leave
you to make up your own minds whether or not there is here a set of technical
aims which we should pursue in the United Kingdom.

I would also like to remind you that hydrogen is an important industrial
commodity now, but primarily for its chemical properties rather than as a
fuel. Individual plants in the UK currently have capacities up to 20,000
tonnes/year, and hydrogen is produced on the multi-million tonne per year
scale world-wide, but largely in plants which are within chemical complexes
providing a captive market, so that the hydrogen is not available for general
distribution. The principal uses are for the manufacture of ammonia,
methanol and petrochemicals and for internal processes within oil refineries
(hydrocracking, hydrodesulphurisation, etc.). In the short-term, i.e. less
than 20 years, the increasing need for hydrogen will arise almost entirely
from the natural expansion of these chemical markets rather than from any
novel fuel uses.

Forecasting the future is a notoriously difficult exercise, especially as far
ahead as the end of the century, but that is the timescale required to develop
a new source of hydrogen. A study by Meadows and de Carlo in 1970[1]
considered that a range of perhaps 2 to 20 times the 1968 rate of production
should be considered for the world-wide production in year 2000, much of the
increase being required for the production of ammonia and for petroleum
refining. We do not need, therefore, to foresee the development of new fuel

uses to render hydrogen important: its uses in the chemical industry will be growing at a rate which is not insignificant in terms of hydrocarbon consumption. This arises because hydrogen is currently made primarily by the steam reforming of naptha or natural gas: as hydrogen demand rises in step with fertiliser and chemical demand so will consumption of the hydrocarbon fuels.

TABLE 1. POTENTIAL ADVANTAGES OF HYDROGEN

Potential Advantage	Comment
1. Cheap, inexhaustible raw material (water).	Energy input from fossil or nuclear sources is needed to make hydrogen from water.
2. Non-polluting fuel.	May require a very large nuclear programme, with its own environmental problems.
3. Easy to store and transport.	Safety hazards may be higher than for its competitors.
4. Flexibility of use.	Economic arguements are likely to be more important than technical flexibility.
5. By substituting for oil it can ease national balance of payments.	Only if the fast reactor is successful in reducing a potentially equally severe problem of uranium supply.

If a new cheap production route through water-splitting by a non-fossil energy source (e.g. nuclear or solar) can be developed there will be a market for the hydrogen in chemical production. To give you an order of magnitude: nuclear power plants are measured currently in outputs of about 1000 MW(e) and to meet today's demand for hydrogen in the UK by electrolysis would already require four such reactors.

In summary, therefore, the main problem before us is not to find new markets but to develop a better method of producing hydrogen, and most of the remainder of this talk will be about this aspect.

HYDROGEN PRODUCTION

Water being the only sensible raw material for our hydrogen, we need a source of energy to split it. Ultimately this might be nuclear, solar, wind or wave energy, but of these the most important on current thinking is nuclear. However, in some senses it does not matter which source is used because all we are interested in for the production process in most instances is a source of heat or electricity.

Electrolysis
I believe that a first expansion of hydrogen production from non-fossil fuel sources is likely to be by the electrolytic route and so I will deal with that first. The reasons will be apparent later. Electrolysis has been with us for a very long time: the definitive original work having been carried out

by Sir Humphry Davy in 1806. In recent years, because naphtha and natural
gas have been so cheap a source of hydrogen – even cheaper than hydro-
electricity – the commercial development of electrolysers has proceeded only
slowly towards the large scale units which would be needed in the future.
Only three per cent of the hydrogen needs of the world-wide chemical industry
is currently produced by electrolysis.

Common features of conventional electrolytic cells are the use of iron as
cathode and nickel as anode, with an electrolyte of 20% NaOH or 30% KOH
solution to give a reasonably high conductivity (acidic electrolytes are
avoided for corrosion reasons). Some of the more fanciful suggestions for
the future use of electrolysis involve siting the plants on or in the sea:
this would present severe materials engineering problems since traces of
chloride in the electrolyte would be highly deleterious. Present technology
would not permit the direct electrolysis of sea water. The lowest practical
decomposition voltage of water at room temperature and normal pressure is
1.48 V, but voltages over 2V sometimes have to be applied to conventional
cells to overcome ohmic resistances and polarisation at the electrode
surfaces. There is, therefore, a considerable gain yet to be made in
efficiency of operation.

The output of individual plants tends to remain low: the larger the required
output the more unattractive electrolysis appears based on present designs
and power costs. The largest electrolytic plant is the one associated with
the Aswan Dam, with a design output of 30,000 tonnes/year. Whilst this is
large by present hydrogen industry standards it is small compared with the
needs of a full hydrogen economy, where hydrogen might substitute for natural
gas, as a fuel for road and air transport, etc. Table 2 shows some recent
estimates of the potential hydrogen requirements to meet present day demands
in some of these areas.[2]

TABLE 2. HYDROGEN REQUIREMENTS FOR SELECTED POTENTIAL
INDUSTRIAL MARKETS IN THE U.K.

	Hydrogen requirement (M tonnes/y)
Ammonia production	0.3
Total chemical hydrogen	0.6
Replacement for natural gas	8.7
Road transport	7.7
Air transport	1.5
Iron/steel making	5.5

A detailed prediction of the cost of very large electrolysis plants has been
made by Schenk et al.[3] The capital costs can be apportioned approximately
as in Table 3, which points up the significant contribution of the
rectification equipment. For plants of small output capacity the capital
charges will be comparable with the electricity costs, but for large plants
the electricity cost is by far the dominant factor.

On present evidence, only a nuclear economy could produce electricity on the
scale required for a full hydrogen economy, and so the success of the latter

will depend not only upon successful development of electrolysers, but also upon the success of new types of reactor system, such as the fast breeder, in reducing the price of electricity.

TABLE 3. DISTRIBUTION OF CAPITAL COSTS OF ELECTROLYTIC PLANT

Electrolytic cell	60 - 63
Water treatment, etc.	2 - 5
Rectifier	28
Installation cost	7
	100%

The efficiency of conventional electrolysers is only about 60 per cent, defined as the yield of hydrogen relative to the yield of an ideal reversible cell. It is possible to improve on this in a number of ways, and exciting new development is being carried out at the General Electric laboratories in the U.S.A., based on a solid perfluorinated sulphonic acid polymer electrolyte which is a good ionic conductor.[4] However, even with such an advance a recent estimate by CEGB has given pessimistic cost figures for hydrogen production: if the hydrogen produced is regarded as a fuel it would be three times as expensive as $10 per barrel oil.[5]

For hydrogen to be regarded as a generally-available fuel, therefore, the prices of oil and gas will have to rise considerably in real terms from their present levels and further advances will be needed to improve the efficiency of hydrogen production. New research on the electrolytic route would be well worthwhile, but it must have fairly radical aims: it will be ineffective to try to tinker about with minor modifications to the well established technology. Some authors have argued that electrolysis will always be too inefficient on energy conversion grounds: perhaps the best that can be hoped for is to generate electricity at 40% efficiency and develop advanced electrolysers to work at a little better than 80%, giving an overall efficiency of 33%.

Water splitting by thermochemical cycles
To split water into its elements by the direct application of heat is theoretically possible, but would require temperatures (about 3000°C) well beyond those currently thought achievable even in the most advanced nuclear reactors. Suggestions have been made that a fusion reactor could generate a very intense source of ultraviolet photons which could be used to cause the decomposition of water in a surrounding blanket.[6] However, separation of the hydrogen and oxygen produced simultaneously as a mixture would probably prove to be an insuperable problem and above all this concept requires the successful development of a fusion system. In fact the use of radiation from any type of nuclear reactor system for the large scale decomposition of water is not a profitable line to pursue.

The use of nuclear heat involves a different set of considerations and is currently the focus of considerable attention in a substantial number of laboratories. If we fix the upper temperature limit at about 850°C for practical reasons, and taking elementary thermodynamic considerations into account, it can be shown that for a purely thermochemical system it is necessary to allow not less than three intermediate steps to achieve overall the decomposition of water.

Many multistep chemical cycles have been proposed in the literature: over thirty from 10 research centres. It is not too difficult to invent them and some laboratories are using computer searching of basic thermochemical data in the quest. A cycle which has received much attention at the Ispra research laboratory of Euratom is based on reactions of the iron chlorides:

$$6 \; FeCl_2 + 8H_2O \longrightarrow 2 \; Fe_3O_4 + 12HCl + 2H_2 \quad (650^{\circ}C)$$

$$2 \; Fe_3O_4 + 3Cl_2 + 12HCl \longrightarrow 6 \; FeCl_3 + 6H_2O + O_2 \quad (200^{\circ}C)$$

$$6 \; FeCl_3 \longrightarrow 6 \; FeCl_2 + 3Cl_2 \quad (420^{\circ}C)$$

All the products apart from hydrogen and oxygen would be recycled and the net effect is simply the decomposition of water. This cycle would have the potential attraction that the required temperatures should be within the range of HTRs now being developed. At first sight is looks an attractive prospect but, as with all other cycles so far suggested, detailed work has revealed many serious problems.

The cycle begins to lose its attractions when looked at through the eyes of a chemical engineer. We know little about the factors controlling the kinetics of the reactions and so the design of a large scale plant can scarcely begin. Moreover, there are considerable problems of heat and mass transfer (this particular cycle carries a heavy penalty in terms of material flow per unit of hydrogen produced), heat recycling, solids handling, etc. Chloride systems are notoriously corrosive and may demand expensive materials of construction. Rather precise control of temperature may be required yet be difficult to achieve, for instance at somewhat lower temperatures the first reaction can take a different route to $Fe(OH)_2$ without the production of hydrogen. Recent work has shown that reaction 2 does not proceed quantitatively as written. Difficulties arise over reaction 3 due to the volatility of $FeCl_3$.

A cycle which sets out to circumvent all the problems of solids handling and which is a hybrid system supplying some of the energy input electrochemically is beginning to attract attention:

1. $SO_2 + H_2O \longrightarrow H_2SO_4 + H_2$ $(E^{\circ} = - 0.17V)$ (cell operated at $150^{\circ}C$ to reduce over voltage).

2. $H_2SO_4 \longrightarrow H_2O + SO_2 + \frac{1}{2}O_2$

It is thought that reaction 2 might be possible in the temperature range 700-800°C with the aid of a catalyst. It is already carried out commercially at 1200°C, but now work is required to attempt to bring down the temperature to a range compatible with nuclear reactor technology.

It will be apparent that thermochemical cycles will be no less difficult to attempt to develop than advanced electrolysers - my guess is that they will be more so - and in addition they will depend on successful development of the HTR as a source of high grade heat. The interfacing of the nuclear reactor to the chemical plant will involve many new problems which are as yet only dimly perceived.

In principle, thousands of multistep chemical cycles will exist that might yield a workable water-splitting process. Many criteria can be set up to sort

out the most promising system, including thermal efficiency, maximum
temperature, materials compatibility, system complexity, safety constraints,
predicted costs, etc. It is at present difficult to arrive at even an outline
cost of producing hydrogen by this route, in view of the uncertainties in both
the chemical and the nuclear plant. However, one of the most important
criteria will be thermal efficiency: unless an efficiency higher than that for
electrolysis can be predicted it will not be worthwhile developing the thermochemical route. Unfortunately the estimates of efficiency vary widely at
present due to lack of knowledge of the chemical engineering aspects of the
plants. A recent CEGB review[5] concluded that an overall efficiency much
over 40 per cent would be unlikely to be achieved. This is only marginally
better than can be predicted for an advanced electrolytic plant. Moreover the
CEGB team pointed out that such an efficiency would be no better than for
conversion of nuclear heat from an HTR into electricity, which many would
regard as more attractive than hydrogen as a secondary fuel.

The subject of thermochemical cycles remains wide open for further investigation - we do not have a winner yet.

Biochemical water-splitting

Before leaving the subject of hydrogen production I should mention the
long-term possibility of water-splitting by biocatalytic action using a solar
energy input. Important pioneering work in this area is being carried out by
Professor Hall and his colleagues at Kings College, London. The scheme
combines the action of light on the chloroplasts which can be isolated from
plants (containing chlorophyll and various electron carriers and enzymes)
with an enzyme (hydrogenase) produced in many bacteria or algae. The process
is complex. In the presence of the chloroplasts, photons produce oxygen,
protons and high energy electrons from water: the hydrogenase subsequently
catalyses the formation of hydrogen by the combination of the protons with
electrons degraded in energy through a series of reactions of the chloroplast
components. This work is still in the laboratory phase, but it presents an
interesting alternative route to hydrogen based on a solar rather than a
nuclear energy input.

DISTRIBUTION OF HYDROGEN

If hydrogen is to achieve an economically competitive position as a fuel then
it will have to be produced on a very large scale, and to make a significant
contribution to the conservation of fossil fuel resources the hydrogen plants
will need to be driven by large nuclear reactors. This implies that the
centres of production will have to be sited remotely from many of the centres
of consumption, probably on coastal sites to allow for sufficient provision
of cooling water. Several authors have considered the problems of transporting hydrogen by pipeline over long distances.

Considerable technological know-how already exists in this area, and
Marchetti[7] has pointed out that a complex large-bore hydrogen pipeline
linking several industrial firms is in operation in the Ruhr and has a length
of some 300 km. A similarly complex, but less extensive, pipeline (14 km)
links various ICI plants on Teeside in the UK.

Hydrogen suffers from the disadvantages of a heating value only about one
third that of natural gas on a volumetric basis, so that a correspondingly
larger volume of gas would need to be transmitted in a hydrogen fuel pipeline.
Gregory concluded that on an equal energy basis and using the same diameter of

pipeline for both gases, hydrogen would cost 60 per cent more to transmit than methane.[8] Most of the increase is due to the higher compression required. Workers at the Ispra laboratory derived a similar figure, but Marchetti has pointed out that for hydrogen the necessary pumping stations can be much further apart[7] and this could be an important bonus for the relatively short transmission distances likely to be required in the U.K. Moreover, if the hydrogen is produced in porous electrode electrolysis plants, part of the initial compression cost is already included in the production figures.

A more pessimistic view of the costs of hydrogen transmission and distribution is to be found in a recent review by CEGB.[5] However, perhaps of greater importance is the relatively small contribution which emerged for transmission costs to total costs: they are completely swamped by the production costs, i.e. by the cost of the primary nuclear energy whether as heat or as electricity. Although there may be interesting scientific and technical problems in the transmission field - for instance hydrogen embrittlement of pipeline materials - they are of secondary importance in comparison with the need to solve the production problems.

The suggestion was made earlier that hydrogen could be a candidate synthetic fuel to replace natural gas when supplies from the North Sea begin to dwindle. Before the advent of natural gas, much of the distribution network, especially that closest to the domestic consumer, was used to transmit town gas which had a hydrogen content of about 50 per cent. The main difficulty in using the same system to distribute pure hydrogen would be associated with the safety implications of leakage (on an energy basis the loss of hydrogen through a given leak will be no worse than for methane despite the higher diffusion rates for the former). It may be necessary to develop techniques for the application in situ of impervious linings to the pipework, or even to introduce a small high pressure hydrogen line inside the existing pipes and to monitor the interspace.

It appears unrealistic to consider a rapid change from methane to hydrogen as the principal gaseous fuel. The feasibility of a gradual change from one to the other requires study, and may well be ruled out. Alternatively, the addition of, say, 10 per cent hydrogen to the natural gas system might be a way of conserving the hydrocarbon fuel, but this requires further evaluation of matters such as compatibility with burner devices.

USES OF HYDROGEN

If a large scale supply of cheap hydrogen based on nuclear or, ultimately, solar energy could be developed, how would it be used? I have just mentioned the possibility of hydrogen as a substitute for natural gas: that would be the largest potential use, but in a situation with a plentiful supply of coal it may be better to use the hydrogen to convert the coal into SNG.

I have also referred to the expanding need of the chemical industry for hydrogen. The metallurgical industry could well demand substantial amounts of hydrogen too, in particular for the direct reduction of iron ore. However, that demand would be smaller than the potential market for hydrogen as a substitute for hydrocarbons as a transport fuel. The imagination of a number of groups has been caught by the potential attractions of hydrogen as a fuel for aeroplanes, both sub- and super-sonic. Whilst some authors claim that this might be the first unconventional use for cheap hydrogen, the awkward fact remains that an extensive redesign of the complete aeroplane system would be needed - not a task to be lightly embarked upon!

The use of hydrogen as a fuel for road transport is certainly feasible
technically and demonstration vehicles have been built in several countries
(e.g. at Cranfield Institute of Technology in the UK). Perhaps the most
intriguing technical problem is that of storage of hydrogen on the vehicle.
The choice lies between liquid hydrogen and metal hydride systems. For the
latter, magnesium hydride is usually chosen for reference discussion on the
grounds of its low density, but other hydrides have more attractive physical
properties. There is a rich field here for the physical chemist to explore.

SUMMARY

The main points I have tried to make in outlining some aspects of the hydrogen
economy may be summarised as follows:

-   Hydrogen is to be regarded as a potential secondary fuel or as an
    energy storage and transmission medium. Energy is needed for its
    production from water and is most likely to be derived from nuclear
    primary fuel, with solar and wave energy inputs as interesting
    longer-term possibilities.

-   Movement towards the hydrogen economy will depend equally on
    solving the problems of producing it and on the commercial success
    of nuclear power to supply electricity (fast reactor) or heat (HTR).

-   The most important technical problem is how to produce the hydrogen
    at a competitive cost. Problems of distribution and storage are
    plentiful and interesting, but are subsidiary in importance.

-   Of the two main methods of production, the technical problems of
    electrolysis on a very large scale look more amenable to solution,
    but thermochemical cycles offer a potential prize of greater over-
    all energy efficiency. That prize will be very difficult to win.

-   Hydrogen can be used as a _fuel_ or as a _chemical_ agent. Expansion
    of the chemical (and metallurgical) demand will probably be the
    first target for cheap hydrogen to substitute for hydrocarbons.

REFERENCES

1. P. Meadows and J. A. de Carlo, "Mineral Facts and Figures", U.S. Bureau
    of Mines, Washington, 1970, p.97.
2. R. M. Dell and N. J. Bridger, AERE, Harwell. Unpublished work.
3. H. Schenk, W. Wenzel, F. R. Black and E. Wortberg. Report EUR-416 (1973).
4. See, for instance, W. A. Titterington, paper to the 8th Inter-society
    Energy Conversion Engineering Conference, Philadelphia, August 1973.
5. P. J. Hampson, A. B. Hart, B. Jones, D. T. Swift-Hook, J. J. Syrett and
    J. K. Wright, "CEGB Research", No.2, May 1975, p.4.
6. B. J. Eastlund and W. C. Gough, Paper 41 at Symposium on Non-fossil
    Chemical Fuels, American Chemical Society, Boston, April 1972.
7. C. Marchetti, Chemical Economy and Engineering Review (Japan), 7th
    January 1973.
8. D. P. Gregory, D. Y. C. Ng and G. M. Long, Chapter 8 of "Electrochemistry
    of Cleaner Environments", ed. J. O'M. Bockris, Plenum Press, 1972.

DISCUSSION

Question: Why is direct electrolysis of sea water not currently feasible?
Answer: This is because of corrosion problems in the present electrolytic cells due to the presence of chlorine in salt water.

Question: Why involve the extra stage of hydrogen production from water in utilising the energy from a nuclear reactor?
Answer: The only reason would be if there were a breakthrough in chemical production which would use the nuclear heat more efficiently.

Question: What wavelength is necessary for direct splitting of water to give hydrogen? Allied to this, would it be feasible to go outside the atmosphere to make direct use of high energy solar photons?
Answer: In answer to this latter question - it would seem to be completely uneconomic, as costs would be astronomical. The threshold wavelength is not known, but one would be using a sledge-hammer, viz. using 40 eV photons to split a 1 eV bond.

Question: The long-term future of the hydrogen economy seems to depend on the development of high temperature reactors; however, we have been told that we need breeder reactors for the future which would not attain the required temperatures.
Answer: Thermal reactors of the present generation probably have no long term future. However, in the future we can visualise HTR's using bred U^{233} from thorium breeder reactors.

Question: 6% of natural gas in existing gas pipelines is lost, mainly due to the drying out of the sealants in the line. Would hydrogen have this effect?
Answer: Electrolytically produced hydrogen is damp, and so this problem should be unimportant.

Question: We have been told that H_2-producing thermochemical cycles are 40% efficient. Is this regarding high-grade heat only?
Answer: Yes. It is conceivable to use the lower-grade heat left over, thus improving the overall efficiency.

Question: The problem of hydrogen embrittlement of metals is not confined to pipelines. What steps are planned to cope with hydrogen on a large plant scale?
Answer: There is already a large body of experience in handling hydrogen in present industrial installations.

Question: What would you say is the future for the hydride systems of hydrogen containment? Are they safe?
Answer: They would seem to be a safe way of transporting hydrogen. As regards their utilisation, they are not very economic, due to their weight. There are not many feasible elements, the most likely being the light metals. With these materials it is difficult to achieve the density of liquid hydrogen. One has to define areas for the use of hydrides, for example, in heavy goods vehicles where their weight becomes less important.

Question: The hydrogen economy would seem to require 3 or 4 major technical breakthroughs. This would seem to be rather much compared with alternative systems.
Answer: I am not sure that other systems will be easier to realise. It

A.E.C. R

must also be borne in mind that a mobile fuel will be the end
result.

Question: Are there severe safety problems in the large-scale use of H_2?
Answer: There is an historical precedent in that a) town gas is 50%
              hydrogen, b) petrol is highly inflammable.

Question: Is there an environmental hazard due to H_2O_2 formation?
Answer: Yes, but outside a very broad range of H_2/O_2 mixtures.

Question: Will the widespread use of H_2 interfere with the hydrological
              cycle?
Answer: I do not know specifically, but energy balance problems become
              important when we produce heat at $\sim 1\%$ of the solar input and on
              this basis the effect will be trivial.

Question: Are there any environmental problems associated in burning H_2
              in an internal combustion engine?
Answer: Only outside a wide range of mixtures. The exhaust emission can
              be less polluted than the intake.

Question: Where will we obtain all the water for an H_2 grid - will there be
              a water shortage in, say, AD 2000?
Answer: Not in the UK - the problem would be one of distribution.

Question: What is the present ratio of hydrogen from electrolysis to that
              from natural gas?
Answer: Hydrogen produced from natural gas is negligible in quantity.

PANEL DISCUSSION:
ENERGY TRANSMISSION AND STORAGE

Panel members: Prof. D.W. Holder, Oxford University (Chairman).
Dr. J.K. Wright, CEGB.
Dr. W.T. Norris, CEGB.
Dr. I.E. Smith, Cranfield Institute of Technology.
Dr. G. Long, ETSU.

[Editors' note: During the course of this discussion some very interesting questions were raised on CEGB policy. These have been extracted and are presented in a separate section.]

Question: Could superconducting cables be used underground by pumping liquid hydrogen along with them?

Answer: A superconducting cable working at liquid hydrogen temperatures or slightly lower say $15°K$ is under consideration, but not as a means of transporting hydrogen. Other advantages are lower refrigeration loads and simpler design of thermal insulation. These points would make superconducting cable look more attractive from economic viewpoint.

Question: It has been said that coal has many uses - one of which could be to produce substitute gasoline. If this were the case, would not coal produced gasoline be preferred to hydrogen as a fuel for cars?

Answer: Two factors must be considered:
1) It wouldn't necessarily be cheaper to produce substitute gasoline rather than hydrogen.
2) Pollution. Hydrogen is less of a pollutant.
Contrasting these two points there is the problem of infrastructure, how would you set up a chain of filling stations for hydrogen?

Remark: It seems to me not only to be a problem of transmission. You have previously said the production costs are very high and yet the technology has not been developed to a sufficient degree.

Remark: Don't forget that the liquefaction of coal requires hydrogen. The option is either to feed the hydrogen in to gasify coal or market it as hydrogen. Hydrogen is still needed, the cost is the same.

Question: Could you discuss storage methods particular to small domestic power plants, say 10 KW down to 1 KW?

Answer: The preferred methods are very dependent on scale. Windmills and solar convertors must have accompanying storage due to their nature. On the domestic scale, safety is very important; nothing specialised or dangerous, simplicity and reliability are important. Sensible heat storage is probably best, with perhaps a few dissolved salts to raise the temperature.

529

There are a different set of considerations for portable storage, eg. motor
car.   I'm very apprehensive about the use of liquid hydrogen and its use as
a fuel for the car; a better application would be as an aviation fuel.

Remark: On the contrary, I can see a definite future for the hydrogen car.
However if the hydrogen economy develops I would like to see the hydrogen
burnt not in the internal combustion engine, but in a fuel cell with an
efficiency of 70-75% with little or no pollution.

Question: Would you not rather see a hydrocarbon burnt in a fuel cell?

Answer: Two points here:
1) Hydrogen fuel cell works at lower temperature and hence is thermodynamically more efficient than the hydrocarbon fuel cell working at a
higher temperature.
2) It is better to conserve our hydro-carbon resources.

Question: Is anything being done to concentrate on simplicity of design
rather than to develop technology for technology's sake, creating more
complicated machines and larger research programmes to increase efficiencies
by 1 or 2%?

Answer: Society generally proceeds by making small advances; the chiselling
effect and not by major breakthroughs.   Britain has a comparable number of
people doing research as the U.S.A. and Europe but only 1/6th the people in
the design field.

Question: Do you think the capital investment for hydrogen storage using
cryogenic stainless steel pressure vessels will be prohibitive?

Answer: No pressure problem.   Aluminium is quite strong enough.

Question: But when the supply of hydrocarbons is exhausted what alternatives
do we have? Battery storage? That is also capital intensive.

Answer: Careful studies in the U.S.A. have shown that hydrogen storage tanks
when mass produced could cost 10% of cost of car.

Question: Some industries suffer from peak demands upon their process steam.
What storage methods could you suggest to overcome this problem?

Answer: This heat load could be stored as molten liquid up to $800^{\circ}C$.   The
only problem regards the rate of heat release during solidification and this
should be taken into account during the design.   The inefficiencies are due
to thermal losses at input and output stages.

Question: With careful design to minimize heat losses through the walls could
heat be effectively stored in a fluidised bed?

Answer: Yes, so long as you can accept the reduction in density and there's
no change in phase, it is possible to make use of the very low ΔT needed to
effect heat transfer.

Discussion on CEGB policy

Question: It has been stated that the installed nuclear capacity of CEGB will be 20 GW by 1990.   This could be made up of 30 - 660 MW plants or 15 - 1300 MW plants.   Can the CEGB hope to achieve this number of completed plants in 15 years?

Answer: There is 10 GW of installed capacity at the present time.   The Winfrith prototype of SGHWR has been working successfully for 7 or 8 years now and there are about 4000 MW of SGHWR plant now on order;   these will have slightly different design features from the prototype.

Question: Even working flat out, employing a great deal more skilled man-power, will the CEGB be able to provide all this nuclear capacity when the demand will be there?

Answer: The essence of the nuclear debate was exactly what you have said. The CEGB wanted a large amount of capacity quickly and to be able to place large orders with confidence, and both of these criteria could be met with the American design.   The British government on the other hand wanted the SGHWR for patriotic and safety reasons.

Question: How do you extrapolate the CEGB demand curves?

Answer: It would be wrong to have a simple extrapolation.   First, we look at the fuel situation in general and estimate our market share. Electricity will inevitably increase its share in the long term but in the short term there will be strong competition from natural gas.   For 8-10 year estimates CEGB consults with the area boards who sell the electricity to predict future demand.   These different estimates taken together provide a general forecast, though unfortunately of low accuracy.

Question: Does the CEGB collaborate with the BGC and NCB?

Answer: Occasionally.   We have consultations about interrelated problems such as coal demand for electricity production.   However we have very few discussions about the market share as ultimately we would never agree.

Question: What chance is there of the Government decision on SGHWR being reversed?

Answer: The Secretary of State will review the situation in 1977/78, and I would imagine that this decision will depend upon the development at that time.   I personally think that a change of policy is unlikely because of the expense of developing a new program in mid-stream.   This decision will not affect SGHWRs on order, only whether we buy more of the same.

Question: There has been no long term strategy in the nuclear program and we are building AGRs from a very small prototype, and it is only now that a long term strategy seems to be emerging.

1) Should we not be considering the building of 1000 MW breeder reactor now
    in order to gain experience?
2) Has CEGB got a policy on energy storage?

Answer: This country has had a long term nuclear policy over the years. The
AEA set its sights on gas cooled thermal reactors and when AGRs looked
unacceptable they changed to SGHWR as the deliberate fallback.
1) As regards fast reactors, as soon as CEGB can order, with confidence,
    the first, full scale prototype will be built. After a long and care-
    ful look at its performance a decision can be made about its future
    development. From the point of view of uranium conservation, it is
    important that the breeder reactor be developed on a world-wide basis
    by the turn of the century.
2) By mid 1990s viable storage methods will be needed and research is
    currently being done in this field. A present, storing hot water bled
    from the turbines seems to be the best method.

Question: Would you like to comment on the concept of a NUPLEX plant
embracing thermal, breeder and reprocessing plant on one site? Would this
reduce the threat of plutonium hijacking?

Answer: There are two aspects here:
1) Ratio and timing of the breeder reactors compared with the thermal
    reactors, and by the end of the century, the world must be capable of
    ordering large numbers of breeder reactors.
2) There is probably more scope for this concept of one site for the thermal
    and breeder reactors and even the reprocessing plant in the USA where
    distances are larger than in Britain. However, I would point out that
    the requirements for reprocessing plants are far more stringent than
    those for reactor sites, and generally they must be sited in remote
    places far away from where the electricity is required.

Question: What would happen if COALPLEX and/or NUPLEX plants were built and
for some reason, one of these complexes shuts down; would this put the grid
system at risk? Are we sacrificing reliability for economy of scale?

Answer: A balance must be made between the size of generating units and
spare capacity. At the moment the biggest single outage that can be made
at any one time on the grid is 1000 MW.

Question: If the CEGB continues building large 2000 MW units, such as Didcot,
does this rule out the possibility of district heating schemes which could
make use of the 2/3 of the power station heat energy which is not used and
is discharged to the environment?

Answer: There are different scales of district heating schemes. The
easiest and most economic way is to install these schemes in new developments. District heating is more suited to smaller stations such as gas
turbine stations. A 2000 MW power plant could supply district heating for
a large city.

Question: What are limitations on the thermal efficiency in fossil fuel
fired boilers?

Answer: There seems to be a general corrosion barrier which prevents working

above 565°C and this is because the salts in the fuel tend to form low
melting point, high corrosive, eutectics which plate out on the boiler tubes.
It is not economic to clad boiler tubes with exotic metals.   Pressure is a
case where we have got to a point of diminishing returns.   At the moment
the working pressure is 2000 psi and further increases lead to negligible
increases in efficiency.

Question: Could we have your ideas on the possibilities of reducing the
demand for electrical energy and the possibility of differential pricing in
the domestic sector to encourage economy?

Answer: Heat pumps would be an effective way of saving electricity but these
are expensive.   If the public were asked to consume less energy and pay
slightly more to do so or continue as they are, they would rather stay as
they are.   People as a whole are not willing to accept a change of this
sort in society.   There are no technical or administrative reasons why
differential pricing of electricity could not be achieved.   Whether this
will be acceptable to the public is open to question.

Question: Could you comment on the thermal efficiency of night storage
radiators and how do they compare with other forms of heating?

Answer: Probably about 90% depending upon the particular design.

Remark: Good thermostatic control of night storage radiators can be achieved
via a fan.   But you are clearly referring to useful heat, one can define
the efficiency as:

$$\eta = \frac{\text{what you want out}}{\text{what you regret putting in}}$$

Also some benefit is gained from heat released during the night by maintaining the thermal inertia of the house.   Otherwise they do have some
disadvantages from one overall efficiency viewpoint.

SECTION 6

POSSIBLE DEVELOPMENTS IN
ENERGY USE

POSSIBLE DEVELOPMENTS IN TRANSPORTATION

S. S. Wilson

Engineering Science Department, University of Oxford

A study of prehistory and history will show that technology is <u>the</u> instrument of change and that no aspect of technology has been more important in changing the pattern of social life than transport.

The Urban Revolution of 5,000 years ago was possible due to the development of the wheel and the sailing boat which enabled food to be bought from the countryside to the towns. The development of the sailing ship in the middle ages opened up the world to trade and empire building. The Industrial Revolution of 250 years ago was dependent on the development of better means of transport through canals, railways and steamships. In the 19th century came what may rightly be regarded as the breakthrough in modern technology, the bicycle; pre-bicycle technology was heavy and inefficient e.g. the steam locomotive, but the bicycle is lightweight and efficient, structurally and mechanically, with its elegant steel tubular frame, lightweight spoked wheel, bush roller chain, ball bearings and pneumatic tyres (Ref.1). It is an excellent example of ergonomics, the machine fully adapted to the man, and as a result it is not only efficient in energy terms (Fig. 1) but also it has brought about immense social changes, by providing personal mobility for all.

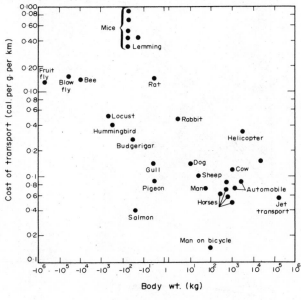

Figure 1

Unfortunately it also led to the development of the automobile; the
year 1885 saw both the final form of the bicycle, the Rover Safety, but
also Daimler's first motor bicycle and Benz's first motor tricycle, both
based firmly on bicycle technology, as was also the Wright Brothers first
aeroplane.

From the time when the motor car was fully developed and became significant in numbers, not more than 50 years or so in the U.K., the motor
car has become the instrument of change above all others and now dominates
our lives to a degree little realised; yet it now faces a number of crises
which may well result in a very different direction of change in the next
fifty years. It is not too much to say that the most urgent questions
facing the people of the 'developed' world are: "Can we learn to live with
the motor car, or can we learn how to live without it?"

The question could be reduced to even simpler terms - whether the
human race can survive without the motor car; to this there can only be
one answer, since we did so for millions of years, the vast majority of
the world's population survived without it and even in the U.K. only a
minority have direct use of a motor vehicle. It is also quite inconceivable
that a much larger fraction of the world population could ever be able to
use motor cars on the scale that they are used in the U.S.A. or even in
Europe.

Since energy consumption for transport is dominated by road transport
(Fig. 2) no apology is made for discussing this matter in some detail,
since no sensible prediction of future energy uses can be made without
facing squarely the role of motor transport in our future way of life.

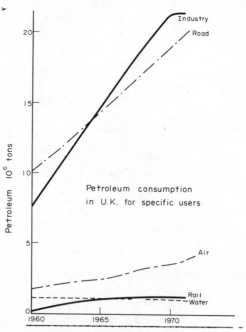

Figure 2 (source: U.K. Energy Statistics '73)

It is no exaggeration to say that our whole life style revolves
around the internal-combustion engined vehicle; where we live, how we
travel to work, to recreation, or holiday, how we move goods (80% of goods
movement is now by road), how we plan our towns and our country side; our
largest export industry is at present the motor car industry, yet this
industry has been a leader in inflation - so long as people are willing to
pay a high price for their cars so long can high wages be obtained, yet
because modern mass-production methods are so inherently unsatisfactory
there is little prospect of long-term industrial peace, resulting in strikes
for higher wages, which prove to be no cure for the basic problem.

Perhaps one of the most serious results of the motor car is that it
has fractured society, by scattering families throughout the country, with
the inevitable results of the loneliness of the aged and losses by the
children of close contact between the three generations, aggravated by the
high proportion of working mothers, with the unhappy social consequences
now so familiar.

On top of this high social cost is the high financial cost of fuel -
our largest import bill - of depreciation, taxation, insurance and all the
hidden costs of accidents, the police, the courts - not just motoring offen-
ces but so many crimes involving stolen cars - road construction and maintenance. Then there is the ill-health associated with too much driving
under stressful conditions with too little physical effort, a major cause
of coronary attacks.

Efforts to plan for the automobile have proved disappointing to say
the least, judging from experience in the U.S.A., Europe and Japan. Now
with the sudden increase in oil prices and the certainty of exhaustion of
oil supplies in a finite time there is evidence of a world-wide slump in
the car industry which the industry itself persists in regarding as merely
temporary but which is more likely to be a permanent recession, though
stability may be achieved at a much lower level of production. Attention
is being directed towards alternative fuels to petroleum e.g. electric
batteries, oil from coal and the production of hydrogen from nuclear power,
the subject of other lectures. But these are far from being certainties
as large-scale alternatives to petroleum fuels and past experience shows
that many years are needed to develop a new technology to the stage where
it can supplant an existing technology - perhaps 30-40 years.

In these circumstances the only sensible policy is to seek to conserve our dwindling resources by all possible means - by fuel economy
measures, by substitution, by changing our patterns of living. In contemplating the changes that may be necessary it is well to remember the
enormous changes which have come about in the last 50 years, largely as
the result of the motor car, and to take note of the widespread belief
that the rate of change is accelerating. Hence whatever technical, economic, political or other changes occur in the next 25-50 years we must
envisage social changes on a scale comparable to those which have taken
place in the last 50 years - whether or not such changes are actively
planned, since the lesson of history is that great changes such as the
urban and industrial revolutions came about without conscious planning and
were unrealised by all but a few until they had taken place.

Despite such discouragements we must attempt to identify the movements

which are today taking place and to seek to promote those which appear to
be generally beneficial, opposing those which seem detrimental.

There is evidence, particularly among the younger generation, of a
tide of opinion turning against the domination of life by the motor car and
all it stands for in the way of high-energy living with the resulting detrimental effects on the environment. Attempts are being made to explore
alternative technologies, based on individual and community efforts. If
these are to succeed they will do so by the development of appropriate
technologies - the instruments of change.

An essential element in any such technology is its relation to the
human scale of values; the same criterion can be applied to any machine,
building, institution - does it tend to increase or decrease the significance of the individual? It is only too clear that many modern developments fail this test; one of the diseases of civilization being megalo-
mania, the myth that "Bigger is Better". Dr. E. F. Schumacher (Ref. 2)
has spelt out the alternative - "Small is Beautiful" - a study of economics
as if people matter.

Let us now examine the possibilities in transport for the following
qualities:

1. Relation to the human scale of values.

2. Economy in energy consumption.

3. Economy in materials.

4. Minimum adverse effect on the environment.

Table 1 shows a useful categorization of transport, showing amongst other
things the complexity of the problem, involving as it does road, rail,
water and air transport, for passengers and goods, for urban, rural, national
or international routes, for door-to-door or trunk services, for developed
or less-developed countries and with a variety of power sources.

Clearly it will not be possible in a limited space to cover all these
aspects, so a selection will be made of those for which developments can
be forseen in the next 25-50 years which may greatly affect energy consumption. A more detailed consideration of the energy economics of urban transport is given in Ref. 3.

Perhaps the most significant contrast is between modes 3. and 4.,
pedal power compared with the motor car. The first fulfills admirably
the conditions suggested of economy in energy and in materials, low impact
on the environment and of relation to the human scale.

In energy terms Fig. 3 shows the power required for cycling at different road speeds; an average cyclist can maintain an output of 75 watts
(0.1 H.P.) for a road speed of 12 m.p.h. but much greater outputs for short
periods, up to 750 watts (1 H.P.). Regarding the body as a heat engine,
using food as fuel, the thermal efficiency is about 25%, which compares
quite well with even the best heat engines for transport purposes. For
example, although a good diesel engine can achieve a peak efficiency of

TABLE 1. CATEGORIES OF TRANSPORT

	MODE	P=Passenger G=Goods	AREA OF OPERATION U=Urban R=Rural N=National I=International	(a) RANGE	(b) SPEED	ACCESSIBILITY D=Door-door T=Trunk	FUEL CONSUMPTION (c) P	FUEL CONSUMPTION (c) G	O=Overdeveloped countries L=Less developed countries	POWER SOURCE S=Simple (Muscle/wind) M=Mechanical (ICE etc) E=Electrical (Mains) B=Battery H=Hybrid
	1 Walking	P G	U R	1	1	D	1?	1?	(O)? L	S
	2 Animal Power	P G	U R	1 (2)	1 (2)	D	1?	1?	(O) L	S
	3 Pedal Power	P (G)	U R	1	1 (2)	D	1	1	(O)? L	S
ROAD	4 Private Car	P	U R N (I)	1-3 (4)	1,2 (3)	D	3-4	-	O (L)	M (H) (B)
	5 Taxi	P	U R	1 (2)	1 (2)	D	3-4	-	O (L)	M (H) (B)
	6 Bus	P	U R N (I)	1-2 (3,4)	1 (2,3)	T (D)	2-3	-	O L	M (B) (H) (E)
	7 Light Van	G	U R	1 (2)	1 (2)	D	-	4	O (L)	M B (H)
	8 H.G.V.*	G	(U) R N I	(1,4) 2-3	(1) 2	(D) T	-	3-4	O (L)	M (H)
RAIL	9 Rapid Transit	P	U	1 (2)	(1) 2	T	2-3	-	O (L)	M E (B) (H)
	10 Rail	P G	(U) R N I	(1,4) 2,3	(2,4) 3	T	2-3	2(3)	O (L)?	M E (B)
WATER	11 Inland	(P) G	N	(2) 3	1	T	(4)	2	O L	M S (B)
	12 Short Sea	P G	I	2 (3)	1 (2)	T	4	2	O (L)?	M (S)
	13 Ocean	(P) G	I	(3) 4	1 (2)	T	(4)	2	O (L)?	M (S)
AIR	14 Air	P G	N I	(2) 3,4	4 (5)	T	4	4	O (L)	M (M/S)
	15 Pipeline	G	U R N I	1-4	1	(D) T	-	1 (2)	O L	M E

(a) Range <10 mls 1
                <100 mls 2
                <1000 mls 3
                >1000 mls 4

(b) Speed <15 mph 1
                <50 mph 2
                <100 mph 3
                >1000 mph 4

(c) Fuel consumption
      Subjective scale
      1-4   1=Good
              5=Bad

* Heavy Goods Vehicle

40% in actual road conditions the average is much lower and in city driving can be as low as 15%; the corresponding figure for a petrol engine is 9%.

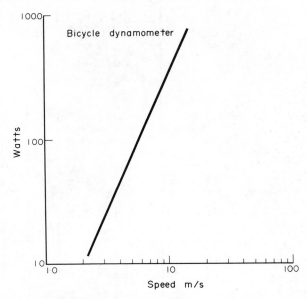

The motor car by contrast is of inhuman scale as regards size, weight, power (from 100 to 1000 times the power of the driver) or speed; as such the tendency for ordinary people to misbehave when driving can perhaps be understood.

Considering the energy efficiency of the motor vehicle, **the basic** difficulty is the excessive range of power demand needed, ranging from zero when stationary to a value several times that needed for average speed on a level road during acceleration, hill-climbing and maximum speed conditions. In fact such conditions are infrequent (Fig. 4), so vehicles are usually grossly over-engined for normal use - in fact engines would quickly fail if run at maximum output continuously. Unfortunately it is the inherent characteristic of a petrol engine that at low loads it is very inefficient, due to the method of power control by throttling. The form of output, i.e. the torque/speed curve is also fundamentally wrong for traction purposes (Fig. 5) and necessitates a multistage gearbox or some form of hydraulic or electric transmission, with accompanying power losses, to overcome this defect.

A consequence is that little saving of fuel results from the imposition of a general speed limit of 50 m.p.h. since most vehicles are designed to cruise at 70 m.p.h. and to have a maximum speed considerably greater; above 50 m.p.h. air resistance becomes significant and increases roughly as the cube of the speed so that power at 50 m.p.h. is so low compared to that at maximum speed that the engine is running inefficiently at low power.

The only way to save an appreciable amount of fuel would be to introduce a universal speed limit of 50 or at most 55 m.p.h., (80 or 90 k.p.h.),

as is already accepted in many parts of the U.S.A. If this could be made
the subject of international agreement then vehicle manufacturers would
adapt their designs to optimise car performance at the new lower speed,
using a much smaller engine and a lighter body; the result would be a significant fuel saving, of the order of 25% with a smaller petrol engine but
with a smaller diesel engine a saving of perhaps 50% should be possible.

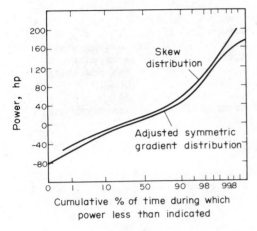

Figure 4

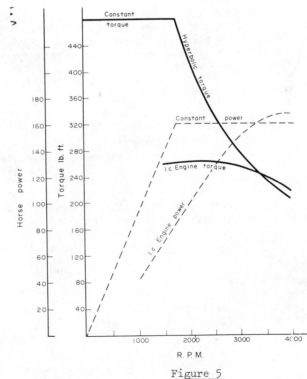

Figure 5

It is worth noting that 50 m.p.h. is in fact beyond the safe speed
for the majority of the world's roads - urban or rural - and is the designed
or legislated limit for most commercial vehicles. Motorways and other new
roads with limited gradients would not be a wasted asset, since they would
enable vehicles of limited power/weight ratio to maintain a good average
speed.

Much attention has been given in recent years, particularly in the
U.S.A., to alternatives to the gasoline engine, mainly in order to combat
the problem of 'smog', especially in Los Angeles, though it is a problem
in most large cities where sunlight is plentiful. Unfortunately most
possible solutions to the smog problem turn out to be more expensive in
energy consumption, apart from other snags, so attention has now shifted
towards energy conservation first, pollution reduction second.

It may be worth while to comment briefly on the alternatives considered:

Wankel Engine

The chief advantages are compactness and low weight for a given output; pollution is inherently worse but can be more easily treated; fuel consumption is about 10% worse, so the further outlook is unpromising.

Stratified Charge Engines

The idea is attractive as an alternative to inefficient throttling
but the practical realisation is different. Although the Honda CVCC system shows low pollution figures its consumption in city driving is worse
than the conventional engine. The Texaco system seems more promising,
especially as it can burn a wide range of fuels, which may be necessary in
the future if there is a large swing towards diesel engines, so upsetting
the refinery balance.

Petrol Injection

This can effect a marginal improvement in economy and low pollution
but at the expense of increased cost and complexity.

Gas Turbine

The performance of small gas turbines tends to be poor due, fundamentally, to the low work ratio. The work ratio is the ratio of nett work to
positive work in the cycle; for a gas turbine this can be as low as 0.25,
so that the effect of any loss of component efficiency due to off- design
operation or small size is magnified in the loss of overall efficiency.
Fig. 6 shows two approaches to good efficiency; either a low pressure ratio
and high performance regeneration or a very high pressure ratio and no
regeneration. The former awaits the evolution of a reliable regenerator;
the latter has a very narrow range of efficient operation, which is not
attractive for vehicle operation,

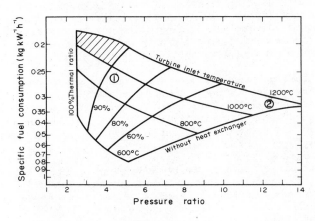

Stirling Engine

Despite the efforts of the Philips Co. of Eindhoven and others over nearly 40 years this type of engine is still some way from commercial acceptance. The reason is the unsound nature of the basic thermodynamics of the cycle, which attempts isothermal heat reception, as in the Carnot cycle. This is a misleading ideal in that it implies the existence of an isothermal heat source i.e. an infinite heat source whose temperature does not change as heat is withdrawn. Unfortunately most sources are finite, so the temperature varies greatly, and it is significant that all other practical heat engine cycles - Rankine, Otto and Joule have variable temperature heat reception processes, which give a better match to the available heat source. The Stirling engine needs a primary heat exchanger able to withstand high pressures at very high temperatures and to be very compact, leading to a difficult fabrication in exotic materials; it also needs a large external regenerator to cool the flue gases and a large-capacity internal regenerator to transfer heat between the heating and cooling processes. Finally, the cooler must be able to withstand high pressures and to be compact. It will be seen that the attempt to achieve isothermal heat reception results in a nightmare of sophisticated heat exchangers such that although laboratory prototypes show promising performance their translation into commercial production is difficult. Load control is also a problem.

Vapour Cycles

In this category we have to consider both steam and organic compounds as working fluids as well as reciprocating, rotary and rotodynamic expanders. Examples of most of these have been developed in the U.S.A. and the most practical appears to be the reciprocating steam engine, working from a high boiler pressure of about 70 bar and condensing at a pressure of about 1.4 bar i.e. at a temperature of about 109°C. This high condensing temperature is necessary in order to reduce the air-cooled condenser to a practical size, remembering that it has to handle about five times the heat of a normal car radiator for the same engine power. It is not therefore surprising that the thermal efficiency is modest, but because of its favourable torque/speed characteristics (Fig. 5) in city driving conditions the performance is better than a gasoline engine for fuel consumption and acceleration. Hence for a purely city vehicle e.g. a city bus, a steam vehicle mught be designed to be effective and efficient.

Steam turbines tend to be inefficient in small powers, while the use of working fluids having a heavy molecular weight, though giving higher turbine efficiency, leads to a lower overall efficiency due partly to a lower work ratio.

Battery Vehicles

Despite all efforts to evolve a battery giving a better power/weight ratio than the lead/acid accumulator no superior battery has yet appeared. Hence battery vehicles are confined to modest performance as regards range, speed or acceleration - unless and until a markedly better battery appears. These battery vehicles are also expensive in first cost and in running cost, since the batteries have a finite life. Also the overall efficiency is quite modest, typically about 18% only, due to the low energy efficiency of the electricity supply system - perhaps only 25% at the users terminals - and to further losses in charging and discharging the battery. Additionally the capital cost of central electricity generation transmission and distribution is excessive compared with the cost of a conventional i.c. engine. Performance is modest because up to a third of the all-up weight is that of the batteries. There is an assured but limited field of application for battery vehicles unless a much better cell is proved in appliction; the likeliest at present appears to be the sodium/sulphur cell, but its life is limited at the moment.

Hybrid Drive

This term implies the use of a small prime mover running continuously and efficiently at the average power level and an energy storage system, usually batteries, to store the energy needed for acceleration and hill climbing. Though attractive in concept it appears to be difficult to make a practical success of this mode, partly because of the need for several forms of energy transformation - chemical/thermal/mechanical/electrical/mechanical.

Fuel Cells

These remain elusive as a practical means of vehicle drive: although in theory they can be much more efficient than a heat engine in practice they fail to achieve anything like as good a performance.

Diesel Engines

We are driven to consider the use of liquid fuel as the most practical means of providing portable mechanical power; the most efficient means of conversion is the compression - ignition or 'diesel' engine, which owes its success partly to its high compression ratio (really expansion ratio, since a high compression ratio actually lowers the work ratio) and partly to the fact that, particularly at low loads, the working fluid has more nearly the properties of air than the products of combustion of the near-stoichiometric

air/fuel mixture used in a spark-ignition engine. The resulting higher
value of γ, the ratio of specific heats of the gas, gives a higher cycle
efficiency. The net result is not only a lower fuel consumption at full
load but a much lower consumption at part load, the normal road condition;
consumption figures as high as 40 miles/gallon are common in 15 cwt diesel
vans and in taxis - roughly twice as good as for petrol engines.

The price paid includes a higher initial cost, due to the cost of
the high-pressure fuel injection system, and to stronger and higher precision engine components; the incidental gain is an appreciably longer engine
life. The noise of a diesel engine is intrusive, mainly due to the 'diesel
knock', an inherent defect which is almost impossible to eliminate. French
and German car manufacturers have produced acceptable designs suitable for
taxis and private cars, but U.K. manufacturers appear to have neglected
this market.

An interesting proposal has recently been put forward by Radwan and
Tee (Ref. 4) for a small diesel engine of 20 H.P., naturally aspirated,
which is economical for use in a normal car in city use but which can be
highly supercharged by means of a turbocharger to give up to 60 H.P. for
high-power, high-speed running on the open road.

If the speed limit of 50 m.p.h. is combined with the use of small
lightweight diesel engine an exceedingly economical vehicle could result.
The author has recently put forward proposals for installing an available
15 H.P., 2-cylinder, air-cooled diesel in a lightweight 4-seater body which
normally uses a 4-cylinder water-cooled petrol engine of 35 H.P. to achieve
70 m.p.h. It is anticipated that at a steady 50 m.p.h. a fuel consumption
of 100 miles/gallon would be achieved, with an average of 70 m.p.g. or more.
Acceleration and hill-climbing would be modest but flexible top-gear
motoring from 20 to 50 m.p.h. should be possible owing to the favourable
torque/speed characteristic compared to a petrol engine.

This vehicle could be the prototype of a new generation of economy
cars (ecocars?) which would achieve a worthwhile reduction not only in
direct fuel consumption but in tyre wear, accidents and wear-and-tear of
the vehicle and driver. The advantages of light weight construction would
be apparent and these could lead to a greater use of aluminium alloys as
the main material giving a much greater body life to match the engine life.
Although aluminum is regarded as a high-energy material its use would save
energy throughout the life of the vehicle and most of the energy of formation would be recovered when eventually the material is recycled.

By measures such as these we might hope to make the motor car more
defensible as a major part of the transport system and to stabilise the
motor industry at a lower level of production.

Alternatives to the monopoly of the private car which are put forward
by substantial lobbies are the two major forms of public transport, bus/
coach and rail. Neither of these is at present a satisfactory substitute,
for a variety of reasons. First, neither is door-to-door transport, so
failing on grounds of convenience. Secondly, the perceived costs are normally higher, particularly for rail journeys, though the true costs of running a car are usually forgotten. Thirdly it is increasingly difficult to
provide a frequent and reliable service, partly due to high labour costs in

an inflationary situation but also due to lack of motivation for staff to
exert themselves to the utmost to make the service as attractive as it is
capable of being. Unfortunately this is a malaise of present society which
afflicts organisation to a greater degree the larger their size.

One possible development which may succeed in recreating a viable bus
service, particularly in rural areas, is the revival of the 'village carrier', in which each village would have one or more passenger-cum-goods
vehicles and would organise its own services to work, school, shops,
outings, etc. to suit its own community. This is, of course, an example
of 'self-help' or community action which represents one of the most hopeful directions of future change in social life; it is applicable to less-
developed countries as well as developed ones.

Turning to the movement of goods by road, there are no obvious means
of combating the ascendancy of the H.G.V. or 'juggernaut', especially with
the growth of roll-on/roll-off cross-channel ferries and the motorway system. Possible alternatives include containers moved by rail, especially
via a Channel Tunnel and the revival of water transport, particularly by
means such as L.A.S.H., B.A.C.A.T. etc., whereby smaller barges are loaded
aboard mother ships for sea crossings. One of the more ambitious schemes,
put forward by Mr. J. F. Pownall is for a Grand Contour Canal which would
cross all the major watersheds in England at the 310 foot contour level
and connect Birmingham with London, Southampton, Bristol, Manchester, Leeds
and Newcastle without locks except near the connexions to the sea. This
would enable barges of 1500 tons to load in Birmingham and sail without
transhipment all over the European waterway system.

Sea transport is likely to maintain its importance in world trade,
probably with a permanent recession in tanker operation and a growth of
container ships and bulk carriers. Passenger liners have almost faded from
the seas due to the success of air transport; these liners were extravagant both in energy terms and in labour.

The large slow-speed diesel engine is likely to remain the main prime
mover at sea, having largely replaced the steam turbine. Gas turbines have
proved acceptable for naval use but are not at present sufficiently economical for merchant service use. A combined gas turbine/steam turbine cycle
could give efficiencies almost equal to that of a big diesel engine but
there remains the problem of providing a clean fuel for burning inside the
gas turbine combustion chamber. One possibility is the use of liquefied
natural gas, especially for tanker use; at present large quantities of gas
are burnt at the oilfields but the techniques of liquefaction, transport
and storage of L.N.G. are now so well advanced that it should be possible
to set up a supply system to bunker tankers which are provided with insulated tanks in their bows, a safe space otherwise little used. Once established as a worldwide bunker fuel its use might spread to, for example,
refrigerated cargo ships where the controlled evaporation could provide a
bonus in refrigeration.

Advanced Transport Systems

These include various forms of monorail, dual-mode vehicles (road/
rail), minitrams, hovertrains, magnetic levitation etc. Many have
reached model or prototype stage but few are being further developed,

mainly because of the probable cost of developing them into a viable full-scale system. A recent six-volume review of some four hundred systems by Ecoplan International states that "at the conclusion of the latest of these investigations we found ourselves with something of a sinking feeling in contemplating what our cities would appear to need and what most 'hard' technology projects seem to be aimed at offering". They decided therefore to begin a study of bicycle use.

Air travel will, hopefully, stabilise at an economical level of operation based on medium-sized aircraft operated at near-capacity and at economical cruising speeds to give minimum energy and operating costs. There seems little justification for supersonic passenger transport; it is the ultimate in technology of inhuman scale, energy consumption and adverse environmental effects.

A better field in which to exercise innovation, ingenuity and individual effort is that at the opposite end of the transport spectrum, categories 1 and 3 in Table I. Apart from walking as a good means of personal mobility there is a scope for development of goods transport even without use of the wheel, e.g. by back packs and carrying poles. The wheelbarrow, although in use in China for perhaps 2000 years and in Europe for several hundred years, is only now receiving recognition as a goods vehicle of great potential; the World Bank is conducting comparative tests in India of different designs; the Rev. Geoffrey Howard recently crossed the Sahara with a neo-Chinese wheelbarrow having a central wheel of 4 feet diameter, while a new design employing a spherical wheel and plastic body has recently appeared in the U.K.

Moving to two wheels, apart from the ubiquitous bicycle, now rightly enjoying a boom in the U.S.A. as well as in the Third World, the potential of pedal power as a goods carrier is relatively unexploited. In addition to the carrier bicycle and carrier tricycle many other forms are possible and the author recently designed and built a basic tricycle chassis, the OXTRIKE (Fig. 7), having a 3-speed gearbox, powerful foot brake and a simple form of differential and designed to carry a payload of 150 k.g. (3 cwt), equivalent to two passengers or a variety of goods. Although primarily intended for the Third World it is clear that many people in the U.K. would like to use it, for a variety of purposes.

A further proposal is for an off-the-road vehicle in which each wheel is directly pedalled, so giving a simple load carrier with good traction.

There is no suggestion that pedal power is the solution to all our transport problems, though as outlined by the author in a recent paper (Ref. 5) much can be done by planning to facilitate the use of bicycles, etc. Where distances are too great several systems of dual-mode may be envisaged - bicycle/rail, bicycle/bus and bicycle/car.

Figure 7   "Oxtrike" chassis. Note 20" carrier cycle front wheel
with hub brake, footbrake and handbrake for rear wheels,
3-speed gearbox driving twin rear chains, body attach-
ment points.

Returning to our starting point of technology as the instrument of change and the role of transport as a major determinant in the form of society, we are faced not just with an energy 'crisis' - present price increase, future shortage - but at the same time a number of other problems - the population explosion, inflation, unemployment, the failure of political, economic and educational theories, disease, violence, crime and a whole spectrum of social problems, many of which stem directly from attempts to live in large cities. There is evidence of rethinking about the answers to all these problems but little sign of agreement on the ability to implement any proposed solution.

In these circumstances of failure of systems which have evolved during the 5000 years or so of civilization consequent on the urban revolution, it is worth while to recollect the slow pace of evolution and to realise that for all practical purposes we are the same creatures now as 5000 years ago - physically, mentally and socially and that we have not adapted to living in large urban masses, nor in 250 years to industrial life nor in 50 years to driving in cars.

Our main hope for the future must be to evolve a way of life better adapted to the human animal i.e. with our institutions, buildings and machines of human scale, in which each individual can play a significant part in the life of the local community. This paper has indicated some ways in which transport of more human scale can play its essential part as an instrument of change.

References

1. S. S. Wilson "Bicycle technology" Scientific American, March 1973.
2. E. F. Schumacher "Small is Beautiful", Blond & Briggs, 1973.
3. S. S. Wilson and N. D. C. Tee "Energy Economics and Transport" Proc. Urban Transport Studies Group, University of Warwick, March 1974.
4. M. S. Radwan and N. D. C. Tee "Highly turbocharged small automotive diesel engine" J. Automotive Eng., March 1975.
5. S. S. Wilson "Bicycle planning" P.T.R.C. Conference Proc., July 1975.

Further Reading

1. V. Gordon Childe "Man Makes Himself", 1936.
2. Ivan Illich "Energy and Equity", Calder and Boyars, 1974.
3. Howard T. Odum "Environment, Power and Society", Wiley-Interscience, 1971.
4. M. W. Thring and R. J. Crookes (ed.) "Energy and Humanity" Peter Peregrinus Ltd., 1974.
5. M. S. Janota (ed.) "Vehicle Engines" Peter Peregrinus Ltd., 1974.
6. R. U. Ayres and R. P. McKenna "Alternative to the Internal Combustion Engine" Johns Hopkins University Press, 1972.

DISCUSSION

Question: Can more be done to encourage the use of bicycles?
Answer: Segregation of traffic helps. In Stevenage, pedestrians, cyclists
             and mopeds are separated from other traffic, and this has been
             successful for over 20 years.

Question: Canal systems are operated very successfully on the Continent.
             Would a revival of canal transport be justified in this country?
Answer: There are some moves in this direction. The cost of the contour
             canal system might be £400 million (1972) but barges would then
             also be able to use the European system, which is widely used.

Question: How can motor cars be designed for the more efficient use of
             energy?
Answer: A 50% saving in fuel could be achieved by using a smaller engine
             and lighter construction materials, e.g. aluminium could be used
             and recycled later.

Question: How does road travel in coaches compare in cost with rail travel
             over long distances?
Answer: If trains were full they would be more economical. In practice
             buses or coaches are slightly better. Both are ten times cheaper
             than private cars.

Question: Could municipal bicycles be provided for free use as in Amsterdam?
Answer: This has been tried elsewhere, but it is doubtful whether it was
             successful.

Question: What is the effect of wheel size on the efficiency of tricycles?
Answer: Large wheels would be more efficient, but 20" diameter wheels give
             a lower centre of gravity.

Question: What are the efficiences of a diesel train as compared with an
             electric train?
Answer: Diesel trains tend to have higher efficiency overall.

Question: How do larger cars compare with smaller cars from the point of
             view of safety?
Answer: No figures are available, but accident rates have been reduced by
             the introduction of the 50 mph speed limit.

Question: Could people be persuaded to do without the motor car?
Answer: Yes, there is a strong possibility that they could be persuaded.
             However, this must be done by offering alternatives, not by bring-
             ing in legislation.

THE INFLUENCE OF ENERGY USE ON FUTURE INDUSTRIAL PROCESSES

M. E. Hadlow and B. Buss

Electrical Research Association Ltd.

1. INTRODUCTION

In the formal papers and discussion during this Summer School a great deal has been said about our Global and National Energy Resources both presently and potentially available. How we acquire, transport and supply these to the consumers who, beyond any doubt, are responsible for depleting resources faster than technology can release them (at least within the presently applicable economic constraints) have also been covered. Leaving aside the large unit of energy consumption that goes directly into our domestic comfort, and the massive contribution to our energy problems that are, more or less, looked after by natural events (e.g. in agriculture), then the largest proportion of our energy in consumed in the indirect satisfaction of our needs through the mechanism of industry. It is to the industrial processes which demand this large proportion of our extracted energy resources that we have been asked to address this paper.

The result of the events which, although not the direct cause, were certainly the alarm bell which focused world attention, is that our energy supplies to industry which until recently have been taken for granted, are now less secure and more expensive. If, therefore, there is a question that it was expected this paper would answer, it must be:

> 'WHAT DEVELOPMENTS IN INDUSTRIAL PROCESSES WILL BE INSPIRED BY THE NEW SITUATION THAT HAS LED TO THIS SCHOOL BEING CONVENED'

It is always satisfying to provide specific answers but to do so in this instance would require either a very narrow thesis of little general relevance to the majority of UK Industry, or the wisdom of Solomon in choosing solutions to apply to problems that are, as we shall see, not yet adequately formulated. As a consequence the subject matter of this paper will, necessarily, take a less direct, less scientifically quantified route and will only indicate the general direction of the answer to the apparently simple but multi-faceted question. To attempt more would be to rationalise the problem beyond the point where the answers would stand up to the practical criteria by which our industrially based society must operate.

The format we will follow is illustrated in Fig. 1 with the starting point objective TO IMPROVE THE EFFECTIVE UTILISATION OF ENERGY IN INDUSTRY. Our terminology has been chosen carefully and is to be preferred to the brevity of ENERGY CONSERVATION. According to the Seventh Edition of Webster's (and we make some apology for this choice of reference in this location); 'CONSERVATION is the careful preservation and protection......, the planned management of natural resources to prevent exploitation, destruction or neglect'. By comparison and from the same reference; 'UTILISE: to make use of : convert to use'.

We submit that the laws of physics, as presently understood, make energy conservation within the boundary of comprehensible systems an unavoidable fact.

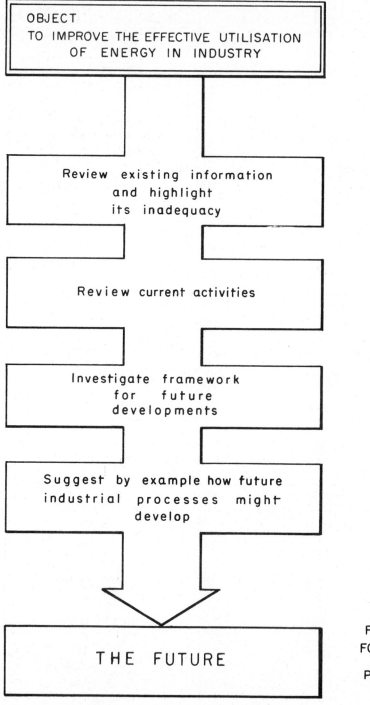

FIG. 1
FORMAT
OF
PAPER

Our human skill and the skill to which this school is addressed is to improve
the EFFECTIVENESS with which we UTILISE our finite resources (both ENERGY and
MATERIALS) to CONVERT them into desired (if not always useful) PRODUCTS.

Returning to Fig.1, we shall discuss briefly the base of information available
as a starting point for our studies of the situation in industry, as it presently exists, both from nationally collected information and from specific
product orientated studies.

Since the crisis of late 1973 a considerable number of initiatives have been
taken to review, survey and, in a few cases, take action towards improving
the situation.   The overall picture is complex but we shall try in the second
section of our paper to indicate the extent of these and the contribution
they make to the total national and industrial problem.

Until the results of at least some of the studies at present getting under way
become available, it is difficult to give a coherent picture of the eventual
impact of energy constraints on industrial processes. We do, however, in the
next section, provide a framework which we find helpful in orientating the
impact of the many possible contributions.

Finally, we shall illustrate our perspective framework with a few examples
drawn from experience within industry over the last eighteen months.

2.   THE INDUSTRIAL USE OF ENERGY

We must, of course, take as our starting point the now time honoured statistics for the UK industrial energy consumption[1] in 1973, classified under
seventeen sectors of the Standard Industrial Classification[2] (SIC) and shown
in the histogram of Fig.2 for the nine largest sectors.

From Fig.2, well known facts such as the energy (coal, coke and oil) intensity
of the steel industry are obvious and from an historic review certain marked
examples of fuel substitution can be noted.   These changes have in themselves
been inspired by the benefits obtainable through more effective energy utilisation.   The recent NEDO report[3] makes particular reference to fuel
substitution:

> 'For example cement has maintained a high proportion of coal (70% in
> 1960 and 66% in 1972), but china, earthenware and glass sector
> almost abandoned the use of coal in the same period (43% in 1962
> and less than 3% in 1972).   The chemical industry has made a simi-
> larly decisive shift from coal to natural gas and petroleum.'

A little further insight can be obtained by looking at the structure of each
sector, although unfortunately, since no comparable data has been assembled
since, we must go back to the 1968 Census of Production[4] to obtain this.

Figure 3 shows the average purchases of energy per company in each sector.
The significance of the relatively small number of very large companies in the
large energy consuming sectors such as chemicals, metal manufacture and
vehicles, is clearly seen.   The examination of the purchases of energy per
employee for each sector in Fig.4, is also a crude indication of how energy
intensive each sector is in relation to its work force. As is to be expected,
the very large energy users such as metals manufacture and chemicals also have
the largest purchases per employee but the high ranking of bricks, food and

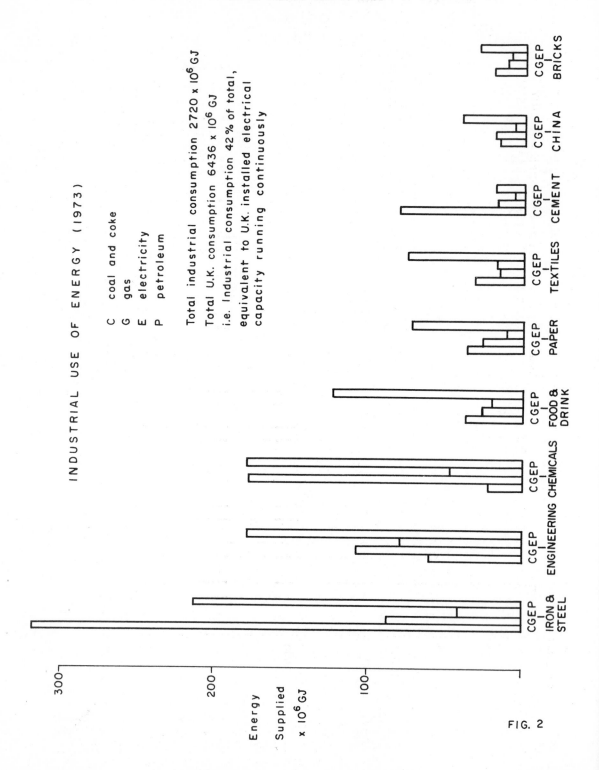

FIG. 2

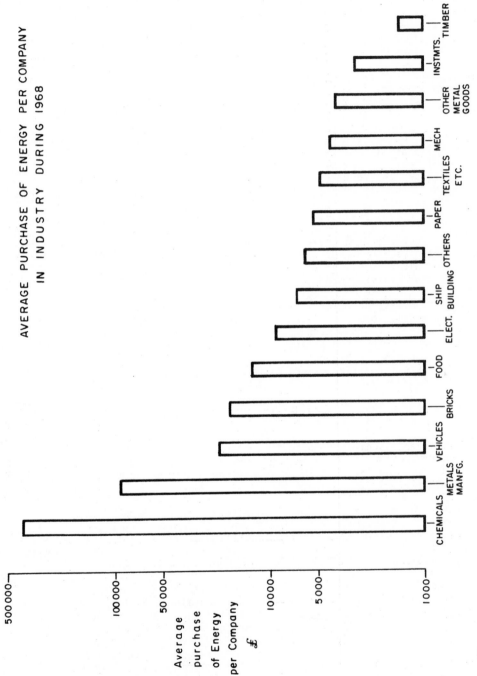

FIG. 3

paper in Fig.4 suggest the importance of automation processes in these sectors.   The extensive use of automation in the vehicles sector might also beg
the question why is it not ranked higher in Fig.4?

From here, centrally collected and published information enables very little
further progress to be made in analysing how effectively industry uses the
energy it consumes.

One way round this problem is to use a detailed knowledge of individual process steps to establish the energy content of particular products.   The art
is known as Energy Analysis and has, through the efforts of Dr Chapman[5] and
others[6,7 & 8] made significant strides forward over the last few years.

Other papers in this school address the subject in detail.   The technique is
essentially that of working from the bottom up and, although there is no doubt
of the contributions Energy Analysis has to make, some considerable effort is
required before the results contribute generally to the problem we are
addressing.   Further comment on this area is made in Section 4 but at this
point we must note that in almost every instance, existing information resulting from the Energy Analysis approach tends to:

(i)   cover products where the manufacturing process can be simply
      analysed and understood, e.g. the production of steel, alu-
      minium, chemicals etc., in the main, raw materials rather
      than finished products.

(ii) conflict, because different assumptions are made as to
      exactly what is included in the energy requirements per unit
      output of production.   Smith[6] exposes the difficulties where
      he estimates the energy content in the buildings and equipment
      required for production.   Chapman[5] has proposed how energy
      costs for different sectors can be analysed but as yet the
      input data has still to be agreed and compiled.

An example of the range of energy requirements per tonne of production for a
number of prominent materials is given in Fig.5.

In a further approach, input-output tables[10] have been compiled from the 1968
Census of Production and have enabled some further manipulation of basic industrial information[11].   Again, however, the breakdown cannot proceed beyond
the minimum list headings in the SIC.

Also, starting from the published information Pick and Becker have, in a
recent paper[12], highlighted the significance of indirect supply of energy
(i.e. in materials) to manufacturing operations.   The work focuses attention
on important aspects but is a little removed from the direct concern of an
industrial company.

Overall, we find that nationally published statistics do not facilitate the
detail of information required to rationally develop and plan either on the
national or individual company scale. Detailed Energy Analysis type studies
are still, in all but a few instances, a stride away from providing industrially relevant results. Within industry, no more than a handful of companies have a clear picture of how well they use their energy purchases.
Many of the present activities are directed towards a resolution of this gap
in detailed information.

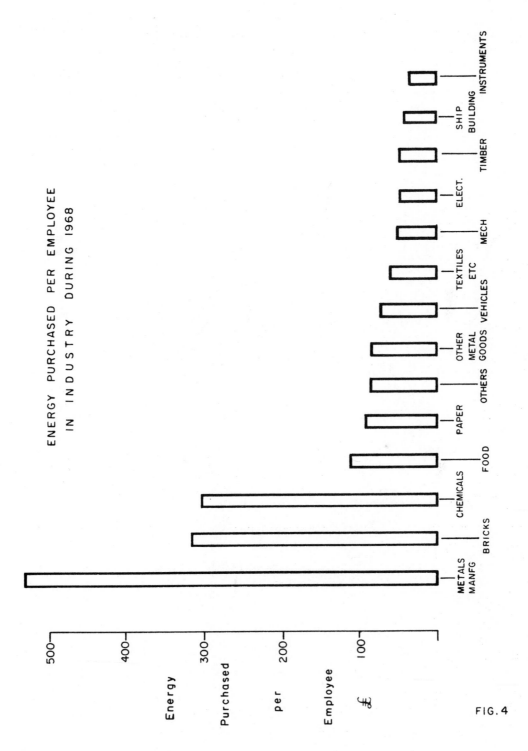

FIG. 4

Material		GJ/tonne		
		Ref. 3	Ref. 8	Ref. 9
Aluminium:	primary	315	230	320
	from scrap			10 – 15
Plastics:			90	80
	+ feedstock			50
	total			130
Steel			50	40 – 200
Paper			18	–
Glass			12.6	25
Cement			4.7	8
Timber			3.4	–
Copper:	primary		–	120
	from scrap			10 – 15

Fig. 5:   Estimates of energy required to
           produce 1 tonne of various materials

3. PRESENT ACTIVITIES

Since December 1973, a great diversity of activity has taken place and in order to gain an appreciation of this, it is useful to briefly review the major landmarks.

In late 1973, the compounded effects of the Arab-Israel war and the indus-trial dispute in coal mining led quickly to a rationing of energy in the form of a three day week.   This rationing, which was brought about by shortages rather than price, induced companies to improvise to make best use of both labour and energy.   Much was quickly achieved and whilst detailed records are scant[13], the claims made of 90% production with 60% energy are indicative of achievements gained.

Many energy saving committees were formed within individual companies and the ideas adopted proved quickly what was possible.   As the crisis passed and supplies were re-established, rapid price increases in fuels took place, but the full impact on financial performance was not apparent until well into 1974.   Industrially, relatively few companies continued corporate programmes that required a more effective use of energy.   The few that did, have con-tinued to demonstate that substantial energy savings are possible.

There followed a phase where both nationally and industrially it was necessary to review and consider what action should be taken to:

(i)   reduce the impact should the crisis re-occur and
(ii)   consider what action to take in the longer term.

The Central Policy Review Staff were asked to quickly respond and what has come to be known as the Rothschild report on Energy Conservation[14] was pub-lished in mid 1974 setting out certain national objectives.   Many other offi-cial organisations also undertook studies during 1974, several of which were directed specifically towards industry.   These included such bodies as NEDO, CBI and professional institutions and learned societies.

All these bodies, and many others, undertook studies and reviews as to the ways in which the particular sector of industry, the professions or science, which they represented, could contribute to this 'new national energy problem'.

Whilst we do not wish in any way to criticise the intentions and results of these studies, it has to be said that from an industrial point of view, the published reports were too broad and generalised to be of a great deal of direct use to industry.

At the same time, and we are really here talking about the first half of 1974, we saw a proliferation of publicity for the impact that individual areas of technology could make on the 'Technology of Efficient Energy Utilisation'. The latter terminology is taken from the title of a report emanating from a NATO Science Committee Conference[15] which contains many high technology ideas. Many technology lobby groups were active at this time making maximum use of the energy vehicle to promote their interests.

Much of this activity finished up at the doors of the newly formed Department of Energy and its Energy Technology Support Unit (ETSU), one of the conveners of this School.

And so by this point in our chronological account we have arrived at much the same position as we have already described in the preceding section.   Simply, it was clear that we knew too little of the detail of how industry uses energy to be able to assess the potential for savings and implement the simple or technological actions that would improve the situation.

There exists in fact a fairly definitive gap in knowledge between the national and industrial level.   This is illustrated in Fig.6 along with some of the more recent developments which we shall go on to discuss.

One of the first initiatives to acquire more detailed information was taken by the Confederation of British Industry (CBI) who issued some 7,000 copies of a questionnaire to member firms in July 1974.   The results of the analysis of the 1,269 valid responses have recently been published[16] and provide some hitherto unavailable detailed information on energy usage within industry although not on the effectiveness of utilisation.   And, as the report points out, it was realised very early in this exercise that the need was for much more detailed information from each industry and the processes within that industry.   We shall return to the activities presently addressing this problem in a moment.

One activity that did come clearly to the fore in the second half of 1974 was the understanding that a great deal was open to be achieved through simple educational campaigns and perhaps modest consultancy assistance directed towards good housekeeping measures.

There was and still is a great deal of promotion by consultants in this field, publication of papers and guide books all overlayed a little later by the 'SAVE IT' campaign launched by the Department of Energy.   One publication of particular note for industry was the Energy Conservation Program Guide published by the National Bureau of Standards[17] in the USA.   This very practical guide is recommended to anyone involved in implementing the first steps of an industrial programme.

A little later, Department of Energy introduced its £3M per annum energy savings loan scheme[18] which seems to have been directed towards stabilising energy savings in the medium term.   With hindsight it must be said that this would seem to have been somewhat premature and not sufficiently specific in its application to bring the benefits that had been hoped for.

Following a period of gestation, we have recently seen more considered and directed plans and actions emerging:

-   within industry from a number of the larger groups of companies
    which now have corporate plans geared towards more effective energy
    utilisation.   Such companies as Petters, Courtaulds, Tate & Lyle,
    BLMC and, as has been reported to this shool, Reeds, have all
    received recent publicity for that energy work.

-   at the academic level, SRC grants have been awarded to both the
    Open University and Newcastle University to study energy require-
    ments; also the publication of the results of energy workshops on
    methodology[19].

-   Internationally we have seen a growing liaison, if not coordination,
    of activities through EEC, NATO, and organisations such as IFIAS.

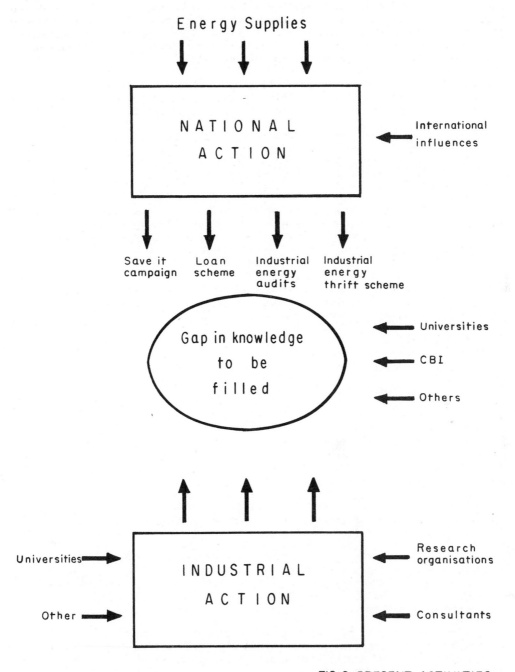

FIG.6 PRESENT ACTIVITIES

By late 1974, both the Departments of Energy and Industry were beginning to formulate plans that would start to fill the national gaps in knowledge on industrial energy use.   First from the Rothschild report and subsequent reviews, many specific areas for potential R & D funding, were identified and reviews, mainly by Government Research Establishments, have been initiated to assess the state of the art and possible national contribution.   Some indication of the range of topics is given in Fig.7.

More specifically looking towards industry, these two Departments are currently embarking on a series of energy audits, initially within the more energy intensive sectors of industry.   Through these, detailed profiles of energy use within different sectors of industry will be established.   Further to this activity an Industrial Energy Thrift Scheme is likely to be initiated to assist companies in exposing and achieving opportunities for immediate energy savings.

Overall, the work by the two Departments will provide a deeper understanding of how industry utilises its energy to expose if processes can be modified or substituted to use energy more effectively.   As the results from this work become available, we shall see a clearer picture emerging as to the actions that can be taken at both the national and industrial level.

The various mechanisms through which the participation of industry in the adoption of energy effective measures is to be achieved, will have to be established.   In this context it is of paramount importance that we appreciate the way in which industry will examine and assess the importance of energy saving measures both in the short and long term.

4.   THE FUTURE IN PERSPECTIVE

In assessing any measures to improve energy utilisation, it is of major importance that the boundary within which the conclusions apply is clearly defined.   The lack of many workers to define this boundary is the reason for a lot of the confusion that exists when it comes down to the critical level of 'who should be taking some action'.

If we may take a gross simplification to illustrate this point with reference to Fig.8.

Primary energy supplies are dominated by global and national considerations. The responsibility for securing and distributing these supplies rests with central government which discharges its responsibility through the mechanism of the nationalised energy supply corporations, legislation, development grants, etc.   Industry is the customer but has little direct control and only in very particular circumstances is its acquisition of primary energy supplies independent of external influences.   Solar and wind energy are two such examples but of strictly limited direct industrial significance in the immediate future in the UK.

All primary energy has to be converted to different forms, refined, distributed and stored.   Responsibilities in this area lie partly in the national arena, where coal, gas and oil can be stored and distributed directly or converted and distributed as electricity, but also, very importantly, in the industrial sector.   After energy supplies cross the 'factory fence' they must be converted into the form required for manufacturing operations.   Here, within the limits of technology, the user controls the effectiveness and cost.

Alternative fuels

Dual fired equipment

District heating schemes

Energy storage devices

        Electrochemical

        Flywheels

Fluidised beds

Heat exchangers

Heat pumps and pipes

High temperature materials

Instrumentation for energy use

Insulation materials

Motor vehicle design

Nuclear process heat

Photo voltaic devices

Wave power

Wind power

Fig. 7:   Potential areas of energy R&D

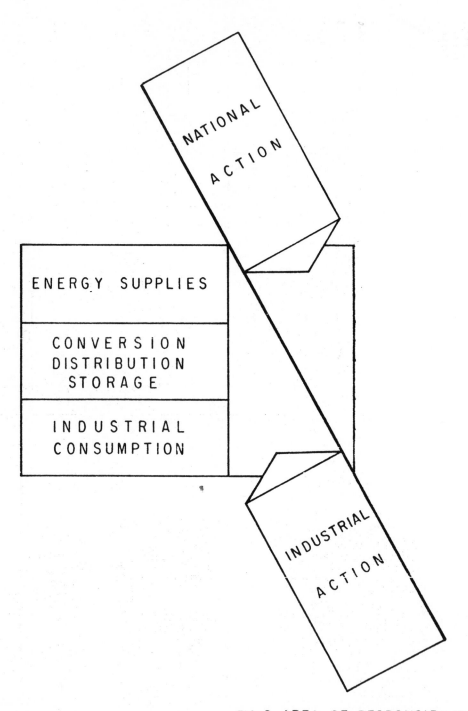

FIG. 8 AREA OF RESPONSIBILITIES

Equally, the end user controls the effectiveness with which energy is used in manufacturing operations and in providing the associated services.

Beyond protecting itself against interruption of a preferred supply of direct energy, industry cannot be expected to shift from one energy form to another or to utilise a presently wasted resource without economic justification.

The arguments apply equally well if the analysis is broadened to include the important indirect energy contribution represented by materials.   Although, if we take the industrial point of view, much of this indirect contribution will fall within the direct energy input to an analytical factory fence boundary at an earlier level in the production chain.

Let us now concentrate on the question of motivation within industry to achieve an improvement in energy utilisation.   In manufacturing industry a capital investment is employed to provide the facilities within which labour, energy and materials are employed to produce goods.   There are many criteria which this operation has to satisfy.   The most important single measure of performance is return on investment and any steps towards improved energy utilisation must be consistent with improving performance against this and/or other criteria against which company performance is measured.   Figure 9 illustrates the point.

In order to achieve the steps suggested in Fig.9, it is a prerequisite that the nature of the need for energy within the industry is fully understood. Whilst the manufacturing processes themselves are of focal importance, all energy consumption within the site must be taken into account.   As shown in Fig.10, energy demand from manufacture, transport and environment must all be considered, particularly if the full possibilities and benefits of integration are to be exposed.

Also illustrated in Fig.10 are the primary goals towards which effective energy management within industry must be directed.

If one accepts the realities of the criteria by which industry must operate within the structure of, at least, the Western World then one must also accept that there will be goals to be achieved in the national interest that must be motivated by a national policy which compensates for any deficiency in justification against purely industrial criteria.   The removal of national subsidies to our energy supply industries and the resultant large increases in the cost of all forms of energy is a positive step in the sense that it affects the economics of any energy consuming process.

Developing Fig.8 into Fig.11 is as close as our graphical skill enables us to get to illustrating the total range of present activities and the overlap between national and industrial areas of action and responsibilities.

Finally, to complete our perspective against which the future can be viewed, we must introduce one further generalised framework as shown in Fig.12 within which future actions are likely to be taken and results will accrue. Whilst there have been many attempts to rationalise the major features of such a framework, we submit that the primary features are:

-        The Timescale   -   over which the impact will be
                                 achieved.

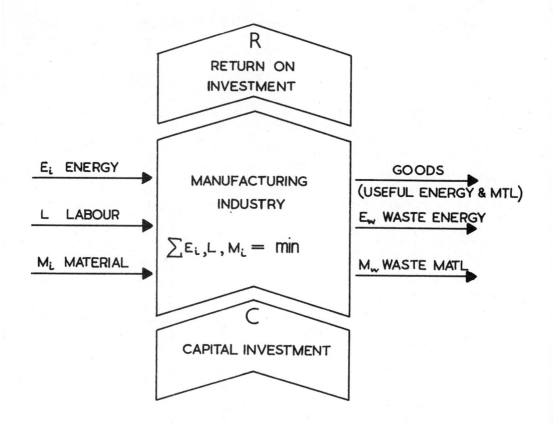

Recycle waste energy or material to reduce E_i , M_i

Substitute materials or processes to increase $E_i - E_w$ & $M_i - M_w$

Substitute energy forms to reduce $E_i - E_w$

ALL TO IMPROVE R

FIG.9 BENEFIT TO
COMPANY PERFORMANCE

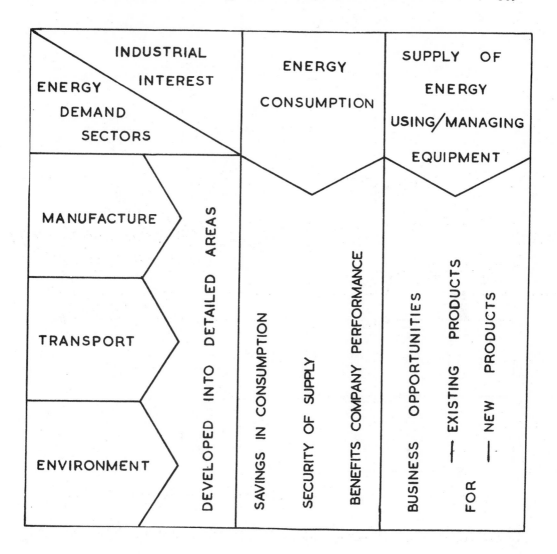

FIG.IO INDUSTRIAL INTEREST
-DEMAND MATRIX

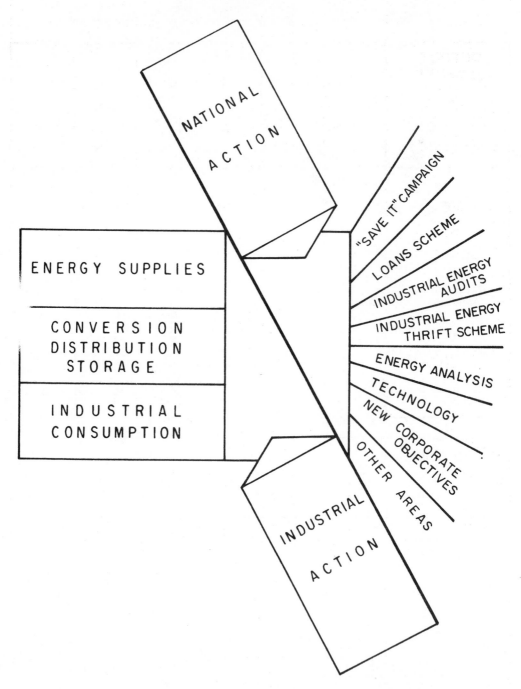

FIG. II AREA OF RESPONSIBILITIES

OUTPUT BENEFIT ╲ FEATURES	ENERGY THRIFT	ENERGY EFFICIENCY	ENERGY ECONOMY
TIME SCALE	TODAY	NEAR FUTURE	LONGER TERM
MECHANISM	EDUCATION	MODIFICATION TO EXISTING EQUIPMENT ╱ OPERATIONS	NEW EQUIPMENT ╱ INVESTMENT
STIMULATION	PROMOTION OF AWARENESS	CRITIQUE OF EQUIPMENT ╱ OPERATIONS	FUNDAMENTAL AND APPLIED STUDIES
	IDENTIFICATION ╱ QUANTIFICATION		OF BENEFITS

FIG. 12 PERSPECTIVE FRAMEWORK

	The Mechanism	-	through which the objectives will be achieved.
and -	The Stimulation	-	that will lead to the decision to proceed.

The action categories that will provide the output benefits are grouped as follows:

ENERGY THRIFT -

understanding the role of energy and maximising energy effectiveness by adopting good housekeeping measures. The timescale for implementation stretches ahead from today, the mechanism requires little more than education and the stimulation must be a continuous promotion of awareness of the opportunities.

ENERGY EFFICIENCY -

relates to programmes designed to achieve the optimum energy performance from existing investments. We look to a timescale of perhaps two to five years for implementation with only minor modifications to existing equipment or operational procedures. The stimulation comes from the benefits identified through an in depth critique of present practices.

ENERGY ECONOMY -

R & D results relate to situations where energy considerations dominate and/or justify new capital investment decisions. The timescale of immediate interest is up to ten years ahead when the results of more fundamental and applied studies can be expected to influence the design of new equipment etc.

From the foregoing, we suggest that new developments in industrial processes will occur as a direct result of fall-out and spin-off from the three action categories just detailed. To attempt to forecast what they may be at this stage could well be misleading, because they could not yet be fully supported by measured argument. That is not to say that very useful early indications are not already appearing, and we would like to conclude with an illustration of an example of what we mean in each of our action categories.

5.   CONCLUDING EXAMPLES

5.1 Energy Thrift

This category of measures is representative of the action rapidly or effectively applied by industry during the UK's compounded problems in early 1974. Individual savings are likely to be very modest, require no added technology, no capital expenditure and result from a motivation through education and self examination. In short, an attitude of 'SAVE IT' in staff at all levels.

Check lists of Energy Saving Opportunities are a useful stimulant in this area and we show in Fig.13 an abstraction of some of these. The SAVE IT possibilities behind many of these ideas are obvious but let us take just one example, that of leakage from compressed air systems, to illustrate the potential.

Area	Recommendation
Space heating	Avoid use of open doors and windows check and repair leaks
Lighting	Light only those areas where needed clean light fittings regularly
Process heating	Retain covers or doors on heated vessels in closed position whenever possible Check and regulate air-fuel mixtures
Drives	Reduce machine idling time Eliminate compressed air leaks
Transport and distribution	Use transport to maximum capacity Reduce unnecessary business travel
Other	Appoint staff responsible for energy conservation Monitor energy use and act where necessary

Fig.13:   List of typical 'thrift' recommendations

Almost every industrial plant has a compressed air installation; the energy
waste through leakage was, at least until recently, generally considered to
be insignificant. However, we can demonstrate by taking the example of an
actual factory site which we have surveyed.   An assessment of the leakage
from the 100 psig compressed air distribution system is given in Fig.14
together with an estimate of the value of electricity savings at the maximum
demand and units supplied tariff rates applicable to the site. The estimated
savings of around £5,000 per annum are certainly not insignificant and provide a good return on the relatively trivial amount of maintenance effort
required.

Just as an illustration, and not to be taken as a valid extrapolation, let us
hypothesise that similar savings could be achieved at about 30% of the industrial companies in the UK.   We find that if this were to be true in practice,
a saving of 1,000 MW would be achieved in the maximum demand for electrical
power and capitalised at £100/kW (a modest sum by present CEGB standards)
would be worth some £10M to our electricity supply industry. Taken one step
further, if all the electrical power were to be generated in oil fired power
stations, this simple measure could produce an £84M contribution to our oil
trade balance.

The above is of course a modest indulgence in arm waving, but does emphasise
the potential aggregate contribution that can be made by giving attention to
individually very modest possibilities for achieving Energy Thrift.

5.2 Energy Efficiency

Within this area we are talking of potential savings that are much more
specific to a particular plant or process than the previous example. Check
lists (Fig.15) can still be a useful stimulant to thinking but generally a
much more detailed critique is required in order to quantify the potential
savings and justify the effort (not necessarily involving capital expenditure)
to implement them.   The effort may be required in changing the method of
operation of production equipment or more generally in the re-scheduling of
the use of equipment to reduce peak demand.

We illustrate the second with the hypothetical example of a plant operating
a group of say ten electric resistance heating furnaces at 40 kW rating
which go through a heat up (40 kW), hold (10 kW) and cooling cycle every
24 hours.

If all the furnace cycles run more or less synchronously, there will be a
maximum demand of 400 kW.   Re-scheduling so that only two furnaces are on
heat up and up to four on hold at any one time would reduce the maximum demand
to 120 kW.

The financial saving will depend on the tariff period in which the savings are
made but, as an indication, if we use the average maximum demand charge
applying to the example given in Section 5.1, the saving could amount to
nearly £6,000 per annum. There will, of course, be no saving in energy used
and this simplistic proposal might well not be consistent with other features
of the production operation.

A further, this time real life example, that we would put in the Energy
Efficiency category, concerns a plant whose main business is heating up steel
stock and performing a relatively simple mechanical deformation. No

No. of leaks	Estimated diameter in	Free air wasted ft³/yr.	Power demand kW	Energy demand MJ	Power £	Energy £	Total
1	1/4	35,667,000	11.14	351,311	235.5	693.5	929
12	1/8	106,629,000	33.26	1,048,887	703.1	2070.5	2273.6
16	1/16	35,467,000	11.08	349,418	234.2	689.8	924
24	1/32	13,280,000	4.15	130,874	87.7	258.3	346
Total		191,043,000	59.63	1,880,490	1261.5	3712.1	4973.6

Fig.14: 'Thrift' saving example

Area	Suggestion
Space heating	Consider use of all forms of waste heat Consider use of control systems
Lighting	Position lamps and/or equipment to maximise use of light Increase light reflectance of walls and ceilings
Process heating	Consider use of heat recuperation Minimise thermal cycling of materials
Drives	Check power factors Investigate production scheduling
Transport and Distribution	Investigate bulk density of product loads Investigate use of containerised loads
Other	Make maximum re-use of packaging Investigate ways of reducing scrap

Fig.15:   List of typical 'efficiency' suggestions

temperature hold or reheat is required so we can do a fairly simple and interesting calculation of the energy input required to heat up the steel throughput of the factory from $20^{\circ}C$ to $800^{\circ}C$. The theoretical energy required to do this is about 360 MJ per tonne compared with the direct production heating energy used during 1973 or 14,800 MJ per tonne. Obviously, one does not expect the practical situation to approach the rather artificial theoretical minimum we have calculated, but a 20% improvement on the 1973 figure to 11,840 MJ per tonne would at <u>1973 prices</u> have represented an annual saving of over £20,000. This would seem to be an entirely achievable target and one that the company concerned is pursuing. The immediate steps necessary are not complex, e.g:

- improve (maintain) furnace door seals and improve operator attention to proper closure and minimum opening times

- provide a simple method of regulating furnace draft

- improve control of furnace combustion. At present, operation is typically at 100-400% excess air.

Longer term on this site and more within our category of Energy Economy, there are substantial opportunities for the direct use of waste heat for factory heating.

5.3 Energy Economy

In this domain we are concerned with situations where benefits will emerge much longer term. More time is available and substantial effort needs to be applied to defining and understanding the detailed nature of the requirement within which we are working to improve energy utilisation.

The resource commitment in feasibility studies, possibly R & D and detailed design, even before a decision to implement is taken, is very much more than the previous two categories of examples. We are particularly vulnerable to taking decisions with a lack of detailed knowledge on the present situation.

Ideas and check lists can still be important but these have tended to appear as an Aladdin's cave of the wonders of science and technology. Before we commit resources, we must be sure that the fit between the technology and the problem is potentially good enough to justify the resource effort that is required to reach the end point.

Some indication of the range of possibilities is given in Figs.16 and 17.

We are only able to take a fairly simple example relating to industrial waste heat recovery to illustrate some of the difficult facts that may emerge on detailed analysis.

The plant with which we are concerned is a fairly complex combination of series and parallel process operations which are integrated with a proportion of site generation. The total energy taken over the factory fence into the site is about 350 MW, expressed in terms of electrical power merely for convenience.

As illustrated in Fig.18, the waste energy contained in the various process effluents represents about 12.5% of the energy intake. After mixing into a common effluent channel losses and quality degradation have reduced this to

Area	Suggestions
Space heating	Investigate use of process waste heat for this purpose Re-evaluate factory layout and heating requirements throughout the year
Lighting	Consider energy efficiency as well as first cost on new systems Consider automatic control of new systems
Process heating	Consider use of cold forming techniques in place of hot Exploit economies of scale
Drives	Consider greater use of preformed sections Understand relationship between energy use and utilisation
Transport and Distribution	Use gravity conveying wherever possible within factory Investigate production, distribution and storage facilities on an energy basis
Other	Consider recycling of waste materials Consider flow processes in place of batch where heating is required

Fig.16:   List of typical 'economy' suggestions

Revaluate design of furnaces to match process need
    low thermal mass;
    radiant reflective barriers

Practical mechanisms for "cleaning" furnace hot exhaust
gases for recycling.

Revaluate corrosion mechanisms:
    Cathode protection;
    Inhibition (negative electro catalysis)
    Reuse with minimum degradation;
    Use of electrochemical potential as a gate
    for selectivity in metal recycling.

Direct routes to final shape to minimise "works" scrap:
    Powder forming;
    Electroforming (increased rates);
    Hot forming.

Organisation (management) of production routes:
    Avoidance of overspecification;
    Management of motive power.

Use of storage to even out variable demands of process
energy (needs careful study of energy needs at each stage).

Heat pumps:
    Development of heat transfer fluids, both two phase
    and single phase - different temperature ranges,
    higher density and sound velocity;
    Better compressor design with ability to modulate
    output;
    Use of Stirling engine on reversed cycle.

Pre-preparation of ores to remove ganque in low energy
state.

Use of plasma reactor to exploit higher temperature thermodynamics

Apply techniques of electrocatalysis to electro winning.

Avoid use of water as a carrier with loss in latent heat.

Match energy resource to process requirement.

Improve efficiency and concentration of power sources.

Improve energy transfer effectiveness.

Fig.17:   Suggested areas of economy R & D[15]

M. E. HADLOW and B. BUSS

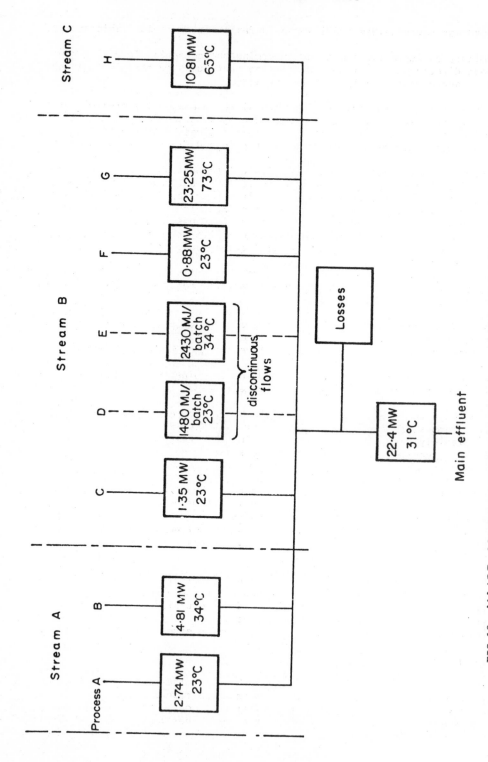

FIG.18: MAJOR COMPONENTS OF THERMAL EFFLUENTS AT SITE

about 6.4% but nevertheless still a considerable amount of available energy.

A preliminary assessment was made to determine the potential for using the waste heat directly or after upgrading using heat pumps to meet some of the process or space heating loads within the site.

Obviously this study was very specific and since it was only a preliminary assessment, the results must be interpreted with some caution. We cannot give full details of the measurements and calculations in this paper but some of the tabulated results are presented in Fig.19 and Fig.20.

These results are all based on electrically driven heat pumps with Performance Energy Ratios that can be achieved with currently practical systems and a maximum delivery temperature at $65^{\circ}C$. The latter arises from the limit set by the refrigerant pressure in the hot part of the circuit.   All the calculations are based on an electric drive although it is possible that, particularly for large heat pump installations, alternative prime movers such as gas or diesel engines may improve the financial comparison.

Considering now space heating, there is more than sufficient total energy available from the effluent streams to meet all the requirements of the site. However, there are two important points:

 - the maximum heat pump output temperature of $65^{\circ}C$ is generally not adequate for space heating without either increasing the radiator surface area or using higher temperature steam or hot water make up;

 - the losses and pumping power suffered in long pipe runs limits the distance from the heat source at which a load can be serviced.

Figure 19 shows the advantages of using the higher temperature ($34^{\circ}C$) effluent source but demonstrates only very modest (or negative) savings in running costs;   certainly insufficient to directly justify the capital expenditure that would be necessary particularly if the seasonal nature of the loads is taken into account.

Process heating is more attractive from the point of view of continuous loading and Fig.20 indicates that savings in fuel cost of between 25% and 30% may be able to be achieved.   A particularly interesting point is that Process B from the $34^{\circ}C$ source requires an input of 4.81 MW(Th) which coincides exactly with the effluent stream conditions from that process. A zero energy process seems at least in principle possible.

Taken in total, the processes examined in Fig.20 indicate an annual saving in energy costs to the site of between £86K and £131K. Clearly the direct financial considerations look to be of an order where significant capital expenditure may be justified. There are, however, many other implications to be taken into account and the full details will only emerge after further studies.

6.   CONCLUSIONS

If the ground we have covered in this paper has left the reader wondering exactly 'what will the impact of expensive and constrained energy supplies be on industrial processes' then we have not failed in our purpose but merely left the answer to the question pending at a level that is representative of the present situation.

M. E. HADLOW and B. BUSS

	Peak load (kW)	Annual consumption (TJ)	Equiv. Fossil Fuel cost pa (£k)	23°C source (PER = 3.25)				34°C source (PER = 3.8)			
				Heat pump rating (kW)	Heat flow from source (kW)	Annual heat pump consumption (TJ)	Annual electricity cost (£k)	Heat pump rating (kW)	Heat flow from source (kW)	Annual heat pump consumption (TJ)	Annual electricity cost (£k)
Building X	283	4.93	4.39	87	196	1.52	4.84	75	208	1.30	4.15
Building Y	274	4.71	4.19	84	190	1.45	4.64	72	202	1.24	3.97
Office block	226	3.90	3.47	69	157	1.20	3.83	59	167	1.03	3.28
Workshop	718	9.24	8.22	221	497	2.84	10.28	189	529	2.43	8.80
Total	1501	22.78	20.27	461	1040	7.01	23.59	395	1106	6.00	20.20

Fig.19: Comparison of current space heating costs with running costs of heat pump systems

Process	Process temperature (°C)	Heat Load (MW)	Annual Consumption (continuous running) (TJ)	Equiv. Fossil fuel cost (£k)	23°C Source					34°C Source				
					PER	Heat pump rating (MW)	Heat flow from source (MW)	Annual consumption (TJ)	Annual elec. cost (£k)	PER	Heat pump rating	Heat flow from source	Annual consumption (TJ)	Annual elec. cost (£k)
A	63	1.46	46.0	51.1	3.06	.477	.983	15.03	39.76	3.54	.412	1.048	13.00	34.36
J	63	3.65	115.1	127.9	3.06	1.192	2.458	37.63	99.50	3.54	1.031	2.619	32.52	86.00
B	66	6.89	217.2	241.4	2.89	2.384	4.506	75.10	198.70	3.31	2.080	4.810	65.60	173.55
K	66	0.50	15.8	17.6	2.89	.173	.327	5.46	14.45	3.31	.151	.349	4.77	12.62
Total	-	12.50	394.1	438.0	-	4.226	8.274	133.22	352.4	-	3.674	8.826	115.89	306.5

Fig.20: Comparison of current space heating costs
with running costs of heat pump systems

We have succeeded, if what has been said has helped to provide a perspective framework into which the contributions and interactions of the many organisations from industry to government, from academic to industrial research workers, can be seen to be pushing forward our understanding of industrial energy utilisation and the various means through which this can be improved to the benefit of industrial and national performance.

Of one thing we are certain, a new and hitherto relatively unimportant aspect of manufacturing procedures has become important in the national context. This is likely to have a lasting and significant impact on the way we produce, package, transport and present our products in the future. We believe that all scientists, engineers and industrialists whose skills are so important to achieving real advances will all profit from an early appreciation of the often conflicting criteria against which the relevance of proposed steps in solution must be assessed. It is as a contribution to this appreciation that this paper is primarily submitted.

REFERENCES

1. UK Digest of energy statistics, HMSO 1974.

2. Standard industrial classification, HMSO 1968.

3. Energy conservation in the United Kingdom NEDO, 1974.

4. Report on the census of production, HMSO, 1968.

5. Chapman, P F: Energy costs: a review of methods, Energy policy, June 1974.

6. Smith, H: The cumulative energy requirements of some final products of the chemical industry. World Power Conference, 1968.

7. Henry G: The Soviet Chemical Industry, Leonard Hill, 1971.

8. Material needs and the environment today and tomorrow, National Commission on Materials Policy. Washington, 1973.

9. Hinde, T, Probert, D: Forget the hot air, just concentrate on conserving energy, The Engineer, 2/9 June 1975.

10. Input-output tables for the United Kingdom, 1968. HMSO, 1973.

11. Wright, D J: Calculating energy requirements of commodities from the input/output table. Imperial College, July 1973.

12. Pick, H J, Becker, P E: Direct and indirect uses of energy and materials in engineering and construction. Applied Energy.(1) 1975.

13. The reaction of industry to the energy crisis. Economic Review. May 1975.

14. Energy conservation. Central Policy Review Staff. HMSO 1974.

15. Technology of efficient energy utilisation. NATO Science Committee Conference. October 1973.

16. A statistical survey of industrial fuel and energy use. CBI. June 1975.

17. Energy conservation program guide for industry and commerce. National Bureau of Standards. NBS Handbook. September 1974.

18. Energy saving loan scheme for industry. Department of Energy. December 1974.

19. Energy analysis. International Federation of Institutes for Advanced Studies. Workshop. August 1974.

DISCUSSION

Question: In what ways can information on improving energy utilisation be communicated to industry? Do you envisage small teams of experts exposing the problems in individual firms?

Answer: The Department of Energy recognises the need for dissemination of information and for some stimulus to this end. Personally I envisage two solutions:-
a) The 'energy-audit' approach to expose the problem areas.
b) Multi-client research programmes conducted by contract - research organisations, such as E.R.A., to determine the beneficial areas of development.

Comment: Energy problems are often specific to particular industries - to this end the British Gas Corporation runs an advisory school on industrial fuel management. The costs of small teams for investigating firms might be excessively high. Organised bodies such as N.I.F.E.S. could be used to generate a brief of practical case studies to highlight problems and indicate the likely magnitude of conceivable savings.

Question: Since 15% of companies consume some 50% of industrial energy, should not these energy-intensive industries be investigated as a first step?

Answer: No, since small-scale users potentially provide as good an opportunity for improving energy utilisation. Also, the larger improvements in energy intensive industries usually come as a direct result of technology changes, e.g. the introduction of basic oxygen plant to replace open-hearth plant in the steel industry.

Question: Discounting short term solutions, should education in energy utilisation come from specialist, suitably-qualified personnel?

Answer: No, although it is important to educate the workforce, this is perhaps better dealt with by technical staff with some knowledge of energy utilisation as a part of their general education rather than by specialist energy management departments.

Comment: I have found, in energy studies, that firms are reluctant to disclose detailed information about energy utilisation which they consider 'confidential'.

Response: This is indeed a problem, but such firms will often impart this data to Research Associations in their confidence or to contract research firms with whom they have confidentiality agreements. The proposed 'energy-audit' schemes will operate through the Research Associations for example.

Comment: With the demise of the detailed information produced by the Census of Production, there only usefully remains the digest from the Business Statistical Office. However, since this takes 4 years to analyse each collection of data, it seems that a lack of up-to-date data exists.

Question: How can industry be encouraged to go beyond just looking for those fuel-saving investments yielding an economic return on capital?

Answer: Judging by the return of domestic consumption trends to 1973 levels it would appear that some financial incentive is required, either by increasing fuel prices or by direct cash incentives. This is particularly true for industries in which energy is only a small proportion of the total cost.

Question: If fuel prices are increased to restrict consumption, will not
             this lead to a direct rise in the cost of living, eliminating the
             higher price effect?

Answer: This may well be true. A more effective means may be to adopt
             one of the three following policies:-

             a) Energy rationing.

             b) Demand - related tariffs related to the current mean consump-
                 tion per capita, i.e. a reversal of the current 'cheaper for
                 more' tariffs.

             c) Industrial energy tariffs related to the effectiveness with
                 which the firm concerned utilises its energy input.

Comment: During the last war, rationing did not lead to inflation but, in
             the current situation, careful monitoring of such effects would
             be required.

TOTAL ENERGY SYSTEMS

C. M. D. Peters

Total Energy Co.Ltd.

I. DEFINITIONS

We define the total energy concept quite simply as follows. "Power Generation with Heat Recovery".

The term derives from the field of commerce rather than that of science. But it has become a ready way of referring to a specialized field of effort that is very relevant to these times. The term has become accepted in governmental, economic, engineering as well as commercial publications. The field might be said to be multi-disciplinary.

The history of the term is quite interesting. 'Total Energy' was originally coined by a gas utility company in the U.S.A., which designed a gas sales promotion campaign around it, emphasising conservation. To make the term sell gas it was defined as "the provision of all the energy service requirements of a site from a single primary fuel". The single primary fuel was of course gas and the sales line "Connect to gas for clean heat and while you are at it generate some electricity too (there was the extra sale) and save fuel by heat recovery (there was the conservation)".

The whole concept of power generation with heat recovery is, of course, as old as the steam engine. So when the U.K. began to find the term total energy useful they adopted a much more sober and rational definition. I quote an Electricity Council paper (Gerald West Revised June 1974). "Total energy has been given many definitions but perhaps the most accurate technical statement is that 'total energy' means an on-site electricity generating system in which the utilization of the energy input is maximized by recovering and using waste heat from the generation process". An admirable definition and I like the inclusion of the word 'maximize' as relating the definition to the commercial exaggeration in the word 'total' and also emphasizing the efficiency and conservation that are inherent in the concept.

It should be mentioned that people also talk of "combined systems producing both heat and electricity". For instance the Central Policy Review Staff publication of July '74 on Energy Conservation talks exclusively of this rather than 'total energy' but technically this phrase could apply to a central power station dissipating all its heat. There should, I think, be mention of deliberate heat recovery to indicate that the heat is to be used.

All in all the fact that there has evolved an internationally accepted two word nomenclature to cover this field, 'total energy' with the simple definition of "power generation with heat recovery" is, I think, very salutary.

It is also worth noting that power generation is not necessarily confined to electricity. An automobile with heat recovery for air conditioning has a total energy system of sorts; an engine driving a pump or propeller and equipped with heat recovery likewise; and here one gets into the whole business of hydraulics and combining that with heat recovery. In this lecture I will confine myself exclusively to the generation of electricity with heat recovery.

II. DESCRIPTIONS

Let me now give an elementary run down of the types of total energy system. There are basically two types, one dependent on internal combustion engines and one dependent on external combustion engines. Then there are subdivisions into reciprocating and rotodynamic as follows

TABLE 1	INTERNAL COMBUSTION	EXTERNAL COMBUSTION
1)	Reciprocating	Reciprocating
	Diesel or Gas engines	(Steam engines) rare
2)	Rotodynamic	Rotodynamic
	Gas Turbines	Steam Turbines

The following are the critical characteristics of these various types of engine that are relevant to the choice of a total energy system.

External Combustion engines

Relatively high pressure boilers are required and there is virtually no restriction on the type of fuel that can be used; coal, heavy oil; even refuse. Heat is recovered as steam direct to process.

Internal Combustion engines

Limited choice of fuels, higher grades generally being needed such as gas or gas oil, though large diesels can use heavy oil. Heat recovery is generally effected through heat exchangers, and waste heat recovery boilers, mostly with direct firing capabilities. With gas turbines however the hot exhaust gases can in certain drying and similar processes be used directly.

Reciprocating engines generally have higher power efficiencies and lower overall thermal efficiencies than Rotodynamic.

The following table (Table 2) is adapted from a recent paper by Andrew Shearer of G.E.C. which shows the distribution of engine inputs in terms of efficiency. Note the Diesels and the Gas Turbines. With the Diesels (Reciprocating) 41% goes to shaft horse power, 32% to exhaust gases, 14% to water. With the gas turbine (Rotodynamic) 20% goes to shaft horse power and 70% to exhaust gases. The approximate temperature of the Turbine's exhaust gas is 525°C, of the Diesel's exhaust gas 325°C. The back pressure steam turbine, which is the relevant one for total energy systems and not shown on the chart, offers a great variety of efficiencies depending on combination of pressures and temperatures and the overall thermal efficiencies can be as high as 85%.

TABLE 2   COMPARISON OF EFFICIENCIES OF PRIME MOVERS

Percentage heat to	Diesel	Dual Fuel Gas	Spark Ig. (gas)	Industrial Gas Turbines
Shaft HP	41%	39%	29%	20%
Exhaust	32%	34%	40%	70%
		60% recoverable (high grade heat)		
Water	14%	13%	21%	Nil
		All recoverable (low grade heat)		
Misc. radiation, etc.	13%	14%	9%	10%
Sizes available kW	Up to 10,000	500 to 3,000	20 to 1,500	500 to 100,000
Fuels usable	Distillate and residual oil	Natural and sewage gases plus distillate oil	Natural propane, butane and sewage gases	Natural propane, butane and sewage gases. Distillate oil
Power/heat	1:5	1:5	1:5	1:20
Best overall efficiency	74%	72%	74%	85%

EXHAUST CONDITIONS WITH FULL LOAD

Diesel		Dual Fuel		Gas Turbines	
Mass flow lbs/bhp	Temp. °C	Mass flow lbs/bhp	Temp. °C	Mass flow lbs/bhp	Temp. °C
$14\frac{1}{2}$	325	13	400	51	525

I will now take the diesel, the gas turbine, and the steam turbine and show a sampling of the kind of efficiencies that can be realized in practice in a total energy system. But what should be emphasized is that every installation is individual and the key is to select the right engine and the right heat recovery

system that will best match the loads of the complex for which you are designing.
Here then are the three main types of total energy system

Fig. 1   BASED ON DIESEL ENGINE

Internal Combustion

(Gas, Gas oil, Heavy oil)

Waste heat recovery boiler

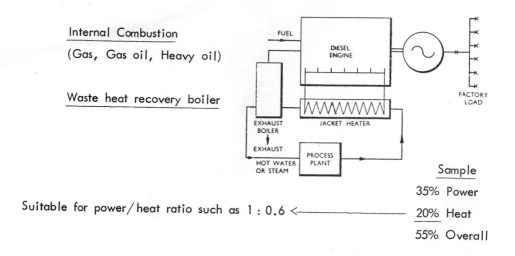

Suitable for power/heat ratio such as 1 : 0.6 ←——————

Sample

35% Power

20% Heat

55% Overall

Fig. 2   BASED ON GAS TURBINE

Internal Combustion

(Gas or Gas oil)

Waste heat recovery boiler

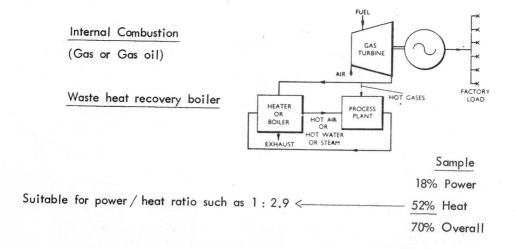

Suitable for power / heat ratio such as 1 : 2.9 ←——————

Sample

18% Power

52% Heat

70% Overall

Fig. 3   BASED ON STEAM TURBINE

External Combustion

(Coal, Heavy oil, Refuse etc.)

High Pressure Boiler

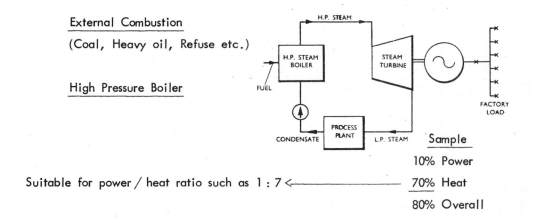

Sample

10% Power

Suitable for power / heat ratio such as 1 : 7 ⟵——————— 70% Heat

80% Overall

The art of designing a total energy system consists, briefly, in the following.

First   analyzing the electricity and the heat loads on a year round basis; the
          temperature and pressure requirements for the heat (the lower the better);
          and the ratios of the respective loads to each other.

Secondly   selecting the prime movers and boilers that will best match the
           ratios of these loads and with maximum economy. The types of
           equipment can, of course, be mixed in a single installation.

On the economics, a quick note. Diesels, as more complicated machines than
turbines, will be the more expensive. High pressure boilers as stouter units
than heat recovery boilers will be the more expensive. Gas and the lighter
oils as premium fuels will be more expensive than coal and heavy oils.
Getting the right design is like solving a very interesting puzzle.

III.   CONVERSION AND CONSERVATION

By definition the total energy concept seeks to convert the primary fuel input
as 'totally' as possible to useful energy in the form of power and heat. It is
thus obviously a 'Conservationist' concept and one can measure its value in terms
of energy savings versus the conventional ways of generating electricity and heat
separately, i.e. electricity in large centralized power stations with massive
dissipation of heat, and heating by decentralized individual on-site boilers.

IV. THE ENERGY RESOURCE CALCULATION

Here are some calculations of the percentage savings in the consumption of primary fuel resources which can be effected under various conditions by substituting the total energy approach for the conventional approach. It in no way affects the consumption of energy. It merely reduces the consumption of primary fuel resources necessary to produce the energy. In the conventional approach it is assumed that the power efficiency of centralized power generation is 35% and that the thermal efficiency of the individual boiler is 80%. In the total energy approach the efficiencies already discussed are assumed.

TABLE 3   SAVINGS UNDER LOAD CONDITIONS SUITABLE FOR DIESEL INSTALLATION

	Load Requirements (1:0.6 ratio)	CONVENTIONAL		TOTAL ENERGY	
		Efficiencies	Primary Fuel Consumption	Efficiencies	Primary Fuel Consumption
Electricity	1000	35%	2857		
Heat	600	80%	750		
	1600		3607	55%	2910

SAVINGS (CASE I) PER CENT			697
	(Overall view)	19%	
	(Electricity view)	24%	
	(Heat view)	93%	

SAVINGS (CASE II with 30% conventional electricity efficiency) PER CENT			1173
	(Overall view)	29%	
	(Electricity view)	35%	

The 19% is from an overall or national view of the savings. One can also look at it from the point of view of central electricity supply and visualize them taking over a heating plant with a consumption of 750 units and saying to themselves - "by consuming an extra 53 units (2910), we could produce our 1000 units of electricity and save that consumption of 750 units for a net saving of 697 units or

$$\frac{697}{2857} = 24\% \text{ ".}$$

One can also look at it from the point of view of the heating plant owners with their consumption of 750 units of fuel and visualize them taking over the central electricity supply plant and saying to themselves - "by consuming an extra 2160 units (2910) we could produce our 600 units of heat and save that consumption of 2857 units for a net saving again of 697 units or $\frac{697}{750} = 92\%$".

One might also visualize the central electricity supply saying to themselves - "we have some older plants that are producing those 1000 units at an efficiency of say 30% involving a consumption of 3333 units. We could retire this old one and replace it with this diesel total energy plant. The resource savings would then be 1173 units or 29% $(\frac{1173}{4083})$ on the 'overall view' and 35% $(\frac{1173}{3333})$ on the 'electricity view' ".

The following tables go through the same calculation under load conditions suitable for the gas turbine and the steam turbine respectively.

TABLE 4   SAVINGS UNDER LOAD CONDITIONS SUITABLE FOR GAS TURBINE INSTALLATION

		CONVENTIONAL		TOTAL ENERGY	
	Requirements (1: 2.9 ratio)	Efficiencies	Primary Fuel Consumption	Efficiencies	Primary Fuel Consumption
Electricity	1000	35%	2857		
Heat	2900	80%	3625		
	3900		6482	70%	5571
SAVINGS (CASE I) PER CENT					911
	(Overall view)		14%		
	(Electricity view)		32%		
	(Heat view)		25%		
SAVINGS (CASE II with 30% conventional electricity efficiency) PER CENT					1387
	(Overall view)		21%		

TABLE 5   SAVINGS UNDER LOAD CONDITIONS SUITABLE FOR STEAM TURBINE INSTALLATION

	Load Requirements (1:7 ratio)	CONVENTIONAL		TOTAL ENERGY	
		Efficiencies	Primary Fuel Consumption	Efficiencies	Primary Fuel Consumption
Electricity	1000	35%	2857		
Heat	7000	80%	8750		
	8000		11,607	80%	10,000
SAVINGS (CASE I) PER CENT					1607
(Overall view)		14%			
(Electricity view)		56%			
(Heat view)		18%			
SAVINGS (CASE II with 30% conventional electricity efficiency) PER CENT					2083
(Overall view)		17%			

The main things that those calculations are intended to demonstrate are

1)   that total energy offers plenty of flexibility. The overall average U.K. ratio of electricity requirements to heat requirements is 1 : 5 i.e. 5 times more heat is consumed than electricity. The ratios for which the samples provide range from 1 : 0.6 to 1 : 7.

2)   that total energy can realistically produce very worth-while reductions in the nation's consumption of primary fuels and without reducing energy use. The samples of the overall view show a scatter of percentages - 29, 21, 19, 17, 14.   Twenty per cent is perhaps a good round figure to keep in the mind.

3)   that looked at from the point of view of individual conventional plants the resource saving argument can be even more persuasive.

V. THE ECONOMIC CALCULATION

Conservation of energy is not the only criterion. There are lots of other vital
commodities like steel. And so one has to do the overall economic calculation
in monetary terms. The saying that British Pounds Sterling are a better guide to
action than British Thermal Units is still valid despite the current sliding nature
of the former. However there are at this time so many distortions in monetary
economics (to which I will refer later) that for the purposes of this lecture I will
try to present the situation just in terms of economic logic.

Let me first present two very general considerations:-

The first derives from the axiom that conversion to power creates a higher value
than conversion to heat. Power derives from heat in the heat engine and by the
laws of thermodynamics there is always a residue of heat left over from the
conversion. The obvious corollary for the total energy man is that every
concentration of heat requirements should be regarded as an opportunity for the
generation of power. Furthermore he should extend his interests into the area of
'District Heating'. Just as the electricity supply industry has used the economies
of scale to justify centralized plants and relatively high cost distribution, so
should the total energy man ensure that he is dealing with the largest and most
economic concentrations of heat generating plant before he considers turning it
into a total energy plant. These, of course, will be on a smaller scale on
account of the different problems of heat distribution. But the latter is
developing fast and has also added some new nomenclature - Telethermics.

The second consideration is that the economics of total energy are marginal
economics. Let us assume that there is a prospective concentration of heat
requirements for an industrial heat process or space heating either in an
individual complex or in a district heating type of plant. The economic
calculation regarding the opportunity offered by the heat requirements for
converting a heating plant to a total energy plant is as follows -

1)   What extra cost will be involved on-site?

2)   What extra value will be created on-site?

The economic logic for the likelihood of extra value exceeding extra cost is as
follows.

The tangible extra value created is the electricity for which the Central
Electricity Supply determines the price. There is also an intangible value that
derives from control of supply and this I will touch on later. The price has
cost components as follows -

a)   a capital component which is based on a highly specialized plant
       complex on a remote site subject to every kind of infra-structure costs
       and subject to a long-line relatively high cost distribution system. In

the league table of distribution costs, oil is cheapest, gas second, coal third, electricity fourth.

b)   an operating cost component which can benefit to the full from the economies of scale.

On the extra cost side there will be similar components –

a)   a capital component, which will include the extra cost of installing the electrical generating capacity. There will be no change in the boiler capacity required for on-site heating but there will be some extra costs for installing waste heat instead of conventional boilers. If external combustion engines are involved boilers will be needed of higher pressures and some capacity over and above that required for heating.

b)   on the operating cost side there will be extra fuel and maintenance costs but they are subject to the economies of fuel conservation and there are no bulk distribution costs.

Table 6 below summarizes the above.

TABLE 6   TOTAL ENERGY ECONOMICS

	Extra Value Created	Extra Costs Involved
	Components of centrally supplied electricity price	Components of on-site decentralized electricity supply costs
a) Capital	Specialized plant with full infra-structure costs and distribution system	Mass produced plant marginal to special complex designed for other objectives
b) Operating	Subject to economies of scale	Subject to economies of fuel conservation. No bulk distribution costs

Comparing the components involved in Cost with those involved in Value, the capital component on the Cost side is manifestly lower. The major extra item, the electrical generator, is mass produced instead of being of specialized design and there is no bulk distribution system. Although the operating component on the value side benefits from the economies of scale, this is offset on the cost side by the very considerable economies of fuel and the elimination of bulk distribution. The logical likelihood therefore is that extra cost will be lower than extra value and that wherever there is an appropriately sized concentration of heat, a decentralized supply of electricity will be economically more advantageous than centralized supply.

VI.   HISTORICAL BACKGROUND

So much for theory and generalities. Let me now offer some concrete experience and historical background with a quick reference to some statistics in the U.S., the U.K. and Germany. I think they are also interesting as reflecting national characteristics - competitive enthusiasm in the first, conservative casualness in the second and extreme thoroughness in the last.   There is as least this to be said for casualness that it gives time to observe the errors of others. But my view is that the crucial time has arrived to cast this off, develop a British approach to total energy with some concentration and sweep the world with sales of the goods and services involved. For conservation methods and equipment are in demand and even in oil rich Saudi Arabia when they build an up to date hospital they build it with a total energy system   (and it is British equipment).

U.S.A.

I mentioned earlier the total energy campaign by the gas utility companies in the U.S. That was started in the early sixties, being almost exclusively directed to commercial and institutional complexes like shopping centres, office buildings, schools and hospitals. Within 7 years about 500 installations had been made, with a total capacity of about 1000 MW - an average of about 2 MW per complex. The concept was oversold, mechanical troubles due to inadequate control of design and operations caused some disillusionment on the part of engine manufacturers supporting the effort and furthermore gas shortages developed and the campaign trailed off.

The total energy concept is, of course, applied extensively in industry and there are, I understand, some 2000 District Heating schemes but I do not have information of how many are combined with electricity generation.

U.K.

There is a long tradition in industry for generating own power - mostly with heat recovery and mostly in Iron and Steel, Chemicals and the Paper industry. There is even an association, the UKIPGG, Industrial Power Generation Group. In 1973 Industry generated 17% of their own requirements.

In commercial and institutional complexes of the type towards which the drive was directed in the U.S. there is a mere handful of examples. But things are beginning slowly to change, with several now in the construction or near construction stage.

There are close to 200 District Heating schemes but only two are in combination with electricity generation under a total energy scheme - namely at Pimlico and Aldershot. The counts of District Heating schemes, incidentally, vary depending on the limits of size selected.

W. Germany

A comparison of the U.K. with W. Germany is highly interesting. As far as
District Heating is concerned a recent statistic indicates that there are close to
500 District Heating Schemes 25% of which are combined with electricity
generation.

In industry the situation contrasts with that in the U.K. even more sharply.
Here are the comparative figures for 1973.

TABLE 7   COMPARATIVE INDUSTRY STATISTICS (1973)

	W. GERMANY		U.K.	
	Kwh x 10^9	Percent	Kwh x 10^9	Percent
Requirements	158.4		102.7	
From public supply	116.0	73.2	85.4	83.1
Net private generation	70.5	44.5	21.6	21.0
Delivered to public supply	(28.1)	(17.7)	(4.3)	(4.1)
Net for own use	42.4	26.8	17.3	16.9
Total Supply	158.4	100.0	102.7	100.0

The interesting feature is the volume of electricity delivered by the private
sector to the public supply system and it is a reasonable assumption that all of
the former has been produced in combination with heat on total energy principles
and more economically than by central supply.   Dr. R. Thiele of the Siemens
company sent me some very interesting official figures and a chart. In Figure 4
I have compressed the latter somewhat and drawn a similar one for the U.K. for
purposes of comparison.

As is evident industry in Germany not only produces a very large slice of their
own requirements but the volume delivered to the public supply system is
equivalent to 11% of all the latter's customer-requirements - including the needs
of the public electricity distribution system.

One further item of interest is that among German blue-prints for the post-fossil-
fuel era is virtually a nation-wide total energy system based on centralized nuclear
power plants from which the waste heat is distributed throughout the country in a
piped network - the ultimate in telethermics and an expression of thoroughness
taken to extremes. It at least indicates that for the uncertain future that lies

Fig. 4   PUBLIC AND PRIVATE ELECTRICITY GENERATION
                                1973

                            W. GERMANY

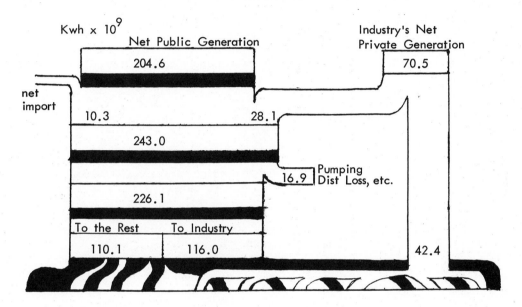

                            UNITED KINGDOM

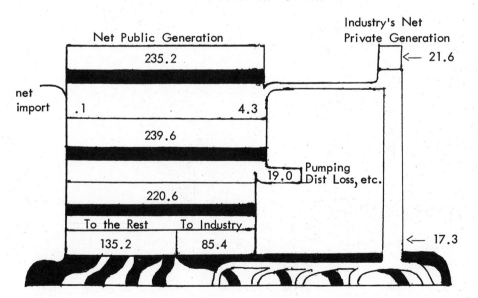

ahead in the energy field, a whole series of total energy stations built throughout the country for the fossil fuel age would be readily adaptable to the nuclear age or whatever is the successor and in the meantime the conservation which they will effect will do more than anything to prolong the fossil fuel era that we know.

As a result of our conservative casualness we have a relatively unexploited field in these islands and have had time to observe the experience of our neighbours. With the time to act becoming overdue, I would select from the German experience, the principle of decentralized electricity generation coordinated with central supply and from the U.S. some of its enthusiasm for application.

VII. U.K. CONSIDERATIONS

The rest of this lecture will be concerned with the problems in making these islands a base for exploiting the total energy concept on a world wide basis, some of the advantages we have, some of the benefits to be gained and some practical precepts.

First the advantages.   These paradoxically derive from political developments, which on the theory of many people, including myself, are only to be recommended as a last resort, namely state monopolies.   But we are singularly fortunate with our Electricity Council, our National Coal Board and British Gas and in two respects these developments can be advantageous to the progress of total energy.   Firstly their nation-wide administrations present a single authority with which far-reaching arrangements can perhaps be made that will expedite development quickly.   Secondly they have nation-wide centrally controlled distribution networks which can rival those of anywhere in the world.   These are a vital adjunct to what I believe should be the specially British type of total energy development and appropriate to the doctrine of the mixed economy i.e. independently conceived decentralized installations that are coordinated with the central supply.

These same state monopolies, however, and the institutional and statutory situations that go with them also offer obstacles to total energy and correspondingly to energy conservation.

Firstly as I have already indicated the key to a total energy scheme is a concentration of heating requirements.   The installation and operation of heat generating plant is almost exclusively the domain of the private sector and of local authorities.   The conversion of a heating scheme to a total energy scheme by a private sector interest or local authority will thus inevitably deprive an Electricity Board Salesman of a sale and will not naturally excite his interest - or that enthusiasm of application which we have mentioned.   And furthermore if that enthusiasm could be generated, an institution devoted to the supply of electricity is not, on the face of it, the right institution to carry forward the interests of a concept for which the recovery of heat for use is the critical technology.   The interest is likely to be on that paradoxical phenomenon of increasing electricity sales by 'degrading' the supreme form of energy, namely power, to a low form of energy such as that for space heating.

Secondly the Coal Board sells about half of its output to the C.E.G.B. and Scottish Electricity Boards. Although a sale to a total energy installation, as opposed to a heat station only, would represent a small increase in sales, this would diminish the C.E.G.B.'s requirements for primary fuel by a greater amount on account of their necessarily wasteful generation. So that apart from the reduction they would cause in their major customer's electricity sales, it might represent a net loss on sales of their own product. So they also are not naturally inclined to embark on a conservation campaign via total energy.

Thirdly British Gas are unencumbered with any large interest in sales to the C.E.G.B. And it is interesting to note that they did give considerable backing for a period to the total energy concept with sales to one or two very interesting gas turbine industrial installations such as John Player's factory at Nottingham and Singer's at Glasgow. But they then had to restrain sales due to their supply position.

The institutional arrangements are thus generally counterproductive for conservation via the total energy concept. There are also statutory constraints such as doubts about the legality of electricity sales to the public by a private sector installation, the laying of cables under roads etc. But these can readily be overcome once the essential has been achieved, namely a framework in which the private sector can 'take off' with the support and backing of the nationalized energy supply industries. What is needed is a 'third force' that can defuse the institutional constraints on conservation via total energy and infuse a concentrated aim. I am confident that out of the various committees that are currently deliberating conservation and structures, an appropriate answer will emerge.

At the beginning I called total energy a multi-disciplinary field and I am talking here of its involvement in politics. You are, I understand, all engineers or physical scientists rather than political and economic theorists or moralists. But it is interesting to speculate that here may be a place where the different disciplines can support each other.

I have described the British approach as being a decentralized electricity generating system operating on total energy principles and coordinated with the centralized system. The conventional wisdom - for which this University is credited with a fair share of responsibility - seems one-tracked in theories of centralized state direction and to ignore the deadening effect of this on freedom, originality and enterprise. The key problem of all administration is the right balance between centralization and decentralization. Perhaps there is here a unique situation where scientists like yourselves can provide a logical answer for this balance based on physical as well as political, economic and moral considerations. The limits for the centralization of heat generation are prescribed by heat distribution factors. This constraint prescribes the limit for a total energy system and the decentralization of electricity generation. Centralized electricity generation supplies the rest.

VIII.   BENEFITS FROM MAXIMUM DEVELOPMENT OF TOTAL ENERGY

Let us assume then that the objective is a multitude of independent total energy installations throughout these islands, firmly contracted with the primary fuel producers for supplies and with the Central Generating and Area Boards for mutual support;   running in parallel, synchronization, standby etc.   I will quickly run over the advantages of this kind of 'scenario' with some efforts at quantification. I will then finish with some practical precepts.

From the Central Electricity Generating Board's point of view –

The money and time available for devising the successor to the fossil fuel age would be increased and the return on capital improved.

Every kilowatt of capacity installed in a total energy plant should with proper coordination represent a saving in the capital required for new and replacement central plant.   The conservation of fuel and stretching of supplies by the total energy plants would correspondingly stretch the time for nuclear or whatever research and development.

Furthermore total energy plants would have their highest opportunities for cheap electricity generation in concert with heat generation during the heating season when electricity also peaks.   Total energy plants therefore would tend to cut peaks to be met by central plants;   lessen the incentive to fill troughs with 'degrading' sales of electricity for heat (which again tend to add to peaks);   improve load factors and return on capital.

From the point of view of the Treasury –

The conservation of fuel would make a sizeable contribution to foreign exchange reserves by either saving in oil imports or increasing oil exports.   There would also be a great fillip to the exports of goods and services to huge markets for energy conservation techniques and equipment in both the developed and underdeveloped countries of the world.

From the point of view of people generally –

Firstly there would be a strategic advantage in not having all our eggs in large baskets.   By dispersing plants in small units, our protection against massive disruption and shut down would be enhanced.   The demand for this has been well illustrated in recent years by the multi-million pound investment by every kind of institution in on-site standby – which is mostly idle and only useful for emergencies. With proper design it could undoubtedly have been productive on total energy principles.

Secondly from the environmental point of view, total energy means less combustion and correspondingly less pollution.

Thirdly from the social point of view total energy means devolution of responsibility, which makes for a sturdier and more independent citizenry and more widespread challenges to involvement and initiative.

Obviously with total energy you can't miss.

On quantifications

These have their limitations when concerned with twenty year projections - except for emphasizing some such conclusions as "The sky is the limit" or "Doomsday cometh". But for what it is worth I have gone through the exercise based on the 'Energy Conservation' reports by the C.P.R.S. of mid '74 and by N.E.D.O. of late '74 and the '73 edition of the United Kingdom Energy Statistics compiled by the D.T.I.

1)   I estimate there is a potential U.K. market for total energy installations by 1990 of 28,600 MW. Following N.E.D.O. I assume that electricity consumption will approximately double by then, that the capacity will also have to about double and that along with the required replacements 28,600 will represent about a quarter of the new capacity to be built.

2)   I estimate that the extra capital requirements for building 28,600 MW in combination with heat generation will certainly not be greater than building it into central power plants and following the economic logic already discussed it would probably represent a capital saving in the £2-3000 million category.

3)   I estimate that the conservation of primary resources would at 28,600 MW of total energy capacity amount to 276,000 B/D oil. If it built up to this level evenly over a 20 year period it would effect at current prices a cumulative addition to exchange reserves of ₤ 12,300 million.

I have no doubt that different approaches would produce different figures but the fact that one approach produces this kind of figure is, I hope, stimulating.

IX.   PRACTICAL PRECEPTS

I am sure that there are among you many prospective managers of industrial, commercial, institutional and domestic building complexes that will be suitable for total energy installation. So let me now offer you six practical precepts including some deflationary ones appropriate to the times.

Firstly   To use a slogan from those hustling American gas utility firms "Think and Talk Total Energy". This means always consider your requirements for heat generation and those of your neighbours as possible opportunities for combination with power generation.

Secondly   Get as accurate profiles as possible of your daily heat and electricity requirements during critical seasons, summer, winter, intermediate. If you have a maximum demand of over about a megawatt, examine the ratios of power to heat and determine the most suitable type of prime-mover on the basis of the various thermal efficiencies available.

Thirdly   Find a sympathetic officer in your Area Board with whom you can discuss mutually advantageous arrangements for sharing the load and make up your mind on a preliminary basis regarding the kind of installation that is likely to fit best - fully independent, running in parallel, peak lopping etc.

Fourthly   Determine your Management's views on security of energy supply, liquidity, and return on money and get down to the economic calculation of the extra cost of a total energy plant versus a conventional plant. If your Management has already invested in on-site emergency electricity generating capacity or think they need it, credit the total energy plant extra cost with either the sales price of the existing standby or the current new price.

Fifthly   Prepare for disappointments in your economic calculations. There are two prime factors in the economic viability of the installation; the load factor and the ratio of the market price of electricity to that of the primary fuel. You can do a little about the former but nothing about the latter.

I have mentioned economic distortions and with the electricity supply industry running at a subsidized loss and primary fuels related to Arab oil, the unfavourable distortions in the price ratio are self evident. Furthermore access to capital is subject to different criteria in the private and public sectors. But have faith, there is a 20 year history of a favourable ratio, the subsidies are in the course of being eliminated and the Arabs occasionally 'threaten' a lower price. The problems of liquidity and the rationing of capital outlay through the imposition by Managements of very low pay out requirements are also problems which every project faces. Economic justification these days is an uphill battle.

Sixthly   Whatever the outcome of your economic calculations, keep faith. If you cannot justify the installation now, orient your designing so that there are minimum obstacles to a future installation. Leave space for future generating plant, consider boilers adaptable to various fuels and to conversion for waste heat recovery.

X. CONCLUSION

This series of lectures is sponsored by the Science Research Council and I have wandered into politics and economics and business concerns of markets and sales. But I do not think I need apologize too much. For the political and economic framework has got to be favourable before Science Research can attract practical interest and vice versa.

Given the right framework there is plenty of scope for research particularly I suspect in the field of coordinating the decentralized system with the centralized one. I have also omitted mention of that other sector, namely Transport, which carries about equal stigmas of fuel extravagence and pollution as centralized electricity generation. And yet it is electricity which is the form of power offering the best hope of removing the noise and other pollution of transport. But as far as concerns waste of fuel and level of pollution, one would be out of the frying pan into the fire unless the electricity were generated on total energy principles. One can visualize the development of a most popular gadget that substitutes for the domestic home boiler a diminutive total energy plant heating the house and charging the electric car in the garage.

This particular lecture also falls under the heading of <u>possible</u> developments in energy use. I hope I have offered you sufficient reason for regarding it as a <u>necessary</u> development and for not only 'talking total energy' but also orienting your designing and building to a future of a decentralized electricity generating system coordinated with a centralized one.

BIBLIOGRAPHY

1)   "Total Energy" by R.M.E. Diamant, M.Sc., Dip.Chem.E., A.M.Inst.F., published by Pergamon Press in 1970 is a useful text book.

2)   The Proceedings of the 'Total Energy Conference' organized at Brighton 29 Nov - 1 Dec 1971 by the Institute of Fuel, 18 Devonshire Street, Portland Place, London W1 N2 AU offers a comprehensive work of reference for most aspects of total energy.

3)   Major engine manufacturers all have useful literature on total energy in the form of manuals or papers to professional societies. In particular may be mentioned G.E.C. Diesels, London, (01) 836-3466; G.E.C. Gas Turbines, London, (01) 580-8439; Mirrlees Blackstone, Stockport, (061) 483-1000; W.H. Allen, Bedford, (0234) 67400; Caterpillar Tractor, Slough, (0753) 38835.

4)   The total energy campaign by the U.S. gas utilities was well documented in a magazine called 'Total Energy' with annual Directories describing and listing installations. It was published by GATE (Group to Advance Total

Energy) and back numbers available from American Gas Association, 1515 Wilson Boulevard, Arlington, Virginia 2209.

5)   Sources for statistics on private electricity generation in U.K. and Germany are -
United Kingdom Energy Statistics, H.M.S.O.
Elektrizitätswirtschaft.   Zeitschrift der Vereinigung deutscher Elektrizitätswerke - V.D.E.M.

6)   Information on District Heating in the U.K. and W. Germany is available from the respective country's District Heating Association. (U.K., Caterham (0883) 42323;   Germany, ARE, Humboldt Strasse 33, Hanover).
Interesting articles are -
"Why Waste all that Heat" by A. Ernest Haseler.   First Chairman of the British Association published in The Consulting Engineer Nov. 1974.

"Germany looks toward nuclear district heating" by Dr. Hans-Peter Winkens, Chairman of the German Association published in Energy International, August 1975.   It includes a 'blueprint of a nation-wide total energy system based on nuclear power'.

DISCUSSION

Question: Could an example be given of a typical customer for a total energy
             system and of how long it would take to recover the capital cost
             of the investment?

Answer: Customers cover the whole spectrum of industry. For domestic
             heating there will probably have to be development of district
             heating systems to make it viable. Total energy systems could be
             used economically in installations where there is 1 MW capacity
             or above. With regard to the time to recover investment, the
             prime consideration is the ratio of the price of electricity to
             the price of primary fuel. The current economic situation is dis-
             torted because you are comparing a commercial system with a sub-
             sidised nationalised industry, which seems to have little diffi-
             culty in obtaining money to cover its losses.

Comment: This is not true for the CEGB, which operates under strict finan-
             cial rules laid down by the Treasury. Total energy systems cannot
             be generalised; any system will be looked at by the financier and
             treated on its merits. Then there is the question of plant relia-
             bility: if it is out of commission for 80 days in the year, the
             cost might be equal to the cost of the generator. This is why we
             go for simple uncomplicated plant in the CEGB. There is also the
             problem of matching demand for heat and electricity – this may
             differ by 10% over the year and might therefore be as large as the
             advantage margins of total energy systems. The value of one joule
             of electricity is not the same as one joule of heat. It is not
             always correct to carry out efficiency calculations that compare
             the two.

Response: The sorry situation at present is that you have high grade elec-
             trical joules degraded to joules for space heating. On the ques-
             tion of reliability the CEGB does have a superb reliability record,
             but what we would like are decentralised, small units with coordi-
             nation between these total energy systems and the public system.

Comment: An important factor in this is the heat exchanger. If you are
             selling electrical power and heat, the best customer may be the
             power man because he can use it to better advantage needing no
             additional heat exchangers and having no heat distribution costs.

Question: Could you explain the difference between the amount of electricity
             supplied by the private sector in Germany (40%) and that in UK (20%)?

Answer: The system in Germany should not be described as a state system
             but as a municipal system where you have a large number of local
             systems not fully integrated. The power supply industry is there-
             fore closer to the scale of private industry.

Comment: The difference between UK and Germany may be geographical. We
             have a large coastline and many rivers, and therefore have few
             problems in obtaining cooling water. Germany has small coastline
             and one major river; therefore they may be forced to look for uses
             for their waste heat. This, in the long run, may be a good idea
             in this country. The problem in UK is centralisation – the answer
             is decentralisation.

Question: Do you envisage a breaking down of centralisation?

Answer: No, because in the present situation there is extra supply capacity
             (14%) which can be used as standby. This would not be possible in
             a municipal set up.

Question: How would the introduction of total energy systems affect the role
            of the C.E.G.B.?
Answer: It would be possible to install about 28 MW of capacity in Great
            Britain over the next 20 years, but this is small compared with
            C.E.G.B. capacity. This is equivalent to 25% of the C.E.G.B. con-
            struction programme but would be extremely useful in saving the
            country's capital and imports. It is important to remember the
            scale for which we would recommend total energy systems.

Question: What happens to total energy systems when fossil fuels have been
            exhausted?
Answer: It is possible to use total energy systems with nuclear power. In
            Germany a large district heating system is being planned using
            waste heat from nuclear power stations.

Question: Would it not be possible to put a large VAT (Value Added Tax) on
            electric space heaters? Until there is some measure like this
            there seems to be little chance of total energy systems being
            introduced.
Answer: Introduction of total energy systems is essentially a financial
            problem. If the financial conditions were favourable they would
            be introduced.

Question: What use can be made of excess heat produced in total energy
            systems during the summer?
Answer: Consumption of electricity also goes down in summer so, if need
            be, electricity could be purchased from the C.E.G.B. during the
            summer rather than dispose of waste heat.
Comment: The Swedes are building power stations which will only be used
            for winter heating. The power station will be close to a town
            and supply district heating. The electricity will be used to
            supply the country districts.

Question: You mentioned various engines - the internal combustion engine e.g.
            diesel and gas turbine - the external combustion engine e.g. steam
            turbine. Could you give the power ranges over which these operate?
            What about other developments such as closed circuit gas turbine
            engines and heat pumps?
Answer: I restricted myself to proven types of equipment and not to new
            developments because I am concerned with practical considerations.
            Mr. Wilson gave a comprehensive review of available engines and
            there is obviously scope for further research and design.

THE AUTONOMOUS HOUSE EXPERIMENT

J. G. F. Littler

Architecture Department, University of Cambridge

Introduction

It should be stressed that the work which we hope to carry out in the
Department of Architecture at Cambridge over the next five years, is
experimental.

In summary the aim of the project is to close the plot boundary to most
of the network services, and to rely on as many on-site resources as
possible.

Such a concept, illustrated in Figure 1, raises many contentious issues.
However, although the social problems involved are to be addressed at the
end of the five year programme, when the house as 'a machine for living' is
functioning, I personally regard the project as a technological and architectural challenge in which the uppermost question to be answered is 'Can
ambient sources really provide acceptable standards in such an autonomous
house, in the U.K., at a price which is not immoderate?'

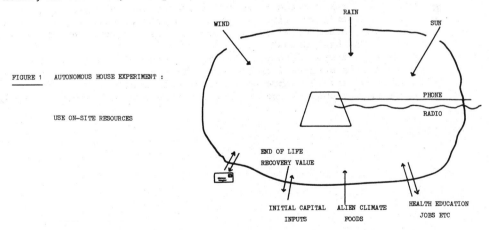

FIGURE 1   AUTONOMOUS HOUSE EXPERIMENT :

USE ON-SITE RESOURCES

Prototype house

The autonomous house will have several (partially interacting)
subsystems:

Table 1

Water supply	Rain collection, purification, storage and recycling
Wastes disposal	Anaerobic digestion ($\rightarrow$ CH$_4$)
Cooking	Methane
Food	Cultivate the 1 ha plot and greenhouse for four people
Electrical demands	Wind generator
Space heating	Solar collectors
Water heating	Solar collectors and excess wind energy
Control system	Initially PDP11. ultimately hard-wired
Building fabric	Integration of subsystems to cut costs

Water supply

Table 2 indicates some of the problems faced in this area:

Table 2

Conventional pattern of water use l/person . day		Comment	Autonomous house l/person . day
W.C.	55	drinking water standard unnecessary	0
Washing	50	1 bath ≃ 110 l 1 shower ≃ 15 to 40 l	17
Dishes	5		5
Laundry	8		8
Drink and cook	5		5
Total	123		35

a = annum
ha = hectare
l = litre

The four person family for whom the prototype house is planned would be
expected to use 140 l/day, and the rainfall in Cambridge is such that even
with an excessively large storage tank (25 m^3) a collection area of
195 m^2 would be needed to guarantee year-round supplies (data calculated
from ten years of local rainfall figures).

The roof of the house will be about 60 m^2, and thus some water ('grey water')
will be recycled using either solar distillation (our computer model of a
simple one pass solar still indicates an output of 1.6 l/m^2 . day from March
to September), or a small reverse osmosis unit. Membrane clogging and pH
adjustment are the principal difficulties associated with the second
method.

Wastes disposal

The human wastes (four people) would provide about 280 kWh/a of methane;
but with the addition of vegetable wastes from 0.2 ha, or wastes from
1 ha growing a complete diet, about 1500 kWh/a may be obtained by
anaerobic digestion.

The sludge from the digester may be returned to the land. In hotter
climates, this effluent could be used to set up a more productive cycle:

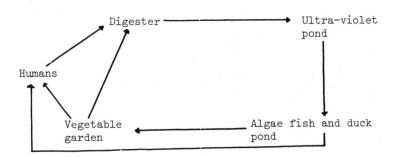

Wind generator

It is proposed to mount a $4\frac{1}{2}$m diameter Darrieus wind rotor on the roof, driving a 2.2 kW generator, and to absorb the energy supplied in winds above the design speed (9 m/s) in mechanical devices such as a 'heat churn' or a heat pump. Table 3 shows the output calculated using local (Mildenhall) wind speed data, and assuming the following cumulative losses:

(a) Betz efficiency 59% (theoretical maximum)
(b) Darrieus efficiency 60%
(c) Half the output stored in batteries with an 80% recovery factor
(d) Generator efficiency 85%, losses in bearings 10%

Table 3

	kWh/month
January	203
February	182
March	171
April	155
May	152
June	104
July	103
August	92
September	122
October	127
November	151
December	209
Total	1771

Typical domestic electrical demand in the U.K. for purposes other than heating and cooking is 1800-2100 kWh/a. The collector pump is likely to add about 200 kWh/a to the normal load; but savings will be made by using redesigned domestic electrical equipment.

Solar heating

Figure 2 indicates the total solar radiation available on a horizontal surface at Kew throughout the year.

Figure 3 shows the solar energy incident on inclined surfaces for the winter (October 1 - April 1) and the summer months.

Figure 4 gives some results of a computer model calculation of collector efficiency for the early days of 1959, and indicates where the optimum tilt lies for collection under the following conditions:

Collector facing south, steel tube in strip fabrication with 14 cm of foam insulation heating a water store at 60°C. Plate emissivity 0.95, steel pipes 2.5 cm diameter spaced at 5 cm - metal 0.1 cm thick. Double glazed with 4 mm good quality glass, pumped circulation of water.

Figure 4 suggests that the efficiency is very low (13%) at the worst time of year. The output indicated by Figures 3 and 4 for the six winter months from a vertical collector is about 7 kWh/m^2 . month.

Table 4 presents the current demands for a four person house in the U.K.

Table 4

	kWh/a
Space heating	12000
Water heating	4000
Cooking	1200
Other	2100
Total	19300

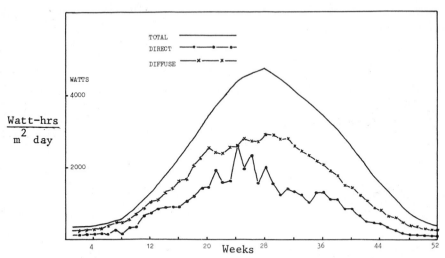

FIGURE 2   RADIATION FOR 10 YEARS AT KEW (1959–1968) AVERAGED WEEKLY

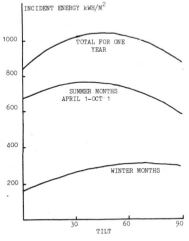

FIGURE 3   INCIDENT ENERGY VERSUS TILT

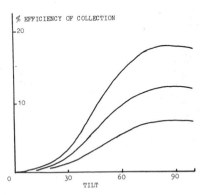

FIGURE 4   EFFICIENCY OF COLLECTION VERSUS TILT

To supply the space heating requirement would thus demand a solar collector area of about 240 m^2. Such a very crude calculation is only intended to indicate that the provision of winter space heating requirements with <u>current</u> collectors is not a practical proposition for existing badly insulated houses in England.

The following suggestions may be made in connection with present day claims made for solar collector efficiencies:

(a) Minutely prescribed standard operating conditions should be laid down

(c) Collector efficiency is somewhat irrelevant, since it is the <u>heating system</u> efficiency which is really significant.

The <u>system</u> includes the heat store and its standard of insulation, the means of heat distribution in the house (which determines the minimum acceptable store temperature), and the energy used for pumping.

In the prototype autonomous house, inter-seasonal heat storage is attempted using a large well-insulated tank of water. Furthermore the controlled ventilation system and the hot water system (both with heat recovery) and the very high insulation standards to be adopted, will cut down the heat demand.

However considerable improvements may be made to collectors which have in the past been designed for hot climates with continuous sunshine, as indicated by the following results of calculations using hourly Kew weather data for the first 20 days of 1959 in our computer model of a solar collector. (Tables 5-8 contain results of computations on south-facing double glazed (glass) collectors, of the steel tube design mentioned previously, at a tilt of 80°, with a water store at 50°C. Table 9 lists some areas soon to be investigated.)

Table 5   Selective coatings

Absorber plate emissivity	% efficiency of collection of	
	diffuse	direct
	radiation	
0.95	7	17
0.20*	12	26

* absorptivity/emissivity of plate = 5

Table 6   Storage temperature

$\overline{\text{T Store}}$	% efficiency of collection	
	diffuse	direct
6°C	35	45
60°C	7	17

Table 7   Heat reflecting coatings on glazing

	% efficiency of collection	
	diffuse	direct
No coatings	7	17
SnO_2/In_2O_3 on internal faces of double glazing (ε=0.25)	7	17

Table 8   Insulation of collector

U value insulation	cm foam	% efficiency of collection diffuse	direct
0.24	14	7	17
0.72	7	6	15
1.44	3.5	4.6	12.4

Table 9   Some other areas awaiting parametric investigation

Put high viscosity, low conductivity gases in the convection spaces

Try anti-convection honeycombs

Interaction of low emissivity cover glasses and a low emissivity plate

Locate dominant points of thermal resistance and redesign collector plate accordingly

Explore effect of a higher viscosity anti-freeze liquid on heat transfer

Compare Thomason and tube in strip collectors

Suggested Further Reading

C. Freeman, J. Littler, A. Pike, G. Smith, R. Thomas, 'The Autonomous House Research Programme', Building Science Special Supplement (Pergamon Press), October 1974.

P. Harper, 'Directory of Alternative Technology Parts 1, 2, 3', Architectural Design 1974 <u>11</u>, 1975 <u>4</u>, 1975 <u>5</u> (useful broad review).

W.A. Shurcliff, 'Solar Heated Buildings: a Brief Survey', 10th ed. Sept. 1975, available from the author at 19 Appleton Street, Cambridge Mass. 02138, U.S.A. $ 9.00.   Details of operation and construction of all solar heated buildings in the industrialised world.

'Energy conservation: a study of energy consumption in buildings and possible means of saving energy in housing', Building Research Establishment June 1975.

G. Smith[*], Economics of Water Collection and Waste Recycling

G. Smith[*], Economics of Solar Collectors, Heat Pumps and Wind Generators

J. Littler[*], Thermal Balance at Multiply Glazed, Heat Reflecting Windows

J. Littler[*], Multi-Layer Evacuated Insulation Panels

P. South, R.S. Rangi, Laboratory Technical Reports LTR-LA-74 and LTR-LA-105 'Wind Tunnel Investigation of a 14ft. diameter Vertical Axis Windmill' Sept. 1972, National Aeronautical Establishment, Ottawa, Canada.

C. Freeman[*], 'Critical Methane Bibliography for Small Systems'.

[*]Member of the Cambridge Autonomous Housing Group, papers available from the Wolfson Industrial Unit, Trumpington Street, Cambridge.

DISCUSSION

Question: How much energy is stored in the batteries?
Answer: Of the order of 200 kWh/month.

Question: How far back does the weather data you have used go, and is there much year to year variation?
Answer: We have used 10 years of data; the only real variation is in rainfall. Some of the data comes from Kew and some is local, e.g. the wind data comes from Mildenhall.

Question: Why is the efficiency of your solar collectors so low when efficiencies of up to 45% can be obtained?
Answer: The efficiency depends on each collector and we are using water stored at $60^{\circ}C$ and pumped circulation. The figure allows for all losses e.g. inclination of the sun, reflection etc.

Question: What kind of foam do you use as an insulator, and why not use ceramic wool?
Answer: Urea formaldehyde is used because of cost. Ceramic wool performs better at high temperature.

Question: What is the water storage tank made from, and have you considered polythene?
Answer: It could possibly be polythene but we are considering trocal.

Question: Why not use the roof as the solar collector?
Answer: Because the roof is used in summer for thermal distillation to recycle water.

Question: Are all windows in the greenhouse double glazed?
Answer: Yes.

Question: Is it necessary to have the windmill on the roof of the house?
Answer: If it was not on the roof a tower would have to be built which would be more costly. There also could be problems with planning authorities.

Question: Have you a cost for the house?
Answer: Yes, £40,000 but this is only the construction cost and does not include the cost of land. It should be remembered that our purpose is to run this house as an experiment at this stage.

Question: Presumably if the house were to be mass produced it would be cheaper. Can you give an estimate of the likely cost then?
Answer: It is not envisaged that this will be produced on a massive scale. At present it is an experimental project to determine whether an autonomous house is viable in this country.

Question: Would it be possible to use wood for cooking?
Answer: Yes, it is possible to use beech hedge without using too much land, 2-3 acres possibly.

Question: Could the house be left for a two week period without any attention, e.g. during holidays?
Answer: In general, yes. The problem won't be the house itself, but the animals and possibly small boys with catapults.

Question: There is extensive use of foam for insulation. Is it possible that thermal bridges may be formed by vapour penetration causing cracking by the freeze/thaw cycle?

Answer: Yes, possibly. It is partly a problem of workmanship but might be eliminated by using vapour barriers on either side of foam.

Question: How often do you need to clean the glass in the solar collectors?
Answer: It is not as important as one might think, losses are only a few%.

Question: What is the cost of the wind generator?
Answer: The generator cost £90. I cannot give a cost for the windmill as none are available commercially.

Question: With the large amounts of foam around, is there a fire risk in the house? Does it comply with fire regulations?
Answer: Because it is an experimental building some regulations tend to be waived by local authorities. I imagine that the CH_4 collector could be a problem.

Question: What is the system of ventilation?
Answer: Mechanical ventilation with heat recovery on the exit via a thermal wheel.

Question: How will food be stored?
Answer: A freezer will be part of the standard equipment.

Question: Do you envisage a strict code of ethics on the part of the occupants? Will there be a precise and ordered way in which they must lead their lives?
Answer: No, the house is designed for a normal style of living.

Question: Am I right in assuming that all the food for humans and animals is produced on the site, apart from luxuries such as coffee, etc.?
Answer: This is the idea, but there may be problems of nutrient depletion. We may require capital input in the form of fertilizer, for example.

Question: Is it physically feasible to cultivate the ground?
Answer: Our calculations show it should be possible. Perhaps we may need a methane- or battery-driven plough.

Question: Is the future for this not a single house on its own, but a plot with 5 or 6 houses combining their resources?
Answer: There are certainly advantages in this, and a number of such projects have been formulated.

Question: How long will it take to reach steady state?
Answer: Our budget is for five years and we would expect to have reached steady state by then. It will take some time after that for economists/sociologists to evaluate the results.

Question: How would the wind generator survive a bird-strike and what about the biological safety of the methane digester?
Answer: We estimate that there should be no trouble with the wind generator in this respect. I agree there may be problems with the methane generator; initially, there will be a control system to monitor the progress of the processes in the digester so that the design of these digesters can be improved.

Question: Will the wind generator produce excessive noise in the house? Have you studied this problem?
Answer: The designer of the windmill claims that it makes virtually no noise. The speeds involved are relatively low and it is possible that it may produce a 'nice' noise.

Question: What happens when the wind speed exceeds 9 m/s.
Answer: The excess electricity generated will be converted directly to heat water in the tank.

ENERGY AND THE DEVELOPING COUNTRIES

P. D. Dunn

Engineering & Cybernetics Department, University of Reading

INTRODUCTION

In this chapter we will consider three aspects of energy use in developing countries.

(i)   The overall situation and the implications of increased energy use in the future.

(ii)   The problem of the provision of power in rural areas, including the consideration of energy resources and energy conversion.

(iii)   A description of some specific examples of small power plant and a brief introduction to the Intermediate Technology Approach to Development.

GENERAL CONSIDERATIONS

In spite of considerable efforts to encourage development over the past two decades, the results, with certain notable exceptions, have been disappointing.   This conclusion is illustrated by Figure 1 in which income per capita is used as a rough measure of development and is plotted against time. The very low value for the developing market economies, whilst increasing, still remains a very low level.   In contrast the gap between these countries and the developed market economies has increased considerably.

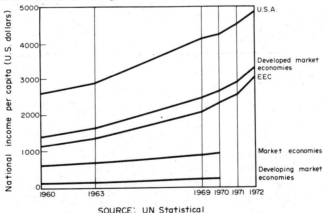

SOURCE: UN Statistical
yearbook 1973 table 183A

Fig. 1. U.N. Estimates of per capita national income at market prices.

In a technological civilisation it is to be expected that there
will be a correlation between wealth and energy use and this is
shown to be the case in Fig. 2, which refers to 1969.

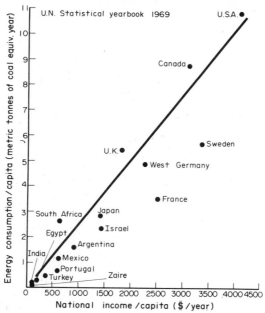

Fig. 2.

The world average per capita use of energy annually is currently
1.9 t.c.e.   This figure disguises the large differences in †
energy use between the developed and developing countries,
shown in Fig. 2 and expressed in histogram form in Fig. 3.

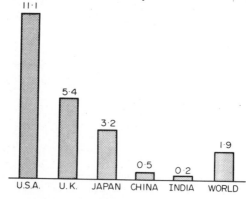

Fig. 3. World energy use t.c.e./capita/annum 1971.
    † Excluding 'non-commercial' energy sources, wood,
    vegetable wastes and dung which may amount to 1 t.c.e./
    cap/yr.

Energy use is one of several essential components for development and in this chapter we will consider how it can be made available in a form appropriate for local needs.

But first we should look at the long term implications of greater energy use by developing countries. The first aspect is the increased drain on energy resources. If, for example, the expectation is to increase the world average to the current US rate of consumption, this implies an increase of a factor of six in energy consumption together with a further factor of three or four for the inevitable population increase before a stable level is achieved. In addition to the drain on resources, such an increase in consumption would result in serious environmental and ecological consequences, together with the increased hazards of pollution and the safety problems associated with a large nuclear fission programme. This is a disturbing prospect. It would be equally unacceptable to suggest that the difference in energy between the developed and developing countries should be the long term solution. It therefore seems sensible and prudent for the developed countries to move towards a way of life which, whilst maintaining or even increasing quality of life, reduces significantly the energy consumption per capita. Such savings can be achieved in a number of ways.

(i)   Improved efficiency of energy use, for example better thermal insulation, energy recovery, total energy.

(ii)   Conservation of energy resources by design for long life and recycling rather than the short life throw-away product.

(iii)   Systematic replanning of our way of life, for example in the field of transport.

These savings should also be accompanied by the development of renewable energy sources, for example solar, wave, geothermal, and an increased effort on controlled thermonuclear fusion.

In the developing countries, on the other hand, we must endeavour to increase energy use, whilst working towards the same ultimate goal. This approach is illustrated in Fig. 4.

The capital and physical resources required for both programmes are immense and hence the time scale will be long. Nevertheless it is important that programmes are set up with these aims in mind. It is not possible to give an ideal energy/capita figure at this stage and much more work requires to be carried out on possible life styles before adequate information will become available.

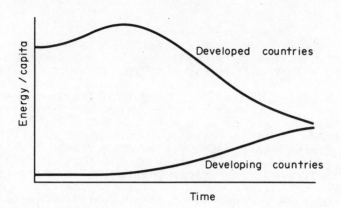

Figure 4.

Turning now to the more immediate future, there are difficulties both of resource and implementation.   Not only is the current world energy use per capita very inequitable but the occurrence of workable energy resources is also highly non-uniform.   For example coal tends to occur in the Northern Hemisphere and Africa (excluding South Africa), South and Central America and Oceania have very little.   Oil is concentrated to an even greater extent and India, Pakistan and Bangladesh are particularly badly off in this respect.   Hydro power is also non-uniform in distribution, suitable sites occurring in North and South America, Africa, South East Asia, USSR and China.

From the point of view of energy, we can divide the world into four areas

A   Developed - Energy Rich   e.g. USA
B   Developed - Energy Poor   e.g. Japan
C   Developing - Energy Rich   e.g. Middle East
D   Developing - Energy Poor   e.g. India.

It is category D which is of particular concern.

Development programmes should take account of energy resources. In this respect the 'Green Revolution' approach is open to criticism since it is dependent on the use of fertiliser which requires a high energy input.   In fact we should examine our agricultural programmes very carefully.   If we move from a traditional agricultural pattern having an energy ratio* of between 10 and 30 to a modern system having an energy ratio around $\frac{1}{3}$ the resulting energy load will be considerable.   For example, in 2000 AD if the population is 7×10^9 and we assume** a food intake of about 1/6 t.c.e. per capita/yr and an energy ratio of $\frac{1}{3}$ we find the energy required to support the programme is

$$3 \times \frac{1}{6} \times 7 \times 10^9 \ = \ 3.5 \times 10^9 \ \text{t.c.e./yr.}$$

or about half our current total world energy use.   A better route from the point of view of energy saving would be the development of nitrogen fixing non-leguminous plants, possibly in association with leaf protein extraction.

We conclude that our long term objectives should be as indicated in Figure 4, this to be achieved by the greater use of renewable energy resources, and at the same time taking account of environmental effects. In the developed countries the aim should be to reduce energy use per capita whilst maintaining quality of life and in the developing countries there should be an increase in energy/capita tending towards the same limit.

* Energy Ratio is defined as the ratio of

$$\frac{\text{Energy Content of the Food Product}}{\text{Energy Input to Produce the Food}}$$

** If we assume a man eats 3200 large Calories per day

$$3200 \times 4.18 \times 10^3 \times 365 = 4.8 \times 10^9 J \triangleq \frac{1}{6} \text{ t.c.e. per year.}$$

POWER FOR RURAL AREAS

Unlike the developed countries where most of the population is to be found in urban areas, the greater part of the population (typically 80%) of a developing country is in the rural areas. In this section we will consider the general problems of the provision of power in such rural areas.

Table 1 lists the most important of these energy needs.

Table 1. ENERGY NEEDS IN RURAL AREAS

Transport e.g. small vehicles and boats

Agricultural Machinery e.g. two-wheeled tractors

Crop Processing e.g. milling

Water Pumping

Small Industries e.g. workshop equipment

Electricity Generation e.g. hospitals and schools

Domestic e.g. cooking, heating, lighting.

Considerations when selecting power plant include the following.

(i)   Power level - whether continuous or discontinuous.

(ii)   Cost - Initial Cost
            Total Running Cost including Fuel, Maintenance
            and capital amortised over life.

(iii)   Complexity of Operation.

(iv)   Maintenance and availability of Spares.

(v)   Life.

(vi)   Suitability for local manufacture.

(vi) is of particular importance in most situations.

Table 2 lists the Energy Sources available to us.

Table 2. SOURCES OF ENERGY

Muscle Power	Human and Animal
Fossil Fuels	Coal Oil Natural Gas
Geothermal	
Tidal	
Solar Direct	
Indirect	Wind Hydropower Vegetation

Vegetation → Direct Use ——→ o
Fermentation o
Animal Dung → Fermentation ——→
Animal →—o

Nuclear	Fission Fusion Radioactive Decay

Muscle Power is given first since it still represents an important source of energy in the developing countries.   It is interesting to compare the total human power available to the installed capacity of i.c. engines in the USA.   If we assume 1/10 h.p. for half the world population, we arrive at the figure of 1.9 x 10^8 h.p., whereas the installed capacity of the i.c. engines in the USA is 2 x 10^9 h.p.   A man is surprisingly efficient as a power converter.   For example if we assume 2000 hours work per year at 1/10 h.p. this will amount to an annual useful work output of 200 h.p.hr. or .1/60 t.c.e./yr.   Since the food intake per year is about 1/6 t.c.e. /yr. the conversion efficiency is around 10%.   This is the net efficiency and does not include any other muscular output such as walking etc.

Currently the 'non-commercial' fuels wood, crop residues, and animal dung are used in large amounts in the rural areas of developing countries, principally for heating and cooking; the method of use is highly inefficient.

The per capita energy figures in Figures 2 and 3 refer to 'commercial' fuel use and do not include the 'non-commercial' fuel component, which may be as high as 1 t.c.e./cap/yr. Thus the figures referring to the developing countries are misleading in this respect. These 'non-commercial' fuels are derived indirectly from solar energy.

Solar energy, both direct and indirect, has considerable development potential. The disadvantages of low power density and variability are often not important in these situations since power level requirements are low and for many applications such as pumping continuity of supply is not an important consideration.

As in the developed countries, the fossil fuels are currently of great importance in the developing countries. Geothermal and tidal energy are less important though, of course, will have local significance where conditions are suitable.

Wave energy is included under wind energy, since wave, energy is not likely to play an important role in the present context. There may be local exceptions, for example, the Mauritian scheme reported by Bott (1).

Nuclear energy sources are included for completeness, but are not likely to make any effective contribution in the rural areas.

Having listed the energy sources we should now consider methods of energy conversion. These are listed in Table 3. It can be seen that a matrix constructed from Tables 2 and 3 will contain a large number of possible systems.

Figure 5 shows the approximate power ranges for the various conversion devices. In the present application thermionic and thermoelectric converters are unlikely to be of interest except in very small sizes due to high initial cost. Most rural power applications can be satisfied by engines of less than 30 kW. An exception is the use of small water turbine driven electric generators; for example these have been used extensively in China and are used for power output of up to a MW for village electrification.

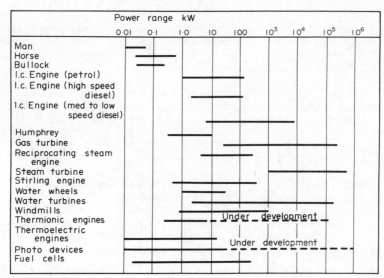

Figure 5.   Power Range of Energy Converters.

Table 3.   METHODS OF ENERGY CONVERSION

Muscle Power	Man
	Animals
I.C. Engines	
Reciprocating	Petrol - Spark Ignition
	Diesel - Compression Ignition
	Humphrey Water Piston
Rotating	Gas Turbine
Heat Engines	
Vapour (Rankine)	Reciprocating* Steam Engine
	Rotating Steam Turbine
Gas (Stirling)	Reciprocating*
(Brayton)	Rotating Gas Turbine
Electron Gas	Thermionic
	Thermoelectric
Electromagnetic Radiation	
Photo Devices	
Hydraulic Engines	
Wheels, Screws, Buckets	
Turbines	
Wind Engines	
Windmills	Vertical axis
	Horizontal axis

Table 3. continued

Electrical/Mechanical
    Dynamo/Alternator
    Motor

---

            * Can be constructed using a water piston

The renewable energy resources are particularly suited for
the provision of rural power supplies, and a major advantage
is that equipment such as flat plate solar driers, windmills,
etc. can be constructed using local resources and without the
high capital cost of more conventional equipment.   Further
advantage results from the feasibility of local maintenance and
the general encouragement such local manufacture gives to the
build up of small scale rural based industry.   In the next
section we give some examples of small scale energy converters
of this type.   Nevertheless it should be noted that small
conventional i.c. engines are currently the major source of
power in these areas and will continue to be so for a long time
to come.   There is a need for some further development here -
not in engine design which is well understood, but in design
to suit local conditions, to minimise spares holdings, to
maximise interchangeability both of engine parts and of the
engine application.   Emphasis should be placed on full local
manufacture.

EXAMPLES OF SMALL POWER PLANT

Muscle power and its effective harnessing to produce useful
work has been described in the talk by Stuart Wilson.   Figure
6 shows a simple manpowered flywheel developed by Weir at
Edinburgh University.   The construction is based on a bicycle
wheel and the moment of inertia of the wheel is raised by
filling with concrete.   This flywheel may be used to power the
simple lathe, Figure 7.

Wind power has been used for several thousand years both for
propulsion and the production of mechanical shaft power.   It
is particularly suited to crop processing such as grain
milling and to pumping applications.   Wind mills can be used
for electricity generation but the requirement for energy
storage raises the cost and complexity of the plant.   Most
windmills are of the horizontal axis type.   An advantage of
the vertical axis mill is that it does not require means of
orientation into the wind.   Figure 8 shows a simple vertical
axis Savonius rotor mill constructed at Reading University from
a 40 gallon oil drum and serving as a prototype for a mill
subsequently constructed in Zambia.   The efficiency of
conversion of this drag type mill is low, but the low speed
high torque characteristics make it suitable for direct

Figure 6

Figure 7

connection pumping applications.   Efficiency of energy
extraction is not of course the major criterion in assessing
a windmill since the energy is free.   What is important is
the work output divided by the capital cost of the plant.   The
Darieus vertical axis design Figure 9 has recently been revived
by a group at the National Research Council o. Canada (2) who
have shown that due to its simple design the initial cost is as
little as 1/6 of a conventional horizontal axis mill of the same
rating.   The Darieus type mill is a high speed high efficiency
life type mill suitable for electricity generation.   There is
some difficulty in starting and the Figure 9 shows how small
Savonius rotors can be used to provide starting torque.   In
spite of its long history, there remains plenty of scope for new
ideas in the windmill field.

The direct collection of solar radiation by flat plate collectors
can be used for water heating, crop drying and distillation,
using simple and cheap plant.   Figure 10 shows a simple water
heater made by a colleague in Zambia from corrugated iron,
plastic sheet and an old cistern found on a scrap heap.   This
heater provided all the domestic hot water required by a small
household.   Suitable designs have been described in detail by
Headley (3).

Higher temperatures and heat fluxes can be achieved by the use
of concentrators.   Such concentrators may be used to drive
small heat engines.   Liquid piston engines are of interest in
association with solar heating (4).   Liquid piston engines are
simple to construct, do not have tight tolerances, do not
require lubrication of the piston, and have long life and low
capital cost.   The efficiency of these engines can be as high
as that of their solid piston equivalent.   Liquid piston
engines can be constructed to operate with either external
combustion or internal combustion.   The simplicity of
construction of liquid piston engines enables them to be
manufactured on a small scale.   This, together with the ease of
servicing, is an important advantage when used in developing
countries.

Figure 11 shows a number of different heat engine configurations
employing liquid pistons.

Internal combustion liquid piston engines are also possible, the
Humphrey engine being the best known example.   The Humphrey
engine used the Atkinson cycle; Figure 12a and 12b show an
actual pump cycle for a four stroke engine.   Figure 12c shows th
the valving arrangement of an engine.   In the early part of
this century, Humphrey (5) constructed several large pumping
engines (6), fuelled by gas, which seem to have operated quite
satisfactorily.   However, for various reasons interest was lost
and development discontinued.   Work on small pumps specifically
for use in developing countries was started at Reading University
about 4 years ago.   Figure 13 shows a 6" (152 mm) bore pump
capable of pumping about 6000 gallons (27000 l) per hour to a

Figure 8

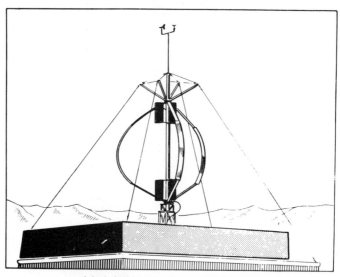

A sketch of a high-speed, vertical-axis wind turbine on the roof of a building in New
Mexico. Built by the Sandia Corporation to a design of the National Research Council of
Canada, the turbine has blades with a symmetrical aerofoil section, which bow out to a
diameter of 14 feet, the shape being designed to eliminate bending stresses. The electrical
output is 1 kW in a 15 mph wind, rising to 8 kW in a 30 mph wind.

Figure 9

Figure 10

head of 20 ft. (6 m).   The total pipe length is about 90 ft.
(27 m) and the frequency of the pump is about 12 power strokes
per min.   The conversion efficiency is at present about 8% but
it is hoped to raise this to 20% with further development.

The water inlet valves are of the cone diaphragm type.   The
pump works on the 4 stroke principle and has a cylinder head
fitted with three 2½ inch valves - exhaust valve, scavange
valve, and mixture inlet valves.   There is a simple valve
locking device arranged so that the exhaust and the scavange
valves open together and then the mixture inlet valve opens on
alternate strokes.

The fuels so far used have been natural gas, simulated town gas,
and propane.   It is hoped that it will also be possible to use
liquid fuels.

Ignition is by a spark plug with a gap of 3 mm to prevent
fouling by water.   The spark is obtained with a 12-volt car
ignition coil;   a piezo ceramic device has also been used with
success, and it is hoped to go over to this device permanently,
thus making the pump independent of batteries.

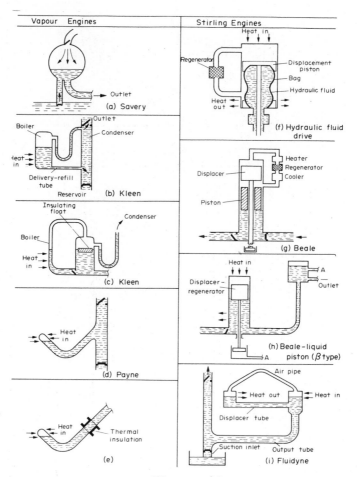

Figure 11.

Methane can be generated by the anaerobic fermentation of dung or vegetation, and small plant are in operation in various parts of the world. Figure 14 shows a simple diagram of such an installation.

The examples in this section were chosen to illustrate the type of power converters which could make an important contribution to development in rural areas. This type of self help or the provision of appropriate soluations is not of course confined to energy. In fact one should not restrict development to a single topic but should consider a total approach to the provision of appropriate solutions to the problems of a community or area. One of the first to express this point of view was E. F. Schumacher who recognised the need for a form of development in parallel with major capital investments such as power stations, steel works, airports, etc. (7, 8, 9).

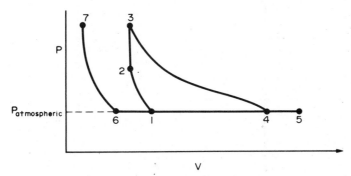

Figure 12a. Theoretical p-v diagram for the Humphrey pump.
1-2: compression, 2-3: ignition, 3-4: expansion,
4-5: free movement of column, 5-6: exhaust,
6-7-6: cushion and bounce, 6-1: intake.

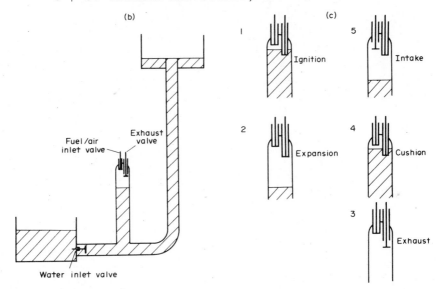

Figure 12 (b and c) The Humphrey Pump

Schumacher proposed that solutions should be found which were
appropriate to

(i)   the needs

(ii)   the skills

(iii)   the financial and material resources

(iv)   the culture

of rural based communities.   In this way a community could
raise its quality of life and remain in its same location.
This is a total system approach to development carried out by
the people themselves.   In order to assist such development,

Figure 13

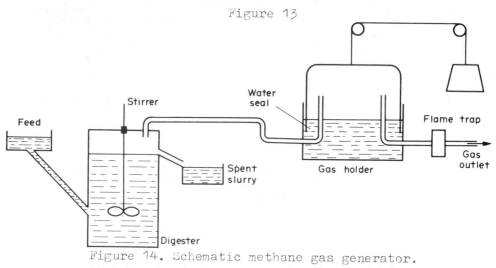

Figure 14. Schematic methane gas generator.

Schumacher founded the Intermediate Technology Development
Group, London, in 1965, to investigate ways and means of
utilising to the fullest extent the resources available to
developing countries through the application of the appropriate
technologies.   Its main aims are:-

to compile inventories of existing technologies which
can be used within the concept of low-cost, labour-
intensive production;

to identify gaps in the range of existing technologies;

to research into and develop by invention or modification
new or more appropriate processes;

to test and demonstrate in the field the results of its
investigations;   and

to publish and make known the results of its work as
widely as possible, so as to facilitate the transfer and
use of appropriate technology.

It is important that all the knowledge and experience gained
through research and practical work in the field is fully
appraised, recorded and made available worldwide.   This
harvesting of experience is one of the Group's vital roles.

A number of voluntary panels brings together a wide range of
people with a high level of professional expertise.   They give
advice to the Group;   introduce new ideas and fields of activity;
widen the Group's range of contacts and expertise;   and lend
authority to the concept of intermediate technology within
governments, funding agencies and professional organisations.
The work of the panels includes the preparation of specifications of available equipment, teaching manuals and technical
reports, and the promotion of new research and development
with universities, technical colleges and professional
associations.

The ITDG Panels are given as an Appendix.   The scope of the
Power Panel is shown in Figure 15, and covers the types of
power converter described in this section.

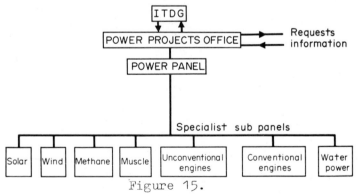

Figure 15.

CONCLUSIONS

The encouragement of greater energy use is an essential component of development.   In the short term we require mechanisms
to enable the rapid increase in energy/capita, and in the long
term we should be working towards a way of life which makes use
of energy efficiently and without the impairment of the environment or of causing safety problems.   Such a programme should
as far as possible be based on renewable energy resources.

Large scale, conventional, power plant such as hydro-power, has
an important part to play in development.   It does not, however,
provide a complete solution.   There is an important complementary role for the greater use of small scale, rural based, power
plant.   Such plant can be used to assist development since it
can be made locally using local resources, enabling a rapid
build-up in total equipment to be made without a corresponding
and unacceptably large demand on central funds.   Renewable
resources are particularly suitable for providing the energy
for such equipment and its use is also compatible with the long
term aims.

There is a need for greater attention to be devoted to this
field in the development of new designs, the dissemination of
information and the encouragement of its use.   International
and Government bodies and independent organisations all have
a role to play in this work.

APPENDIX

The ITDG panels are as follows:

Agriculture	Chairman: Dr. H. S. Darling, CBE, Principal, Wye College.
Building and Building Materials	Chairman: J. P. M. Parry, MBE.
Chemistry and Chemical Engineering	Chairman: P. S. Reid, B Sc, MBIM.
Co-operatives	Chairman: Lord Taylor of Gryfe.
Forestry and Forest Products	Chairman: Keith Openshaw, B Sc (For), MA (Econ), International Forest Science Consultancy.
Power	Chairman: Professor P. Dunn, Dept. of Engineering & Cybernetics, University of Reading.

Rural Health   Chairman: Dr. Katherine Elliott,
                                          Ciba Foundation.

Transportation   Chairman: Dr. J. D. Howe,
                                          Transport & Road Research
                                          Laboratory.

Water Treatment   Chairman: Professor D. J. Bradley,
                                          London School of Hygiene
                                          & Tropical Medicine.

Communication   Chairman: John Bowers, University of
                                          Reading.

Other fields in which the Group has access to specialist advice
include: textiles, food technology, ferro-cement, glass
manufacture.

REFERENCES

1. Bott, A. N. W.   Power Plus Proteins from the Sea, Journal
   of Royal Society of Arts, July 1975.

2. South, P., Ranji, R.   The Performance and Economics of the
   Vertical-Axis Wind Turbine developed at the National
   Research Council, Ottawa, Canada.   Paper PNW 73-303.
   Annual Meeting of Pacific North West Region of American
   Society of Agricultural Engineers, Calgary, Oct. 1973.

3. Headley, O. St. C.   Cascade Solar Still for Distilled
   Water Production.   Journal Solar Energy. Vol. 15, pp.
   245-258. (1973).

4. Dunn, P. D., Walpita, N.   Liquid Piston Engines, International Mechanical Engineering Congress, Shiraz, Iran.
   Apr.-May 1975.

5. Humphrey, H. A.   An Internal Combustion Pump and Other
   Applications of a New Principle. Proc. I. Mech. E. Nov.
   1909.

6. The Engineer. March 14. 1913.

7. Schumacher, E. F.   Small is Beautiful.   Pub. Blond and
   Briggs, June 1973.

8. McRobie, G.   Technology for Development - Small is
   Beautiful.   Journ. Royal Soc. of Arts. March 1974.

9. Dunn, P. D.   Intermediate Technology. E.R.A. Journal,
   Autumn 1971.

DISCUSSION

Question: You said that ITDG* is designing technology to fulfil the needs of developing countries. How do you know what these countries want?

Answer: The ITDG develops a systems approach in any one area. Problems are usually solved in the UK but development is carried out with the people of the country.

Question: Are people from ITDG going out to these countries?

Answer: One aims for a technology transfer. Thus one mechanism is to link up with an overseas university. An example of this is a solar pump being developed in the UK and a solar collector being developed in Trinidad. On completion, these projects will be linked. An involvement over there is essential.

Question: You were the first lecturer, in this 10 day summer school, to show a time-projected graph where you suggest that the developed world might try to decrease its energy consumption. How realistic is this?

Answer: The developed world can decrease its energy consumption by reducing its "energy wasters", e.g.
1) Your car is a visible waster of energy (15% efficient). There is no reason why the size of the car could not be halved. A car could be dispensed with by providing good long-range public transport, plus many small, short-range, self-drive taxis.
2) Industry and the private house are also concealed wasters. There is much room for decreasing waste, without decreasing our quality of life.

Question: Is it not impossible to reduce the "big machine" of the developed countries?

Answer: This is simply a time-scale problem. The long-term objectives should be to maintain an optimum quality of life.

Question: Do not politicians in the developing countries require highly developed status technology and thus regard IT** as an insult?

Answer: When E. Schumacher first proposed ITDG many treated it as second-rate. However it is now rapidly gaining ground. Note also the example of Tanzania.

Question: What would be the capital cost of an IT energy development which yields, say, 1 ton of coal equiv/year?

Answer: IT is based on local resources so that the capital cost is within the range of the local community. Capital costs can be offset by involving the local community in a project e.g. the building of a rainfall catchment device by the teachers and pupils of a school in Botswana.

Question: So far you have talked about small localised projects in these countries. Is ITDG concerned with larger operations, e.g. exploiting hydropower?

Answer: No, ITDG is concerned with small equipment primarily for rural areas.

Question: How do you see the future of developing countries in terms of increasing population?

Answer: There is a great need for the practice of birth control in these

*
** Intermediate Technology Development Group, London.
   Intermediate technology.

countries; it should be encouraged. It is not widely practised
because:
1) children provide security for old age,
2) large families are a social tradition,
3) muscle power may often be the only form of energy available,
   requiring large families to work the land.
The problem is that in these countries increasing population means
lower GNP and lower life expectancy. A rate of change of GNP of
3% per year and population growth of 4% per year is a dynamic
condition with usually adverse consequences.

Question: The problems here are enormous. How much real influence does IT
            have?
Answer: ITDG has a lifetime to date of 8 years; give us time!

Question: Surely real development can only come within the developing coun-
            tries themselves. Does IT just represent an increase in trade only?
            Would it not be better to influence governments to initiate research
            in the countries themselves?
Answer: ITDG workshops are sited in these countries. These, with the co-
            operation of local governments and local universities, grow within
            the locality in accordance with the needs of the people. Hence it
            does involve the local population at all levels. As regards trade,
            ITDG has 2 beneficial effects:
            1) it considers internal substitution for imported goods,
            2) it encourages development of local industries and production
               with a view to export markets.
            For example a bolt making factory and a weaving industry have
            been introduced in Kumansi, Ghana.

Question: Are your fluid piston engines made in the developing countries?
Answer: When the research is completed in this country, they will be
            developed and produced there.

Question: Since the Arab oil embargo, the costs of fertilizers and associated
            chemical products have increased drastically; what can be done
            about these agricultural needs?
Answer: Both the Agriculture and Chemical Panels of ITDG, are looking at
            IT agricultural methods, such as nitrogen fixation etc., but it is
            difficult to encourage a less energy intensive style of agriculture.

SECTION 7

ENVIRONMENTAL AND SOCIO-ECONOMIC ASPECTS OF ENERGY USE

ENVIRONMENTAL ASPECTS OF ENERGY CONVERSION AND USE

C-G. Ducret

Geneva, Switzerland

1. Environment

"What does the word 'environment' mean to the people I will be speaking to?"
This was the first question that came to my mind when I started to prepare
this lecture. Then I spent half-a-day in the library searching for a good
definition, found complicated descriptions but no suitable answer: environment has no precise limits because it is in fact a part of everything.

Indeed, environment is, as you probably already know, not only flowers
blossoming or birds singing in the spring, or a lake surrounded by beautiful
mountains; it is also human settlements, the places where people live, work,
rest, the quality of the food they eat, the noise or silence of the street
they live in. Environment is not only the fact that our cars consume a good
deal of energy and pollute the air, but also that we often need them to go to
work and for holidays. It is also the fact that a new car is in effect a
more or less carefully planned future waste representing an impressive
energy investment. Environment is also the way in which developing countries
will develop, the fact that they should have the choice between clean and
polluting development, that their population is growing very fast and too
often is underfed, that we at present do not know where we will find the food
necessary for nourishing six thousand, five hundred million people who will
populate the earth in 25 years' time (half being in Asia). Environment is
also the fact that a delicate climatic balance controls the growth of crops
and that human consumption of energy might, according to some climatologists,
quite easily alter this delicate climatic balance and consequently the
feeding capacity of the planet.

I could easily continue along these lines for quite a long time, but this is
not the purpose of this lecture. What is fundamental here is the fact that
environment, from the energy point of view, is characterized by a permanent
flow of energy. The source of this energy is, as you know, that fusion
reactor sitting in space that we call 'the sun'.

This flow of energy through the 'natural' environment is very well organized
and controlled; it is a little bit like a complicated electrical machine
working continuously and having fuses for all its numerous sub-systems. If
the load is too high, the fuse of the overloaded sub-system blows; the
machine is still working but one of its sub-systems stops functioning. This
is naturally an illustration and we will use it with the care one should
always exercise when using images.

What about Man? Obviously Man uses energy just as plants, bacteria,
mushrooms, bees, fish and rats do. He largely uses solar energy – food,
hydro-power, wood – and thus participates harmoniously in the natural flow
of energy through the environment. But he also uses oil, gas, coal and
nuclear power which are not included in the present natural energy flow
through the environment. By using such sources of energy, Man is thus
modifying his environment.

But does Man have the right to modify his environment?

Being a biologist by training, I suggest we turn to wild life to answer the
question. Let us take beavers. Beavers, as you know, build dams and often
flood fairly large areas; they modify their environment in order to make it
safer and more easily usable. Nobody could claim that beavers do not live
in harmony with their environment. Similarly, Man has the right to modify
his own environment, like any other living being. But let us now imagine
that, in a given place, ambitious beavers have built a particularly large
dam which floods a huge area. As they are not aquatic plants, all the trees
will die, and as a consequence, the beavers will no longer be able to repair
their dam. In time the dam will break and the place will dry up. But our
ambitious beavers would have to wait 30 to 50 years, i.e. the time required
by trees to grow, before they could again have the raw materials needed for
re-building their dam.

Similarly, Man's freedom to modify his environment is contained within
certain limits beyond which the fuses blow; the environment is then
drastically affected by too great an impact which then does a disservice to
Man and his activities. As a general rule we can conclude that Man has the
right to modify his environment but only within limits established by the
smooth functioning of the environment. These limits can be called the
acceptance or the restorative capacity of the environment.

This applies to energy as well as to mercury in the flesh of fish. What we call pollution or harm to amenities, and what economists call externalities are in fact the effects of an environment, the acceptance capacity of which has been saturated.

Why is it so? Why do we have pollution problems? I see two main reasons for this. The first is the fact that the smooth functioning of the environment is considered as granted free of charge and therefore economically non-existent. A cubic mile of air has no economic value, one billion gallons of sea-water have no economic value. Clean air and transparent sea water cannot be included in national accounts; that is the source of most of our problems. Fortunately the tourist industry might improve this situation.

The second reason lies mostly in the growing size of the sources of impact on the environment. Take coal-fired power plants as an example; their size grew from a couple of hundred kilowatts to several thousand megawatts in seventy years, with a similar increase in size of the amounts of burnt coal and consequently the effluent gases and particulates that can be released into the atmosphere. This evolution is mostly due to the fact that the bigger you build, the cheaper it is per unit of power produced. It also allows a better technological control of the quality of the effluents released into the environment. However, we do not know very much about environment and its acceptance or restorative capacity; we are just beginning to have some general ideas about what is happening when it is saturated as   seems to be the case, for example, with rains falling over Scandinavia containing high amounts of sulphuric acid resulting from SO_2 emitted, among other sources, by coal and oil-burning power plants in the United Kingdom and Central Europe.

2. The energy hierarchy

As you know, all energy sources are not equal and we can list them in order of hierarchy. The king of energy is electricity. With it everything can be done; we can run computer programmes, telephone to friends, light houses, play records, produce mechanical work to move trains or pump water, for instance, and generate heat to make our morning toast. All these operations can be performed with electricity.   In addition, electricity can be transported practically instantaneously and is almost pollution-free at the consumer level. In the environment, electricity is sometimes found in large quantities (lightning) but mostly in minute amounts in many biological processes like nerve transmissions which in larger amounts would make it dangerous to living organisms.

The next grade in the energy hierarchy is mechanical energy; mechanical energy can achieve two tasks that electricity performs: it produces movement and heat. Mechanical energy can also generate electricity but the efficiency of its conversion is 90-95%, the remainder being lost as diffused heat. Mechanical energy is almost untransportable and occurs most widely in the environment as wind, flowing water and tides, all three being used by Man.

The third level in the energy hierarchy is dense heat, i.e. high amounts of heat in a confined space. It is generally produced by burning oil, gas, coal and wood or by nuclear fission. Because of entropy, dense heat sources do not exist in nature with the exception of volcanic phenomena (including geothermal energy). Per se, dense heat cannot, because of entropy, do very much except produce diffused heat. It is widely used for domestic applications not requiring high grade energy like home heating, cooking, etc.

Dense heat can also be converted into mechanical energy and, if the latter is in its turn converted, it can generate electricity.

Thermal power plants are designed to accomplish this conversion. They perform it by burning oil, gas or coal, or by controlling a fission reaction. As you know, this is not a pollution-free operation. Interesting comparative estimates have been made by the U.S. Council on Environmental Quality* on the environmental impact of standard 1000 MWe power plants with a load factor of 0.75. These estimates include not only the releases occurring at the power plant itself but also cover fuel extraction and treatment, as well as the storage of waste and the area of land required for all these operations. These estimates are:

---

* Council on Environmental Quality, Energy and the Environment - Electric Power, Washington, D.C., Government Printing Office, August 1973 (tables 3, A-11 and A-12).

(figures in thousands of metric tons and in km^2)

Primary source of energy	Emissions		waste	area
	atmosphere	water		
coal	380	7–41	60–3,000	120
oil	70–160	3–6	negligible	70–84
gas	24	1	–	84
nuclear	6	21	2,600	77

The atmospheric emissions of fossil fuelled installations are mostly aldehydes, carbon monoxide, nitrogen oxides, sulphur oxides and particles (i.e. ash), as well as, obviously, carbon dioxide. With the exception of CO_2, all these chemicals can in theory be removed with present technology, the only limitation being the expense of this purifying; indeed, the cost of cleaning increases exponentially with the purity reached at the outlet of the power plant stacks.

For nuclear power stations, the problem of chemical pollution is negligible, but this statement does not apply to the remainder of the nuclear fuel cycle. Indeed, the enrichment step of this cycle requires the chemical conversion of uranium oxide into a gaseous fluorinated uranium compound, uraniumhexafluoride (UF_6). Isotopic enrichment through gaseous diffusion further requires a large investment in electricity - about 5% of the energy which will be produced at the nuclear power station itself. When the fuel is enriched, UF_6 is again chemically converted into metallic form and some fluorine, a very toxic pollutant, is released into the environment. Fluorine, which is the most electronegative chemical element is not, however, a typical nuclear pollutant as other industries also release large quantities into the environment.

The reprocessing of spent nuclear fuel rods is the drawback of nuclear power mostly because this step requires sophisticated chemical reactions involving strong acids, releases several radio-active gases into the atmosphere and results, among other things, in the production of relatively compact but extremely toxic wastes that will be stored over a very long period of time, i.e. probably hundreds of centuries.

I would like here, and without entering into the nuclear safety debate, to draw your attention to the fact that storage does not mean disposal; we are used to associating the concept of disposal with wastes but high level radio-active wastes should, and must, be associated with the word 'storage' and thus 'management'. This very fact raises two major questions which, up to now, have not been clearly and fully answered by anybody.

The first one is: "What are the consequences for the socio-political environment of this very long-term management of highly toxic wastes?" Some people, and among them the most ardent supporters of nuclear power, Dr. Alvin Weinberg* for instance, argue that, due to the nature of its wastes, a nuclear-based economy would necessarily lead to an extremely long-lasting social commitment meaning no war, no revolution, absolutely stable societies and almost perpetual institutions. This is obviously something quite different from what we know of the history of humanity.

The second question is: "What is, in the long run, the overall efficiency of a nuclear economy?" If this question seems strange to you, just think about this illustration:** "... if a power station produces 1000 MW for 25 years but leaves waste materials which require a power input for maintenance of 100 kW for 250,000 years, then the net energy output will be zero". And what are 100 kW? One-seventh of the power required by one single floor of the World Trade Centre***; the energy required in 1970 by 10 average Americans or by 29 average Europeans.

One of the conclusions that can be drawn from what has preceded is that one can, with good reason, argue that our non-renewable resources are being rapidly consumed by our growing energy requirements and that we do not have the right to bequeath to our successors the strange heritage of our highly toxic wastes which they will be obliged to manage carefully for centuries and force upon them the necessity of establishing excessively stable and long-lasting social institutions.

---

* Weinberg A., "How Can Man Live With Fission?", Proceedings of IIASA Planning Conference on Energy Systems, Laxenburg Castle, Austria, 17-20 July 1973.

** Chapman P. and Mortimer N.: "Energy Inputs and Outputs for Nuclear Power Stations", ERG 005, December 1974, (Milton Keyes, Bucks., England).

*** The World Trade Centre is wired for an electric consumption of 80 MW.

3. The "waste" heat problem

Now let us be optimists and assume that, through a technological wonder, all environmental problems have been solved and fossil fuelled power plants are 100% clean, as well as the whole of the nuclear industry. Would there be any problems remaining at the conversion level?

The answer is yes because, as already mentioned, power plants cannot operate at high efficiency; as long as they "climb up" the energy hierarchy, using the sequence dense heat → mechanical energy → electricity, they will release into the environment diffuse heat, which represents the last and lowest level in the energy hierarchy. Quantitatively the amount of this energy is high: it currently represents about 65% of the total energy produced by power plants. Qualitatively it is such a low-grade energy that it is called "waste heat" which is discharged like any other waste either into rivers or the sea or into the atmosphere.

The environmental effects of this waste heat, particularly on rivers and the sea, are little understood mostly because they result from the combination of physical, chemical and biological parameters, the latter varying enormously from one place to another and depending upon various factors, such as climate, behaviour, developmental patterns, etc. To cut a long story short, I will just mention that these effects can range from a beneficial improvement of the situation - a rare case however - to the complete asphyxia of a water body - a not so rare case - with all possible steps in between. Nevertheless, I would agree with the idea that, in the case of a non-polluted and large river, the thermal impact due to waste heat released by a clean, medium-sized power plant, is negligible. This apparently falls within the acceptance capacity of a non-polluted river. But what are we to think of projects aimed at building a dozen or more 1000 MWe power plants along the banks of the Rhine or the Rhône rivers, both already heavily polluted?

When water is scarce and clearly the acceptance capacity is already saturated, one uses cooling towers. The purpose of these cooling towers is to accelerate a normal process in rivers and in the sea, i.e. the transfer of heat to the atmosphere. There are two types of cooling towers, wet ones which use the "waste" energy for evaporating water, and dry ones which directly transfer heat to the air without evaporating water. Dry cooling towers are very expensive to operate and consume electricity for pumping

water and for creating a forced air draught, but save large quantities of
water and have practically no environmental impact except noise and the fact
that they release heat into the atmosphere. Their maintenance and operating
costs, as well as the capital investment they require, are responsible for
their rare use when water is abundant.

The usual wet cooling towers are huge hyperbolic structures measuring more
than 300 feet in height and using the natural air draught which is created
by their height and shape for cooling down the water which evaporates and
falls in small droplets inside the structure. Under unfavourable meteorological conditions, they are the cause of fog and cloud and some meteoro-
logists have claimed that they could alter the climate of the region where
they are erected. In addition, their size creates aesthetic concern,
particularly in rural areas, although some people find them pleasing to the
eye.

All these waste heat problems are due to the fact that, in thermal power
plants, dense heat is converted into electricity with a low efficiency
coefficient. Contrary to hydro-electricity, the environmental "price" of
this electricity is high. Indeed, for one unit of electricity we have two
units of wasted energy discharged into the environment, making fossil-fuel-
generated electricity an environmentally precious and expensive asset, the
uses of which should logically be restricted to applications that only
electricity is able to perform. In effect, isn't it stupid to use
electricity for heating purposes? In fact, for each house you heat with
electricity, you spend the energy that could be used for heating three
houses; for each kettle of water you boil electrically for making tea, you
add two kettles of boiling water to rivers or the atmosphere, and so on.
I really rebel against some current policies aiming at adopting all-electric
systems, just because it is politically or economically the easiest way. In
fact, these policies are wasting large amounts of non-renewable resources,
like oil, gas, coal or even uranium, a wastage which is not environmentally
acceptable as long as the efficiency coefficient of the electric conversion
process is not much higher.

4. <u>Does climatic balance represent the ultimate limit to energy consumption</u>?

Mankind is hungry for energy and, in the decades to come, this hunger cannot
but grow rapidly, particularly in the developing regions of the world.

Moreover, if industrial output is to increase, even larger inputs of energy will be required for pollution control, particularly in developed regions, if environmental degradation is not to reach alarming proportions. The question we will now try to answer is whether the environment and specifically the climatic environment would permit the consumption of very large quantities of energy.

I will first try to summarize the situation by saying that solar radiation is undoubtedly the most fundamental climatic factor. About 23% of it is absorbed on the Earth's surface corresponding, on a world average basis, to 95 watts/m^2. Most of this energy is used for keeping up the water cycle, while about 1 watt/m^2 is required to drive winds and sea currents and for maintaining waves and convection. From this, one can claim that a region producing approximately 1 watt/m^2 of energy in the form of heat, releases into the environment a quantity of energy comparable to the amount needed to govern winds and ocean currents. In view of the climatic and agricultural importance of winds (monsoons, etc.) and sea currents (the Gulf Stream), it is tempting to infer that this order of magnitude marks a limit that should not be exceeded if the risk of causing non-negligible climatic disturbances is to be avoided.

But how fragile is the climatic equilibrium? Various studies have been undertaken in order to answer this question. M. Budyko* in the USSR has calculated that a 1% increase in the solar constant would result in a polar icecap meltdown and W. Washington** in the U.S.A. has reckoned that a slight change in the direction of high altitude winds would have a powerful warming effect on the Greenland region (if the ice on Greenland were to melt, the level of the ocean would be raised by 7 metres) and at the same time a strong cooling effect on northern Europe, southern Siberia and the north-west part of Canada.

Thus it seems quite reasonable to estimate that 1 watt/m^2 represents approximately the heat acceptance capacity for our climatic environment.

---

* Budyko M.: "The Future Climate", Transact. Americ. Geophys. Union, (1972), page 868.

** Washington W., "Numerical Climatic Change Experiments: The Effects of Man's Production of Thermal Energy", Journal of Applied Meteorology, Vol.11 (1972), page 768.

Using the data for energy consumption as published in the United Nations
Statistical Yearbook — a very official document — I have calculated the
energy released in 1970 per m^2 of land in various areas of the world. The
result of these calculations is given below:

Area	Energy $(watt/m^2)$
Africa	~0.00
Asia	0.03
Central America	0.05
South America	0.01
Western Europe	0.32
Eastern Europe	0.42
Oceania	0.01
USSR	0.04
Canada	0.02
U.S.A.	0.24

It can be seen that Eastern and Western Europe are not far from the
1 watt/m^2 zone; indeed, one or possibly two doubling periods of energy
consumption would make them come within this zone. U.S.A. would follow at a
somewhat later stage. It is striking to think that if the past trends in
energy growth continue and consumption doubles, say every ten years, the
whole European region might therefore be in a delicate situation as regards
climate before the end of this century, although some diluting effect can be
expected, due to the fact that Europe is partially surrounded by oceans. The
"energy crisis" which was in fact a "price of energy crisis", i.e. an
economic problem far more than an availability shortage, might give us some
time to think about these questions before definitely orienting our future
economic activities towards highly wasteful energy policies.

Per capita consumption is often regarded as an indicator of the well-being of
a given population (although many other factors also enter into consideration)
and it has been estimated* that a 20 kW per capita energy consumption represents the energy situation of a characteristic post-industrial society living

----

* H. Weinberg and A. Hammond, Proceedings of the 4th Conference on the
Peaceful Uses of Atomic Energy, Geneva, September 1971, Vol. 1, page 171.

in a materially static civilization. It might be asked what the world
situation would be like if - through a technical wonder - every human being
living in the year 2000 consumed the energy equivalent of 20 kW. The hypothetical result would be:

Region	Population[†] (millions)	Watts/m^2
Africa	818	0.54
Asia	3,778	2.74
Central America	180	1.44
North America	333	0.34 [*]
South America	472	0.52
Western Europe	441	2.24
Eastern Europe	127	2.57
Oceania	35	0.08
USSR	330	0.29 [**]

[*] Includes Canada   [**] Includes Siberia

On the basis of the results of this hypothetical situation one can infer that

(a) an extensive energy-consuming development of some regions of the world
could not take place without causing changes in climate. These should
be associated with the various indirect effects they could have on
climate-related processes, amongst others, on food production, water
régime, etc.

(b) since the control of population growth appears to be a characteristic of
the major energy-consuming nations, the later the poor countries
experience high individual consumption, the more their populations will
have grown and, consequently, the greater their national consumption
will have to be if they wish to provide the standard of well-being their
inhabitants desire.

5. Conclusions

Several general conclusions can be drawn from what I have tried to discuss
under the title "Environmental aspects of energy conversion and use".

[†] United Nations Statistical Yearbook, 1972.

The first conclusion is that energy is a normal constituent of the environment but its excess above the environmental acceptance capacity is, like any excess, detrimental not only to the environment but also to Man.

A second conclusion is that Man has at his disposal a wide variety of energy resources ranging from energy resources which are almost unlimited, renewable and which are included in the normal flow of energy through the environment, to limited, non-renewable sources which are not included in this flow.

A third conclusion is that most energy sources are more or less clean as long as they are used on a reasonably small scale falling within the acceptance capacity of the environment.

A fourth conclusion is that Man, because of something arbitrary he has invented which he calls "money", is tending to choose for the future development of his civilization, sources of energy which are limited, non-renewable and not included in the normal energy flow through the environment.

A fifth conclusion is that Man uses large amounts of these non-renewable energy sources for producing environmentally expensive electricity which he uses only in very limited amounts for purposes that only electricity can meet. He should therefore not be surprised if he finds his stocks of non-renewable energy resources diminishing more and more rapidly, particularly if he replaces, by electricity, functions of energy which were performed by sources of energy situated in the lower grades of the energy hierarchy.

A sixth conclusion is that the planet's climatic balance has a limited acceptance capacity and therefore the extent of the use of energy sources not normally included in the energy flow through the environment is necessarily limited.

A seventh conclusion is that Man is proliferating very rapidly when the feeding capacity of his planet is apparently growing less rapidly, and more and more energy is required for increasing or sustaining the production of crops. This crop production is highly dependent on the climatic balance and Man should take particular care not to alter it. This is a very frustrating situation, particularly for scientists; we like to know precisely where the limit is, we like to be able to translate it into figures. However, in this particular instance this is impossible because, when a climatic balance starts to be modified, we cannot ensure that even if we stop altering it immediately, it will revert to its former state. It is thus better to keep well away from the climatic limit.

My eighth conclusion is that our general energy policy should be oriented towards a strategy aiming at using the 1 watt/m^2 acceptance capacity of the environment wisely; as we have it to use, let us use it for emergency purposes, particularly the development of developing countries. The growth of energy consumption in countries which already consume high quantities of energy should not produce a situation where developed countries use the greater part of the acceptance capacity of the environment. In other words, if already developed countries want a good deal more energy, they should get it from the sources already included in the energy flow through the environment.

The last conclusion I would like to draw is that it would be advisable to concentrate our research on the better utilization of our currently available energy resources and on the development of renewable energy sources which are included in the normal energy flow through the environment. These, if properly harnessed, will certainly not modify in any way the climatic balance nor ever, obviously, be exhausted. This would require only very limited amounts of electricity - the electricity circulating in the brains of several hundred physicists, engineers, chemists, biologists and planners. It is very simple, maybe too simple to be adopted by some decision makers.

My final word is that we are living in an incredible age and not for anything would I have chosen to live in another century; indeed, never before has it been so clear that Man holds his future in his own hands. Our hands. The alternative before us is quite simple: either we build higher and higher dams like ambitious beavers, simply projecting into the future our historical and instinctive hunting behaviour of the animal exploiting its environment, but with means at our disposal that history never saw, or we behave as our Latin name would appear to indicate, like "Homo sapiens", a new species no longer exploiting his environment but managing it to his own advantage and for the well-being of his successors.

Let us hope that we will choose the right way!

DISCUSSION

Question: As solar panels have a higher absorptivity than the Earth's sur-
            face, will their use add to the net atmospheric heating?

Answer: Yes, but the efficiency of collection will determine the amount.
            It is unwise to have large solar collection plants transmitting
            energy to regions where the heat dissipation is already high.

Question: Whilst it was stated that for domestic heating one unit of elec-
            trically produced heat represents three units from fossil fuel,
            in practice this will be closer to a ratio of 2:1 as the conver-
            sion efficiency of fossil fuel appliances is less than the elec-
            trical counterpart. The current energy consumption in Western
            countries is about 20 kW per person; is there any reason why this
            should increase in the future?

Answer: The actual figure is nearer 7 - 11 kW per person at the moment.
            There will be a need for an increase in energy consumption to
            cover environmental control, recycling of metals, desalination of
            sea water and transportation.

Question: We currently have more comfort in domestic situations although
            with less primary energy consumption than in the past, indicating
            a better efficiency of use. Perhaps this will continue to improve
            The London area has a population of 10 million, each emitting 100w,
            within an area of about 10 km x 10 km giving an energy density of
            10 w/m^2. This density occurs with sedentary human activity. Now
            this figure is greater than the 1 w/m^2 you mentioned although the
            climate in London has not apparently changed over the last 100 years.
            What, therefore, is the area over which this density is significant?

Answer: Gibraltar currently has an energy density 50 - 60 w/m^2, but because
            of its small area it has an insignificant effect of climatic modi-
            fication. Even the 100 w/m^2 density found in Manhattan does not
            appear to have an effect. However, a large number of such regions
            may have a significant effect.

Question: Pollution due to noxious materials will tend to increase with both
            population and energy consumption. Can you comment on the problem?

Answer: The problem was not discussed because the technical solutions are
            already available. There is a race between consumption and cli-
            matic disturbances. Technologically we can overcome the problems,
            but at a price. For example, strip mining is a very energy inten-
            sive problem. The solutions exist but will require the will to
            pay the price for a clean environment.

Question: Can we do this in time? Will the Mediterranean become another
            Lake Erie?

Answer: This is a very complex problem. The Mediterranean is bordered by
            countries at different stages of industrial development and the
            research that has been done is not easily comparable. Another
            illustration of this type of international problem is the acid
            rains falling in Scandinavia, where governments have put pressure
            on the UK and Germany to reduce power plant sulphur emissions.
            Here there was an immediate incentive to solve the specific problem.
            International pressure is the best method.

Comment: At the Stockholm conference it was concluded that the sulphur pol-
            lution in Sweden had migrated from Poland.

Response: Agreed, but sulphur pollution in Norway comes from the U.K.

Question: Acidic pollutants are a common problem; instead of removing this
              at source, could we not inject ammonia at source in order to neu-
              tralise it, and wouldn't this be cheaper?
Answer: This is possible, but ammonium sulphate particulates will be emitted.
              The reflectivity of these will cause greater radiation from the
              earth, thus we should try to remove problems and not merely
              replace them.
Comment: Regarding thermal pollution, the energy density in w/m^2 is an
              average value and surely local excesses will be evened out by
              thermal diffusion? Where is the boundary? In your table of energy
              densities, Oceania had a value of 0.08 w/m^2 which, when included
              with countries of high local value, could produce an acceptable
              mean.
Response: Energy density can be calculated on a global scale but this would
              assume immediate diffusion of heat release. This is not the case
              and only the land should be used for calculation purposes. Also
              the limit of 1 w/m^2 is not of great accuracy, it could be 0.7 or
              1.3 w/m^2. It would be preferable to keep the oceans as a reserve
              to keep us under the limit.

Question: How does the thermal energy output of a typical country compare
              with climatic systems such as cyclones or other weather systems?
Answer: A direct comparison is difficult because a cyclone represents a
              stock of meteorological energy accumulated over a long period,
              whereas thermal energy is immediately released. The 1 w/m^2 limit
              is a controversial point and others would disagree with it. Indeed
              the climate has changed radically already in this country with no
              apparent connection with Man's activities. However we should still
              take steps to reduce the effects of these activities.

Question: Would you elaborate on the environmental problems involved in strip
              mining in North America, for example, where 50% of coal is pro-
              duced by this method?
Answer: There is a problem in Europe as well, large holes are left after
              supplies are depleted. The usual solution is to fill the hole in
              later. However in East Germany a coal-pit has been rendered with
              concrete and made into a lake suitable for leisure activities.
              This is very expensive but only legislation makes the mining com-
              panies reclaim the land. Limits other than financial must also be
              considered. For example, water availability affects the ability
              to reclaim both strip-mined land and oil shale rocks.
Comment: Reclaiming the land should be a legislated requirement for the
              companies involved.

Question: You used a 1 w/m^2 limit as this was the energy that goes into wind
              and ocean currents, but the planet receives 90 w/m^2 from the sun
              which it uses to evaporate water; why not use this figure for a
              limit?
Answer: 1 w/m^2 could be an important climatic factor for, for example, the
              Gulf Stream. If we produce hot spots in that area we may change
              its situation. In Peru deep ocean currents bringing in the anchovy
              catch have stopped in the last two years, reducing that catch by
              some 90%. Research is continuing, but no-one really knows why
              this occurred.

Question: Concerning oil spills, the Mediterranean is not as pure as it used
              to be. Is there an effect of pollution on plant and fish life near

the pollution source and what are the effects of oil spills?

Answer: There are several effects involved:
1) the albedo of the ocean is altered;
2) the oil spreads over the water forming a film, thus slowing gaseous exchange between atmosphere and water which affects animal life;
3) the effect on fish is unknown but fisherman have noted an oily flavour which is not good for business. However, this does no apparent harm to Man.

To tackle this problem we can use bacteria which can degrade oil. There is a problem when a detergent is used as this merely disperses the oil, forming small detergent-covered globules of oil which sink. Fish may eat these as they sink, otherwise the oil will sit on the ocean floor and pollute local animal life. Suez is an illustration of this - pollution was decreasing after the canal was closed but now it will increase.

RADIOACTIVE WASTE MANAGEMENT, REACTOR SAFETY AND SITING

E. C. Williams

Nuclear Inspectorate, Health & Safety Executive

The main statutory basis for the safety regulation of nuclear reactors and of
the processing of nuclear fuels and storage of radioactive waste is the Nuclear
Installations Act 1965. The Act requires that no body, other than the United
Kingdom Atomic Energy Authority or a Government Department, may build or operate a nuclear installation without a Nuclear Site Licence. In respect of
detailed regulation of safety, the Act gives wide discretion to the Nuclear
Inspectorate to attach conditions to a nuclear site licence which makes it
possible to frame such conditions as may be most appropriate at a particular
installation for the protection of operators and members of the public from
ionising radiations at or from any of the licensed sites.

Recent UK legislation (The Health and Safety at Work etc. Act 1974, governing
the safety and health and welfare of persons at work and the protection of the
public from the risks arising from work activities) has led to some amendments
to the Nuclear Installation Act 1965. The principal effect of these amendments
is to transfer the functions connected with the safety aspects of licensing
and inspection of nuclear installations from the Secretary of State for Energy
and the Secretary of State for Scotland to the Health and Safety Executive of
the Health and Safety Commission.

The Nuclear Installations Inspectorate, which was formed in 1959 to assist in
the administration and execution of the Nuclear Installations Act, now forms
part of the Health and Safety Executive. The Executive was set up on
1st January 1975 by an Order in Council. An outline of the organisation of
the Health and Safety Commission and Executive is shown in Figure 1. Although
relevant provisions of the 1974 Act will impose some additional obligations on
both employers and employees in the interests of safety, these and other amendments to the 1965 Act do not introduce any changes of immediate significance
to the main principles of safety control imposed on licensed nuclear installations.

At the present time there are nine commercial nuclear power stations of the
'Magnox' type in operation and five stations of the Advanced Gas-Cooled Reactor
type in various stages of construction and commissioning. Several other installations have been licensed under the Nuclear Installations Acts, including
seven research reactors, two of which have been subsequently dismantled and
the sites abandoned. In April 1971 the main fuel processing plants and isotope preparation units were detached from the UKAEA and thereby became subject
to the licensing and inspection regime of the Nuclear Installations Acts.
Figure 2 shows the location of the major nuclear establishments in the United
Kingdom.

Dealing first with the management of radioactive waste, it is worth noting
that almost all the radioactivity produced in the nuclear power programme ori-
ginates from the nuclear power plants, mostly in the fuel elements but also
from other materials activated in the nuclear reactors. The bulk of the radioactivity is transferred in the fuel to the reprocessing plant of British Nuclear
Fuels Limited at Windscale where the fissile and fertile materials are recovered
for subsequent use as reactor fuel and almost all of the other radionuclides
are extracted for long term storage as potentially hazardous wastes.

A small proportion of radioactivity is disposed of to the environment. The
main statutory basis for the control of these disposals is the Radioactive
Substances Act 1960. The Act is administered by the Department of the Environment and the Ministry of Agriculture, Fisheries and Food (or their counter-
parts in Scotland, Wales and Northern Ireland). No disposals may be made without prior authorisation by the DOE and MAFF. The controls are exercised to
ensure that disposals of radioactive waste do not exceed the minimum necessary
and in any event do not add significantly to the radioactivity in the general
environment or give rise to a hazard to health. Close liaison is maintained
between the Nuclear Installations Inspectorate and the authorising Ministries.

The radioactive waste stored on nuclear power station sites includes fuel element debris, reactor components and materials arising from the treatment of
the spent fuel element cooling pond water. The principle adopted in the design
of the storage facilities is that all the waste should be recoverable. It is
of course necessary to provide suitable facilities for the temporary storage
of the irradiated fuel elements. This fuel is usually stored in ponds for a

minimum period of 90 days before being transported to Windscale where it is
stored for a further period – up to 240 days – before reprocessing. There is
no particular difficulty in designing suitable cooling ponds but wet storage
does give rise to a further source of waste such as filter sludges and spent
ion exchange resins. High activity solid waste, in particular canning materials from irradiated fuel elements, is stored in concrete silos at Windscale.

The most important category of radioactive waste arising from fuel reprocessing
is the highly active liquid waste which contains over 99% of the fission products and a high proportion of the transuranic elements (other than plutonium).
The long half-lives of the radioactive materials and their biological hazard
require them to be isolated from the environment for very long periods of time.
From the point of view of external radiation (due mainly to fission products)
these wastes could remain exceedingly dangerous for several hundred years.
Thereafter the level of radiation (due mainly to transuranics) will decline
slowly over some thousands of years and could still be sufficiently high to
warrant continued isolation from the environment. In addition, there is a
potential ingestion hazard from long lived radionuclides entering the water/
food chain.

At present, highly active liquid waste is stored at Windscale in high integrity
stainless steel tanks with secondary containment and shielding. This system
of storage requires controlled ventilation arrangements for heat removal and
facilities for transfer of the liquid waste to spare tanks in the event of
leakage. It is clear that these structures have an expected useful life considerably shorter than the duration of the radioactive hazard. Although
storage in liquid form has proved to be satisfactory it is recognised that, in
order to reduce the requirements for surveillance and tank replacement, a
method of waste solidification should be developed which would facilitate long
term storage or ultimate disposal of this waste.

Research is under way in a number of countries into possible methods of ultimate disposal, and burial in stable geological formations seems to be the most
realistic approach. However, a generally agreed solution is some way off and
in the interim most countries are developing methods of waste solidification.
This requires the incorporation of the waste into a matrix material having
resistance to leaching, radiation and thermal degradation. The most favoured

system is a form of solidification of the waste - e.g. as a glass followed by canning in stainless steel containers.

It is expected that a solidification programme for highly active waste will be introduced in the U.K. by the 1980's. The process it is proposed to adopt should allow immediate solidification thus reducing the need for liquid waste storage and eventually allowing all existing liquid waste storage tanks to be emptied. The solidified waste would then be stored in suitable facilities such as cooling ponds which would provide shielding as well as easy access for any inspection or action thought to be necessary.

So far as the potential hazard from the storage of highly active waste is concerned, assured means of isolation from the environment are required for very long periods of time - perhaps thousands of years. The amount of highly radioactive waste generated in the U.K. nuclear power programme on the scales currently envisaged should not lead to any unmanageable problems. In liquid form all this waste could be accommodated on a site no larger than two or three acres and there should be no difficulty in designing storage structures to an acceptable standard of safety. However waste solidification should reduce the potential hazards and simplify storage.

Dealing now with the procedure for granting a nuclear site licence for power reactors, an applicant must provide sufficient information to enable the Inspectorate to satisfy itself as to the safety of the proposed plant and the suitability of the site.

Am important characteristic of a power reactor site is that the population distribution, topography and so on should be such that emergency action could be taken to deal effectively with any accidental release of harmful amounts of radioactive material. In respect of the reactor itself, an assessment would be made of the overall design and of the proposed commissioning and operating procedures.

It is appropriate at this point to review briefly the development of U.K. siting policy over the past 20 years. The development of a siting policy for nuclear power reactors in the U.K. was made necessary by the decision of the Government in 1955 to undertake a major programme of construction of power

reactors based upon the Calder Hall design. Engineering requirements (for
cooling water supplies, suitable foundations etc.) already imposed some small
constraints on site selection, but on safety grounds it was considered right
to take a strict view of siting, while at the same time recognising that the
most remote sites available (in the North and West of Scotland) were much too
far from the major load centres in England to be of practical use. It was,
therefore, necessary to consider sites available in regions of relatively high
population density, but where sizeable towns are widely distributed.

In England, the average population density is 860 per square mile and the
population density of the maritime counties ranges from about 3,000 persons
per square mile to a few hundred per square mile. In England and Wales, with
a combined area of about 60,000 square miles, there are some 500 towns with a
population greater than 10,000 and 55 in excess of 100,000.

Population is concentrated in six major areas. In these areas the local population density in towns may rise to 40–50,000 per square mile, but over a
wider area, the average population density falls to around 10,000 per square
mile or less.

It is apparent, therefore, that the degree of isolation available in England
is limited because of the high population density and it follows that the
factor of safety to be gained by the remotest practical siting is also limited.

The White Paper "A Programme of Nuclear Power" published in 1955, stated that,
although properly designed nuclear power reactors should present no more danger
to nearby population than many existing industrial plants, it was the Government's intention that the first nuclear power stations should not be constructed in heavily built-up areas. Following this statement, the sites actually
chosen for nuclear power stations were situated in comparatively remote and
sparsely populated areas. In each case, arrangements have been made to ensure
that residential development in the vicinity is controlled so that the original
character of the site is preserved.

This was a sensible precaution to take in relation to a developing technology.
As experience of nuclear power grew, however, it became clear that a policy
of remote siting created problems of its own. In the first place, remote

sites could often only be found in parts of the country valued for amenity
reasons. Secondly, since such sites were distant from centres of electrical
load, extensive overhead lines had to be constructed to transmit power from
the stations. These lines constituted a further intrusion upon amenity and
were also costly. They would be many times more costly if they had to be placed
underground. Of the first eight sites selected, six were the subject of Public
Inquiries during which strong objections were made to the intrusion of power
stations and the associated transmission lines. Since 1955 much experience of
power reactor operation has been gained, and the technology has generally
advanced. The introduction of massive pre-stressed concrete vessels and the
development of a greater confidence in methods of safety analysis are examples
of matters which led to even greater confidence in plant safety than was felt
in 1955.

For all these reasons the Government concluded that it would be right to reconsider reactor siting policy, on the basis of an intensive study of the contribution which siting could make to public safety in practice. This review was
carried out in consultation with the Nuclear Safety Advisory Committee, a body
appointed by the Secretary of State for Energy and the Secretary of State for
Scotland and which consists of distinguished independent experts. As a result
of this review, in which it was reaffirmed that the major safeguard for the
public is the achievement of high standards in the design, construction and
operation of nuclear power stations, a new siting policy was formulated which
allowed Advanced Gas-Cooled Reactors in pre-stressed concrete pressure vessels
of approved design to be sited nearer to urban areas than hitherto. The change
of policy was announced by the then Minister of Power on 6th February, 1968.
He stated that because of advances in the technology, the safety of nuclear
stations was such that it was possible to allow the building and operation
of gas cooled reactors in pre-stressed concrete vessels much nearer to urban
areas than had been previously permitted. Each site proposed must be considered on its own particular merits but the sites at Hartlepool and Heysham
were given as examples of acceptable sites. Population density close to a
site has to be sufficiently light to enable effective emergency countermeasures
such as the evacuation of people from the area to be taken in the very unlikely
event of an accidental release of radioactivity having effects beyond the
station boundary.

The class of site now considered suitable for nuclear power stations of the
AGR type is defined by criteria which specify the acceptable density and distribution of the surrounding population. These criteria which limit the proximity of urban populations to a nuclear site are called limiting 'site' and
'sector' characteristics. The suitability of any proposed site is first tested
by comparing its characteristic curves, derived from an analysis of the density
and distribution of the surrounding population including projected development,
with the limiting characteristics. The derived characteristics of the proposed
site should not exceed either the limiting site or sector characteristics. The
method used to derive the characteristic curves for any site and the basis of
the selection of the limiting characteristics is briefly outlined in the
attached Appendix which also sets out the AGR siting criteria.

An important aspect of siting practice is the provision which is made to mitigate the effects of a radioactive release just beyond the site boundary. At
short ranges from the site, even if an activity release were small, the risk
to an individual might be unacceptably high. Therefore, special consideration
is given to persons up to two-thirds of a mile from a reactor site. Prepared
and rehearsed schemes for monitoring, warning, evacuation and medication are
required and provide safeguards for persons in this zone.

Since the level of risk to members of the public does not change sharply at
two-thirds of a mile, it is a requirement of the revised siting policy that
the population distribution, the communications, and the general nature of the
terrain should not preclude the extension of the emergency arrangements further
out, should that ever prove to be necessary. To satisfy this requirement controls are exerted on population development up to a distance of two miles.

The extent to which emergency measures are implemented to protect members of
the public is based upon a balance of the risk to persons from exposure to a
radioactive plume and that involved in removing them from its path. The choice
of action levels in the United Kingdom is based on the recommendations made by
the Medical Research Council on the permissible limits of human exposure to
radioactive substances.

However, if emergency measures are to be effective, they require urgent applications in zones of highest risk. To this end, the scheme for evacuation of

persons living within two-thirds of a mile of a reactor site is required to be
capable of full implementation within two hours of the declaration of an
emergency.

Although they may be significant at much greater distances, the levels of
radioactivity in milk, food, agricultural produce, and public water supplies
are not likely to create any immediate public hazard and can be dealt with on
a much longer time scale. These factors do not influence the selection of
nuclear sites, but, nevertheless, require detailed consideration by the responsible authorities and the maintenance of adequate facilities for a monitoring
programme should the need ever arise.

The relaxation of siting restrictions for gas-cooled reactors of approved
design permitted by the revised siting policy reflects the increased confidence
gained from the extensive experience with this reactor system. In principle,
a reactor can be designed and constructed so as to be free from siting restrictions imposed on public safety grounds. The Inspectorate are not yet satisfied that such a plant is available at present though it is clear that a remote
siting policy could not be maintained for a large scale nuclear power programme.

The Government has now chosen the Steam Generating Heavy Water Reactor as the
system to be used for the next programme of nuclear power stations in the U.K.
The sites chosen for the first few of these stations are of the remote type.
The siting policy for any further nuclear stations of this type will be considered in the context of the experience gained.

Returning to the procedure for granting a nuclear site licence, if the preliminary safety assessment proves to be acceptable, the applicant is directed to
publicise the proposal and give notice to specified public and local authorities who have the right to make representations regarding the proposal within
three months. The procedure to be followed is that laid down for all power
stations by the Electric Lighting Act 1909. Under this Act the Minister's
consent is required before any power station can be built or extended. A
nuclear site licence will not be granted by the Health and Safety Executive
until the Minister gives his consent to build a station under the Electric
Lighting Acts.

When all interested parties have been given an opportunity to comment or object
to the proposed station, the Minister decides whether or not the proposals
affect their interests to the extent which makes it desirable to hold a Public
Inquiry. If the local Planning Authority objects, the Minister is obliged to
hold such an Inquiry.

Under these arrangements Public Inquiries have been held on eight of the applications for consent to build a nuclear power station. In only one of these
cases was the Minister obliged to order an Inquiry because of objections by
the local Planning Authority which is quite a new experience in the United
Kingdom. This was a proposal to build a nuclear power station as a site where
the Planning Authority felt that the consequential restrictions on residential
development, an integral part of U.K. siting policy, would inhibit the development potential of the region. This was the only nuclear power station to be
refused following a Public Inquiry and indicates the difficulties which might
be encountered in siting nuclear power plants near urban developments where
there is a conflict between the need to preserve a relatively thinly populated
area near a nuclear power plant and the long-term development plans for that
area.

Strictly speaking, the Inspectorate has no formal powers over any proposed
plant until a licence has been granted but, in practice, no licence would be
issued until the Inspectorate was fully satisfied that the proposals submitted
were to a satisfactory safety standard. Once the licence has been granted the
Inspectorate has full powers to maintain adequate control over the design,
construction and operation of the plant, and as already indicated, the licence
conditions may be varied at the discretion of the Health and Safety Executive
which makes it possible for them to be amended, added to or revoked at any
stage. The flexibility of the licensing system has advantages for both the
Inspectorate and the licensee in that it is appropriate to the progress of the
work. For example, in the early stages of the project the conditions might
merely require the licensee to submit officially to the Inspectorage only those
details of the site, the design and the development and testing programme
already discussed in detail before the issue of the licence. At a later stage,
when the construction is approaching completion, the licence can be amended
and new conditions imposed regulating the commissioning and operation of the
plant, which again, will have been fully discussed previously and which cover

the operating and maintenance procedures, the storage of radioactive waste and
the emergency arrangements.

The conditions on which a nuclear site licence is issued are designed to ensure
that all possible steps are taken to protect both the public and the operating
staff. The safety assessment of the design of the nuclear plant is backed up
by a continuous inspection during the actual construction and the subsequent
commissioning and operation of the plant. At all stages the Inspectorate must
be satisfied with the demonstration of the safeguards provided to prevent an
escape of radioactivity or the emission of ionising radiations. It should,
however, be emphasised that the licence conditions are, as far as possible,
framed to make sure that every stage of the process, from the initial fuel
loading through the commissioning to the operation of the plant, is properly
thought out and put down in writing in advance. Also subject to formal approval are the station operating rules which specify the safe operating limits,
the availability of essential safety equipment as well as the proposed testing
procedures. It is, of course, inevitable that during the long life of a nuclear
plant some alteration of the rules will become necessary, but it is naturally
desirable that such changes should be carefully considered before being
implemented.

This task is undertaken by a Safety Committee, which can advise on the sound-
ness of any proposed changes in the operating rules or of any suggested modifications to the plant which might affect the safety standards. Any alterations recommended by the Committee must be approved by the Inspectorate before
implementation.

Similar regulatory control procedures are applied to installations processing
nuclear fuels or handling and storing radioactive wastes. On such sites the
licensee is required to produce a site plan showing every building or plant
on the site that may affect nuclear safety together with a specification of
the materials and operations of those buildings or plants. The specifications
are relatively simple and descriptive and do not go into technical detail of
the plant nor detailed safety provisions. The identification of plants that
may affect safety can include non-nuclear plants where a malfunction of such
a plant could affect a nuclear operation - e.g. a gas pressurising plant or a
flammable liquid plant. Such a site plan must be kept up-to-date in consultation

with the Inspector for the site and must also be fully reviewed each year.

The site plan concept provides an overall view of plants requiring safety
inspection and establishes an early warning system for possible changes of use
or process. Within this general control detailed controls may be elaborated.
If it is desired to introduce a new plant on the site then a more detailed
plant specification must be produced. This specification must indicate the
type of process to be introduced, the nature of the materials to be processed,
the nature and volume of likely wastes, the main nuclear hazards and a design
safety report which is a statement of the principles and means for the control
of safety. This information must be submitted to the Inspectorate before the
licensee begins to construct the new plant on his site.

Once construction of the plant has commenced field inspection concentrates on
the quality of materials of construction, fabrication and inspection methods
and component testing. By this time a plant commissioning programme is formally required which details the plant tests important to safety. No product
material may be introduced into the plant other than for the demonstration of
a commissioning of test procedure. During this phase of the development of
the plant the main operating rules relating to safety must be developed. In
the case of plants processing fissile materials these operating rules will
include the issue of Nuclear Safety Certificates which incorporate both a
criticality analysis of a process vessel or vessels and a statement of operating procedures and control limits. Upon the satisfactory completion of all
the commissioning tests and the production of safety documentation the operator is then formally authorised to commence production operation of the plant.

As with nuclear power plant operators, so also are fuel processing installations required to set up a Safety Committee to consider the safety of any modifications to the process or the plant. Specialist sub-committees or panels -
e.g. criticality safety - may also be established within this committee
structure.

The Nuclear Installations Acts place the licensee under a strict duty to avoid
harming anyone. The Health and Safety Executive has direct responsibility to
ensure that all necessary safety precautions are taken by the licensee. In
the final analysis it is the maintenance of high safety standards in the

design, construction and operation of nuclear plants that provide the best
protection for the public and it is the task of the Nuclear Installations
Inspectorate to ensure that the rules and regulations drawn up for this purpose are not only strictly observed, but that they are amended and brought
up-to-date as experience dictates. The very wide discretionary powers given
to the Health and Safety Executive under the Nuclear Installations Act to
enforce any conditions in the interest of safety, make it possible to be confident that technological changes in the industry can be dealt with under the
existing law.

FIGURE 1   **HEALTH AND SAFETY COMMISSION / EXECUTIVE**

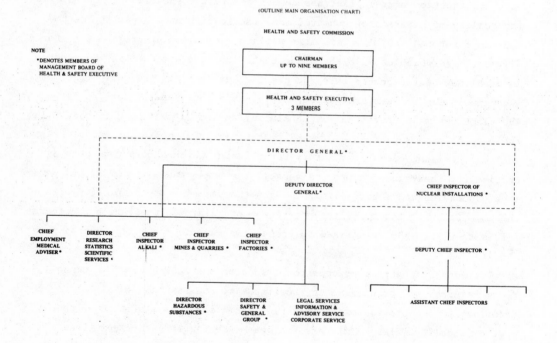

NUCLEAR POWER STATIONS
OTHER NUCLEAR SITES
APRIL 1975

FIGURE 2

APPENDIX

SITING CRITERIA FOR NUCLEAR POWER STATIONS

Under the new siting policy, a review of the suitability of a proposed site
proceeds by first deriving two characteristic or risk factor curves, one for
the site and one for the most densely populated 30° sector. These curves
include a measure of the population distribution around the site and an assessment of the potential thyroid dose, i.e. the risk, to the population in the
event of an incident resulting in the release of radioactivity to the
environment.

The region within a twenty mile radius of the plant site is divided into seven
radial zones with the following zonal distances, in miles, from the centre of
the plant: 0-1, 1-1$\frac{1}{2}$, 1$\frac{1}{2}$-2, 2-3, 3-5, 5-10 and 10-20. The population, P_i, in
each zone is then multiplied by a zone weighting factor W_i, and the <u>site</u>
characteristic curve is obtained by plotting

$$\sum_{i=1}^{n} P_i W_i \quad \text{against the outer radius of the nth zone.}$$

The weighting factor for a zone represents the potential thyroid inhalation
dose that a hypothetical "standard man", located at the centre of the zone,
would receive from a ground level release of iodine-131 – the most important
isotope as it is the most volatile of the fission products and one to which
the human thyroid is extremely sensitive. The release is assumed to occur
during Pasquill type "F" atmospheric conditions with a wind speed of 2 metres
per second. The Pasquill atmospheric diffusion model (Pasquill, 1961) is used
to determine the radioactive material concentration downwind from the point of
release. The model assumes the plume to exhibit a Gaussian characteristic,
decreasing in concentration with increasing distances because of changes in
the lateral and vertical dimensions of the cloud.

A procedure similar to the above is followed in order to obtain weighting
factors for the calculation of the characteristics of the most densely

populated 30° sector, i.e. the sector characteristic curve. The only difference
is that whereas before, in the calculation of the weighting factors for the
site characteristic the lateral spread of the cloud was assumed to vary with
the distance, the lateral spread is now assumed to be a constant 30°.

The site and sector characteristic curves for a proposed site are then compared with the corresponding limiting curves (see below). A site whose
characteristic or risk factor curves are higher than the limiting curves is
deemed unacceptable. On the other hand, a site whose characteristic or risk
factor curves are lower than the limiting curves is subjected to further
examination.

Factors which are taken into consideration in this further examination include
the topography of the area, local communications, population mobility and any
other special features which might affect the emergency procedures which will
have to be put into operation in the event of an incident. In addition, the
siting policy aims to approve only sites which would not restrict likely
development around them. The implications of all likely large scale developments in the area are accordingly considered when the suitability of a proposed site is assessed.

The selection of the limiting risk factor curves was based on the need to dif-
ferentiate between those sites which were acceptable and those which were not.
Risk factor curves were derived for a number of practical plant sites, ranging
from those selected for the first phase of the UK nuclear power programme to
a number of hypothetical sites located in or near London. Examination of the
resulting curves revealed that the risk factors varied by three orders of magnitude; sites within each order of magnitude of risk were considered to be of
comparable risk. The sites chosen for the first nuclear power programme lie
in the lowest order of risk factor and metropolitan and near-metropolitan sites
in the highest order of risk factor. In between lies a range of sites considered suitable for gas-cooled reactors in concrete pressure vessels.

The limitations on population distribution around AGR sites are given in the
next paragraph, but it must be emphasised that where a choice of sites exists
preference is given to sites in less populated areas.

E. C. WILLIAMS

SITING CRITERIA FOR ADVANCED GAS COOLED REACTORS (AGRs)

Table 1 specifies the number and area of each zone, and the values derived for the limiting site and sector characteristics, respectively. The limiting 'site' and 'sector' characteristic curves for the AGR are shown in fig. 3.

Reference

Pasquill, F., "Atmospheric Diffusion" published Van Nos-rand-Reinhold, Princeton, New Jersey, 1961.

Table 1. Site and sector factors

Zone Number n	Zonal Distance from Reactor Site centre in miles	Site Zonal Weighting Factors W_i	Limiting Site Characteristics $\sum_{i=1}^{n} P_i W_i \ (\times 10^{-6})$	Sector Zonal Weighting Factors W_i	Limiting Sector Characteristics $\sum_{i=1}^{n} P_i W_i (\times 10^{-6})$
1	0 - 1	40.0	0.190	32.4	0.063
2	1 - $1\frac{1}{2}$	22.5	0.411	17.6	0.137
3	$1\frac{1}{2}$ - 2	13.75	0.595	10.5	0.198
4	2 - 3	8.75	0.917	6.4	0.305
5	3 - 5	4.75	1.448	3.3	0.482
6	5 - 10	2.0	2.436	1.31	0.812
7	10 - 20	0.75	3.823	0.46	1.274

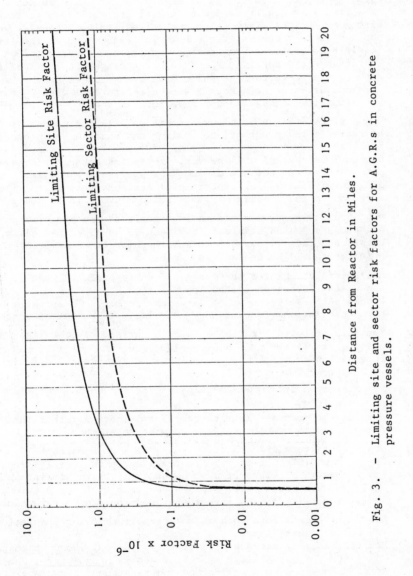

Fig. 3. - Limiting site and sector risk factors for A.G.R.s in concrete
pressure vessels.

DISCUSSION

Question: It is clear that the waste products of a nuclear reactor necessi-
            tate long term storage and isolation. This implies a location of
            equally long term political stability. What would be the dangers
            of a site in a country where political instability occurred? We
            also process waste from other countries (Italy, Sweden). What is
            the danger to these materials both in transit and at the site
            itself from either a bomb during war or from terrorists?

Answer: The danger to a specific "pond" (i.e. storage site) is small.
            With regard to terrorists there have been studies of what security
            is necessary and what steps are required to deal with terrorists
            but naturally, these are not made public.

Question: How serious are the problems associated with the sudden dispersal
            of waste in the event of the cooling and containment processes
            failing?

Answer: Obviously dispersal will be very widespread because of the enor-
            mous heat generated. We do not know precisely and clearly would
            not contemplate making experimental tests of this sort. The
            solution to the problem of waste is to convert it to solid form.

Question: How much radioactive waste is in transit at any one time and what
            are the associated hazards? How much is in pipelines being pro-
            cessed and thus vulnerable to sabotage (or accident)?

Answer: A fair amount. It is kept to a minimum.

Question: Has anyone calculated the energy of these products when in transit
            and what will happen if they're released?

Answer: Of course these calculations are carried out. The problem with
            such calculations is that they are always taken out of context.

Question: You talk of solidification by vitrification. A glass is notori-
            ously unstable (it is a supercooled liquid and will be subject to
            intense ionisation). Do you think it will remain solid for, say,
            1,000 years?

Answer: U.K.A.E.A. Harwell says it will remain solid.

Question: Surely not?

Answer: I think the process might be one of calcination rather than vitri-
            fication and hence not subject to your objections.

Question: Would containers withstand 100 years or more erosion?

Answer: There is a basic safety rule that all waste must be (a) monitored,
            (b) recoverable. Thus it must be possible to put it into new
            tanks whenever necessary. This also deals with the previous point
            about de-vitrification since the liquid too can be re-enclosed.

Question: Since terrorists form such a small fraction of the population,
            why not eliminate them?

Answer: Perfect security at each site is the solution. However one must
            allow for the possibility of terrorists gaining entry. The very
            worst case would be a well-armed and very knowledgeable group,
            but this is unlikely and one probably over-estimates the dangers.
            As an indication of how secure a site could be made, remember
            that Fort Knox has never lost any gold!

Question: Is a Fort Knox approach sensible? Won't stronger fortresses
             provoke stronger attacks?
Answer: For safety it is worth trying to have maximum security. Even if
             terrorists penetrate a site they need to have specific and specia-
             lised knowledge.

Question: To remove the chance of terrorists reaching the waste can we dis-
             pose of it geologically?
Answer: Although recovery remains a basic policy, stable geological regions
             are a possibility for storage. The best suggestion seems to be
             storage in salt mines; geological sites might require international
             coordination, however. The present accumulation of waste in UK
             needs about 30 MW to give the necessary cooling over hundreds of
             years. People assume that technology will remain static (in its
             ability to cope with the disposal problem). This is not sensible
             and probably new ideas will come along. Remember that nuclear
             technology is comparatively young (30 years or so).

Question: Will salt-mines continue to be stable?
Answer: The geologists' advice is that they will.

Question: Although the UK standards of safety are high, might not some
             countries be tempted to take short cuts?
Answer: The US is also worried about this. Perhaps the vendor-country
             will insist on "follow-up" safety control and security. The US
             does this to some extent already. Remember that a reactor is
             designed to shut down if it is not operated and maintained
             properly.

Question: Recently Germany sold (or is in the process of selling) a reactor
             system to Brazil. Although some regard this as a "coup" by
             Germany, one should realise that the US is not allowed to sell
             systems which include reprocessing plants. In this light would
             you regard this sale as potentially dangerous, since it gives
             Brazil the ability to make bombs?
Answer: As a British Government official I cannot comment.

Question: Will the UK process Denmark's waste? What will be the procedure?
Answer: We will solidify and then return it.

Question: The figure of 30 MW for cooling of wastes was given. Does it
             include manpower, administration and buildings?
Answer: The number of men involved is small, of the order of a dozen and
             therefore unimportant in terms of energy.

Question: How does the danger of genetic effects from chemicals compare with
             that from radiation? Could you compare Porton Down with Windscale?
Answer: The risks are comparable. One might add that 100 miners die each
             year; 22 people die on roads each day; 400 people can die in a
             single air crash; i.e. we are used to mechanical or visible
             damage to the body. Radiation however evokes emotion, perhaps
             because the damage is not visible at once. We must learn to alter
             attitudes.

Question: How serious is genetic damage from radiation?
Answer: Chemical poisoning produces genetic damage too, mercury for
             example.
Comment :   It is not sufficient to say that nuclear deaths are less than (say)
             road deaths and thus can be simply added to our list of "tolerated

deaths"; because chemicals can produce genetic damage we can't
justify genetic damage from nuclear effects. In other fields
safety is not stringent, e.g. only recently have lorries carrying
dangerous materials been properly labelled.

Response: Comparison with road deaths, for example, is at least an attempt
to obtain a perspective view. Air-line passengers can suffer far
higher radiation doses than exist at ground level. There is no
evidence of genetic damage in Hiroshima and Nagasaki victims, so
we must not over-estimate the dangers. Genetic engineering may
pose a far greater hazard than radiation.

Comment: Regarding genetic damage due to radiation, perhaps insufficient
time has elapsed for this to become apparent.

Question: What would happen if a commercial organisation asked for a licence
for a nuclear site, (a) from a legal point of view, (b) from the
point of view that it would be near a population centre?

Answer: In principle there is no legal problem e.g. several Universities
have (admittedly small) reactors. Rolls Royce have one and
Vickers had one. The same restrictions would apply as for the
C.E.G.B. One doesn't have to put a reactor near a large population, one could build it in a remote area.

Question: If the public insists on remote sites for the C.E.G.B. reactors,
how would this, necessitating as it does an extended national
grid, affect the economics and cost benefits of nuclear power for
electricity?

Answer: I am not really sure, but the UK is fairly small and quite well
covered by the grid so it's probably not important. In this context it is interesting that a small village in Scotland actually
preferred a nuclear power station to a chemical works or an oil
plant because the nuclear station would be clean and quiet.

Question: What will be the policy of the Nuclear Inspectorate between now
and 2000 AD (say) regarding waste? Will waste simply be buried
and monitored continuously?

Answer: This is not a situation which is satisfactory. The Inspectorate
has asked for active steps to allow waste to be converted to
solid as soon as is possible.

Question: Could armed guards become necessary to safeguard transport of
waste and materials. If so, whose responsibility would that be?

Answer: Transit is not dangerous. It is carried out in double skinned
vessels. (After the 90 day cool-off, danger from radioactivity
is small.) The vessels have been tested thoroughly, e.g. by
dropping 200' from a helicopter; by dropping from a truck at
60 mph; in fire at 600°C. After a further 200 days cooling the
contents are kept in Windscale thus obviating transport problems.
If guards were needed, it would be a matter for the Home Office.

Comment: 500,000 years storage has been mentioned. Over this length of
time the actual radiation from the original uranium ore could be
comparable.

Question: How does total radiation in bomb tests compare with radiation
released by power stations?

Answer: The latter is much less although a bomb does disperse the debris
so that the situations are not strictly comparable. Remember,
although you have been very critical, there is no real alternative
to nuclear power.

Comment: There is an alternative to nuclear power, and that is to achieve
a state of zero growth.

ECONOMICS OF RESOURCE USE

D. L. Munby

Nuffield College, Oxford

My purpose is to outline some very simple general propositions about
resources which economists might want to make.

Resources and their value

I refer in the first place to pieces of matter lying about in the world
which have not been deliberately shaped by human beings for some purpose
or other. Are they resources, or are they waste? The answer to this
question does not depend on the physical characteristics of the materials.
There are for example up to 100,000 hectares of derelict land in Great
Britain, and about a third to a half of this is covered with industrial
waste much of which could be used as aggregate in building or
construction of one kind or another. There is a stockpile of colliery
spoil of about 3,000 m. tonnes, which increases at some 50 m. tonnes p.a.,
and there are also stockpiles of china clay waste in Cornwall and slate
waste in Wales, each of about 300 m. tonnes, the former growing at some
20 m. tonnes p.a. Most of these are technically suitable to replace the
200-250 m. tonnes of sand and gravel and crushed rock dug up each year in
(among other places) National Parks and the Thames Valley with growing
concern from environmentalists of all sorts. But we go on digging holes in
places where people do not want them dug, and making mountains in other
places where people do not want this sort of mountain. The simple fact is
that these materials are required at a particular place and the cost of
transporting the "waste" to where it is wanted is much greater than the cost
of digging up other materials.

Material resources only have value in the places and at the time they can
be used to meet economic demands. Their value depends on the valuation
society and individuals put on the final services they perform. This
depends among other things on the pattern of consumption, the technology
of a society, and the total costs of transforming the material into
services demanded. Total costs in their turn include the costs of using
land and capital, and the labour and material costs of production, as well
as the transport and distribution costs. All costs in the last resort turn
out to be valuations, the values society or individuals put on the use of
all these resources (in the wider sense) to provide alternative services
or benefits.

The simplest final breakdown of costs is the traditional economists'
division between the costs of land, capital and labour of all kinds, not
forgetting the peculiar kinds of labour required to organise production
at all levels. None of these costs can be taken by itself as a final
yardstick, much less any other cost, such as energy cost. We are not on an
energy standard, no more than we are on a labour standard. There is no
alternative standard for economic decision-making - which means, with all
the necessary qualifications, society's decision-making - other than
overall costs. (The qualification relates to the extent to which market
costs or apparent money costs do not adequately reflect all true social

costs, and we have to replace the market, if we cannot "rig" it
appropriately, with some form of political decision-making.)

If one could in a very simplistic way reduce all costs to one standard,
the best to choose would be labour costs, not energy costs, because one
fact about the kind of growing world we live in is that labour becomes
relatively more expensive each year. If output per head grows at 1% p.a.,
labour costs grow relatively at 1% p.a.; this is just the other side of the
coin. To put it another way, if real output per head grows at 1% p.a.
and prices remain stable, wages will still grow at 1% p.a. (If prices don't
remain stable, wages will still grow 1% faster than prices.) Thus outputs
which require a given quantity of labour (roughly "services") become
relatively expensive.

Economic growth almost inevitably involves some things becoming relatively
more expensive and others relatively cheaper. The most obvious example of
this is the price of domestic service in the last seventy years. In general,
this leads to some shift of demand to the relatively cheap and some
substitution of other ways of achieving the same end. Thus people buy more
"goods", and they use washing machines etc. in place of servants. But, of
course, as incomes rise, people may shift their demand even more to these
sectors, so that the prices rise even more and a greater proportion of total
resources of all kinds are devoted to these sectors (e.g. to education or
civil servants). Pure price effects and income effects may lead in different
directions.

What we need to note is that these relative price changes are a fact of life.
They may come about because of shifts in the availability of resources (in
the narrow sense) as well as from changes in other supply factors, changing
technology or changes in demand. They may come about relatively
suddenly or fairly slowly. The evidence of economic history does not suggest
that we should take too seriously the "crises" beloved of politicians every
time they are suddenly struck with the impact of some change.

How much do we know about resources?

Resources are of no use if we do not know about them. Surprisingly to the
layman, we find that we know very little about the availability of physical
resources. It is fairly easy to take some estimate of known resources, say
coal, at a given date and show that the rate of use of these resources since
that date has led, and will continue to lead, to their fairly rapid
diminution. (The famous predictions of the economist, W.S. Jevons, of more
than a hundred years ago that the rate of growth of the British economy could
not continue beyond about 1930 because of the shortage of coal should make us
hesitate.) All these statements really do is to tell us how little we knew
in the past. It is not surprising that after the war we found a great new
coalfield in Nottinghamshire, or recently in South Yorkshire. Similarly,
with North Sea oil and gas.

The facts seem to be that even in a very well surveyed country like Britain
where the Institute of Geological Sciences has been surveying for a very long
time, we know very little. How much more must this be true of countries
which have not been surveyed nearly so continuously or so thoroughly. The
present boring programme of the I.G.S. to sink one borehole in every square
km. will not tell us much we want to know about many minerals, and this

programme itself will take some time.

An economist is a layman in this field, but he can note the difference
between known resources and usable resources. He can note that resources
at one time valueless become valuable at another because of shifts in
technology (e.g. uranium), or because of changing production techniques
which enable low grade materials to be used more cheaply. He can note how
the discovery of new reserves reduces the value of old reserves which were
not so well-placed. Above all he notices how rises in costs lead to more
prospecting and new discoveries. The converse of the doom hypothesis would
be the proposition that the more we use the more we discover, a proposition
whose general truth is not disturbed by the fact that it is obviously wrong
in the case of particular minerals in particular places. Whether there are
any general limits is a matter for other expertise than that of an economist.
But an economist may doubt if there are any limits of any importance for any
future we would be wise to worry about, even if we may have to suffer before
the end of the century shifts in availability and costs perhaps more radical,
and certainly of a different nature to those we have already undergone in
the last 25 years.

The future

The basic fact is that we live in a world of uncertainty whose future we
cannot predict with any likely degree of success, though for all sorts of
purposes we have to make our best guesses about the future. Here are some
general propositions about the kind of society in which we live and its
future:-

(a) Our economy grows and will continue to grow, even if at the rate of only
some 1-2% p.a. (which is relatively slow as compared with what other developed
countries achieve). From this it follows that (quite apart from inflation)
real incomes (wages) will rise on average at this average rate of growth of
productivity of the economy as a whole. People in future will be richer than
they are at present;

(b) At the same time, people will have more leisure in the future. They
will prefer to take some part of the possible increase in output in the form
of leisure rather than goods and services;

(c) People will be better educated and their tastes will change as a result;

(d) Knowledge grows and will continue to grow. The results of future
knowledge are unpredictable (technical progress), and we can only make
guesses as to what possibilities may be open to the future, and these will
often be wrong. (e.g. Did anyone in the great ages of railway-building
anticipate the bulldozer? Did anyone seriously in 1939 anticipate the use of
atomic energy for the production of electricity? Did anyone for that matter
in the great days of coal anticipate the substitution of electricity for coal
and gas?);

(e) Technical changes resulting from greater knowledge will make some
minerals now thought worthless worth a great deal and will downgrade others;
how rapidly and in which directions is subject to all the uncertainties of
(d);

(f) Productivity will increase relatively faster in some industries than others, depending on changes in relative real prices and technical change. The nature of these changes is subject to every kind of uncertainty. The only fairly certain thing one can say is that productivity will increase faster where it is possible to make less use of labour as labour costs rise, and that this will probably lie in the production of goods rather than services.

Interest and uncertainty

£100 next year is not worth as much as £100 today. There can be much discussion as to why this should be so, but one basic reason follows from the fact that people will be richer in the future than in the present; the richer one is, the less one values a given addition to one's income, e.g. a man who thinks his children will be very much richer than he is will be less concerned to hand on to them a given amount of wealth than he will be if he thinks they will be only as rich as himself. Some rate of interest must exist in a growing economy, but there is no reason to think that the rate at which people or firms actually discount the future represents the rate at which society as a whole should discount it. Individuals and firms may suffer from short-sightedness, while society as a whole must rightly assume that it will continue for ever. For this and other reasons, many, if not most, economists consider that the socially right rate of interest would be below actual rates. In other words society should rightly put a higher value on future resources than individuals and firms do; e.g. firms will tend in this case to use up natural resources faster than is warranted by society as a whole. The seriousness of this discrepancy is a matter of judgement. At the moment the Treasury regards a 10% rate (ignoring uncertainty and inflation) as the rate at which the private sector discounts the future. One might, for example, think that the proper social rate was 5%. The difference can be expressed as follows:-

VALUE TODAY OF £100 IN FUTURE YEARS

Years	10%	5%
5	62	78
10	39	61
15	24	48
20	15	38
25	9	30

This table not only illustrates the marked difference between interest rates of 10% and 5% (at 5% £100 in 20 years is worth as much as £100 in 10 years at 10%), but also how the distant future is devalued at any rate of interest (even at 2% £100 in 25 years is only worth £61 today).

The only thing we can be really sure about is that we cannot foretell what will happen. This is, of course, no argument for not doing the best we can on the basis of the best knowledge we have today, as we have to act today with implications for tomorrow. But it does imply that we should act with some humility about our expectations for the future. Whatever we do will almost certainly turn out to be wrong, or, if right, right only by accident. It also implies that we should be ready to spend something more to leave options open for the future, if such flexibility is possible. Equally it is worth spending something to reduce uncertainty, by improving our knowledge of the future or spending money on research.

Research involves costs, the costs of fairly scarce manpower. At some point it is not worth engaging in research either to reduce uncertainty or to help decision-making in other ways. Even if there is no clear way of deciding priorities, we have to decide where priority should be given in terms of possible benefits (even if we do not know the relative costs of different areas for research).

The price system

Society in fact adapts to all the changes we have been talking about. One of its means of adaptation - and not less powerful for not being always noticed - is the price system. As things become dearer people tend to buy less of them. Producers try to produce substitutes, and alternatives that were not open in the past become viable. There are incentives to look for mineral resources in places where people haven't looked before. There are inducements for research to be redirected to find alternative ways of achieving the same ends. The results of all these pressures produce new situations and new configurations of prices.

It would be foolish to pretend that what happens is necessarily better than what has happened before, or that the adaptation is always easy or without shock effects which seriously affect certain groups in society. Nor need we look for some perfect pattern of adaptation or equilbrium which may sometimes attract some economic theorists. The world is a messy place. But some kind of adaptation does take place, and the human race does (usually) "get by". And in this the price system plays its part.

In general there are three reasons why a market system may not work satisfactorily in the field of mineral resources:- (a) the market may discount the future at too high a rate; (b) the market may take an incorrect view of future price trends; (c) the market may fail to take account of environmental and other social costs to which market prices cannot be given or are not given.

(a) The effect of too high a rate of discount is to stimulate current production at too low prices, and to use up unreplenishable resources too quickly. With a lower rate of discount, current prices would be higher relative to future prices , e.g. assuming no costs of holding land, if the value of a given quantity of minerals is believed to be £100 next year, and the rate of discount is 10%, the present price of the minerals will be at least £90, while at 5% the price will be at least £95. If the present price to owners is £92 (i.e. the present selling price less costs of extraction and sale), the minerals will be extracted if the market rate of discount is 10%. But if the "proper" rate of discount is 5%, they should not be extracted, and if the market used this rate, it would hold off supplies until the present price rose to £95. There are thus grounds for saying (following the earlier argument) that we extract more than we should today at the expense of the future .

(b) Does the market tend to hold a wrong view of future prices? Indeed it may. Most markets are too influenced by present prices. But who might know better? Most politicians and most civil servants, and even so-called experts, whether scientists or economists, are also biassed in the same way. It is not obvious that there is any source of pure knowledge which is so clearly untainted with the world as to offer consistently better answers.

It may be that markets are unduly influenced by uncertainty, and that caution leads to unwise decisions. But bold decisions by governments often also lead to equally unwise decisions. This may be particularly true where the uncertainty lies in the future of technology. The classic case of this is the assumption in the early 1960s that atomic energy costs would soon fall, as technological know-how grew, below those of conventional power stations; at a time when the capital cost per kW. of conventional power stations was falling dramatically, a dramatic fall in costs which was observed too late, probably because it was a "mere" matter of improving an old technology and not a "break-through".

(c) It is clear that there are many environmental and other social costs which do not enter into the market and these are the most important reason why the market has to be subjected to planning.

Conclusion

Far more could be said on all these issues. I have skimmed the surface of vast oceans without even being able to claim a bird's-eye view of the whole. Perhaps the most important thing is to stress how little we know and how much we live in a world of uncertainty. How nice it would be, the layman may think, if the scientists could tell us what the future of technology is to be. But we know they can't. All the economist perhaps can contribute is to assert firmly that uncertainty is what life is about, and that we had better stop trying to avoid the uncomfortable fact.

Select Bibliography

Penguin Modern Economics Readings:- Public Enterprise (ed.) R. Turvey.
                                     Cost-Benefit Analysis (ed.) R. Layard.

G. L. Reid, K. Allen and D. J. Harris, The Nationalized Fuel Industries.

J. N. Wolfe (ed.) Cost Benefit and Cost Effectiveness - Studies and Analysis.

Reports of the Royal Commission on Environmental Pollution.

W. Beckerman, 'Economists, Scientists and Environmental Catastrophe',
  (Oxford Economic Papers, Nov. 1972).

War on Waste (Cmnd. 5727, Sept. 1974).

DISCUSSION

Question: What would be the interest rate in a 'steady-state' society?
              Would you explain 'discount rate'?

Answer: I assume you mean by a 'steady-state' society one where the aver-
              age income per head is constant rather than a society with no
              changes. In such a society there might be an interest rate. It
              all depends on the extent to which people wish to save (and there-
              fore on the distribution of income etc.). There could be a zero
              interest rate but probably not, since gross investment still takes
              place (exactly offset by disinvestment in other assets). The'dis-
              count rate' is merely the interest rate looked at from the future
              backwards rather than from the present forwards (e.g. I pay £90
              today to get £100 in 1 year's time - discount rate of 10%, interest
              rate of 11%).

Question: Is the existence of an interest rate something unique to a capi-
              talist society?

Answer: I do not think so. The interest rate exists because people want
              to save and do not regard £1 in the future as equal to £1 today
              and because capital investment produces returns. All this occurs
              in all forms of society. Uncertainty is not unique to capitalism
              if for no other reason than that technological progress is
              unpredictable.

Question: How do you assess discount rates on works of art or national
              security?

Answer: Most people would argue that society should use one discount rate
              for all activities. Some would like to use a zero discount rate
              for works of art but even a high discount rate will have little or
              no effect if their value rises rapidly over time. The same applies
              to national security. It depends on what value you put on defeat
              in the future. If high, you will form a large army but, if low,
              you will have no army or police force.

Question: In a project evaluation, how do you evaluate risks?
Answer: Commonly a higher rate of discount is used to allow for risk but
              risks do not necessarily increase with time as this practice
              implies. A better way is to look at the actual figures for the
              periods when risks are high and put appropriate lower (or higher)
              values on them. This brings out the real incidence of risk.

Question: The Western World seems to control the world prices of both the
              manufactured goods it sells and the primary products it buys from
              the rest of the world. Is it not true that the underdeveloped
              countries suffer from this? Should not these prices be controlled?

Answer: Without accepting that the Western World has control of the market,
              I would agree that some redistribution of wealth to the under-
              developed countries is necessary. How this should come about is
              a much more difficult question. Adjusting the price system (except
              to deal with monopolies and to improve it) is not sensible. The
              alternative is to ration but this does not take account of different
              tastes and is inefficient.

Question: Can we rely on the economics of a situation to protect us from
              using up all our resources?

Answer: "All our resources" cannot be used up. Nothing in economics can
              protect us from using up some resources. It may be economic to

use them up but, with private enterprise control of the market, our oil, for example, is likely to be used up more quickly. This may not be a bad thing because we could find another fuel to use and would be forced to live without cars as we know them today, for example.

Question: You have been optimistic about the future but have stressed the uncertainty in any economic forecasting. Are the two incompatible?

Answer: All our plans may be wrong but it is still often necessary to make an economic forecast. We cannot predict anything exactly but an order of magnitude calculation is possible and sensible. My optimism rests fundamentally on the belief that knowledge increases, and that there is nothing inherent in our economic system that will bring disaster. Such a view is perfectly compatible with a pessimistic view of human nature and of human political behaviour, in particular.

Question: You have made an assumption for the future that technical knowledge will increase but will we become wiser?

Answer: Indeed not. People are not very wise and I expect they will often make unwise decisions.

ENERGY AND SOCIAL ECONOMICS

M. Gaskin

Political Economy Department, University of Aberdeen

Energy and the systems by which it is obtained and delivered have all sorts of economic effects - effects, that is, on that side of our lives concerned with producing and with consuming the fruits of production. Energy is one of those resources that is so basic to our systems of production, to our whole way of life in fact, that it invites the dramatic touch. Projected changes in its supply, or its systems, are presented in millenial or cataclysmic terms - the opening of a golden age if favourable, the crack of doom if not. For the economist energy is certainly a central resource in the country's economy, but his view of it is coloured by his disposition to look at process and change in incremental (or decremental) rather than global terms, and also by his knowledge of the mechanisms by which the economic system adapts to changes in the resources available to it, and of its very great capacity to do so.

All this tends to make an economist regard energy and its manifestations in less distinctive terms than one meets in some other approaches. But nevertheless it is important, and through its economic effects it has important social consequences. In this contribution I shall look at those economic aspects of energy use and energy systems that bear closely on society through the part they play in our living standards and in determining the pattern and development of our industries (this is what I mean here by 'social economics'). To put it very baldly, I shall look at energy in relation to man as consumer and man as producer - though restricting my view to a selection of elements in those two roles.

ENERGY AND LIVING STANDARDS

In a very general sense our present standard of living - as measured, for example, by the sum total of the goods and services we enjoy - depends in very large measure on the effective and massive use of energy. In what is mistermed 'productive industry', meaning those industries that produce or process tangible goods or commodities, present-day levels of output, either in total or per man employed, have clearly required the development and application of power-driven or energy-using processes. And in recent

decades, the direct consumption of large quantities of energy within or by
the household has become a mark of the affluent society.

The predictable consequence of these trends is that the use of energy has
increased very greatly over those periods of time for which we have
adequate records.   But the inter-relation between the growth of income,
and the increased use of energy, is a complex one and cannot be summarised
in simple statements.   For example, the figures we have show a considerable
disparity between the growth in energy use, as measured by the consumption
of primary fuels, and the long-term increase in the national income.   It
should be said, of course, that our attempts to measure both these
quantities over long periods of time are extremely hazardous, and the
figures we use for the purpose cannot be taken as more than suggestive
indications.   But taking them as they are, and looking at the stretch of
time from around 1850 to the present day, it is plausible to suppose that
the total national income of the United Kingdom has grown by about ten
times.   The consumption of primary fuels on the other hand, aggregated on
a heat equivalent basis, seems to have increased during the same period by
only about one half that amount.   If we look at both trends on a per capita
basis, the rise in income has been about fivefold, while the use of energy
has grown only by a little over half that amount.   The corresponding
figures for the U.S.A., from the early 1870s to 1950, show a rather wider
disparity - a sixfold growth in per capita income against an energy use
that has little more than doubled; but the pattern is broadly the same[1].

To a large extent this difference of movement is, of course, spurious.
We know that over this period there have been vast improvements in the
efficiency of fuel use, so that to equate energy use with quantities of
primary fuels consumed is progressively to understate the amounts of
effective energy used.   Historically there is some confirmation of this
in the fact that the growth trends of income and energy use diverge most
markedly during the forty years 1910-1950, a period when there were notable
advances in fuel efficiency in some major energy-using industries[2].   But
there is at least one important trend factor tending to cause energy use to
rise more slowly than income, in the way in which consumption evolves as
income rises.   As countries increase in affluence, they consume
proportionately more services - education, health, administration and so on -
and these are patently non-energy intensive.   Of course, this is a long-

term trend, and there are certainly shorter periods when it is partially,
even largely, offset by upsurges of other more energy-intensive modes of
consumption.   Rising car ownership and an increasing use of domestic
heating and other appliances (encouraged by a marked fall in the relative
price of energy) probably explain a steepening of the upward curve of energy
use in the United Kingdom during the 1960s.   But in the long-run, the swing
to services and perhaps other less energy-intensive products, is probably
dominant, though predictions of this kind must be seen for what they are -
extrapolations of empirical trends, not deductions from natural laws.

Energy and living standards: the quantities involved

But however fast or slowly the curve of its consumption will rise, energy
is bound to remain a central element in our economic and social life.
Equally inevitably, the conditions of its availability, especially its
cost, will change as the world supply situation evolves, and as world demand
grows, not least as those less affluent than ourselves become better off.
These prospects lead us to ask, what are the likely effects of changes, up
or down, in the relative cost of energy, on our living standards?   This
is not a question that we can answer in precise terms but one can at least
put it in a quantitative perspective.   Dr. Johnson once said, "A thousand
stories which the ignorant tell, and believe, die away at once when the
computist takes them in his grip."[3]   My claims as a computist are modest
indeed; but even a little crude figuring is useful here.

In 1972 total expenditure on fuel by final users (these exclude the energy
industries themselves) was rather more than 8 percent of gross national
expenditure, which is simply another measure of gross national product[4].
Roughly speaking, this means that, as a society, the energy we consumed
both directly and indirectly (e.g. through the products of energy-using
industries) amounted to one-twelfth of the sum total of goods and services
that were at our disposal in that year.   This is not a negligible figure,
but neither is it of towering importance, and its very scale places limits
on the consequences of changes in energy supplies and use.   Assuming that
energy usage continues at about this level it means that a 50 percent change,
up or down, in the average real costs of our primary fuels would make a
difference, unfavourable or favourable, of about 4 percent in our total
consuming power.   Again, 4 percent is not negligible; but given the return
of more favourable economic conditions it would represent something like

15 to 18 months' economic growth in the U.K.'s national income.

But let us explore a topical example, the effect of the fourfold increase
in the imported price of crude oil after October 1973.   In 1972 petroleum
products had supplied just under half the energy needs of this country, so
that the effect of a sudden change of this kind was bound to be marked.
There has in fact been a mixture of effects, including a severe worsening
of an already difficult balance of payments position and a strong additional
impulse to domestic inflation, and it requires a considerable feat of
abstraction to distinguish the effects of the change on living standards.
But given the quantities involved - and here we assume, quite unrealistical-
ly and against the facts, that levels of energy use remained unchanged -
the crude effect of the rise in oil prices should have been to raise
expenditure on energy from about 8 percent of national expenditure, to
14-15 percent, with a consequent tendency to cut the level of national
consumption in real terms by about 6-7 percent.   Such a reduction, coming
so suddenly, would be bound to impart a severe jolt to any society.   But
simply in point of scale, as a cut in living standards of a rich country
it would not be the end of the world, and in our case no more than could
be made good by about three years' growth in our national income.   In
fact, as a nation we have not had to take this cut since we have so far
been able to borrow the means to buy oil at the higher prices, and no
reduction in the living standards of the ordinary consumer has been imposed
from this quarter.   (It may be added that when we get it, North Sea oil
will only partly, if at all, offset this unfavourable effect of the oil
price rise - which must catch up with us eventually - since compared with the
pre-1973 cost of Middle East oil it will be very expensive).

To sum up at this point, more costly energy must tend to depress our living
standards, but within the ranges it is sensible to consider (which fall well
short of a catastrophic decline in availability) the very scale of energy
use within the total annual use of all productive resources puts restraints
on its power to do so.   By the same token, the gains to be had from the
opposite change are similarly moderated: even 'free' energy - energy
costless and unlimited - would, on present relative usage, only make us
about one-twelfth better off than we now are.   This conclusion may come
as a surprise to some, and perhaps we should look into it further for
possible qualifications.

Questions of growth and investment

Could it be argued, for example, that a really dramatic cheapening of
energy, such as we seem to be promised in the hydrogen era to come, would
release springs of technological progress that would greatly accelerate
the growth of our real income? To ask this question is, implicitly, to
get one thing right: the overriding importance of knowledge and technical
innovation as a factor in the long-term improvement of living standards.
But whether or not technical progress would be speeded up by the availability of very cheap energy is a point on which we must remain agnostic, if
only because as economists we have very little to offer in the way of firm
propositions about the <u>ultimate</u> forces which determine economic growth.

But there is one point worth making in this connection, not least because
it leads to another of wide significance in the whole field of energy
system development.   Sudden changes in technology, such as might produce
a dramatic cheapening of a major product, are historically unusual.   The
typical progression, especially in difficult areas of technology such as
nuclear power clearly presents, is one of a long, slow development, with
uncertainty about outcomes and heavy costs for those who do the developing.
The high costs of development, coupled with the large amounts of capital
investment demanded by modern energy systems, points to a different and
quite separate way in which energy impinges on our living standards.   The
process of extending modern energy systems to provide for future growth in
energy supplies, as well as developing them in a technical sense, make
very heavy demands on <u>present</u> resources, in advance of the fruits which
their subsequent operation yields.   The simple growth of energy use in
society as real income advances requires a continuing investment of present
resources if supplies are to be available in the face of growing demand.
The generation and transmission of electricity are particularly voracious
in this regard: in the ten years 1963-72 the installation of new electrical
capacity claimed £5,800 m. (at the prices current in each year of that
period), or 7 percent of the total capital investment undertaken in this
country during those years.   But the development of an entirely new
energy source over a short period can be equally demanding as we are
finding with North Sea oil.   The total capital expenditure required to
produce an annual flow of 100 m. tons of oil (roughly our present consumption) from the offshore fields by 1980 will probably exceed £6,000 m. (at

1975 prices)$^{(5)}$.   Of course, it is the case that society is continuously
devoting a proportion - currently 18-20 percent in the U.K. - of its
productive power to such forward-looking investment; and our living
standards in the present, as well as our expectations about the pace of
their ongoing improvement, are adjusted accordingly.   But even were we
to assume, quite wrongly as it happens, that the relative scale of capital
formation is or should be a constant, we have to recognise that there are
very strongly competing claimants on the supposedly given stream of
investible resources - not least housing and the social services.   But in
fact the proportion of present resources devoted to extending the productive
assets of society can and does vary, and in doing so can make significant
marginal demands on these resources, and thus on their availability for
securing our consumption needs in the present.   But there is an important
qualification to note here: it is always open to any one country to
persuade foreigners to provide the resources needed to increase its stock
of domestic productive assets - by borrowing from them.   This is what, on
a vast scale we are doing (I include the operating companies in the 'we')
to exploit North Sea oil.   And indeed, in recent years, we have been
doing it on an increasing scale to finance the capital investment of the
nationalised energy industries.

ENERGY AND PRODUCTION

Apart from their effects on our standards of living, energy systems have
wide economic-cum-social influences by virtue of the industries which grow
up to extract or transform energy from fuel sources, to supply these basic
energy industries with equipment and services, and to produce all the further
equipment and appliances which employ the energy forms produced.   It is a
moot point where one should sensibly draw the boundaries round all these
energy-related activities.   On a conservative definition, grouping the
energy industries with their more obvious suppliers and satellites
(including vehicles) they employ 8 percent of the country's workforce; a
more extended view bringing in those who make the basic materials like
metals and chemicals, or supply the services (including construction),
needed by these industries would raise this figure to at least 10 percent.
But besides their influences on employment the effects of energy systems
on production are felt in other dimensions of the social economy, including
the location of industry and population which we shall discuss later.

The interest for us in these industrial aspects of energy centres very much on situations of change.   What effects are changes in energy systems likely to have on those industries which in some way are related to them?   How readily can the economic system absorb or adapt to such changes?   To what extent, if any, would shifts in the balance of energy systems induce movements of industry and population?

Energy and the pattern of employment

As far as the employment created by energy systems is concerned, the clearest picture is of those engaged in the direct production and delivery of primary and secondary fuels.   In 1973 these numbered 632,000, broken down as shown in Table 1.

TABLE 1.   EMPLOYMENT IN THE FUEL INDUSTRIES, IN 1973, IN GREAT BRITAIN

Coal mining	315,000
Electricity	186,000
Gas (manufacture and distribution)	107,000
Petroleum and natural gas (incl. refining)	24,000
	632,000

Source: Department of Employment Gazette, March 1975, pp.197-202.

This figure accounted for only 2.8 percent of the total employees in the country in that year.   Of course, at the moment the basic material for our single most important energy source, petroleum, is imported, and to that extent the significance of the energy sector as an employer of labour is depressed.   But petroleum, like the other energy industries,with the exception of coal, is a very capital intensive industry.   When the North Sea is producing as much oil as we are now consuming domestically the numbers employed on production itself will only add about another 4-5,000 to the fuel industries' employment at that time.   Of course, between now and then many more than that will be involved in getting the oil to flow ashore; but that takes us into a different layer of the employment associated with an energy system.

It hardly needs saying that all the energy producing industries, whether
primary or secondary, depend on a back-up of supplying industries which
produce the materials, plant and structures that they require.   The extent
of this industrial sub-structure varies with the capital intensity or at
any rate the capital spending of the individual energy industries.   The
extent to which this spending varies between them is shown for the three
state-owned fuel industries in Table 2.

TABLE 2.   GROSS FIXED CAPITAL FORMATION IN THREE FUEL INDUSTRIES, IN 1973

	£ millions
Coal	61
Electricity	524
Gas	116

Source: National Income and Expenditure 1963-1973 (HMSO), Table 58.

In the case of North Sea oil and gas we have no comparably firm figure to
offer, but a recent estimate has put capital expenditure on exploration and
development at £1,250 millions during 1975[6] though probably only half of
that amount is at present being spent in Britain.

Leaving North Sea oil on one side, it is the great capital expenditure by
the electricity supplying industry that stands out.   It has typically been
a heavy spender on electrical equipment and process plant.   In 1968, a year
for which we have more detailed information on the industrial sources from
which it drew this equipment, the nationalised electricity industry spent
£250 millions on electrical machinery, and this was 49 percent of the total
output of the electrical machinery industry which at that time employed
200,000 people[7].   In fact, in that year the number of people making
equipment of one sort or another for the electricity industry was more than
half the number employed in the industry itself.

As the figure quoted earlier suggests, the electricity industry is now
completely overshadowed as a capital spender by the offshore oil and gas
industry of the North Sea;   this is currently spending at an annual rate of
about £5-600 millions within the British economy, though this figure will
certainly rise.   However as a creator of employment the expenditures of the
offshore industry are more modest in their effects than those of electricity.
Table 3 gives an approximate indication of the level and structure of

employment in the offshore industry in Scotland, in late 1974 (when oil
production had not yet begun).   If England and Wales were brought in the
total would of course be higher, but its scale in comparison with other
energy industries would not be altered.

TABLE 3.   ESTIMATED EMPLOYMENT IN OFFSHORE OIL AND GAS IN SCOTLAND,
                        NOVEMBER 1974

	No.
Exploration; support bases and other back-up services; administration.	7,500
Platform and module fabrication.	7,000
Pipe coating; pipe-laying (on- and offshore) and associated terminals and site works.	4,500
Other site works, terminals and oil-related construction.	1,000
Other, including equipment manufacture and mobile rig construction.	6,500
	26,500

Note: As indicated in the text the distribution between categories is
        approximate.

Of course, with petroleum we are faced with an energy system of which the
most substantial employing components are those industries which make
products that consume the fuel itself, above all the motor car and tractor
industries, but with aerospace also coming into the picture.   In 1973 in
Britain there were 536,000 people employed in making cars and tractors,
with a further 195,000 in the aerospace industry – vastly more than are
actually needed to extract and process the energy form that powers their
products.   Compared with this, the electrical appliance industry with
65,000 employees is small beer (though arguably one could add to this
the workforce in all the electronics industries).

Apart from being very large employers, the energy industries and all their
related industries have other features which make them notable elements in
the present-day economy.   For example, as a group they are high-paying
industries: their workers – even in coal-mining, after recent pay settlements – are, on average, among the better-paid members of the country's
workforce.   To an extent, these high incomes reflect high levels of skill
or technical expertise, though this is by no means universally the case.
Some of the industries we have been mentioning, electrical equipment

manufacture and aerospace in particular, are pre-eminently high technology
industries. With the prominent exception of coal-mining, and at certain
periods the gas industry, all the industries concerned have been among the
rapid growers of recent decades - that is, so long as we measure growth in
terms of output, since coal-mining has not been alone among the energy
industries in seeing its labour force decline, as Table 4 shows.

TABLE 4. EMPLOYMENT IN SOME ENERGY AND ENERGY-RELATED INDUSTRIES, IN
                                GREAT BRITAIN

	thousands			
	1960	1965	1970	1973
Coalmining	696	556	346	315
Electricity	209	244	215	186
Gas	127	121	123	107
Motor vehicles and tractors	455	513	538	536
Electrical machinery	191	204	154	133

Source: Department of Employment Gazette, March 1975, pp.197-202.

But so much for the facts of employment and industrial structure in the
energy sector. What does all this signify for the social economy? At
the outset one of the important questions we posed was how readily can
changes in energy systems be absorbed by the industries concerned, and by
the economy as a whole. The kind of changes that spring to mind in this
context are those longer-term ones that flow from marked rises or declines
of particular energy systems. But we should first note some shorter-term
problems of change that arise within the wider energy sector. For some
prominent elements within this are markedly subject to short-run instability,
as a consequence of corresponding movements in the wider economy. The
motor car industry is an obvious case in point: along with other industries
making durable goods - including plant and equipment - it suffers
disproportionate movements of its output in the course of fluctuations of
the national economy. This happens especially when the government,
pursuing a policy of restraining total national spending,clamps restrictions
on credit, especially hire purchase credit. In these circumstances the
capital goods industries, including those making equipment for the
nationalised fuel industries, also suffer, since cuts in the capital
spending of the nationalised industries are another handy weapon for a
government intent on a policy of restraining total demand. The capital

goods sector - the industries making plant and equipment, and also the
construction industry - is in any case one of the less stable parts of the
economy since capital expenditure is itself an unstable item in total
spending.   But the energy sector broadly defined is particularly subject
to destabilizing influences, paradoxically because it forms such a
convenient channel for government policy aimed at stabilizing the economy
as a whole.

Where a change in an energy system of longer term nature is involved, say a
run-down of oil or coal as primary energy sources, and their replacement by
nuclear electricity, there would clearly be consequent changes of a rather
different order.   We have seen this happen strikingly in the case of coal
during the 1950s and 60s when a large labour force was reduced by 50 percent.
This was accomplished with remarkable smoothness, though not without problems
within the industry itself, partly because the age structure of the workforce
lent itself to a relatively painless reduction.   But it has to be said that
the run-down of the coal industry has had very serious regional effects:
it has been a major factor in the economic difficulties with which Scotland,
South Wales and North East England have had to contend because of the high
concentration of the more costly (because older) sections of the industry
in these regions.

In Scotland with the rise of a new and exciting extractive energy industry
there is a combination of advantages and disadvantages.   The advantages,
which are powerful ones, consist in the very considerable injection of new
jobs which has brought the Scottish unemployment rate into a more favourable
relation with the national rate (though still higher) than it has ever been
before.   The disadvantages lie in the fact that much of the employment
created by offshore oil and gas, which is still rising, will not last
indefinitely since it is being generated largely by activities like
exploration, platform fabrication and pipe-laying which are once-for-all
operations.   The period of high employment may last quite a long time,
depending on the scale of the reserves discovered; but much of the
employment now being generated has a definite term to it.   And it may be
added, parenthetically, that even at its high point it will not meet the
total needs of Scotland for new employment to offset the continuing loss of
jobs in traditional industries, including coal.

The problems posed now and in the future by the development of Britain's new
hydrocarbon sources are not so fundamentally different from those on which
attention is more frequently focussed: the repercussions of changes in
energy sources on industries, such as the motor car industry, which are
closely linked with specific energy forms. Looking now at this wider
issue of the effects of possible changes in primary energy sources one
presumes that the group of industries least likely to be affected, except
by growth, are those involved in the manufacture of electrical equipment –
though even here some changes may be called for in the process plant
manufacturing sector. But quite apart from the possibilities of nuclear
sources, the convenience of this form of energy guarantees its continued,
and indeed increased, use regardless of changes in primary energy sources.
On the other hand, an eventual exhaustion of the less costly sources of
petroleum could in principle, have far-reaching effects on the automotive
industries. But there are two points to keep in front of one's mind here.
One is that employment in the vehicle industry is likely to decline anyway,
with the increased use of automated processes of production. And the
second is that an increasing pressure on supplies of petroleum is unlikely
to reduce the demand for transportation, even personal transportation. The
development of the electrically driven vehicle, which will surely come,
could obviously absorb much of the industrial capacity and skills at present
employed in making petrol-driven cars.

This exemplifies the general point that in many of the modern energy, or
energy-complementary,industries it is easy to underestimate the extent to
which the skills of the workforce at present engaged may be transferred to
other lines of activity. In much modern industry, the people employed are
far from inflexibly tied to the processes of the particular section of it in
which they work. Indeed to an extent that is not always recognised – and
can be obscured by the quotation of global figures of employment – they are
hardly specialised at all to these industries. In the motor vehicle
industry, for example, no less than 28 percent of the recorded workforce are
'administrative, technical and clerical workers', while another 15 percent
work in departments like stores, transport and canteens. In the electrical
machinery industry all these categories lumped together make up 48 percent
of the total employment. But even among the other workers in these
industries, among the skilled craftsmen and the semi-skilled production

workers, the transferability of skills to other industries, particularly
those of a generally engineering character, is very high.   Here we are far
removed from the position in some areas of the coal industry, where a
highly skilled but narrowly specialised workforce has been difficult to
redeploy as the industry has contracted over the last half century.   In
the case of mining, where difficulties arose they were compounded by the
comparative isolation - social and industrial - of many mining communities.
These conditions are not notably present in any of the other energy-related
industries.   A sudden, and permanent decline in demand for motor cars would
set Coventry back on its heels;   indeed the repercussions would be felt
throughout the engineering and metalworking industries of the Midlands.   But
those who predict that this region could become one of the depressed areas
of the future have no conception of the immense flexibility of its industry
and workforce.

But in any case, short periods of instability apart, I cannot see changes
in energy systems descending upon us with intolerable suddenness.   This
being so, it is unlikely that the changes that will occur will be
distinguishable from, or less tolerable than, the general and continuing
need for redeployment of labour and industrial capacity consequent upon the
whole process of industrial change in our society.

Energy systems and the location of industry and population
Finally, how far do energy systems impinge on social and economic life by
affecting the geographical location of industry and therefore population?
The most superficial awareness of the economic history of the older
industrialised countries points up/relevance of this question.   In Britain,
above all, the great concentration of industry and people on the coalfields
during the 19th century - a pattern still much with us today - is a
striking example of the powerful influence of an energy source on the
geography of the social economy.   But do the forces which operated so
strongly in the early coal age continue into the present day?   And will
they, or similar forces, so operate in the future?

Energy affects the location of population and industry in two quite separate
ways.   One, the more obvious, is simply through the business of extracting
a primary fuel.   Thus, in the heyday of coal large populations had to be
assembled on the coal measures because the getting of coal was (and,

relatively, still is) a labour intensive operation, needing large numbers
of men.   Today, among the newer sources of energy, primary or secondary,
there are none that compel the location of population in such numbers, in
particular areas.   To take the outstanding example at the present time, in
Scotland the exploration and development phases of the North Sea oil
operations are reinforcing an eastward drift of population which was already
present, and they will certainly produce higher permanent levels in areas
such as Aberdeen and some of the islands which have long-term roles to play
in the offshore industry.   But two-thirds of Scotland's five million people
live in the western half of the Central Belt (a legacy of coal);   and
considering that in its peak employing phase North Sea oil may generate
something like 50-60,000 jobs in Scotland - a fair number of them in the
industrial west - it is clear that its effects simply as a new extractive
industry, on the centre of gravity of the Scottish population, will be
scarcely perceptible.   Nevertheless, locally these effects will be very
significant for some areas, and we shall return to one class of them
shortly.

The other way in which energy systems influence the location of industry is
through transport costs - of the energy inputs themselves and their relation
to the costs of moving outputs to markets.   To simplify a fairly complex
matter, the geographical sources of an industry's inputs exert a locational
pull on it according to the presence or strength of three conditions.   One
is simply that the cost of the input should vary with distance from its
source.   The second, is the relative importance of the transport cost
element within the total cost of the input;   the measure usually quoted
as providing an index of this is the value-to-weight ratio.   And the third
is the extent to which the input transfers its own weight to the output.
In the case of coal these conditions were such as greatly to enhance the
attractions of a coalfield location to many industries: its costs did, and
still do, vary with the distance transported;   its value-to-weight ratio,
even in these expensive days, is relatively low;   and as an energy source
it transfers none of its own weight to the product.

However in the present century the influence of energy sources on industrial
location has been greatly diminished by a number of developments of which
three may be mentioned here.   One is the great improvements made in the
efficiency of fuel use: this has substantially reduced the input of energy

(especially in a 'raw' form like coal) needed per unit of output in all industries.   The second is the growth since 1900 of many industries which are markedly less energy-intensive than the old staples.   The third, the one of most interest here, is the development of cheaply transportable energy forms: electricity, oil and gas.

The cost of moving the newer energy sources over distance is far from negligible, but on a unit basis it declines sharply with the volume of throughput and is comparatively low in relation to the cost of the fuels themselves.   Furthermore, in the cases of electricity and gas (and, with modification, petroleum also) systems of geographically uniform prices have become the rule, with the consequence that over much of industry the energy input has ceased to have a significant effect on the locational pattern of production.   What has happened, in fact, first with electricity following the institution of a national grid, and recently with the natural gas pipeline network, is that the transmission systems have been framed to the existing distribution of industry and population.   In the case of electricity, the siting of generating capacity has been similarly influenced, though here regard has also been paid to primary energy sources. Petroleum is still moved largely by surface transport;   but prices to consumers though varied on a zoned basis, show comparatively small geographical differences.   Furthermore, given the fact of seaborne supplies of crude oil and the consequent desirability of coastal sites for refineries, it is again the existing distribution of industry and population that has largely dictated the location of refinery capacity in Britain.

Energy developments in remote areas

There is one problem of a locational kind, not unique to energy, but arising in connection with energy system developments, especially petroleum, that calls for comment.   Some of the operations connected with exploiting offshore oil and gas involve developments in places where, relative to local population and labour supplies, they are very large.   Some of the sites, actually in operation or proposed, for the fabrication of oil production platforms – especially concrete structures – fall into this class;   and a single major example of another kind is the Sullom Voe oil terminal in the Shetlands.   In the field of electricity generation, a continuing need to site nuclear power stations in areas remote from large populations contributes to the same problem.

Essentially this is a problem of scale of impact, and the effects of
imposing very large developments on small communities.   Economically such
developments may pose severe threats to local industries, usually not very
robust anyway, by competing away their labour.   Socially they create the
problems that arise when labour and population are brought into an area
from outside, frequently in numbers that swamp the original inhabitants -
problems, it should be emphasized, for the incomers as well as for the
aborigines.   In some cases the developments concerned have a limited life
and there will eventually arise a problem of transition to other activities,
or of run-down of employment and population.   Oil platform fabrication is
a notable example of this since eventually the demand for these structures
in home waters will be satisfied, and the possibilities of exporting them
are, generally speaking, limited.

One is faced in these cases with a set of problems extending over time -
problems in the first place of choosing the optimum location having regard
to all the social and economic costs and benefits;   problems in the end of
planning their transition, possibly to other activities.   As a general
socio-economic problem thrown up by all such large developments in remote
areas, it is limited in terms of the number of people it affects (though
 environmental considerations may enlarge the size of this group).   It
demands attention because projects of this kind will surely continue and
may become more numerous with the passage of time, and because no responsible
society can neglect the difficulties imposed by economic change simply
because the communities affected are small.   But 'responsibility' here does
not mean preventing such changes: given the difficulties, social and
economic, which face many of these areas in their undeveloped state (and
about which they frequently complain) it is altogether too simple to say
that no such large-scale developments should take place. But the issue
does need to be approached with a clear appreciation of all the consequences
involved, even those remote in time.

CONCLUSIONS

'Social economics' is a rather ill-defined term which usually connotes those
economic questions that bear most closely on social conditions.   In relation
to energy systems I have used it as a cue to direct attention to the
implications of changes in energy supplies and systems for living standards,
and for the structure of the industries in which men actually get their

living.

In considering the implications of energy and its changes for living
standards we saw that it was necessary to keep these in a quantitative
perspective and one that stressed marginal changes rather than all-or-
nothing type situations.   For vital though energy may be to our society,
in some total sense, its contribution to our material well-being reckoned
in terms of the quantities involved is finite and comparatively modest.
Because of this the power of quite large changes in energy costs to affect
our living staddards has its limits.

Within the productive system we have seen that there are important linkages
between energy industries and their equipment suppliers on the one hand,
and those industries whose products depend on particular energy forms on
the other.   There are possibilities here for the generation of disturbance
and some energy linked sectors have suffered in the past from instability
emanating from variations in energy supply programmes, and from other
aspects of short-run economic policies.   But the ability of the system to
adjust to not-too-sudden changes in the balance of energy forms and systems
is considerable.   The labour force in the energy related sector, in
particular, is more versatile than is sometimes realised, and with the
components less heavily localised than in earlier phases of industrial
history the possibilities for creation of new problem areas are the less.
It is, for example, wrong, in my opinion, to think that the industrial
system would not be able to adjust to very considerable changes in
automotive systems, provided these did not descend too rapidly upon us.

Finally, energy systems today neither are, nor show any likelihood of
becoming, the moulders of industrial geography that coal was in the 19th
century.   Several distinct reasons were advanced for this, but the effect
of all of them is to cause modern energy systems to accommodate themselves
to a geographical pattern of industry and population which is, or has been,
created by other forces, not least those of history itself.

REFERENCES

1. Statements about relative trends in U.K. national income and energy use are based on national income series in Mitchell, B.R. and Deane, Phyllis, Abstract of British Historical Statistics, (Cambridge, England, 1962) and successive editions of National Income and Expenditure, (HMSO), and energy (or fuel) series published in Mitchell and Deane and the Digests of fuel or energy statistics published by a succession of government departments, from Fuel and Power of the early 1950s, to the present Department of Energy. Before 1913 coal is assumed to have been the sole source of energy; after that other primary sources are included, and total energy use calculated on a thermal equivalent basis. The statement about U.S. trends is taken from A Time to Choose: A Report to the Energy Policy Project of the Ford Foundation, (Cambridge, Mass., 1974), p.17.

2. Consumption of primary fuels appears to have risen by only about 10 percent over this period, against an increase in real national income of perhaps 50 percent. Among other industries iron and steel is said to have greatly increased its fuel efficiency during these years.

3. This delightful quotation has been resurrected for us by Professors Richard Stone and Alan Brown. It appears in their paper, A Computable Model of Economic Growth, (Cambridge, England, 1962), p.vii.

4. In calculating this percentage expenditure on energy is taken as the total expenditure on fuels by final consumers, less duty on hydrocarbon oils, as given in United Kingdom Energy Statistics 1973, (Department of Trade and Industry), Table 6. 'Final consumers' exclude secondary fuel industries.

5. Gaskin, M., North Sea Oil and Scotland: The Developing Impact, (The Royal Bank of Scotland, 1975).

6. MacKay, D.I. and Mackay, G.A., The Political Economy of North Sea Oil, (London, 1975), p.75.

7. Input-output Tables for the United Kingdom 1968, (HMSO, 1973), Tables B and Q.

DISCUSSION

Question: You say that 8% of the gross national income is spent on energy
and if energy prices went up by 50% this figure would increase
to 12%. Presumably large energy users would increase their prices
by more than 4%?

Answer: Yes. The increase in price is the rise in energy cost multiplied
by that proportion of the price that is energy, so high energy
users will increase prices more than low energy users.

Question: What would be the effect of an energy price increase on goods
composed of components?

Answer: Let me give an example. Suppose we have an object whose price is
made up of 30% for components and 10% for energy in the final process while the components' price includes 10% for energy. If the
cost of energy doubles the price will increase by 13% for the final
product.

Question: If there was a decline in the car industry, what would car workers
do?

Answer: Other industries would build up over 20 - 30 years and this is
what happened during the decline of the cotton and coal industries.
Industries in the Midlands tend to be flexible and they would
soon adapt to a new situation.

Question: Could they be making an alternative to the car?

Answer: Possibly. The price of petrol does not kill the desire for transport and something like an electric car could be developed.

Question: When energy costs increase, is there not a need for capital investment, for example in the development of North Sea oil or of a new
technology to combat the increase, with the community having to
find money for this investment and for the costs increase?

Answer: On the whole, yes. When prices dropped in the fifties there was
investment to increase the consumption of oil but when prices
increase there is a two-fold squeeze on both price and added
investment.

Question: What criteria affect the investment in North Sea oil?

Answer: The main criterion is to have a balance of payments large enough
to at least pay the interest on the loan, which is what we have
assumed for 1980. The balance of payments has to be much greater
to pay off the debt and I think we may have to export oil to do
this.

Question: What are current world interest rates?

Answer: I would say world interest rates are about ten per cent or so and
as long as the country's benefit is greater than this ten per cent
then it is worth it.

Question: Do you think North Sea oil will save us from the current crisis?

Answer: We will be alright for the next twenty years only. The North Sea
is only a minor producer on a world scale, but it does have the
advantage of being politically safe.

PANEL DISCUSSION:
SOCIAL AND
ENVIRONMENTAL LIMITS TO GROWTH

Panel: Prof K Mellanby, Monks Wood Experimental Station (Chairman)
        Dr J Davoll, Conservation Society
        Dr C Ducret, United Nations
        Prof M Gaskin, University of Aberdeen
        Prof F M Martin, University of Glasgow

Mellanby: It is quite clear that we live in a finite world and that growth
of population, of energy and of resource use cannot continue indefinitely.
There is a tendency today for many people to believe that our environmental
situation is much worse than it has ever been but that is nonsense. One tends
to be told that the increase in energy use must cause greater environmental
destruction and a higher degree of pollution but the City of London uses much
more energy today than 100 years ago and its air is now much less polluted.
One is also told that the control of pollution must be very expensive but
this is nonsense, too. Certain forms of pollution are cheap to control,
others expensive. London's cleaner air has been achieved mainly as a by-
product of the use of cheaper fuel - it is much cheaper to heat a building
with an efficient oil furnace than have skivvies carrying scuttles of coal to
a fire in each room with 95% of the energy lost up the chimneys.

I think, then, that there is no simple answer to pollution. Indeed, perhaps
there should be a moratorium on the word "pollution" because the subject is
infinitely complicated. Almost every type of pollution has different
properties and has to be controlled in different ways, some easy and some
difficult. It is cheap to control the grossest pollution but it would become
infinitely expensive to remove all pollutants. Yet, of course, some
pollutants such as nitrates and sulphur are necessary in our environment
so total removal would be undesirable. A balance must be struck.

Question: We have been told that the CO_2 level in the atmosphere is rising
continuously. We see the CO_2 consuming forests being destroyed and we know
that there is much fossil fuel yet to be burned. Should we worry about this?

Answer: It is true that the CO_2 concentration in the atmosphere has been
increasing continuously (about 10-15% over the last 100 years or so) and is
now expected to double by the year 2000 compared with its present value
which is only a few parts per million. Presumably the increase we have seen
is due to the burning of fossil fuel but, of course, there are other sources
such as plants and volcanoes and the CO_2 cycle is not well understood. We
are not sure at what level the greenhouse effect will become important; some
might say that an increased average temperature would be welcomed here yet
increased sea level due to polar ice melting might not be! (At present, mean
annual temperatures appear to be generally falling). Apart from the greenhouse effect, a general increase in CO_2 level might stimulate food production
as does artificially-introduced CO_2 in greenhouses. Also, marine life could
benefit as more CO_2 dissolved in the sea would become available to plankton
for its photosynthetic processes.

709

We should be uneasy about the progressive change in the amount of an
essential constituent of the atmosphere without our being clear about its
results. This rapidity of change does not usually occur in nature; we have
overridden the homeostatic mechanism which was holding a steady state and we
should not be apathetic about it. The major effect of cutting down our
forests is not likely to be on the CO_2 balance (plankton is the main recycler
of CO_2 in the biosphere) but on the water balance. An incredible amount of
water is being evaporated by the Amazon forest, for example, which will not be
if the area is planted with crops. Indeed, experience suggests that the land
will be useless after a few years' cropping.

Question: Regarding water resources, one of the limitations on energy use
will be the availability of water for power station cooling or extraction of
oil from shale, for example. In the latter, it is suggested that one barrel
of water will be needed for each barrel of oil extracted. Can you comment?

Answer: It is partly a matter of economics. The net intake of water to
power stations with cooling towers is a small fraction of the total used
because vast quantities are recycled. The more you recycle, the costlier
it becomes and presumably this would apply to oil shale refining, too.

Question: An economist named Jevons wrote in 1870 that, by 1970, we would
have run out of coal. He had not foreseen the dominance of oil so can we be
optimistic about the future? What sociological changes will be necessary
to enable us to adjust to the changing energy situation?

Answer: I think people will adapt; it is very easy to underestimate the
flexibility, the adaptability and the toughness of the human being. If you
take a long historic perspective, we have undergone many difficulties but
have still emerged as recognizably human. We have become accustomed to
certain standards of living and softness of life and expect these to continue
to advance and many would find it difficult and painful if this did not
happen. Unfortunately, some of the simple outlooks, assumptions and values
that people had in the past which enabled them to overcome difficulties have
been casualties of improving material standards of life. It is possible
that it could only be difficult and stressful and, perhaps in certain respects,
potentially socially destructive to come to terms with energy assessments
which imply major changes in the standards of living. At the end of the
day, some kind of human society will re-emerge but it may be very different
from any we have been accustomed to. Human beings are tough and resilient
but their social and political systems are much more fragile. Many of our
institutions rest on quite fragile foundations and we perhaps over look the
extent to which they are buttressed by our excessive resources.

Question: One reason advanced for the desirability of nuclear power is that it
would allow more fertilizers to be available to third world countries. Will
these fertilizers be too expensive?

Answer: It is very doubtful that nuclear power will be of any advantage to
third world coutries in the term which really matters to them, namely the next
20-30 years. We would do better to cut down our use of cars and give the oil
to these countries so that they could make their own fertilizers now. We are
not terribly interested in the third world and we really rather hope they will
fade away. Of course, the nuclear facilities some of them are and will be
receiving will eventually force us to pay rather closer attention to their

requirements.

Question: What is the most useful practical step we should take to use energy more efficiently in the short term?

Answer: Raising the price of fuels is the best way. In addition, the present tariff structure should be reversed so that each person can buy a fixed amount of fuel at a low price and pay extra for any more fuel he wishes to use. In the present system for gas and electricity, there is a fixed standing charge or high initial rate to cover capital costs followed by a lower rate to cover running costs. This method is designed to recover costs and not to encourage greater consumption, the two being only coincidental. The difficulty of introducing a higher rate after an initial lower rate is that of deciding the point at which this should be introduced, as there will be no clear-cut economic basis as is the case for the present tariff structure. What would determine the amount of cheap fuel allowed per person? It is surely not beyond the wit of man to devise such a system; indeed, the Japanese have just introduced the reverse tariff structure we have been talking about.

Question: Surely raising the cost of fuel would simply create wage demands and inflation? Would not rationing be better?

Answer: The cost of fuel should be raised relative to the cost of other essential items so that ways of saving fuel become economic. At the moment, it is easier to use fuel than imagination.

Question: What would be the best way of persuading the public that they should save energy, apart from raising prices?

Answer: It is very difficult to persuade people to take actions unless the reason for taking these actions are manifestly obvious at the personal level. Human life is replete with examples, cigarette smoking being an obvious one. If a crisis is really looming, only draconian measures forced on the public will have an effect. Even if people see difficulties, they tend to feel that problems have always been solved and always will be. Energy saving campaigns seem to have an effect but only marginally. In Switzerland last year, the electricity utility company switched its advertizing from promoting consumption to promoting saving and a 2% fall in consumption followed even with no price rise.

Question: The price of fuel might have relevance to social and environmental limits to energy growth in the short term but it seems a long way from the perspective we should be seeking. Industrial societies all over the world are based on greed and I feel very gloomy about the long term. Our greed expresses itself in many ways, not just not in consumerism but also in rising violence in our societies. What can we do about it?

Answer: Certainly our propensity to consume energy is just one manifestation of our general propensity to consume all manner of goods and it can be described basically as greed. I am not necessarily talking about original sin but I do believe that greed is a deeply engrained human weakness and not one to be easily remedied. In a curious way, the efflorescence of democratic society has itself giving a huge impetus to this relatively greedy consumer behaviour. It would be easier to reverse this in an autocratic society but, as we do not have this at present, it will be difficult to achieve. Society will have to change — one could imagine one which paid less attention to competition and found virtues in conservation and sociality. Whether this can be done by a peaceful and continuous transition is doubtful as there are very powerful people doing very nicely out of the present system who will certainly oppose changes. Our society relies on growth and the promise of more growth to smooth over the moral problems of inequality and exploitation. Industrial societies of institutionalized growth cannot run their economies at zero growth - they rely on positive feedback. Zero growth would produce collapse so we strive for growth. On the world scale, this can only widen the gap between us and the majority who will soon be getting plutonium and an ugly situation will develop. The future does look very dark but one should not stop trying. We seem to be left without ideals, without guidance in an existential vacuum. The past seems meaningless. What is more important than the technological, social or political solution is the moral choice we have to make regarding our personal relationships with the natural environment and with other people. If we choose to continue our existing relationships we may be unable to solve our problems. We have many and energy use is only one of them.

SECTION 8

ENERGY ANALYSIS

PRINCIPLES OF ENERGY ANALYSIS

P. F. Chapman

Energy Research Group, Open University

The aim of this paper is to introduce a number of important
perspectives into the consideration of energy questions.
At this point in the proceedings I want to concentrate on
some important general principles, leaving the details of
methods of analysis and results until the sessions devoted
to particular aspects of energy analysis. (See later papers).
Essentially I want to explain some of the reasons why I feel
energy analysis, or similar studies, are worthwhile and also
show how the results of these studies can be incorporated in
more conventional techniques of evaluation. To accomplish
this I will be taking issue with the conventional economic
view that all the information needed to resolve a production
or consumption choice is contained in prices. I want to
demonstrate that whilst this is true in the abstract
'perfect market economy' it is not true in the real world.
Although I will be comparing 'energy analysis' with
'economic analysis' there is a very important distinction
between them which has to be made clear from the outset.
This is that energy analysis is a <u>descriptive</u> method whereas
economic analysis is a prescriptive (or evaluative) method.
Energy analysis attempts to tell you what will happen to energy
consumption if you make certain choices ; economic analysis
attempts tells you which options you <u>should</u> choose. Thus I
am not trying to put forward an "energy theory of value",
indeed I am not in any way trying to impose a value system
or tell you what you ought to do. However by using a
refined descriptive tool I think I can demonstrate that some
of the things that economists say we ought to do are downright
silly in that they are counterproductive.

The first distinction between energy analysis and economics
is in the size of the "standard system". In much (but not
all) of economics the system studied is an individual firm

or household.   The transactions between this firm
(or household) and the rest of the world are described
in terms of the quantities and prices of the commodities
exchanged. For the prices of commodities to give
information for an efficient use of resources it is
necessary to assume that the prices are determined in a
'perfect market'. It is recognised that there may be
imperfections in the structure of the market, that there
may be social costs excluded from prices and that
externalities such as pollution or waste disposal costs,
may not be included. In addition there may be uncertainties
about the future, restricted information about possibilities
and technological time-lags all of which might cause prices
to deviate from those which would lead to optimal
investment decisions. But, the argument goes, there is no
other method available and provided these imperfections can
be taken into account then the price system produces
satisfactory results.

In fact by considering sub-systems in the economy it is
possible for two firms to make 'optimal' investment
decisions which lead to an overall sub-optimal behaviour
of the entire system. Some such examples can be brought
to light by energy analysis since, unlike economic analysis,
energy analysis looks at the entire system. This is
illustrated in figure 1 which shows the sequence of
production, from resource stocks to some final good,
including not only the final production industry but also
those industries which supply it with raw materials, fuels,
equipment and so on.   Provided all the stages of production
have been included in this diagram then the gross energy
requirement of the output is defined as the sum of the
energy which is sequestered from the fuel stocks in the
process of producing the output. Note that this gross-
energy-requirement (ger) does <u>not</u> include any contribution
from labour since

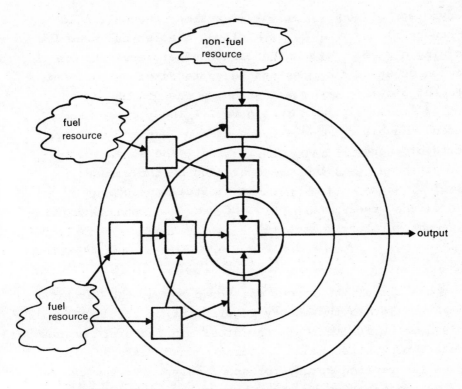

Figure 1.   Economic analysis tends to deal
with the transactions across the inner circle
representing the purchases and rates of the
producer. Energy analysis traces the sequence
of production steps back to the primary
resource stocks shown outside the outer circle.

(a)   to count all the commodities consumed
       by labour as a ger of labour would
       leave no system output.

(b)   the change in fuel consumption due to
       employment is marginal since our social
       system allows unemployed to continue
       to eat, keep, warm etc.

There are also pragmatic grounds for keeping labour analysis
separate from energy analysis in that by constructing two
more or less independent variables it is possible to
subsequently examine the trade-offs between them.

The advantage of looking at larger systems than a single firm can be illustrated by examining a theoretical example concerning a steel works and a car manufacturer. The operation of steel furnaces can be represented by material inputs, of pig iron and steel scrap, a fuel input and an output of steel, as shown in figure 2. The pig-iron is produced in a blast-furnace which, let us assume, consumes E_p units of energy per ton of pig-iron. The steel scrap is not given any ger, but the operation of the furnace requires E_f units of fuel per ton of steel throughput. Thus from the steel makers point of view the total energy requirement per ton of steel (E_o) is given by

$$E_o = E_f + E_p (1 - ß)$$

This suggests that the larger the scrap input, the larger ß, then the smaller is the energy requirement of steel. Thus in this hypothetical example the steel manufacturer decides to install more scrap handling furnaces and increases the price he is prepared to pay for steel scrap. (In the cause of energy conservation he might even get a government grant to help pay for the capital investment!)

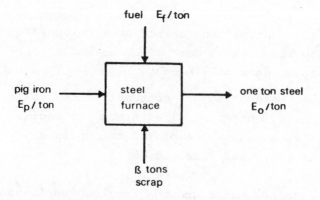

<u>Figure 2.</u>   The material and fuel inputs to a steel furnace. Note to produce one ton of output the inputs are *ß* tons of scrap and (1-ß) tons of pig iron.

At about the same time a car manufacturer is faced with
a choice between two steel presses. Press A consumes
10 kWh per sheet pressed and rejects 10% of the plates
as scrap. Press B requires 12 kWh per sheet, but doesn't
reject any scrap. In the cause of energy conservation,
and with the price of electricity rising (and the price
of scrap rising) our energy conscious car manufacturer
installs Press A.

Unfortunately the net result of these two investment
decisions is to increase the energy required to produce
motor-cars. This is illustrated in figure 3 which shows
the fuel input to the press as E_m per ton throughput
and all the car manufacturers scrap being used by the
steel plant. Now clearly if the output car requires 1 ton
of steel then, by conservation of mass, the pig-iron input
must be 1 ton. However the fuel consumed in both the steel
furnace and car press is proportional to the total mass
throughput which includes β tons of scrap in addition to
the 1 ton flowing from input to output. Thus the total
energy requirement of the car, E_c, is

$$E_c = E_p + (1 + \beta) \cdot (E_f + E_m)$$

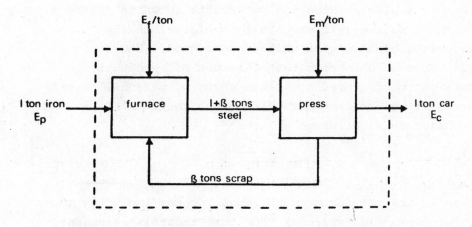

<u>Figure 3</u>.   The enlarged system including a
car manufacturer who generates the steel scrap
consumed by the furnace.

This shows that increasing the quantity of scrap generated
and used, increasing β, increases the energy requirement.
The market 'imperfections' involved in this example include
the ability of a manufacturer to declare a higher price
(showing control of a market by a manufacturer), the car
manufacturer's uncertainty concerning future fuel prices ,
and the inability of prices to show up a "system" effect
before the investment decisions are taken. Eventually
the price system should produce a higher cost per car -
but by then both manufacturers are locked into a
combination of bad investments, possibly for 20 years!

This leads me to the second feature of energy analysis
in that it may be possible to decrease uncertainty about
the future by anticipating certain types of technical
changes or the consequences of such technical changes.
It is obvious that no-one has perfect knowledge of the
future and so there is no mechanism for present prices
to contain information about the future. However
investors and manufacturers clearly need some knowledge
of the future if they are to make rational decisions.

In order to understand how energy analysis is of any use
in this matter let me first relate energy analysis to
economic analysis using the bar-charts shown in figure 4.
The chart on the left shows all the factor inputs
for the production of a unit of some good. Now it is
well known in economics that this set of production
costs can be factored down into a set of personal incomes.
Representing the cost per unit output as C, the factors
purchased as x_i at prices p_i then

$$C = \sum_i x_i \, p_i \equiv \sum_j P_j \tag{1}$$

where P_j are all forms of personal incomes including those
for property rights (rent), for investment (shareholders)
and labour. In energy analysis this factor breakdown
is not extended to the payments to the fuel industries

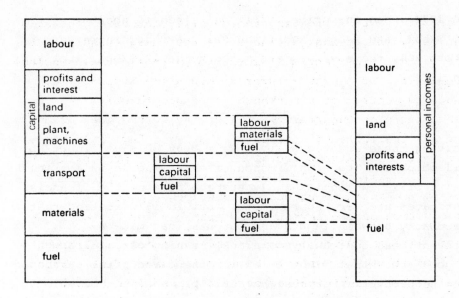

<u>Figure 4</u>.   Showing the division of factor
inputs into payments to personal incomes and
payments for fuels.

(although these could also be represented in terms of
incomes).   Instead the total costs are represented by

$$C = x_f \, p_f \; + \; \sum_{j \neq f} P_j \qquad\qquad (2)$$

where

x_f = total quantity fuel = g.e.r. of product

p_f = price of fuel

$\sum_{j \neq f} P_j$ = personal incomes (excluding those
described from the fuel industries.

To illustrate the utility of this breakdown let me consider
the production of oil from oil-shales. Now in 1970,
when the price of oil (p_f) was 2.5 $/bbl, it was claimed
that it would be possible to produce a barrel of oil from
the shales at a cost (C) of 6 $/bbl. i.e. in 1970

$$6 = x_f . \; 2.5 + \sum_{j \neq f} P_j$$

In 1974, when the price of oil had risen to about
10 $/bbl, and labour and financial costs had increased
about 20%, the cost was estimated to be 12 $/bbl. i.e.
in 1974

$$12 = x_f . 10 + 1.2 \sum_{j \neq f} P_j$$

Solving these equations gives x_f = 0.69 bbls oil input/bbl
oil output.   The significance of this example is not in the
value of x_f (which is highly suspect because the financial
estimates are very approximate) but in illustrating that a
feedback, between inputs and outputs, can make a
conventional cost analysis almost meaningless. Although
a good accountant might allow for the direct fuel purchased
in the production process he would probably not be able to
incorporate the indirect fuel inputs associated with the
production of equipment, chemicals etc. which are bought
in.   In terms of the energy analysis framework the true
break-even price of oil from shales is when the production
costs equal the market price of oil i.e. when

$$C = P_f$$

Putting this into equation 2 gives the result that the
process breaks even when

$$P_f = \frac{\sum_{j \neq f} P_j}{(1 - x_f)}$$

$$= \frac{\text{non fuel costs}}{\text{(net energy yield)}}$$

This is a very direct way of accommodating the feedback
in the situation (c.f. formula for gain of feedback
amplifier) and shows that the net energy yield of a fuel
production process may be significant in determining
when it will become "economic". This may be particularly
important for other fuel sources such as the tar-sands
and low-grade uranium ores.

Before describing other applications of energy analysis
I should like to establish another reason for carrying
out these kinds of studies. It is obvious that energy
is a universal input to production and that the recent
rise in the use of fuels has been largely responsible
for increases in productivity per man. Although many
people studying energy are doing so because of potential
resource problems, i.e. potential scarcity, I am involved
in this area because of the potential problems of abundance.
This concern is based on the idea that man's use of stored
fuels represents a net input to the heat budget of the
earth atmosphere system which has the potential of
significantly altering our climate. This has been
extensively discussed in the literature (1,2,3,4,5,6) so
here I will only outline the case.

In its simplest form the argument says that the only way
for the earth to increase the amount of heat it radiates
into space is to increase its surface temperature.
According to the Stefan-Boltzmann radiation law the annual
energy input,E, equal to the energy output in equilibrium)
is related to the surface temperature, T, by

$$E = \sigma A T^4$$

where

σ =   Stefan-Boltzmann constant

A =   radiating surface area.

Using an annual solar input of 10^{18} kWh/annum this gives
a surface temperature of 280K, which is reasonably close
to that which we observe. This also leads to a
relationship between the change in energy input and change
in surface temperature, namely

$$\frac{dE}{E} = 4 \frac{dT}{T}$$

Thus if the energy input were increased by 1%, a net
addition of 10^{16} kWh/annum, then the rise in surface

temperature would be 1/4% or 0.7°C. This is serious since
it could lead to a melting of the polar ice-caps.
Before describing some of the factors which complicate
this simple analysis let me just establish that this level
of fuel consumption is within our time-horizon. Two
extrapolations of world fuel consumption are shown in
figure 5. That marked (b) assumes a constant growth of
5% p.a, the present growth rate. That marked (a) assumes
a continued increase in the rate of growth. Both put the
1% solar level within our own time horizons.

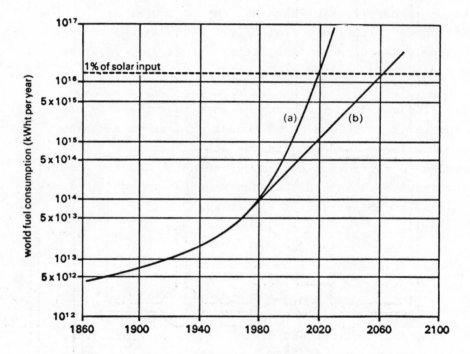

Figure 5.   Two extrapolations of the past
trends in world fuel consumption.

Now the 'black-body' theory would be exactly correct if
the world could be viewed as some kind of barren planet.
However the thin skin of air and water surrounding the
surface on which we live significantly alters the situation.
The complex flows of energy in the atmosphere are shown in

figure 6 which also identifies four 'partition coefficients'
which describe the reflectivity and absorbtivity of
different parts of the system. All these coefficients can
be altered either by an additional heat input or by extra
carbon dioxide inputs (due to combustion of fossil fuels)
or by extra water vapour (due to cooling systems) or by

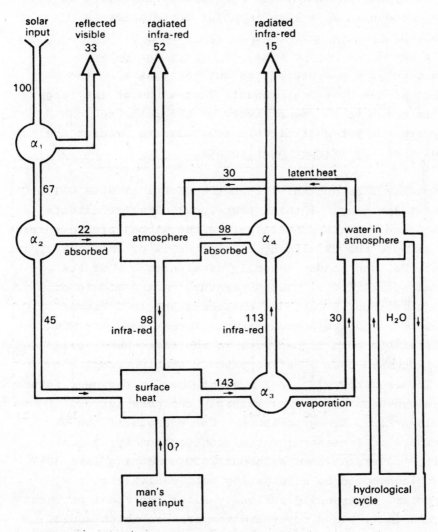

Figure 6.   The flows of energy in the
atmosphere. Note that a significant amount of
infra-red radiation is trapped between the
surface and atmosphere; this is the
'greenhouse' effect.

extra dust. There has been some success in modelling the
role of carbon dioxide [7] and dust [8,9] in the atmosphere.
However even the largest climatic models are not yet able to
predict the results of combinations of inputs. One of
the larger general circulation models has been used by
W.Washington [10,11] to perform 'heat input experiments'.
He found that an additional heat input equivalent to 5%
solar caused a $28^{\circ}C$ rise in polar temperatures -
a catastrophe. A more refined experiment with a
distributed heat input equal to 0.5% solar caused
a $+5^{\circ}C$ polar temperature rise but this was also the
effect of a - 0.5% heat input! Thus although the large
models point to the sensitivity of the polar regions
they are not yet sufficiently sensitive to predict the
results of significant heat inputs.

Personally I do not think that the global effects will
be most limiting, I think that local climatic effects
will sound the alarm bells before the global problems arise.
For example the UK already dissipates 1% of the solar input
to the UK, and London annually dissipates 20% of its
annual solar input. The temperature in the middle of
London is about $5^{\circ}C$ higher in summer and $10^{\circ}C$ higher
in winter than its surroundings.[*] Recently a whirlwind
in Essex has been traced back to the time when a cold
front crossed over a refinery/power station complex on
the Thames Estuary[13]. I suspect that in the near future
these types of effect may multiply and pose serious
problems for U.K. agriculture. For example if the UK
continues to increase its fuel consumption by 2% p.a.
(historical rate) then by about 2030 something like 30%
of England could be dissipating heat equivalent to
50% solar. This could seriously alter the local patterns
of rain fall, cloud cover and temperature which would
pose serious problems for agricultural production in the
UK.

[*] See Appendix.

In summary there seems to be a prima facie case for concern
(but not for alarm). Climatology is an extremely difficult
subject because there are large natural fluctuations in
climate extending over decades. However sooner or later
we must constrain our heat release, and hence our use
of stored fuels, if we are to retain a climate
approximately the same as that which we presently enjoy.

The idea that we may face a future of constrained
fuel use has serious implications for future productivity.
As shown in figure 7, recent increases in productivity
have been associated with increases in fuel use, often
replacing labour by fuel. There are grounds for
suggesting that simply to maintain the present high
agricultural yields may require continuing increases in
fuel inputs to accommodate declines in soil structure or
the evolution of resistant insect and fungal pests. In
general there is a case for suggesting that many basic
production technologies are approaching the point at which
they may need increased fuel inputs simply to maintain
present production.

This rather sweeping assertion can be illustrated by
considering the production of copper from ores. Detailed
analyses have been published elsewhere [14], here I
simply want to offer evidence in support of the above
argument. The simplest form of the argument runs as
follows. As time proceeds then the grade of copper ores
worked declines. Around 1900 the average grade worked
was 2.5% Cu. Now it is about 1%, and the average grade
of copper ore worked in the U.S.A. is 0.55%. The
theoretical amount of work needed to mine enough ore to
produce one ton of metal is inversely proportional to
the ore grade. This theoretical work is required to
lift the material out of the ground, transport it to a
mill, crush it (thereby creating free surface area) and
separate the mineral particles from the dirt. Thus, as

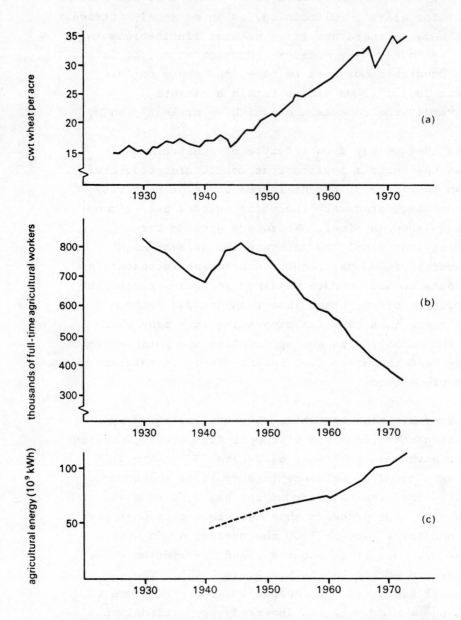

<u>Figure 7</u>.   A typical example of the increasing
productivity per unit land and per man due to
increases in fuel consumption (data from refs
17 and 18).

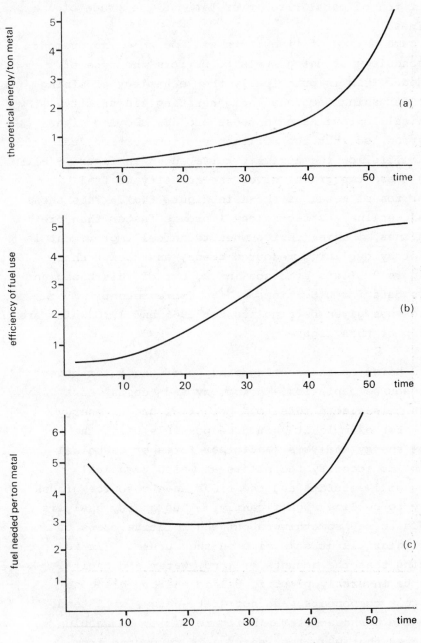

<u>Figure 8</u>.   The time trends in a) the
theoretical minimum of work needed to mine one
ton of metal b) the efficiency of converting
fuels to work and c) the fuel needed per ton
of metal. (arbitrary scales).

shown in figure 8(a) the theoretical amount of work
needed per ton of metal rises over time as the grade of
ore declines.

Over the same period of time there will be some improvement
in the technology of using fuels to perform the type of
work needed. This is principally the technology of mining
equipment, crushing machines and generating electricity.
The technical improvements in these are now showing signs
of saturation, so that the efficiency of converting fuel
to work will follow the logistic curve shown in figure 8(b).
Combining these two graphs gives the quantity of fuel
needed per ton of metal as shown in figure 8(c). This shows
an initial decline, as technology improves faster than ore
grade declines, a level period when technical improvement is
just offset by declining resource grade, and then a sharp
rise. Figure 9 offers evidence in support of this based on
Lovering's data for U.S. mines [15]. (Note recent
information has given data points for 1965 and 1970 which are
off the top of this figure [16]).

Apart from being an interesting example of the interaction
between resource depletion, technology and "costs" of
production this also points to a potential use of energy
analysis. For example it should be possible using the
methods of energy analysis (and other forms of technical
evaluation) to identify the points at which technical
efficiency will saturate and the times when resource grades
are likely to decline significantly. Thus energy analysis
may be able to say something about the relative costs of
producing materials at some time in the future. (It is
worth noting that the amounts of land, water and capital
will also be inversely proportional to the ore grade so
that all cost components, and not just energy, will tend
to follow the J-shaped curve in figure 8(c). It should
also be noted that when the resource is recovered from
sources close to the average crustal abundance of the
element then this J-shaped curve will also show a
saturation effect).

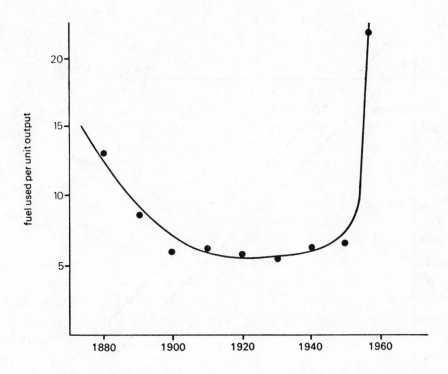

<u>Figure 9</u>.   The fuel used per unit output of
<u>all US</u> mines (source: ref 15).

A similar argument seems to apply to fuel resources.
Although there is no difference in the "grade" of oil
recovered from under the North Sea to that obtained
from the Middle East, it is still "more difficult" to
recover oil from the North Sea. This is reflected in its
lower net-energy yield (or higher energy requirement per
unit fuel output). For some fuel resources, such as
uranium, there is clearly a defined grade and, as shown in
figure 10, there is a grade of uranium below which there is
no net energy yield. (Details of the construction of this
graph will be given in a paper later in the conference).

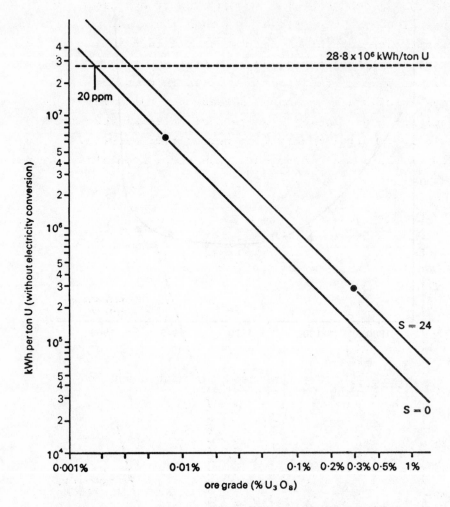

Figure 10. The fuel needed to mine one tonne
Uranium as a function of ore grade (without
any conversion losses due to electricity
generation) compared with the electricity
produced from one tonne of Uranium in an SGHWR
reactor shown by the horizontal line. The two
lines correspond to mines with different
ratios of overburden/ore.

There is thus a good case for saying that if our technology
continues along its historical path then there may be a
significant increase in the total heat put into the
atmosphere simply to maintain our material output. This
is amplified by the simultaneous decline in both fuel
resources (giving a larger heat release per unit fuel
produced) and a decline in the grade of material resources
(giving a larger fuel requirement per unit of material
produced). There are undoubtedly many areas remaining

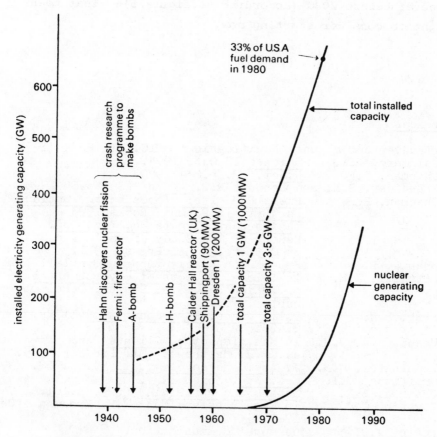

Figure 11. Showing that even with the very
intensive development of nuclear power in the
USA it is at least 50 years between the
'invention' and the time when nuclear power
provides a significant fraction of total fuel
supplies. (official projections of electric
capacity from ref. 19).

for technical improvements and technical changes, but time
is running out. Figure 11 shows that even with crash
research and development, spurred on by war and potential
oil shortages, it will take 50 years for a nuclear
technology to make a significant impact on total fuel
supply. Now if the thesis I have been mounting is correct
then we ought to seriously consider developing solar and other
renewable fuel sources since only by using these can we
maintain our present lifestyles and not seriously upset our
climatic system. But we need to have accomplished this
changeover before 2030 (according to figure 5). That means
we ought to consider starting now.

References

1. 'Inadvertent climate Modification' SMIC. MIT Press 1971
2. Weinberg, A.M.   Science 18 Oct. 1974.
3. Chapman, P.F.   N.Sci. 47. p.634 1970
4. Schneider, S.H. and Dennett R.D. Ambio 4(2). p.65 1975
5. Chapman, P.F.   Fuels Paradise: Three Energy Options
                       for Britain Penguin Books 1975
6.   -   Inadvertent Climate Modification.
                       Report of The Study of Man's Impact
                       on Climate.   MIT Press. 1971.
7. Machta, L.   Proc. 20th Nobel Symposium. p.121. 1971
8. Mitchell, J.M.   in 'Global Effects of environmental
                       pollution' ed. S.F. Singer: D.Reidel.
                       publ. 1970.
9. Lamb, H.H.   Phil. Trans. Roy. Soc. A266(1178).
                       p.426. 1970
10. Washington, W.M.   in'Mans Impact on Climate' ed.
                       ed. W.H. Matthews. MIT Press p. 265. 1971
11. Washington, W.M.   J. Appl. Meteorology. 11(5). p.768. 1972
12. Griffiths, C.I.   Met. Mag. March 1974.
13. Chapman, P.F.   Metals and Materials 8(2). p.101. 1974.
14. Lovering T.S.   in Resources and Man W.H. Freeman & Co.
                       1969.
15. Lovins A.B.   personal communication.
16.   -   'A century of agricultural statistics'
                       HMSO. 1966.
17. Leach, G.   private communication.
18. WASH - 1139   'Nuclear power growth 1974 - 2000'
                       U.S.A.E.C. Publ. US. Govt. Print Office.
                       1974.

Appendix   Urban climates

The aim of this appendix is to present some evidence in support
of the thesis that heat release in urban areas is capable of
bringing about substantial temperature changes. The figures
below show the patterns of isotherms around London both for a
particular time and for the annual average. The best general
review of man's impact on climate is the SMIC report (1); on
urban climate is Peterson's review (2) and publication by
Chandler (3,4).   The evidence for very large urban 'heat
islands' is irrefutable but the magnitude of the man-induced
change depends on what is assumed about what the local climate
would be without the presence of the city.

Fig. 1. Distribution of minimum temperature in London, 14 May 1959. Broken
lines indicate some uncertainty of position. Isotherms are numbered
in $^{\circ}$C with $^{\circ}$F in brackets (after Chandler "The Climate of London",
Hutchinson, London).

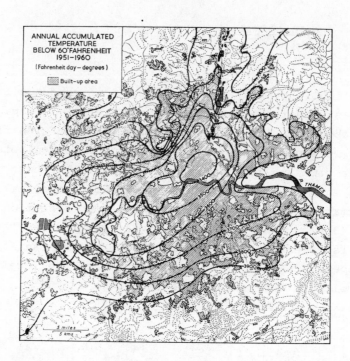

Fig. 2. Distribution of annual accumulated temperature below 60°F in
         London, 1951 - 60 (after Chandler "The Climate of London",
         Hutchinson, London).

References

1. Report of the Study of Man's Impact on Climate :
    'Inadvertent Climate Modification' MIT Press 1971.
2. Peterson. T.J. "The climate of cities : a review"
    Nat.Air.Poll.Control Admin.Publ.No.AP59
3. Chandler T.J. 'The Climate of London' Hutchinson & Co.1965
4. Chandler T.J. 'Mankind's impact on the atmosphere'.
                  Geog.Mag.43(2). p.83. 1970.

DISCUSSION

Question: Regarding the concept of a climatic limit to energy release, can
            you estimate the relative magnitudes of geothermal energy released
            to the atmosphere and the energy released from industrial sources?
            Is all industrial energy input released to the atmosphere as you
            suggest, i.e. is not some energy stored in a chemical form?

Answer: Solar input to the Earth's atmosphere is many orders of magnitude
            larger than the geothermal input. The proportion of energy stored
            in transformed materials is only a small fraction of one per cent
            of the energy expended. The rejected heat from industrial pro-
            cesses cannot be stored in plant life, indeed it is not possible
            to "store" heat at ambient temperature.

Question: I am troubled by your exclusion of labour inputs. If they are
            excluded from the factor inputs in energy analysis, should they
            not at least be quoted?

Answer: By including manpower in the system under consideration the system
            would have no output. This is because the people supplying the
            labour also receive the output from industry. If the inputs are
            kept separate the scope of energy analysis is increased. An ana-
            lysis where manpower is obviously important is the agricultural
            example given in my paper.

Question: Should you include energy inputs expended overseas in an energy
            analysis of a product produced in the U.K.?

Answer: It is important to state the objective of any energy analysis and
            to make clear all conventions used. There is no sense in which
            the conventions chosen are right or wrong, they are merely chosen
            to be appropriate to the chosen objective. The inclusion or
            exclusion of energy expenditure overseas obviously depends on
            whether U.K. or global energy consumption is being considered.

Question: Should energy analysis be on a global scale or should it be carried
            out within, say, an individual firm?

Answer: In my paper, I demonstrated that optimisation of sub-units could
            defeat the overall objectives. In a few years it will become more
            obvious where energy analysis will be most useful.

Question: Why is there a large difference between the theoretical and present
            practical energy requirement for manufacture of products? Have
            car manufacturers learned anything from the results of Berry and
            Fels concerning Detroit car production?

Answer: To produce something at theoretical efficiency it would have to be
            produced infinitely slowly, which is not attractive to car manu-
            facturers.

Question: In one of your figures, why was the minimum useful grade for
            uranium ore 20 ppm?

Answer: The data referred to SGHWR burner reactors and will presumably
            be better for breeder reactors.

Question: How useful is energy analysis as a decision making tool?

Answer: Energy analysis does not make the best choice between alternatives,
            it simply reveals the implications of choosing each of the alter-
            natives. It is not the job of the energy analysist to make the
            decisions; that is the job of the politician. The more informa-
            tion the politician has and the more he is aware of the conse-
            quences of his decisions, the better his decisions should be.

METHODS OF ENERGY ANALYSIS

P. F. Chapman

Energy Research Group, Open University

An earlier paper in these proceedings has described the general principles of energy analysis and discussed some of the applications of these analyses in a very general way. In this paper I want to describe one energy analysis in some detail so as to show how such analyses can be performed and the types of problems involved. The particular example used in this paper is the energy analysis of nuclear power stations which has been described in detail elsewhere [1]. Here I want to concentrate on the methods used to obtain 'energy requirements' and the ways in which these numerical results can (and cannot) be used. I hope to establish that whilst the evaluation of an 'energy requirement' for a commodity can be done relatively objectively, the interpretation of the results require the addition of certain behavioural assumptions. I also hope to demonstrate that these assumptions are not 'arbitrary' but can be ascertained by a careful examination of the system under investigation.

There are a number of methodological problems involved in energy analysis which should be carefully considered in formulating the procedures appropriate to any particular investigation. Without any doubt the most serious problem is that of obtaining enough accurate and reliable data on energy and material flows in the system being examined. There is some data, such as the fraction of domestic electricity used for lighting, which can never be accurately ascertained. At the outset of any investigation it is worthwhile considering what data is needed and the likelihood of obtaining sufficiently accurate information.

The methodological problems are conventional problems which have to be resolved by reference to the aim (or purpose) of the investigation. Briefly these are

(i)   definition of the system being examined;
       i.e. where is the system boundary
(ii)   attribution of energy requirements to the
       input to the system
(iii) partitioning energy inputs between the outputs
       of joint production processes.

All types of analysis have these conventional problems, the
difficulty with energy analysis arises because until recently
no one has considered it necessary to analyse the flows of
energy.   In economic analysis these conventional problems are
not significant because, as a matter of course, every producer
attributes prices to his outputs and pays cash for his inputs.
The convention used by a particular producer to price the
outputs of a joint production process may be quite arbitrary,
but once determined is no longer open to question.   The
conventional problems will be illustrated by a few examples which
are pertinent to the nuclear analysis later.   The simplest way
to illustrate the problem of defining a system boundary is to
consider the problem of imported commodities.   Here the system
boundary is the geographical boundary of the UK, but there are
still a number of possible conventions which can be used to
accomodate imports.

(i)   imports can be given a zero gross-energy-requirement
       (ger) on the basis that no energy is expended on
       their production within the system boundary. This
       may produce anomalous results if comparing production
       processes with different imported components since
       it will show the process which imports most as
       having the smallest ger (see ref.2 for an example).

(ii)   since imports have to be purchased by selling
       exports, which do involve energy use within the
       system, we could attribute the ger of £1 worth of
       exports to £1 worth of imports. This is difficult
       to accomplish (how do you know the ger of an
       exported motor car without first establishing the

ger of imported iron ore); it presumes something
about the balance of trade and it does not allow
any investigation of the "real" energy exchanges
involved in trade (does the UK make an energy
profit on its trade?).

(iii) imports could be given a ger which would arise
if the same goods were produced within the UK.
This corresponds to a 'replacement' cost and could
be useful for comparing some policy options.
However it doesn't work for activities such as
copper mining or growing bananas since these are
not activities carried out in the UK.

(iv) imports could be given the actual ger of their
production in the country of origin. This gives
further information about net energy flows into
(and out of) economies but is, in effect,
extending the system studied to include overseas
production processes.

In the nuclear analysis the convention used is (iv) on the
grounds that if any of the imported items are energy intensive
then this will be reflected in relatively high prices which will
require the export of similarly energy intensive goods or a
larger volume of low energy intensive goods. Thus if a large
part of the energy inputs do arise as imports I am assuming that
the UK will have to expend a similar amount of energy in
producing goods which can pay for the imports.

The problem of attributing energy requirements to the system
inputs is partly covered by the above discussion of imports.
However there are other types of problems which can arise in
counting "fuel" inputs. These can be illustrated by considering
the ger to be attributed to uranium. In the Digest of Energy
Statistics the nuclear power input to the UK is counted on a
'coal-equivalent' basis i.e. the nuclear electricity is counted
as equivalent to the quantity of coal which would have to be
consumed in a coal-fired station to produce the same output.

This is a useful convention for making international comparisons
but precludes any possibility of comparing the efficiencies of
nuclear and coal-fired stations. If one was interested in
comparing the actual and theoretical performance of a nuclear
system then one might use the theoretical fissile energy
contained in uranium. If the investigation were concerned with
evaluating substitutions between fuels, or in documenting heat
release, then one might count the heat generated in the reactor
as the ger of uranium since ultimately it is this heat which can
substitute for coal or oil. However the analysis described later
has none of these objectives. Instead it aims to evaluate the
total quantity of fossil fuel needed to produce nuclear
electricity. This requires a convention which attributes no ger
to the uranium in the ground. It is important to note that
provided it is quite clear which of these conventions is chosen,
and for what reason, then they are all "Legitimate". (But not
arbitrary! It would be inconsistent to attribute a 'coal-
equivalent' ger to uranium in order to evaluate the fossil fuel
required to produce nuclear electricity!)

The final conventional problem is the partitioning of energy
inputs between co-products. All energy analysis is based on the
"conservation of ger" such that

$$\sum_{in} x_i E_i = \sum_{out} y_j E_j$$

where

x_i = quantity of input i
E_i = ger of input i
Y_j = quantity of output j
E_j = ger of output j

This conservation law guarantees that at the end of an analysis
all the primary fuel input is attributed to the outputs.
However even when all the inputs have been evaluated there
remains the problem of partitioning them between the outputs.
For example a nuclear power station produces outputs of both
electricity and plutonium. How much of the energy input should
be attributed to the plutonium?

There are a number of methods which can be used. For example
the energy could be partitioned according to the financial
value of the output so that all outputs have the same energy
per unit value. This is used in studies based on financial
data (e.g. input-output based studies) but cannot be used here
since plutonium is not a marketed commodity. The energy could
be partitioned according to some physical parameter such as mass,
volume or calorific value such that each output had the same ger
per unit mass (or volume etc). However plutonium and electricity
do not have any obvious physical property in common! In some
situations one of the products is made in a single production
process. Under these conditions one product can be given its
ger when produced alone, leaving the remaining energy inputs
to be attributed to the other co-product.

This is a useful procedure when trying to arrive at an overall
picture of a system with a wide range of production technologies.
However it precludes the possibility of <u>comparing</u> the joint
production process with the single production processes.
Furthermore in this particular example it would result in a large
negative ger for plutonium, which is not a helpful conclusion.
In this example the convention used is to attribute a zero ger
to plutonium in order to be consistent with the convention used
for uranium. (Both uranium and plutonium are thereby regarded
as useful <u>materials</u>, not as fuels.)

Before embarking on the details of the nuclear analysis it is
also necessary to say something about the principle methods of
obtaining ger information. A powerful method, known as the
input-output, or I/O, method, uses the published input output
tables of an economy to derive energy intensities for various
sectors. The I/O table can be regarded as a matrix, $\underline{A}$, each
coefficient of which, a_{ij}, gives the £'s worth of commodity i
required to produce £l's worth of commodity j. Thus the direct
inputs needed to produce a vector of commodities $\underline{x}$ is given by
a vector $\underline{y}$ where

$$\underline{y} = \underline{A}\ \underline{x}$$

However to produce commodities $\underline{y}$ requires a set of inputs given by $\underline{z}$ where

$$\underline{z} = \underline{A}\, \underline{y} = A^2\, \underline{x}$$

Thus the total set of commodities required to produce $\underline{x}$ is given by $\underline{s}$ where

$$\underline{s} = \underline{A}\, \underline{x} + A^2\underline{x} + A^3\underline{x} + \ldots$$

$$= (\underline{I} - \underline{A})^{-1}\, \underline{x} \qquad .$$

$$= \underline{B}\, \underline{x}$$

where the matrix $\underline{B}$ is known as the Leontief inverse matrix and is the sum of the infinite geometric series. The entries in matrix $\underline{B}$ give the total quantity, including all the indirect inputs, of commodity i (in £'s) to produce £1's worth of commodity j. It is a fairly simple matter to then insert the quantities corresponding to these purchases so as to deduce the quantity of input required to produce £1's worth of output. Using the appropriate calorific values of the primary fuels leads to the evaluation of the energy input per £1 value output for each sector documented in the I/O table. This method provides aggregated data on the entire industrial system but suffers from the fact that

a) it depends upon the level of aggregation of the
    I/O table. For inhomogeneous sectors (e.g. "chemicals")
    the energy intensity can only be used when a
    representative range of products from the sector are
    involved

b) it cannot take into account preferential prices for
    fuels and materials.

The other method of energy analysis, referred to as process analysis, avoids the aggregation problems of the I/O method, but is very much longer and requires a large accumulation of data. The method examines a single process for producing some commodity and identifies all the major material inputs, transport requirements, machine parts, buildings etc. are

included in the network of inputs. Each input is then given
an energy requirement (often using data from the I/O studies)
and the total ger input evaluated. Table 1 below shows the
inputs required for just one stage in the production of copper
from sulphide ores [3]

Table 1.   Energy requirements of inputs to crushing and
           flotation of copper ore.

		Energy per ton ore MJ(thermal)
Electric power	15 kWhe @ 14.4 MJ/kWh	216
Lime	4 lb @ 2.14 MJ/lb	8.6
Oil	0.015 lb @ 22.9 MJ/lb	0.4
Flotation chemical	0.1 lb @ 8.2 MJ/lb	0.8
Water	750 gall @ 0.04 MJ/gall	27.0
Metal loss from crusher (due to abrasion)	0.2 lb @ 98 MJ/lb.	19.6
Machinery (approx 8000 tonnes) (amortised over 60×10^6 tons ore)		10.1
Buildings (amortised over ore body)		0.3
	TOTAL	282.8 MJ(th)

This example is fairly typical in that it shows that the direct
fuel input is the most important term, but that all the indirect
inputs amount to 30% of the direct fuel and therefore cannot be
ignored. The answer is relatively insensitive to any one of the
indirect inputs which can therefore be estimated from statistics
or deduced from I/O studies. It would appear that provided an
item has a large vector of direct inputs, such as a crushing
machine (which has inputs of steel, copper, aluminium, wire,
nuts and bolts, insulation etc) then the I/O results are
accurate to $\mp$10%. However for a product which has only a few
inputs (such as lime or the flotation chemical) the I/O results
should not be used.

We are now in a position to work through the energy analysis of
nuclear power stations.   Throughout the following analysis the
data is for a nominally 1000 MW(e) installed capacity station ;
uranium (in the ground) and plutonium are given zero ger and
imports into the UK are given the ger required for their
production in the country of origin.

The capital inputs to a nuclear power station consist of the
electrical machinery, the buildings, the nuclear steam system
and the initial core of the reactor.   For the Steam Generating
Heavy Water Reactor (SGHWR) analysed here there is also a heavy
water inventory to be provided.   The first three items,
electrical equipment, buildings and the nuclear steam system,
are characteristic products of homogeneous sectors documented
in the I/O table.   (Nuclear steam systems are simply very
elaborate bits of plumbing!)   Furthermore these all satisfy
the criteria of having a large vector of direct inputs, so
their ger's are calculated using the I/O table energy intensities
and documented costs.

Uranium and heavy water are not characteristic products of any
I/O sector (they are actually included in "general chemicals"!)
so their ger's are calculated by process analysis.   This is also
sensible because the initial core and heavy water inventory
account for a large fraction of the total ger.   The production
of 160 tonnes of 2.1% enriched reactor fuel (which is the initial
core of a 1000 MW(e) SGHWR) requires 643 tonnes UF_6 to enter the
enrichment plant, which requires 750te of U_3O_8 to enter the
conversion plant, which requires mining 250,000te of 0.3% ore,
which also requires the removal of 6,000,000te of overburden!
Table 2 gives details of the energy inputs to each of these
stages for both 0.3% ore and 0.007% ore. (For details see ref 1.).
The energy requirement for the heavy water inventory (250te) was
deduced from data provided by the manager of the heavy water
plant in Canada.   All these energy requirements are set out in
Table 3, the two columns representing different conventions for
counting the electrical inputs (discussed later).

Table 2.   Energy requirements for producing
            the initial core for 1000 MW(e)
            SGHWR (160 te 2.1% U.)

	Uranium from 0.3% ore TJ	from 0.007% ore. TJ
Mining:	306	3612
Milling	302	15588
Conversion	317	317
Enrichment (328 SWU)	11545	11545
Fuel rod fabrication	130	130
	12600	31192

Table 3. Energy inputs to 1000 MW(e) SGHWR

	Electricity from fossil station TJ	Electricity from nuclear stations TJ
electrical equipment (£52m)	6,177	5,510
buildings and services (£30m)	3,124	2,810
nuclear steam system (£50m)	7,236	6,520
heavy water (250 te)	7,740	6,000
initial core (0.3% ores)	12,560	3,600
	36,838	24,400

The next stage of the analysis involves evaluating the power
(and energy) outputs of the nuclear station. Although the
installed capacity is 1000 MW the station will have a life-
average load factor of about 62%, which means that the output
is equivalent to a continuous power of 620 MW. From this must
be subtracted distribution losses (46.5 MW or 7.5%), electricity
used by area boards, electricity offices, showrooms etc,
(23.25 MW or 3.75% *), the energy required to make up the loss
of heavy water (3.43 MW) and the energy needed to refuel the
reactor during its life (23.40 MW of uranium from 0.3% ores).
This gives a net output to final consumers of 523.4 MW, which
over a 25 year life is equivalent to 412,650 TJ(e). This is
more than ten times larger than the energy input required to
build the station, so one nuclear power station produces a
handsome profit.

However no-one is planning to build just one nuclear power
station. Nuclear power has been advocated as a substitute for
fossil fuels and most industrialised nations have plans to build
up a large stock of nuclear stations. In building up the stock
of power stations problems may arise since all the energy to
construct a power station has to be invested before any output
is available. Furthermore the output is spread over 25 years
whereas the input may be required in a five year construction
period. Thus although 'in the end' the nuclear programme may
show a handsome profit, it may run into a temporary net-energy
deficit. In order to analyse this let us assume that a nuclear
station takes 5 years to construct, during which time all the
energy inputs occur, that there is a one year commissioning
period, and that it then operates for 25 years. As shown in
figure 1 we can consider two ratios, either the power ratio,
P_R, defined as P_o/P_{in}, or the energy ratio, E_r, defined as
E_o/E_{in}. (Note P_R depends upon the assumed construction time,

---

* This assumes that the use of electricity by area boards,
  showrooms etc., is directly proportional to the total
  installed capacity of the system.

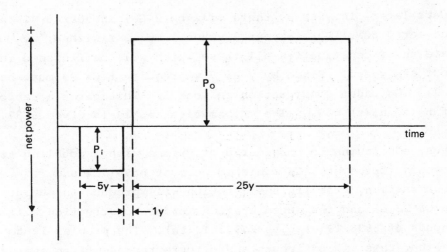

Figure 1. The power inputs and outputs for the
construction of one nuclear power station.

but is independent of station lifetime ; E_R depends on
the lifetime but is independent of the construction time.) From
the times assumed

$$E_R = \frac{E_{out}}{E_{in}} = \frac{25\ P_{out}}{5\ P_{in}} = 5\ P_R$$

and for the SGHWR discussed above $E_R = 11.2$, $P_R = 2.24$.

At any time, t, in the programme of building the reactors, let
the number finished be denoted by $n_f(t)$ and the number under
construction be $n_c(t)$. Then the net power output will be given
by

$$\text{net power} = P_o\ n_f(t) - P_i\ n_c(t)$$

Now if the rate of building reactors is exponential then the
ratio $n_c(t)/n_f(t)$ is approximately constant (as shown in
Appendix 1). Its value can be deduced from the mathematical
analysis, or by a simple calculation based on doubling time.
For example if a programme is doubling the number of reactors
every two years then in year 1 there will be, say, 1 GW, two

years later, in year 3, there will be 2 GW, in year 5 4GW and
in year 7 8GW. But in year 1, when 1 GW is finished, we must
have under construction all those which will be finished in the
next six years. Since by year 7 we require 8 GW we must have
8-1 = 7GW under construction in year 1. This gives a ratio
of $n_c(t)/n_f(t) = 7$. (The mathematical analysis gives 6.57.)

Thus, continuing to consider a programme with a doubling time of
2 years, for every 1GW finished we must have 7GW under
construction. The finished station will produce an output of
16506 TJ(e) and the seven stations under construction will require
inputs of 5896 TJ(e) plus 27972 TJ(th). The problem is to decide
whether this mix of inputs and outputs represents an energy
deficit or profit. I will describe three conventions for
resolving this problem.

Convention A

This starts by observing that one nuclear power station does not
significantly alter the overall efficiency of generating
electricity in the UK. Furthermore the nuclear electricity is
evenly distributed to all consumers (via the national grid).
Hence the electrical input to the nuclear construction
industries (5896 TJ(e)) requires 23584 TJ(th) of fuel, which
combined with the 27972 TJ(th) input gives a total input (of
fossil fuel) equal to 51556 TJ(th). Assuming that the nuclear
output is used as a fuel (and not exclusively for work) its
output, 16506 TJ(e), is subtracted directly to given an energy
deficit of 35050 TJ(th) per year per GW completed. Whilst the
building schedule grows exponentially the fuel deficit gets
larger and larger as more stations are completed.

Convention B

The starting point for this convention is that the electrical
inputs to the nuclear construction industries are assumed
to come from nuclear power stations. Thus subtracting the
5896 TJ(e) input from the 16506 TJ(e) output leaves a net output
of electricity of 10610 TJ(e). Now it is also assumed that this

electricity is purchased by consumers who presently use fossil
fuels for performing work. Since electricity is about three
times more useful than a fossil fuel for performing work the
consumption of 10610 TJ(e) <u>saves</u> 31830 TJ(th) of fossil fuel.
(If the consumers substitute nuclear electricity for fossil fuel
then this is actually the fossil fuel released for other
consumption.) This fossil fuel saving is larger than the required
fuel input (27972 TJ(th)), so according to this convention the
two year doubling programme actually shows an energy profit of
3858 TJ(th) per year per GW completed.

Convention C.

This is the convention which I feel is most realistic in the
sense that it appears to be closest to the way in which fuels
are actually consumed. The above conventions produced
apparently opposite answers because they made extreme
assumptions about consumer behaviour. Convention A is
unrealistic because some nuclear operations (particularly
enrichment) are quite likely to get "special" electricity,
quite possibly from a nuclear station. Convention B is
unrealistic because it assumes that all future users of
electricity will only use it for work hitherto performed by
fossil fuels. In fact a large proportion of electricity is
already used for heating, perhaps as much as 25%. Thus
convention B would be a good approximation if for every 1GW of
nuclear capacity finished 1GW of fossil-fired capacity were
closed down (thereby guaranteeing the substitution). Without
this assumption it would seem more reasonable to subtract the
electrical input to the nuclear construction industries from
the nuclear output (leaving 10610 TJ(e) net output as before)
and then subtract the thermal inputs (27972 TJ(th)) from this
net output. This gives a net energy deficit of 17362 TJ(th)
per year per GW completed which is almost exactly mid-way
between the conclusions reached by conventions A and B.

If convention C is applied to a programme where the number of
reactors is doubled every 4 years, which gives a ratio of

$n_c(t)/n_f(t)$ close to 2, then it is found that there is a net
energy profit of 10825 TJ(e) per year per GW completed (roughly
half the gross output of the finished reactor). Although this
makes it appear that this slower programme is preferable it
should be   borne in mind that the slower programme takes longer
to reach any total capacity. Figure 2 shows how the two
programmes compare assuming a starting capacity of 5GW and a
final capacity of 80 GW. The period when the faster programme

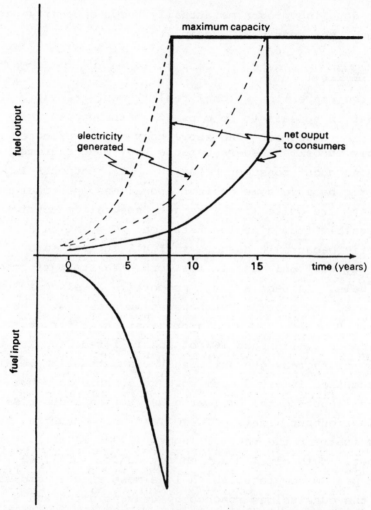

Figure 2.   Comparison of two nuclear programmes
which have a maximum capacity of 80GW, one of
which has a doubling time of 2 years, the
other 4 years.

is a net consumer of fuel is almost exactly balanced by the
period (from year 8 to 16) when it produces more output than
he slower programme. Thus the choice between these programmes
depends upon whether the deficit can be 'afforded' and when
the output is needed. Again this emphasises that whilst energy
analysis may help by providing information on the consequences
of certain policies alone, it cannot tell you which policy is
"preferable".

The original study of nuclear power stations was not instigated
in order to try to resolve the debates about growth rates but
to try to investigate the effect of uranium ore grade on the
energy balance of nuclear burner reactors. This work is still
in progress, but some results can be presented in order to show
the types of effects involved.

Earlier we found that a 1000 MW(e) SGHWR station produced an
output of 412650 TJ(e). If we do not subtract off the energy
required for refuelling this increases to 430977 TJ(e).
Subtracting the energy required to build the station and enrich
the initial core gives a net output of 406699 TJ(e). To produce
this output requires (over the 25 year period) 3743 tonnes of
uranium, giving a net output of 108.66 TJ(e) per ton U. Now if
ever the energy required for mining and milling uranium equalled
108.66 TJ/te U then there would be no net output from the
system. Since the energy required to mine and mill a material
is inversely proportional to ore grade (see earlier paper on
energy analysis for this) this implies that there is a cut-off
grade of uranium ore below which nuclear burner reactors are
not net energy producers.

In fact the energy required to mine one ton of a material
depends on both the ore grade and on the stripping ratio (S)
which is the ratio of tons of overburden to tons of ore. High
grade ores tend to have a high stripping ratio whereas very low
grade ores have no overburden at all (since they are themselves
not much more than dirt). Assuming that for a given stripping
ratio the energy is exactly proportional to 1/G gives the

results shown in figure 3 which suggests that the minimum ore
grade is around 20 ppm.   (Note the energy to mine uranium shown
in figure 3 assumes that all electricity comes from nuclear
power stations; i.e. there is no electricity conversion loss
included.)

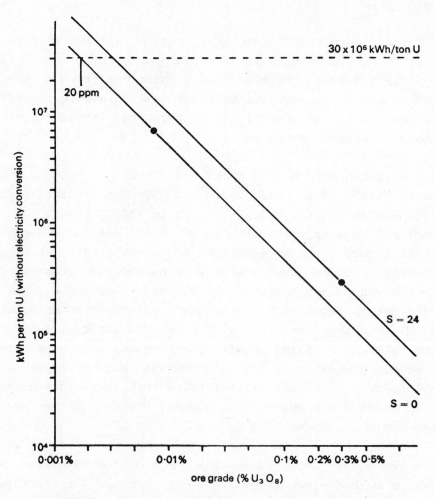

Figure 3.   The energy input to produce a ton of
uranium as a function of ore grade and stripping
ratio(S)   (Note 10^6 kWh = 3.6 TJ).

Table 4.   Uranium reserves

ore cost $ per lb U_3O_8	concentration (ppm)	reasonably assured (million tons)	awaiting discovery (million tons)
Less than 10	700-2500	1.0	1.0
10-15	450-1600	0.7	0.7
15-30	250-800	1.0	1.0
30-50	140-500	0.5	1.0
		3.2	3.7

Source.   Vaughan 1975.

Now if the cut-off grade is 20 ppm then mining uranium of, say
100 ppm, would give a poor energy return.   (The power ratio for
an SGHWR fuelled from such ores would be about 0.4, corresponding
to an energy ratio of 2.)   Thus for practical purposes the
cut-off grade for burner reactors may be about 100 ppm.   If this
is the case then the projected total uranium available is, as
shown in Table 4, only 6.9 million tonnes.   (Note this includes
a very large quantity "awaiting discovery".)   According to OECD
this total reserve is to be divided three ways between the USA,
Europe and the rest.   Of the 2.3 m.t. which is Europes share
the UK can expect to get about a fifth, i.e. 460,000 tonnes.
Since 1 GW (for 25 years) requires 3743 te this gives a total
capacity of 122.9 GW or a total capacity equivalent to 3072 GWy.
This allows either 122.9 GW for 25 years or 61.5 GW for 50 years
or 30.7 GW for 100 years.   If nuclear burner reactors are to
make a significant contribution to energy supplies over say the
next 50 years this restricts the maximum capacity to 60-80 GW
(depending upon rates of growth and decline).   Since the UK's
present generating capacity was 69 GW in 1973 this implies that
nuclear burner reactors cannot provide a large fraction of our

total energy needs. This type of conclusion, which rests upon
the identification of a boundary condition on a technology,
seems less prone to problems of interpretation than those
discussed earlier and is likely to be the area in which energy
analysis proves most useful.

References

1. Chapman, P.F. and Mortimer, N.D. The energy inputs and
    outputs of nuclear power stations : Open University Rsch
    Rpt. ERG005 Sept.1974.
    see also. Chapman P.F. N.Sci.64: p.866. 19th Dec.1974.

2. Roberts. F. (paper in this volume).

3. Chapman. P.F. Metals and Materials. 8 (2). p.107. 1974.

4. Vaughan, R.D. 'Uranium conservation and the role of the
                   GCFBR.' paper presented to
                   British.Nucl.En.Soc. London. Jan 1975.

Appendix: Mathematical analysis of construction programmes

Let the number of stations started in year t be
given by $n_i(t)$ where

$$n_i(t) = a \exp(at)$$

Then the cumulative number of starts up to year T is
given by $n(T)$ where

$$n(T) = \int_0^T n_i(t) \, dt = \exp(aT) - 1$$

This function is zero when $T = 0$ which is convenient.
The number finished at the time T is given by

$$n_f(T) = \int_6^{T-6} n_i(t) \, dt = \exp(a[T-6]) - 1$$

and the number under construction, $n_c(T)$, is given by

$$n_c(T) = \int_{T-5}^T n_i(t) \, dt = \exp(aT) \left\{ 1 - \exp(-5a) \right\}$$

Provided that we are well into the programme then
$n_f(T) \gg 1$, so that $\exp(a[T-6]) \gg 1$, hence

$$n_f(T) \simeq \exp(a[T-6])$$

Then the ratio, $n_c(T)/n_f(T)$ is given by

$$\frac{n_c(T)}{n_f(T)} \simeq \frac{\exp(aT) \left\{ 1 - \exp(-5a) \right\}}{\exp(aT) \cdot \exp(-6a)}$$

$$= \frac{1 - \exp(-5a)}{\exp(-6a)}$$

This shows that provided $n_f(T)$ 1 then the ratio of station
under construction to those finished is independent of T.
Evaluation of this ratio for doubling times of 2 years and
4 years gives 6.57 and 1.64 respectively.

DISCUSSION

Question: How is the first assumption, that the fraction of nuclear elec-
            tricity generated is small, affected by an increasingly large
            number of reactors?

Answer: The assumption becomes less valid. The third assumption, that
            nuclear electricity supplies a large fraction of total electricity,
            is the most valid, especially when considering all-electric
            consumers.

Question: How is the analysis affected by changing the reactor life, say
            doubling or halving?

Answer: The effect is serious only when the life of the reactor becomes
            less than the total construction period.

Question: What is the effect of waste heat utilization on the analysis?

Answer: Another different energy analysis would have to be done to deter-
            mine its significance.

Question: The energy value of the plutonium produced by the reactors has
            been assumed to be zero. Is this correct, as plutonium is cer-
            tainly a useful material?

Answer: It is not meaningful to attribute an energy value to the plutonium
            unless a value is given to the original uranium ore. This analysis
            attempts to determine the fossil fuel requirements for nuclear
            electricity.

Question: Should the energy used in fuel reprocessing and fuel storage be
            considered?

Answer: At least 2 people have studied the storage of waste. The conclu-
            sions reached were that the waste would consume 10 to 100 times
            the total energy generated by the reactor. However they consi-
            dered storage times of up to 10^8 years. Energy analysis is pro-
            bably the wrong tool for studying this problem.

Question: Is it not possible that surplus hydroelectric power in certain
            countries could be used to mine and process uranium, thus freeing
            the electrical output of the reactors for other purposes?

Answer: It is likely that the cost of processed uranium imports would rise
            to the extent that our corresponding exports would require a simi-
            lar amount of production energy as the uranium. It is also doubt-
            ful whether such surplus hydro-power exists.

Question: What is the effect of a 500 MW or 2000 MW reactor size on the
            energy analysis? Is there some economy of scale?

Answer: The analysis considers the cost per MW of installed capacity. It
            is probable that present reactor sizes have been optimized for
            minimum financial cost. Energy analysis may suggest a different
            optimum size.

ENERGY ANALYSIS IN MODELLING

P. C. Roberts

Systems Analysis   Research Unit, Department of the Environment

Background

Energy analysis is a tool, to be employed by the analyst; it is not an end
in itself. The proposers of an energy theory value like Odum[1] have not
made their case. On the contrary, as I shall show later, the value which
we attach to particular objects and services is certainly not linearly
related to the flow of energy necessary to provide them. However, energy
intensity figures are extremely useful as the building blocks in assembling
models of resource depletion.

When we contemplate the problems of modelling resources in the long term it
is clear that there are several individual factors which must be taken into
account. At least the following are relevant:-

  1) technical improvement
  2) substitution
  3) resource depletion
  4) consumption profile

Technical Progress

It is commonplace to think in terms of technical progress occurring through
a spasmodic series of discoveries which carried into the industrial machine
yield steadily increasing overall efficiency. For the macro production
function the rate of increase can be deduced from a regression analysis of

---

* The views expressed in this paper are those of the author and do not
necessarily coincide with those of the Department of the Environment.

the time series for capital, labour and output. In the general relationship:

$$Q = \exp(mt) \cdot f(K,L)$$

The growth rate m can be found for a single industry or for the total output
of an entire economy. Typical rates are about 3 or 4% persisting steadily
over decades[2].

For such a macro function all contributions to increased efficiency are
subsumed and we have no means of understanding the process, but only of
observing it second hand through a number falling out of the regression
procedure. In stark contrast to this picture, the efficiencies of primary
processes do not grow exponentially for more than brief periods of time.
Consider time plots for the following examples of conversions or manufacturing processes:-

    1) Synthesis of ammonia
    2) Pig iron production   (figures 1 - 3)
    3) Soda ash manufacture

In each case, the acceleration of improvement gives way to a linear phase
and then a deceleration as we appear to approach an asymptotic limit. There
are good reasons based in physical law why such limits should exist. The
ultimate efficiency of a thermal power station depends on the upper and
lower working temperatures and no real device can do more than approach the
theoretical Carnot cycle performance.

The reason for the logistic shape of these curves probably derives more from
the "market penetration" effect than from the pattern of innovations underlying the efficiency increase. The discoveries curve of illumination is far
from logistic (fig. 4), whereas all of the market penetration cases examined
illustrate logistic form with striking precision (figs. 5-7). The two
effects operating together are likely to generate apparent logistic shape
for the overall observed efficiency time path.

If the time path is logistic, then a plot of log $f/(F-f)$ with time should be
straight (f figure of merit, F ultimately obtainable value of f at $t = \infty$).

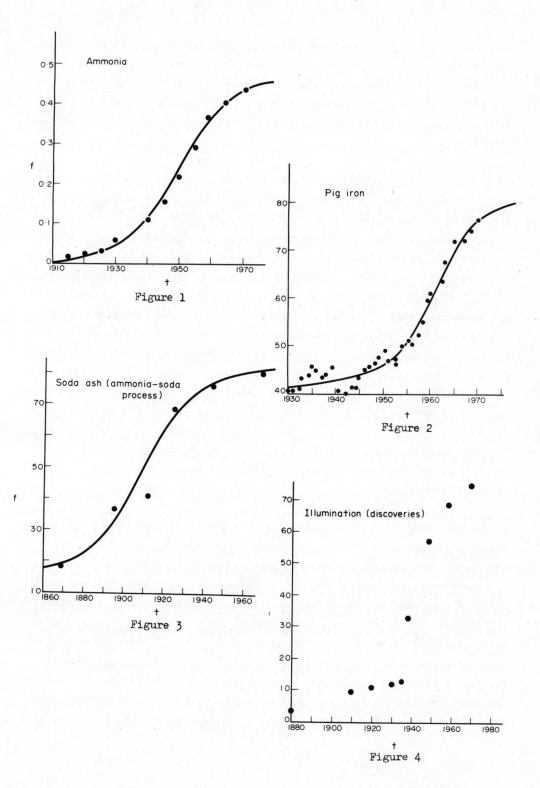

Figure 1

Figure 2

Figure 3

Figure 4

The log plots for power stations and ammonia synthesis suggest the validity of this suposition, (figs. 8-9). Ultimate efficiencies for these cases have been derived by taking the F value for highest correlation observed in the log plot (fig. 10). This method is only appropriate where there is good a priori reason to suppose that current practice is close to the best that will be obtained.

It is clear that there is the possibility of deception by deriving an asymptote that is too low and to guard against this the nature of the physical limit must be considered. A listing of current efficiencies for a variety of processes indicates that there are radical differences dependent on the mechanisms involved (fig. 11). For conversion of high grade energy (mechanical/electrical, hydraulic/mechanical, electrical/mechanical) efficiencies greater than 90% are easily achievable. For heat engines and many chemical processes the range 40 to 60% is likely to be the upper bound, and for processes involving radiation (photosynthesis, solar photo-voltaic cells, discharge lamps) the efficiency limit is in the lowest range. Although there is more work to be done in particularising the categories, a consistent picture already emerges of the constraints which operate on the progress paths that are observed.

For many important cases the efficiency trajectory has passed the point of inflection, and yet the econometric analysis reveals steady growth in value of final product.

For primary industries, the energy use is a good indicator of overall cost. Handling and processing bulk commodities requires energy pro rata to the mass being treated, and capital equipment adequate to allow such energy flow is itself expensive pro rata to the power needed. The prices of metals averaged over the post war years are closely correlated with the quantity of rubble that must be processed to yield unit mass of metal. An even better price predictor is obtained by taking not only the bulk effect, but also the Gibbs free energy associated with release of the pure metal into account. Finally, a third term associated with rarity of deposit, defined by ratio of ore concentration to crustal abundance, completes a three term linear predictor which over some five orders of magnitude (from iron to platinum) relates

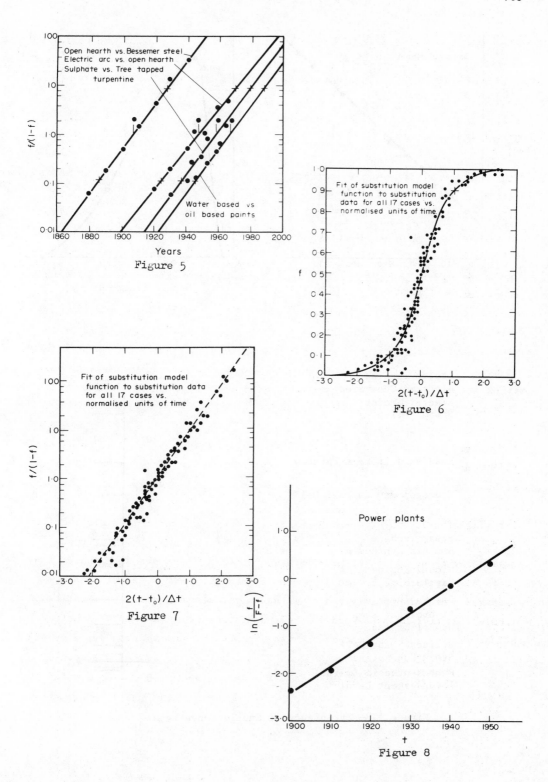

Figure 5

Figure 6

Figure 7

Figure 8

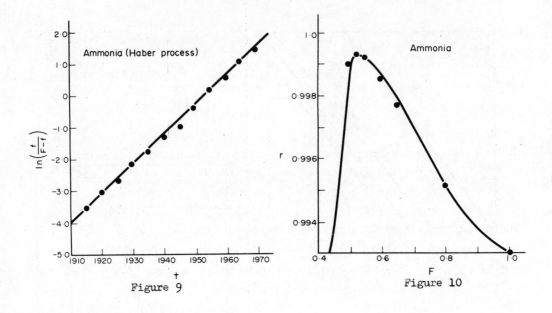

Figure 9

Figure 10

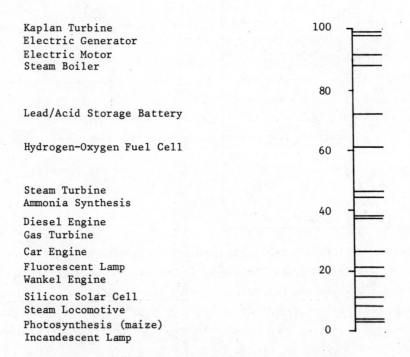

Figure 11 - Efficiency of Energy Converters

the relative prices of metals for the majority, with an error of less than
3dB (fig. 12).

Substitution

As we move to lower ore grades, so the energy bill mounts. There can be
further improvements in efficiency and the efficiency rate change will
slacken - as it must for all engineering progress towards physical limits.
There can also be substitution, but it should be noted that the tonnage of
the various metals used is inversely related to the price or the energy bill.
We already use the cheapest materials for the task. Silver is a better conductor of electricity than copper, but copper is much cheaper. Silver
halides are used for photography because we know of no other material capable of such high resolution. The task of finding substitutes that are actually cheaper is a formidable one.

These points could all have been made early in the century, and yet economic
growth has proceeded fairly steadily. The nature of this growth is not apparent when GNP is taken as a scalar measure in one dimensional £ or $.
However, if we make use of energy analysis some light is shed on the means
by which economic growth can proceed despite the limits of efficiency growth
and the rising difficulty of expolitation. Cross section study of energy
intensities demonstrate that lower energy intensity correlates with high
income.

Energy intensity related to income level

From the energy intensities of the separate categories of goods and services,
we may proceed to calculate the energy intensities of particular classes of
consumer. The household expenditure survey provides us with a breakdown of
the proportions of categories of expenditure which consumers make. Thus,
the input/output analysis of the UK economy for 1968 reveals that about 36
kw-hours of primary energy (coal, gas, oil) have been used overall in order
to provide one pair of shoes. We also know the rate of purchase of pairs of
shoes by consumers in various income groups. By taking the full range of

commodities and aggregating consumption into the main groups we arrive at the striking conclusion that the energy intensity of consumption falls off quite markedly with rise in income. As people get richer, their consumption preference moves towards the goods with lower energy intensity. Table 1 and figs. 13 and 14 illustrate this point. In particular the energy intensity falls from about 90 kw-hours per £ of expenditure for the lower income groups to about 55 kw-hours per £ in the higher income groups.

TABLE 1 — RELATIONSHIP BETWEEN HOUSEHOLD INCOME AND PRIMARY
ENERGY CONSUMPTION, 1968

| EXPENDITURE SECTOR | HOUSEHOLD PRIMARY ENERGY CONSUMPTION: MWh/ANNUM | | | | | | | | | | | | AVERAGE |
| | WEEKLY INCOME BANDS: £/WEEK | | | | | | | | | | | | |
	6	8	10	15	20	25	30	35	40	50	60+		
No. OF HOUSEHOLDS	189	374	273	671	728	963	965	814	631	365	385	426	7184
AVERAGE: PERSONS/ HOUSEHOLD	1.12	1.07	1.61	1.99	2.54	3.06	3.30	3.42	3.42	3.48	3.94	4.14	2.96
HOUSING: CONSTRUCTION	1.33	1.78	1.98	2.29	2.45	2.65	2.75	3.04	3.22	3.43	3.78	5.12	2.92
HOUSING: REPAIRS	0.23	0.24	0.53	0.84	0.68	0.81	1.46	1.19	1.55	2.18	1.93	2.27	1.24
FUEL: GAS	2.47	2.32	2.76	3.48	3.05	4.06	4.21	4.50	4.64	5.08	6.67	9.86	4.41
ELECTRICITY	4.00	5.72	6.68	9.40	11.80	13.40	16.00	17.60	18.40	20.00	23.10	27.20	14.33
COAL	12.20	16.30	13.20	13.70	15.90	12.90	12.20	12.80	11.90	15.20	14.30	13.20	13.55
OIL	2.29	1.83	1.39	1.97	2.34	1.58	1.31	2.25	1.73	1.62	1.70	3.12	1.87
FOOD	3.50	3.54	4.88	6.00	7.99	8.45	9.40	9.89	10.27	11.04	12.47	14.46	8.87
ALCOHOL	0.22	0.20	0.26	0.49	0.87	1.22	1.49	1.64	1.98	2.31	3.25	4.13	1.55
TOBACCO	0.10	0.12	0.24	0.31	0.47	0.60	0.66	0.70	0.70	0.83	1.00	1.05	0.61
CLOTHING	0.42	0.61	0.95	1.23	1.81	2.37	2.89	3.36	3.98	4.39	5.08	7.21	2.98
CARS: PURCHASE	0.06	0.08	0.14	0.17	0.53	0.79	1.10	1.73	2.03	2.10	2.87	4.28	1.34
RUNNING	0.48	0.24	1.06	1.35	3.68	4.70	6.63	8.68	10.90	11.60	14.93	17.91	7.12
RAIL	0.02	0.27	0.16	0.33	0.36	0.64	0.63	0.90	1.17	1.64	2.24	4.20	1.01
BUS	0.18	0.21	0.26	0.43	0.58	0.72	0.83	0.78	0.89	0.90	1.12	0.98	0.71
OTHER	0.04	0.06	0.03	0.24	0.09	0.17	0.27	0.38	0.41	0.44	0.52	1.12	0.32
SERVICES	0.47	0.58	0.66	0.86	1.06	1.27	1.40	1.70	2.00	2.45	2.92	6.16	1.77
GOODS: DURABLE	0.82	0.78	1.32	1.77	2.52	2.85	3.90	4.47	4.73	6.17	7.52	9.21	3.97
OTHER	1.31	1.25	1.65	2.20	2.85	3.62	4.24	4.60	5.21	5.79	6.84	9.23	4.22
DISTRIBUTIVE SERVICES	2.35	2.63	3.43	4.37	5.64	6.78	7.93	8.96	9.89	11.17	13.25	17.87	8.14
ENERGY TOTAL	32.49	38.76	41.84	51.43	64.67	69.58	79.33	89.17	95.60	108.34	125.49	158.58	80.93
ENERGY/EXPENDITURE KWh/£	86.9	92.3	76.6	73.8	71.7	64.2	62.7	62.4	60.7	60.8	59.4	55.7	62.4

(TRANSPORT label at left of CARS/RUNNING/RAIL/BUS/OTHER rows)

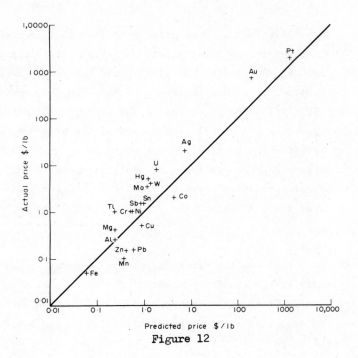

Figure 12

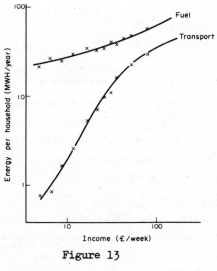

Figure 13

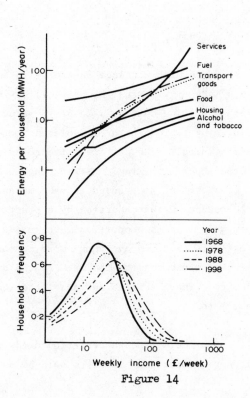

Figure 14

As a result of this shift in the consumption profile with income change, one would expect that, other things being equal, the energy coefficient to be .73. By this I mean that a rise in GNP of 3% would correspond to an increase in the consumption of primary energy of 2.2%. The 1973 edition of the Digest of Energy Statistics quotes values of the energy coefficient both for every individual year (e.g. 1966-67) and average of the preceding 1, 2, 3... years providing series from the commencement of records in 1957, until 1972. Every such series starting between 1957 and 1966, extended to 1972, shows values of this coefficient lying between .7 and .8. Shorter series exhibit greater year to year fluctuation and the values up to 1972 range between .9 and -.2.

Although it is clear that one can use such results in order to make forecasts of the primary energy requirement and of the effects of changes in conversion efficiency, it is not my purpose to consider these aspects. Instead, I wish to explore the deeper implications which lie behind the energy intensity figures.

Mass intensity

Associated with a change in energy intensity is a corresponding change in mass intensity. Mass intensity is defined as the mass of produce associated with one value unit e.g. kg per £ value. Apart from occasional surges stimulated by rapid growth of isolated sectors of the economy, mass intensity has been falling over time. The relation between mass intensity and energy intensity is striking. Over a range of three orders of magnitude in mass intensity the value added to base material is related to the energy used in the course of that addition by a power law (fig. 15). Total energy required per unit mass is proportional to $(value\ per\ unit\ mass)^{\frac{3}{4}}$. The data needed to test this relation over many more commodities is difficult to acquire. For most of those chosen, the mass of the output is recorded in industrial statistics. Where this is not done, the mean mass per unit of output has been estimated.

It is unlikely that the relation holds good for extreme values and indeed it is not obvious what 'mass' should be associated with service activities.

However, even for the range considered, the paradoxical character of economic growth is resolved. As individuals, our patterns of consumption move so
as to raise the value per unit mass of the goods we consume. The epitome of
this progression is seen in recent additions to the shopping list: cameras,
tape recorders and now pocket calculators. Moreover this effect is not of
recent origin but has moved in parallel with the other more often noted
aspects of industrial advance - technical improvement, rising labour productivity and accumulating capital stock.

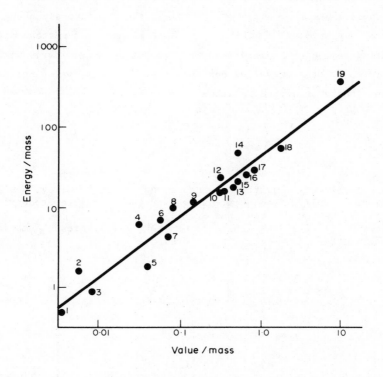

1	Bricks	8	Paper	14	Man-made Fibres
2	Cement	9	Cans and Metal	15	Copper
3	Stone (Quarrying)		Boxes	16	Electrical Machinery
4	Fertilizer	10	Cocoa	17	Textile Machinery
5	Grain Milling	11	Aluminium	18	Footwear
6	Iron and Steel	12	Paint	19	Tobacco
7	Sugar	13	Motor Vehicles		

Figure 15.

Modelling

To provide an adequate model of an economy facing progressive resource depletion, it is not enough to know the reserve quantities of metalliferous ores,
fossil fuel deposits and areas of cultivable land. The probable time path
of improving efficiencies must be represented – but not as exp(mt) multiplying the capital and labour function to yield a monetary value of output,
because in subsuming all sources of improvement in one portmanteau term, the
inherent constraints are concealed. Finally, the continuously changing
character of output must be built into the model, so that instead of
extrapolating demand on primary resources to astronomical levels within a
century or two[3], the pervasive tendency towards higher value per unit mass
and per unit energy required is taken into account.

References

(1) Odum H.T. (1971), "Environment, Power and Society", Wiley, New York.
(2) Solow R.M. (1957), "Technical Change and the Aggregate Production
     Function", Review of Econ. and Stats.
(3) Meadows D. (1972), "The Limits of Growth", M.I.T. Press, Cambridge,
     Massachusetts.

DISCUSSION

Question: You pointed out that when income increases, the value per unit
            mass of goods consumed increases. Is that not contradictory to
            the fact that the value per unit mass of some items (e.g. pocket
            calculators) decreases over time as they move into mass production?

Answer: This decrease in value per unit mass is just the result of a
            learning curve in the early days of production of that specific
            item.

Question: You have claimed that there is no theory behind the economists'
            way of modelling technological progress. In what you presented
            to us about technological progress I see no theory, but only a
            lot of measurements. Are there any theoretical ideas behind your
            data gathering?

Answer: The mechanisms behind technological progress are very complicated,
            so we start our investigations by gathering all kinds of informa-
            tion about the mechanisms. This will hopefully help us to form
            theories later on. We are, however, beginning to get some ideas
            about how to construct production functions. We think that capi-
            tal should be disaggregated into several components to obtain a
            closer correspondence to the physical reality. Capital should be
            broken down into at least two categories. One category is bulk
            capital or necessary capital, e.g. land and rain in agriculture.
            This capital cannot be substituted by anything. The second cate-
            gory of capital is concerned with control of processes, e.g. ferti-
            lizers or computers. This second category is substitutable with
            labour. The main function of labour in our society is a control-
            ling function, not as a source of energy.

Question: Would it not be possible to make forecasts about the future
            energy intensity in U.K. by comparing with U.S.A.?

Answer: That method would be too gross. You must consider that there are
            many factors involved in the change of energy intensity with time.
            Technological change is one component, but there are other compo-
            nents involved e.g. future substitutions between raw materials
            and goods, future costs of producing energy etc.

Question: Considering the complexity of the world, does your model have any
            chance of providing us with good predictions?

Answer: We do not want a uniform accuracy in all parts of the model. Often,
            sensitivity tests will show that great accuracy of some data is
            not essential.

Question: You suggested that catastrophies may happen not only because of
            resource constraints, but also because of social disruptions. I
            would suggest that in most cases social disruptions can be traced
            back to resource problems.

Answer: In the past we have had many wars due to ideological and religious
            differences which had nothing to do with resources. There seems
            to be a positive correlation between wealth/capita and social
            disorder. Eskimos have no time to spare for fighting.

Question: We have seen models based on exponential growth, and you suggested
            a logistic growth curve. Why does nobody make linear growth models?

Answer: We tend to want exponential growth in wages because of marginal
            utility considerations. If your wage is £60 a week you would appre-
            ciate an increase of £6 per week. But if your wage is £600 a week
            you would hardly notice an increase of £6 per week. Thus, it would
            seem that our perceptions of growth are consistent with logistic
            behaviour.

ANALYSIS OF SELECTED ENERGY SYSTEMS

F. Roberts

Energy Technology Support Unit, Harwell

This paper illustrates some general features regarding the flows of energy through the industrial sector of the UK. It shows the newcomer to the field of energy analysis some of the problems which are met when studying real-world situations in industry. The examples selected come mainly from the more energy-intensive industries - iron, steel, cement and petroleum refining - but a brief examination of the motor car industry is included, as being a good example of a low-energy-intensive industry which consumes a good deal of energy in aggregate.

It is evident that a single-minded approach seeking to minimise all the energy requirements for our various industrial process systems would not be acceptable because of various interlocking effects with other important aspects of the economy.

1. Introduction

For the many reasons given at this Summer School, it has become necessary to focus greater attention on why and how energy is used in supporting the economy. Greater knowledge in this field could well lead to a better appreciation of the scope for energy conservation in the medium to long-term future. It could also help us to make more realistic forecasts of future energy demand and study the effect of the changes in the mix of fuels used.

It is a characteristic of all developed countries that industry is the largest energy-consuming sector and the UK is no exception to this rule. Over 40% of our total primary energy consumption may fairly be attributed to the needs of our industries. This can be compared with roughly 31% for the domestic sector, 16% for all transport and 6% for the public sector, as shown in Figure 1, based on 1972 data. This paper is solely concerned with the industrial sector of the UK. Figure 2 shows the relationship between industry and the economy as a whole, indicating the flow of fuels (dashed lines) and materials - which represents the main flow of sequestered energy

773

F. ROBERTS

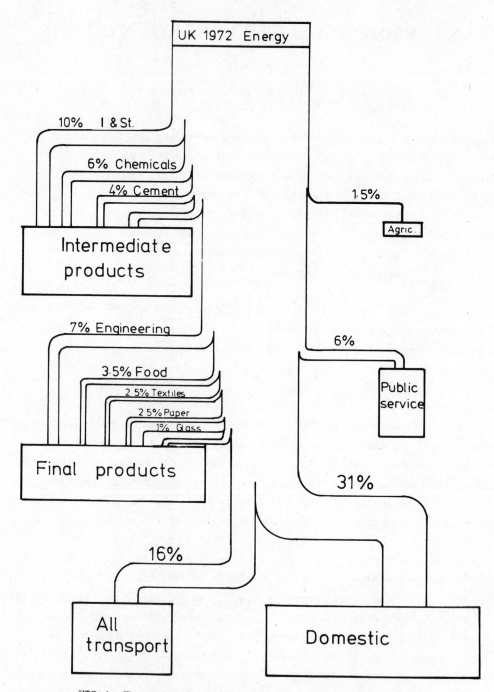

FIG.1 Flow of fuels into various UK sectors

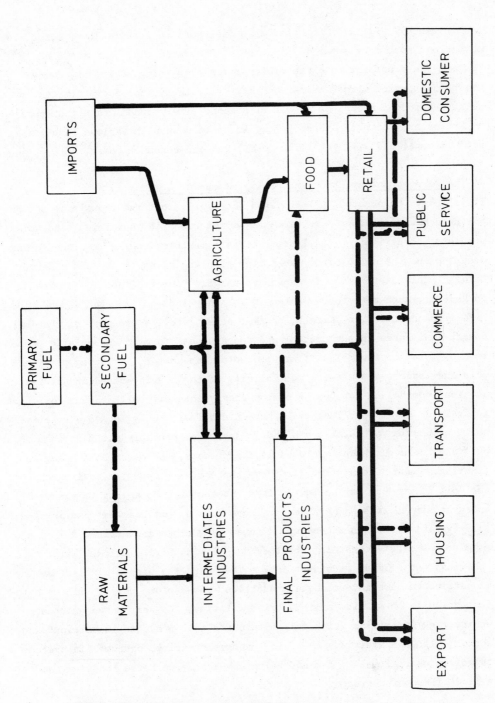

FIG.2 Flow of fuels and materials between sectors of
the UK economy

(solid lines).   It will be noted that fuel goes to every sector, materials move along much more specific pathways.   Note also that "imports" and "exports" are brought into the picture; for everything we import, energy has been expended abroad on our behalf.   Similarly, for everything exported, we have expended energy on behalf of someone else.   Food and agriculture is a particularly complex system, tying in as it does with various industries, with imports, and with retailing to all the end-consumer boxes.

2. The Largest Energy-Consuming Sector of the UK - Industry

It will have been noted in both Fig.1 and Fig.2, that the industrial sector has been divided into two sub-sectors - "intermediate industries" and "final products industries".   Fig.2 depicts raw materials flowing into the former grouping and the output from the latter grouping going to the end-consumer boxes.   This distinction is drawn because the "intermediate industries" are, in general, much more energy-intensive.   They include important industries like iron and steel, chemicals, bricks, cement etc., where much of the fuel bought-in is used in the process and energy costs form an appreciable proportion of total costs.   The "final products" sub-sector, including paper and printing trades, textiles, a multiplicity of engineering-based industries and food, drink and tobacco is of low energy-intensity with regard to bought-in fuels.   For these, fuel costs have historically only formed a relatively small proportion of total costs and in any case, probably not more than about half the fuels purchased are used in the process, the other half going to lighting, space heating etc.[1]   Nevertheless, although the fuel purchases per unit of product may be low in this sub-sector, the aggregate amount of energy consumed is considerable.   Figure 3 is a 'pie diagram' showing the relative total amounts of energy consumed by various industries.   Iron and steel comes first, consuming about 24% of all industrial energy, with 'chemical and allied' second at 14%.   These are the biggest fuel-consuming industries in our "intermediate industries" sub-sector, but it will be noted that owing to sheer size of industrial activity, the very large group of engineering industries which fall into our "final products industries" sub-sector, although individually of low energy-intensity, consume, in aggregate, 14% of industrial energy demand, being comparable with the 'chemical and allied' industries group.

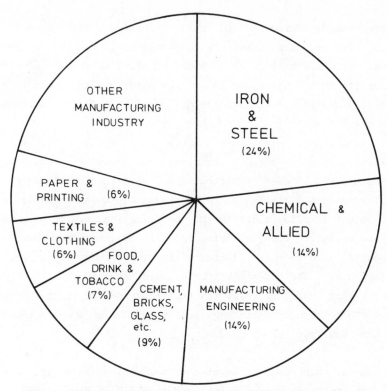

FIG. 3 Relative industrial sector energy consumption

A useful but simplified model of the flow pattern of energy in the industrial sector of the economy, valuable to energy analysts, can now be described.   Firstly, raw materials of almost negligible energy content are
processed by the use of considerable inputs of secondary fuels, which themselves have required an energy input to convert them from the raw or primary
fuel state to a secondary state which is of premium value to the processor.
The materials thus produced may be said to have a high energy requirement,
or energy content.   These materials are then passed on to other manufacturers, who process them to produce further products with a high added value
in cash terms, but only a small proportion of the latter will be accountable
to fuel.   Nevertheless, although process fuel consumption may be low, there
may well be considerable expenditure on fuels for non-process uses such as
lighting, heating and transport purposes.   Also, the rate of capital investment may be high, and this involves inputs of fuels in diverse ways.   From
the energy conservation standpoint, we move from processes where bought-in
fuels dominate the picture and where process fuel conservation is vital,

down through manufacturing systems where the energy content of materials
consumed is high and hence materials conservation is vital in achieving
national energy conservation.   The sheer quantity of energy in aggregate
terms which flows into the lower end of the manufacturing chain calls for
careful energy analysis, although the data will be difficult to obtain from
available statistics.

This sub-division of industry is important in a second respect, the diversity
of its products and product specifications.   Intermediate industries charac-
teristically produce a limited range of products, in large quantities, for
which the weight is often a sufficient specification, e.g. PVC, mild steel
and Portland cement.   Thus, studying one product (and often there are only
a few processes used to make it), may well cover a large fraction of industrial energy use.   For end-product industries there may be thousands of
products for each industry, (e.g. the plastics products industry makes teething rings, floor tiles, plastic bags, car interiors etc. which cannot be
studied individually).

The remainder of this paper is devoted to describing, in brief, some selected
energy systems in UK industry, to illustrate a few of the general points
made in the introductory sections and also reveal how process energy analysis
can throw up some strange, perhaps unexpected, yet vital information to assist
decision-makers in the future.

3. The Largest Industrial Consumer of Energy - Iron & Steel Industry

It has already been pointed out that this industry uses, overall, about 24%
of all industrial energy measured in primary energy terms.   Put another way,
this is about 10% of all primary energy in the UK.   Apart from the secondary
fuels - coke, oil and electricity - the main physical inputs to this system
are iron ore, scrap iron and oxygen.*   We start by looking at the significance
of iron ore grade with respect to energy consumption.

---

*Other inputs are also of importance, such as limestone, refractories,
graphite electrodes, capital plant, but these are perhaps of secondary
interest for our purpose.

3.1 Extraction and Processing of·Iron Ore in the UK

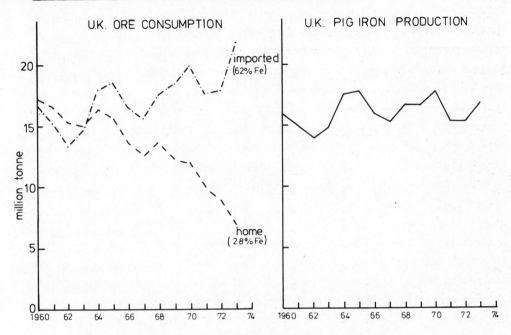

FIG. 4 Showing variation of inputs of home and
imported iron ore, from 1960 to 1973, to UK pig
iron production.

In order to produce about 20 million tonnes of iron and steel products a
year, the UK industry has to obtain its ferrous inputs from both scrap metal
and ores.   Iron ore is partly produced from home resources, partly imported.
The accompanying graphs (Fig.4) show the increasing tonnage of imported ore
used, at about 62% Fe on average, accompanied by a decreasing tonnage of
indigenous ore, at about 28% Fe on average, over the period 1960 to 1973.
The tonnage of pig iron produced in the United Kingdom over the same period
of time is also given.   Now according to Laws of British Steel Corp.[2],
the difference in energy requirement for extracting iron from about 60% high
grade haematite as against low-grade ores in the region of 28% is about 10GJ
per tonne of pig iron produced.   Taking the latter figure of 10GJ, using
Fig.4 one can easily show that the UK must by now be saving itself roughly
130 million GJ of energy per year; but whilst it is true we are saving home
energy equivalent to roughly two million tons of coal a year by this procedure, the trade-off must come in the balance of payments.   Now, since the

actual import cost for ore at present is fairly low, overall it seems a very
sensible thing to do, but it does make us increasingly sensitive to world
price variations in iron ore and even the continued availability of this
world commodity.   One might speculate on what could happen when countries
like Venezuela, Mexico and others in the developing world commission their
massive steel production programmes which they are currently developing.
From the energy analyst's standpoint, the difficult question is the energy
requirement to be ascribed to imported ore.   Superficially, we in Britain
do not start to expend energy on it before we receive it, but we do have to
expend energy in producing exported goods in order to obtain the foreign
currency to buy it!   (It is all a question of where one draws the system
boundary).

A general point can be made; it is easily demonstrated how the energy require
-ment per tonne of steel produced in Britain has been steadily falling over
the years, but we have to remember that not all this saving accrues from
introducing energy conservation measures - at least some of it has been
obtained by steadily raising the quality of the main raw material input to
the system from abroad.   Although this may be a good thing to do, it does
introduce complications in assessing data with regard to our performance in
reducing the energy requirement for steel.

3.2 Crude Steel Production - 3 Routes examined

The pig iron, produced in the blast furnace complex of plants, along with
ferrous scrap metal, is turned into crude, or raw steel, in any one of three
main steel-producing processes.   These are the Open-Hearth, the Basic Oxygen
Steel (BOS) and the Electric-Arc.   The first of these is being phased out
and the second two are rapidly increasing in importance in the UK.   Process
energy analysis may be applied to these three systems in order to see which
one is the most economical in the overall use of energy.   This may seem a
simple thing to do, but in practice it turns out to be complex.   Very different proportions of pig iron to scrap iron are used in each process, and it
therefore makes a considerable difference to the final result what energy
content is ascribed to these material inputs.   Also, the Open Hearth uses
direct heating from liquid and gaseous fuels, the BOS process doesn't receive
any direct fuel but the energy for operating the process has been 'sequestered' in the oxygen being fed to the process vessel; the Electric-Arc uses

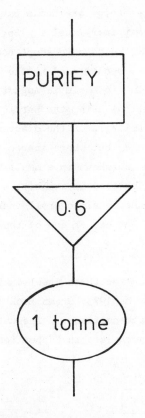

Name of the process stage.

Process energy requirement for the
process just named.   The vertex
points in the direction of flow.
Figures in these triangles should be
consistant on a heat supplied or
primary fuel input basis.

Name and quantity of product.   Stages
are of course linked in sequence in
the flow diagram so the product of
one is the input to the next.   By
convention the quantities are adjusted
to provide 1 unit of output of the
desired final product.

Explicit presentation of the
electricity contribution to the
process energy requirement is by a
box on the top of the triangle.
This is frequently necessary as the
inefficiencies in the primary fuel
utilisation in electricity occur outside the scope of the flowsheet.

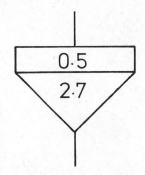

Other symbols are available for the energy content of capital, raw materials
which could be used as fuels and products which could be used as sources of
energy, e.g. combustibles.

FIG.5 Definition of symbols as used in steel energy
analysis flowsheets (based on IFIAS Workshop 1974
recommendation as in Ref.3)

electrical energy efficiently but considerable energy overheads have been
expended in producing this form of energy from primary fuel . So, to be
able to usefully compare these process systems, we need to adopt some conventions.   The conventions used here are those recommended ny the IFIAS
International Energy Analysis Workshop, held in Stockholm in August 1974[3].
Data presentation is by the flow-sheet system, but using the particular
symbols as shown in Fig. 5.   Each triangle on the diagram contains
the energy cost of operating the <u>preceding</u> stage, but where appreciable
electrical energy inputs are concerned, these are shown in a small super-
imposed rectangle.   Masses of materials are enclosed in ovals.   A summing-
up of the figures in all triangles will give the total, overall energy
requirement for producing the material at the bottom, or end of the flow
sheet.

A first process energy analysis was run out for the three steel-making proces
-ses referred to, using essentially data from the 1973 "Brown Book" of the
industry,[4] and adhering to the Workshop recommendations.   The data are
shown in Figures 6, 7 and 8.   The units of energy are in GJ per tonne of

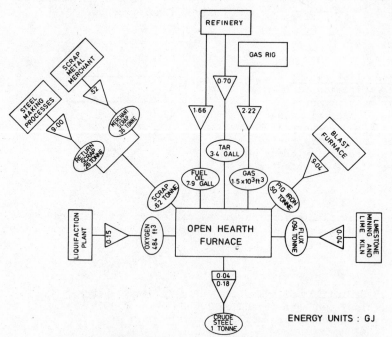

FIG. 6   Energy Analysis of Open Hearth Steel Manufact-
         uring Route

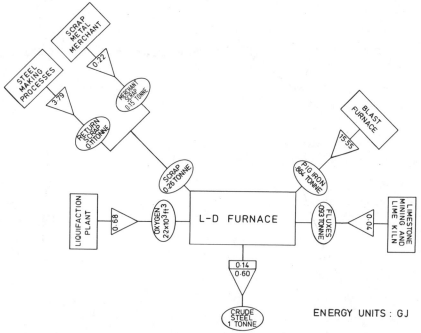

FIG. 7 Energy Analysis of L-D Steel Manufacturing Route

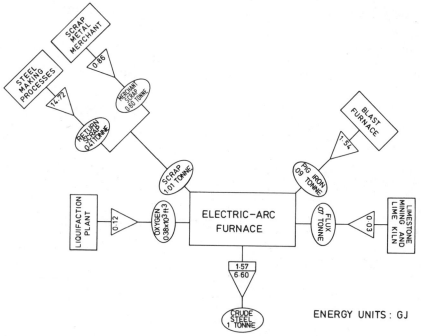

FIG. 8 Energy Analysis of Electric-Arc Steel Route.

crude steel produced.   (In Figure 7, it will be noted that the actual steel-producing furnace is called 'L-D', which is one of the Basic Oxygen Steel processes in use).

It will be noted that the energy sequestered in the pig iron is 18GJ per tonne of pig and also the scrap metal is considered to have an energy value. The latter needs some clarification.   Scrap arises from the various process stages en route to the production of finished steel goods.   Further scrap arises from the fabrication of steel in the production of plant and machines. In the interests of good materials conservation much of this is recycled to the steel-making furnaces.   For the present exercise, it has arbitrarily been given an average energy content of 35 GJ per tonne of scrap; this is perhaps an over-simplification since all the various arisings will have different amounts of energy sequestered in them in accord with the amount of energy expended in their production.   Also, there is what is often termed 'merchant scrap' or 'old scrap'; this is the material collected from old scrap heaps.   The convention has been adopted that this has a nil energy content, apart from the relatively small amount of energy used to collect and sort it, say 1.5 GJ per tonne of scrap.   (These conventions on scrap were adopted following discussions with P.F. Chapman of the Open University). The accompanying Table 1 collects up the total energy given in the attached flowsheets, for the three crude steel-making routes, for comparative purposes, and also includes the relative proportions of pig iron, recycled scrap and old scrap used in each of these processes.   It will be seen that the lowest value is obtained for the BOS route, being about 21 GJ/te steel, compared with 23.5 for Open Hearth and 24 for Electric Steel.   (The average figure for the UK, for what it is worth, comes out at about 22.3 GJ per te).   Now, it would be a gross over-simplification to draw the naive conclusion that we should instal BOS steel-making processes to the exclusion of others merely because this route appears to consume the least amount of energy.   (There are many diverse reasons, some technical, some economic, why we should also have a good deal of Electric Steel making plant in the country, but this is not the place to pursue this argument).   From the energy conservation standpoint alone, such a conclusion could be challenged.   This is because of the convention we adopted with regard to the energy sequestered in scrap.   Some iron and steel manufacturers prefer to give all steel scrap, from wherever

it arises, an energy value of nil*.   If we run the energy analysis once more
putting in this new convention, then we find that the Electric Steel route
is by far the best with regard to energy conservation; at 8.3 GJ per te.
Open Hearth comes out at 14 and BOS the worst at 17 GJ per te.   A very
different picture emerges, solely because of the very different proportion of
scrap which can be handled by the three routes.

TABLE 1

Energy Sequestered in the Production of 1 tonne of Crude Steel by 3 Routes

	Open Hearth	BOS	Electric -Arc
Recycled scrap used, tonnes Merchant scrap " " Pig iron " "	0.26 0.36 0.50	0.11 0.15 0.86	0.41 0.60 0.09
Energy sequestered in scrap, GJ	9.5	4.0	16
Total energy per tonne of steel, GJ	23.5	20.9	23.9
Total energy per tonne of steel, GJ if all scrap has "nil" energy value	14.0	16.9	8.3

In the author's opinion, it is unhelpful to allocate an energy content of
nil to all recycled scrap; in fact, this can conceal energy waste.   Very
large fractions of process-generated scrap could be going round and round the
system, using up energy each time they are heated up, and causing the use of
unduly large furnaces which themselves radiate more heat.   We need to adopt
conventions in energy analysis which can help us to spotlight where there is
unnecessary use of energy.   This example has been chosen simply to highlight
some of the problems met with in carrying out process energy analysis.

---

*They say the production of scrap in the industry is inevitable and unavoid-
able, it is an endogenous thing.   In fact, if one draws the boundary for the
systems being studied around the whole of the iron and steel making system of
processes, then there are no inputs and no outputs of scrap metal, it all re-
cycles inside the boundary.   This ignores merchant scrap, which has a very
low energy content anyway.

3.3 Steel Finishing Processes

The foregoing analysis only took us up to the production of crude, or raw
steel. A number of finishing operations, all requiring heavy inputs of
energy have then to be carried out to produce steel in the required state of
fabrication for the consumer.   There are many forms of 'finished steel',
ranging from sheet to tubes, and on average, roughly 40 GJ of energy are
required to produce a tonne of such products starting from iron ore and
scrap.   Thus it appears that a further 18 GJ of energy are required to
'finish'raw steel.   This obviously does not involve chemical reactions, the
energy is expended essentially in re-heating and mechanical working operations and further quantities of scrap material are produced which are then
recycled.   A NEDC report estimated that on average, the finishing of 1 tonne
of steel yields 0.36 tonne of scrap[5].   This report points out that signifi
-cant losses of energy result from the production and recycling of this scrap,
some of which may be avoidable by new techniques such as powder metallurgy.
Since the amount of energy associated with steel finishing operations in the
UK must be of the order of 4.5% of our total primary energy consumption, a
careful energy analysis of this area would seem to be worthwhile.

4. Cement Industry

Cement production uses 4% of the total primary industrial energy in the UK,
equal to about 5 million tonnes of coal equivalent.   By far the greatest
amount of this energy is accounted for by rotary kilns, which have now
reached a high degree of technical development and large-scale operation.
This technology is widely known and is widely applied by industrialised
countries.   One might therefore expect to find little variation between the
kiln energy consumption per unit of output of various countries, but this is
not so; indeed there appear to be wide country variations, ranging from 6.8 GJ
per tonne of cement for the United States to 4.2 GJ for West Germany.   Now
it is a well established fact that the so-called "wet process" of cement manu
-facture requires a good deal more kiln energy than the "dry" or even the
"semi-dry" process.   So, according to the proportion of cement made by "wet"
or "dry" routes in different countries, we might well account for some of
the recorded variations in kiln energy.   There are other variables though
which can affect average kiln energy consumption, notably the mix of cement
types being manufactured in a particular country.   Portland cement production
requires the most energy, Pozzolanic and Slag cements require a good deal less

energy for their production.   Furthermore, there are wide variations as
between countries in the mix of fuels used in cement kilns.   Table 2,
accompanying, summarises the data which were presented jointly by the
Building Research Establishment and the UK Cement Industry to the NATO/CCMS
Industrial Energy Data Base Project, June 1975 meeting, (the data for the
USA were provided at this meeting by representatives of the cement industry
in the USA).

TABLE 2

Variations in Cement Kiln Energy Consumption between Countries

Country	Average Kiln energy	% Coal used	% Wet Plants	% Portland Cement
USA	6.8 (GJ/te)	40	60	100
UK	6.1 "	80	70	100
Italy	4.6 "	2	13	53
Holland	4.7 "	20	67	75
W. Germany	4.2 "	6	6	75
France	4.5 "	4	30	40

The two questions which come to mind after studying Table 2 are these.
Which of the three variables has most effect on the average kiln energy
consumption for a given country and, are there other variables of equal or
perhaps greater significance?   Although the data are sparse and possibly
imprecise, if it is assumed that a linear additive model fits, then multiple
regression analysis leads to the following equation relating the variables:-

$$E = 3.2 + 0.873C + 0.816W + 1.89P,$$

where E is the cement kiln energy consumption in giga joules, C is the fraction of cement kiln energy provided directly by coal in a country, W is the
fraction of cement produced in "wet" plants and P is the fraction of the
cement output which is of the Portland variety, (the standard error of
estimation came to 0.91 and the multiple correlation coefficient was 0.84).

It is evident from this relationship that there may well be other major factors operating which affect kiln energy consumption and account for differences between countries.   The amount of Portland cement produced seems to
have a great effect on the energy consumption, with the mix of fuels and mix
of plant types having subordinate effects.   But one may well be dubious
about accepting the results of this statistical examination of the available
data.   The important point here is the difficulties inherent in comparing

average energy consumption data for specific industrial processes carried
out in different countries.

If one accepts the evidence presented, can anything be done to improve performance in the USA and UK?   Actually it might be bad policy to plan to
reduce the proportion of coal used to make cement (the Americans are believed
to be planning to push up the amount of coal they use, to conserve oil), there
is nothing in cement manufacture which <u>demands</u> the use of premium quality
fuel.   On the other two variables, we are tied to the raw materials available within a country and also the specifications set by concrete manufacturers and civil engineers for the product.   "Wet" or "Dry" plants have to
be installed to suit the indigenous raw materials - chalks, clays and lime-
stones.   Only certain countries have the materials to hand for Pozzolanic
cements; most developed countries can obtain slag or fly ash as feed material,
but the cement produced might not meet the specifications insisted on by the
home market.

The general point which the author wishes to make is that energy conservation possibilities in the energy-intensive industries, where heavy inputs of
fuel are employed to process raw materials, are often very dependent on the
types and quality of the material inputs and outputs of the process systems;
these may not be variable, except in the very long term.

5. "Final Products" Sector of Industry

When one turns to look at the "final products industries" as defined in
Section 2, it is a characteristic of them that their energy consumption when
viewed in terms of 'value added' is usually very low.   However, since the
energy content of the feed bought-in materials from the "intermediate product
industries" is frequently high, the total energy content (or requirement) of
the final product will be even higher when the energy sequestered in inputs
is taken into account.   Motor car manufacture is a good example to take to
illustrate the general features of this sub-sector of industry.   Until recent
-ly, the motor vehicle industry only spent 1.5% of its income on bought-in
fuels and on this basis it would not be considered an energy-intensive industry.   However, the total energy requirement to produce a medium-sized motor
car was high, of the order of 160 giga-joules.   A revealing way of presenting
the flows of energy necessary to produce motor cars is by way of a flow chart
and this is presented in Figure 9.   It is based on 1968 data for the UK

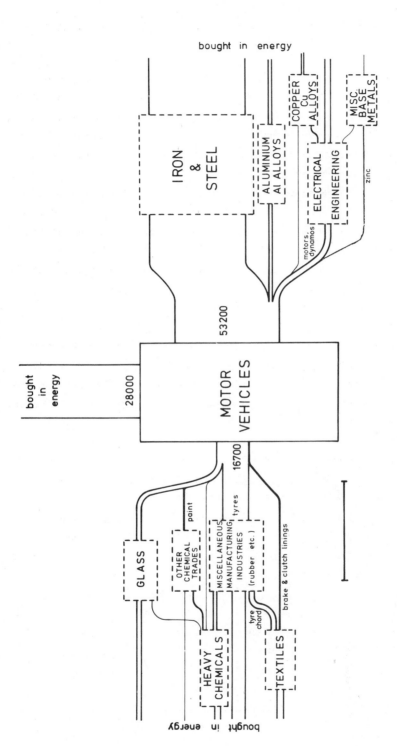

FIG. 9 Flows of Energy into Motor Vehicle Manufacturing in the UK (1968 data)
Notes: (i) Band widths are drawn in proportion to energy content of fuel and
materials.

(ii) Scale; bar equivalent to 50,000 million kWh of energy, on a primary
equivalent basis.

motor vehicle industry.   The various energy inputs have been drawn as bands
flowing into the system, the width of the bands being drawn in proportion to
the respective amounts of energy involved.

It will be noted that iron and steel consumption accounts for most of the
energy (54%).   For some industries which supply the motor vehicle industry,
the energy in the product is provided mainly by the bought-in fuels of that
industry, e.g. glass, whilst for others, it is in the materials bought-in
by that industry, for example, paint, in which most of the energy is contained
in the chemicals and pigments used to make it.   Some industries make a substantial contribution to the motor industry indirectly, but not directly, for
example textiles, which makes its major <u>direct</u> contribution through clutch
and brake linings and its main <u>indirect</u> contribution in the tyre chord which
is sold to the rubber industry.

The analysis on which Fig.9 was based used only bought-in energy costs in the
UK and so does not take into account further energy costs incurred abroad,
such as mining and extraction of copper ore, iron ore and bauxite and smelting of aluminium.   In 1968 only 38,000 tons of aluminium was smelted in the
UK and the nett import of aluminium was 359,000 tons (crude).   Aluminium
production is extremely energy intensive and the total energy cost of the
aluminium input to motor vehicles is in reality far larger than Figure 9
indicates.   To put the data into perspective, in 1968, road transport used
17.32 million tons of petroleum which is equivalent to 3.2×10^{11} kWht.   The
total energy used in motor vehicle manufacturing for that year was 9.8×10^{10}
kWht, i.e. 0.31 times the energy used for road transport!   Finally, in order
to get an up-to-date picture of motor vehicle manufacture in the UK from the
energy analysts standpoint, we would now need to take into consideration the
large number of cars imported each year.

From the point of view of energy conservation, it is obviously imperative for
the motor vehicle industry to reduce wastage on all material inputs, as well
as saving on bought-in energy.   Indeed, design should aim at reducing the
amounts of materials used, consistent with safety.   (And of course, lighter
cars would themselves require less fuel to operate).   Motor designers need
to become aware of which of the materials they use have high energy requirements, so that they can pay particular attention to these.   Also, in the

national context there would of course be an energy saving in making the
body shell of the car last much longer...

6. Industrial Energy Conservation & Business

Energy analysis in general helps us to understand the overall pattern of
energy flow through the industrial system and process energy analysis in
particular can reveal opportunities for saving energy in a given business.
But optimising on energy conservation may run counter to the interests of
profit maximisation in the long run.   Clearly, economic assessment has to
run parallel to energy analysis - if it doesn't pay, an energy conservation
option will not normally be taken up.

To illustrate this important point, an example is taken from a recent
American study of the petroleum industry[6].   In this report based on USA
refining practice, it was shown that the curve of minimum energy requirement
plotted against motor spirit yield followed closely the trends in refinery
capital costs, and the minimum fell around 40%/50% motor spirit production.
If 60% motor spirit were to be produced from a refinery, then a high penalty
had to be paid in terms of internal energy consumption, and if the motor
spirit output fell below about 40%, then internal energy usage increased
again.   The exercise was then extended to include three case studies, based
around this optimum energy consumption at 50% motor spirit yield.   Study 'A'
was based on running a refinery, admittedly a somewhat hypothetical refinery,
to minimise energy usage as a prime target.   Study 'B' removed energy conservation as the primary objective, kept the refinery configuration constant
as in 'A', but optimised on typical product prices and operating costs; the
model thereby obtained was used to determine the 'maximum profit' situation,
allowing the product mix to vary.   The third and final study, 'C', was
similar to the second, but in addition, freedom of choice with regard to
refinery configuration in aiming at maximum profit was included.   The results
came out as shown in the accompanying Table 3.

TABLE 3

	Study 'A'	Study 'B'	Study 'C'
Profit margin in £/million per year	12.5	17.3	17.3
Internal energy use, BTU x 10^6 per barrel of crude	0.489	0.555	0.648

Strictly within the context of this study, there appeared little incentive to
run a refinery according to 'C'.   There is no greater profit margin than
with case 'B' and energy consumption actually is higher than with case 'B'.
The interesting feature is the increased profit margin of case 'B' over 'A'
is about 38%, at the expense of about 13% more energy being consumed within
the refinery.   This is a case, not uncommon, where what is best for energy
conservation is not necessarily best for business.

7. Conclusions

The organisers of the Summer School on Energy asked the author to select
and discuss some actual energy systems taken from the industrial sector, in
order to illustrate a few important features from the standpoint of energy
analysis.   One of the greatest difficulties energy analysts working on the
industrial sector have to face is lack of useful data.   The examples discussed in this paper have necessarily had to employ what published data was
to hand and each example has been treated fairly superficially.   The actual
figures arrived at should not be considered definitive and should certainly
not be quoted out of the present context.   The aim has simply been to illustrate some important general points.   These should have emerged as follows:-

  (a) There are considerable variations in the amounts and patterns of
      use of energy throughout our industrial system.   The model sketched
      in Section 2 provides a useful backdrop against which to approach
      industry - the largest sector in the UK in respect of energy con-
      sumption.

  (b) It is impossible to pursue energy analysis to any meaningful extent
      without constantly being concerned with the material inputs and out-
      puts of the particular industry under study.   Indeed, a study of
      the use of energy resources is meaningless without a full apprecia-
      tion of all the physical resources involved in the system.   This
      was evident from the iron and steel industry examples, where raw
      material grades and scrap metal inputs emerged as very important.

  (c) Where an 'intermediate industry' uses almost all its energy in one
      major process stage, as in the case of cement, then it would seem
      at first sight that 'energy analysis' and 'fuel analysis' tend to
      become the same thing.   But here we have found we are led into
      complications relating to the quality and availability of raw
      materials and also the product markets and specifications.   This

makes it very difficult to compare efficiency of operations as
between different countries.

(d) When studying an ' end product industry', <u>process</u> energy consumption
may be a very small proportion of total costs.   Here, one has to
look for major energy savings in terms of materials economy, and
perhaps in the use of energy for non-process uses; lighting, space
heating and transport of materials and goods.

(e) Optimising on energy consumption doesn't necessarily run in the
direction of optimising the cash return on investment, however
energy-intensive the product.   If an energy conservation measure
does not pay, it will usually have to justify itself on some other,
additional grounds to be introduced.

(f) The most important thing to do when commencing an energy analysis
of an industrial system is to define the system boundary across
which the inputs and outputs to be measured will flow.   This can
only really be done after first deciding the purpose of the analysis
- its objectives.   It may be necessary to carry out several
analyses of the system, using the same data bank but different process boundaries to satisfy all the objectives of the study.

Acknowledgements

Thanks are due to Mr. P. Love of the Programmes Analysis Unit at Chilton for
the multiple regression analysis of the cement data, to Dr. K.S.B. Rose of
ETSU at Harwell for preparing the steel industry energy flowsheets, and to
Dr. P.S. Harris, also of ETSU, for the motor vehicle industry energy analysis.

REFERENCES

1. "A Statistical Survey of Industry Fuel & Energy Use," Confederation of
British Industries, June, 1975.
2. Laws W.R., Iron and Steel International, April, 1974, p.106
3. "Energy Analysis Workshop on Methodology & Conventions, 25th/30th August,
1974", pub. International Fed. of Institutes of Advanced Study, Nobel
House, Sturegatan 14, Box 5344 S-102 46, Stockholm, Sweden - (Workshop
Report No.6).
4. Iron & Steel Industry ANNUAL STATISTICS 1973, pub. British Steel
Corporation, on behalf of the Iron & Steel Statistics Bureau.
5. "Energy Conservation in the UK," A report by the National Economic
Development Office, HMSO 1974.
6. US FEA;" The Data Base, Potential for Energy Conservation in Nine Selected
Industries."   Vol.2, Petroleum Refining, 1st Edition, January 1975,
(Washington D.C.)

DISCUSSION

Question: Why is the steel industry, apparently knowingly, recycling 25% of
            its output as internal scrap?
Answer: This is a problem inherent in current steel making processes.
            Future developments such as continuous casting could reduce this
            figure. More attention is required to reduce the energy wasted
            by internal recycling. A large amount of scrap is not generated
            by the steel company but is generated by the steel processing
            company who can obtain a good price for it. Thus it is a problem
            of split responsibility.

Question: In the energy analysis of the motor car industry, how far back
            have the energy requirements of the various inputs been traced?
Answer: The energy used by the supply industries has been taken into
            account including primary fuel inputs.

Question: What has been left out?
Answer: The capital energy requirements of the supply industries.

Question: Would you clarify whether the electrical input was calculated as
            electrical energy or primary energy?
Answer: All inputs were regressed to the second level. The electrical
            inputs were multiplied by a factor of 4 (which includes conversion
            and transmission losses). The steel industry regards this figure
            as too large since it does not include tariffs and off-peak supplies
            and uses a factor of 10/3.

Question: Does the import of scrap from other countries have an effect on
            the energy analysis carried out on the iron and steel industries?
Answer: Since the energy was not supplied in the U.K. this can either be
            treated as old scrap or costed against the energy used to produce
            the exports which paid for the purchase of scrap. The tonnage of
            scrap bought from overseas is a very small proportion of the total
            scrap used. The problem associated with relying too much upon
            scrap metal supply is that it is a variable supply and the plant
            used for recycling scrap must be kept on full load.

Question: What are the areas of relevance of energy analysis to industry?
Answer: The Department of the Environment's energy analysis unit is con-
            centrating on the metal industries and is using well established
            research associations to collect the data. The Energy Technology
            Support Unit uses consultants from the chemicals industry and the
            British Ceramic Research Association from the brick and refractory
            industries. Ideally the whole of the industrial sector should be
            subjected to energy analysis.

PANEL DISCUSSION:
IMPLICATIONS OF ENERGY ANALYSIS

Panel: Dr. P.C. Roberts, Department of Environment (Chairman).
        Mr. L. Brookes, UKAEA.
        Dr. P.F. Chapman, Open University.
        Dr. P.S. Harris, ETSU.
        Mr. F. Roberts, ETSU.

P. Roberts: I think it might be useful for you to hear about two IFIAS work-shops on Energy Analysis held recently in Sweden. A handbook was produced after the first workshop in Stockholm last year when conventions were thrashed out and the energy analysts who were there agreed how the discipline should develop, what conventions should be used, the way in which system boundaries should be defined and the units which should be used. In the second workshop, which only finished a week or two ago, there was an effort to find common ground between economists and energy analysts. The 'conventional' economists have noted the rise of the discipline of energy analysis and are interested to know what it has to contribute to their work. It seems to me that each has a great deal to contribute to the other.

Chapman: I think that probably the most important point established at the second Stockholm workshop was that the practitioners of energy analysis who were there had no desire to try to establish an energy theory of value, which was one of the aims that the economists feared we were trying to achieve. It took a long time to establish some sort of communication between the economists and energy analysts; if you're a scientist or a technologist, as almost all energy analysts are, you believe that every single process in the world requires some consumption of energy, whereas economists believe that only capital and labour matter. There were clearly some areas where strong and immediate inter-action was possible; for example, econometric models to forecast fuel demands or the availability of fuels at different prices have great use for energy ana-lysis. It also seemed likely that the type of data that energy analysts un-covered (not just energy data but also material flow data and investment tech-nological data) would be useful for econometric studies.

P. Roberts: There were discussions on some very fundamental issues; for example, should you take a strictly commercial view towards a resource which is exhaus-tible? In a paper written by Murlees and Kay of Oxford University entitled "The Desirability of the Depletion of Natural Resources", it is argued that it would be better if we left to our grandchildren not a stock of resources lying in the ground but a stock of useful productive capital equipment made out of those resources. Most people have views about this. They either think that it is a good thing or a bad thing that there should be some conservation. It all depends on the relative valuation of the present and future stream of bene-fits and on your personal discount rate compared with society's discount rate (or the Government's discount rate which is 10%).

Question: The advisability or otherwise of conservation of a resource surely

795

depends on the possibility of substitution. Knowing the long lead times that
are sometimes involved, would it be wise to leave this process to the play of
market forces?

Answer: The various energy planners in the public bodies are well aware of
this problem of long lead times. They recognise that, although technological
fixes have tended to work in the past, the problem is now that lead times of
the technological fixes are tending to be rather longer than the period over
which we can anticipate problems so we have to try harder to see what the
problems will be, what the imperfections of the markets are likely to be and,
therefore, what intervention needs to take place. To some extent it might be
argued that our nuclear power strategy was developed knowing that early nuclear
stations would be uneconomic but on the basis that we ought to be preparing
for some future option.

Question: Are the prices that the Japanese are offering for uranium based on
an examination of the situation similar to the sort of examination you have
just described?

Answer: The Japanese have ordered some supplies of uranium to be delivered
over a quite lengthy period; they have offered to pay twenty dollars a pound
while the current market price is about eight dollars a pound.

Question: Economists tend to regard energy as just another raw material. You
say they will have to change their views. Did you succeed in doing this at
the Stockholm meeting?

Answer: Several of the economists there had not appreciated that the thermodynamic limits of which physicists speak have a kind of absoluteness. They
could see the power of this and this point had not been appreciated. They
said that they would look to the energy analyst to provide the inputs relating
to these extra constraints.

Question: Are systems models subjective because the modellers insert their
own value judgements?

Answer: It is inconceivable that anyone could ever construct a model of anything without making a selection and that selection would reflect his own feel-
ings about the world. This does not have to remove the objectivity of what he
does. It means that he will be thinking about a part of the world and his conclusions about it will be partial, but not necessarily untrue.

Question: Will this situation be alleviated if several groups from different
countries produce models because they would have different biases and different
assumptions?

Answer: There should be several groups involved because otherwise there can
be no criticism by peers. There can be no proper discipline developed unless
it is possible for there to be criticism.

Question: How do you introduce cost-benefit or risk-benefit into models?

Answer: Most of the Atomic Energy Authority's models are not energy analysis
models but are cost-benefit and risk-benefit models. Some of them simply look
at large electricity systems, and insert the various cost ingredients to see
what overall costs emerge. These are, in a sense, cost optimisation studies or

at least cost simulation studies. They look at certain predetermined mixes. The problem with this type of model is in predicting future costs. This was what we talked about earlier, viz. that future markets are not well known. We have to try to look forward at future costs. This may involve one in a crude energy analysis in the sense that one looks at the shortages which might occur given present practices. On cost-risk analysis, we like to think in the AEA that we have carried out much pioneering work on this and have ante-dated Rasmussen's work by some years. Certainly, cost-benefit analysis is something which is done by most of the energy industries.

Question: How do you insert the results of energy analysis into cost-benefit or risk-benefit analysis?

Answer: Energy analysis, as rigorously practised by the energy analyst, has not been used for this. It has been argued that we need never run out of uranium because we can always extract it from the sea. We are energy-analysing this process to see whether it is sensible from the energy balance point of view.

Question: On the cost of extraction of uranium from sea water, I have seen figures ranging from about 30 dollars a pound to over 100 dollars a pound. Do you have any idea what the cost is likely to be?

Answer: We have looked at this and I think that it will be near your higher value but I think that the price of uranium is going to get pretty near that anyway.

Question: What is a reasonable social discount rate to use in cost-benefit analysis?

Answer: About 3% in real terms.

Question: Is energy analysis of the iron and steel industry worthwhile because it looks as if one can make many models and arrive at different conclusions depending on the input assumptions?

Answer: If that is your conclusion, I am pleased because this was one of my aims. Clearly I obtained two answers for the best steel-making route depending on the assumptions I used for scrap, but this is still a useful study because it focusses attention on the energy sequestered in scrap.

Comment: At high grades of uranium ore, all the reactor systems look very similar but for low grade ore, they looked different in energy analysis terms. By making this analysis you can make sure that you choose the best design of reactor if you ever have to use uranium of low grade and there is no other way of making this decision.

Response: There are many factors influencing this, e.g. reliability, safety and whether a reactor will produce plutonium in sufficient quantities and at the right grades to stockpile for the fast breeders.

Question: The Rasmussen project has aroused a considerable amount of controversy. There are a number of regions in which really fundamental data are desperately needed because the Rasmussen inputs were no more than wild guesses. Nevertheless, on the basis of this crude model, the United States government seems to have concluded that the light water reactor is sufficiently safe for

the future to start making a very considerable investment. If the model is
based on inadequate data, could this lead to a catastrophe?

<u>Answer</u>: I'd like to feel that other industries have done as much for their
industries as has been done by Rasmussen for the nuclear industry.

<u>Comment</u>: The consequences of mistakes are much greater with nuclear plant.

<u>Comment</u>: We have learned in the nuclear business mainly from near-misses and
careful analysis of these rather than from actual accidents. In industry generally, accidents actually have to happen before lessons are learned.

<u>Question</u>: In the assessment of any kind of risk, surely one must consider the
magnitude of a risk and whether it is voluntary? It is quite clear that nuclear
power involves an involuntary risk to very large numbers of people. Do you have
a way of comparing lives on one hand with technical precautions on the other?

<u>Answer</u>: You should not get this out of scale. We were talking about subjective
judgements earlier. I would much rather live next to a nuclear reactor than to
most chemical works as I think the risks would be very much less for the former.

<u>Question</u>: Is the probability of action by terrorists orders of magnitude grea-
ter than that of a reactor accident?

<u>Answer</u>: Part of the difficulty in assessing the probability of terrorist action
is that it is a problem we have dreamed up for ourselves. If you really wanted
to create havoc you could probably do it very much more easily with something
other than nuclear materials.

<u>Question</u>: What are the methodologies that you could use in assessing the mag-
nitude of public fears against the probable number of lives lost in an accident?

<u>Answer</u>: There certainly are methodologies for evaluating how far you should go
in taking counter-measures and the cost that is justifiable for different risks.
These are now quite well established methodologies and the AEA makes this service available to other people.

<u>Question</u>: Is there not a much more difficult question involved here in the use
of cost-benefit analysis when the benefits are borne by one generation and the
risks are delivered to another?

<u>Answer</u>: It is not unlike, in reverse, the judgement that you make when you
decide to forego some benefit now for the benefit of future society. There
are quite a number of examples. Almost all investment takes this form.

<u>Question</u>: Do we have a suitable, acceptable technique for deciding this issue
other than a value judgement?

<u>Answer</u>: You do not just produce a benefit for yourself; you may also be pro-
ducing a benefit for future societies. How do you evaluate the balance between
these two or the dis-benefits to future society? I do not know of any really
good methodology of doing this, however. The best you can do is to say that
you will minimise the risks as far as you possibly can and you will not proceed
if you think that the risks are inordinate.

<u>Question</u>: I'm not objecting to the technique being used as far as it can be

used, but do you not think we should be very clear that data may contain value judgements and so may not be objective?

Answer: Cost-benefit analysis is an extremely inexact art.

Comment: It is not inexact, it is wrong in principle for many types of decision. It is not capable of being made exact.

Comment: If we made a straight cost-benefit analysis, we would not take anything like the trouble that we are over radioactive waste storage, because if you apply the discount rate to these risks, they virtually disappear. We do depart from standard cost-benefit analysis when we decide what we ought to do about this.

Question: In principle, cost-benefit analysis rests on the single value judgement that people's preferences matter; once you make that value judgement, can you be objective?

Answer: Not when the costs and benefits fall on different people.

Question: If you are not prepared to admit that people's preferences should count, what alternative is there?

Answer: I think it can be settled at the tribal level because there is a generally accepted set of values that makes comparisons valid. In a modern pluralistic society, it may not be possible but this could be an argument against such a society.

Question: Surely the coal industry implies social consequences possibly quite as serious as those for the nuclear industry?

Answer: People outside both industries have made analyses of the relative hazards of coal and nuclear power and heavily favour nuclear power as being much less damaging environmentally.

Question: Surely much more money has been invested in the nuclear industry to make it safe?

Answer: If we invested the same amount of money in the coal industry to try and make it safe, we would never make it as safe as the nuclear industry appears to be at present. We certainly could not diminish sufficiently the hazards of sulphur dioxide contamination, for example.

Comment: The statistics on deaths and accidents in the coal industry are based on a long historical record. When you say that nuclear power hasn't killed anybody, it hasn't killed anybody yet.

Comment: By all means keep the nuclear industry on its toes and keep nagging at it to provide plenty of information, but do keep the risks in perspective.

Question: Is it true that energy models may be more applicable to the world as a whole, because they deal with physical laws, than economic models which appear to deal with one type of society or system?

Answer: There are several respects in which energy analysis will depend on the nation being considered. Consider the United Kingdom. You can use energy

analysis to determine the energy required for all the goods and services we
produce in this country. When you energy-analyse a process in India for
example, where much of the work is manual, the energy analysis is going to
depend on how you energy-cost labour.

Question: In energy analysis, it is crucially important to know your objectives. Could you say a little more about the application of energy analysis
to steel production, where the object was to explore the possibilities for
energy conservation? Do you define energy conservation from the point of view
of the manufacturer, the country or even of the whole biosphere?

Answer: The Energy Technology Support Unit is carrying out energy analyses to
determine the effects on the total demand for primary energy in this country
now and in the future. We want to look at the present pattern of energy flows
through industry to get some idea of how this is likely to change in the future
and what might be done to constrain those patterns. When we looked at the
specific energy requirement for a ton of steel by three different models, we
were interested in whether there would be any startling differences and we
found that the result depended on how scrap was handled. This was mentioned
here because it does highlight an important point.

Question: If you wish to obtain information on the flow of energy through the
industrial economy, what energy value do you assign to the scrap flowing between
one process and another or within a process?

Answer: There is an arbitrary convention that the content is zero as scrap is
derived from a scrap heap. It is a convention which seems to make a useful
starting point.

Question: If energy analysis is never going to become a theory of value, can
it ever supersede economics?

Answer: There are people like Howard Oldham and Bruce Hannon who take the view
that the old monetary economics is being superseded and that energy will take
over as the unit of value. The case does not seem to have been made, however,
and it is extremely unlikely that there will be any such take-over. We do
need different tools in the varied problems that face us and it seems to me
quite sensible to use now one, now another or combinations.

Comment: Peter Chapman has said that energy analysis is not a broad enough
tool for resource allocation. Economics is the right art - I say art you note -
for this but only if it absorbs new tools to help its resource allocation
studies.

SECTION 9

CONCLUDING PAPERS

ENERGY CONSERVATION AND GOVERNMENT INFLUENCE ON ENERGY USE

Lord Balogh

Department of Energy

There are very few occasions in economic history which are remembered, as are
vast volcanic eruptions, for changing the whole landscape. Professor Marshall
in his classic superiority chose as his motto "Natura non facit salta". Since
his old age we have seen at least three and perhaps four such brutal changes:
the victory of communism; the great Depression which gave Hitler his power
and which led inexorably to the Second War; the Japanese assault on the colo-
nial possessions of the US and the European Powers which heralded the collapse
of Empires. Finally and more recently there is the OPEC revolution.

No doubt efforts have been made before to control commodities. Rubber, tin,
and coffee were subject to restriction on production to achieve high prices.
They all collapsed and were never as all-pervasive and violent in their effects
as the oil price crisis. Oil has no substitute in the short run. The price
increase is so patently profitable for all producers that a collapse of the
cartel cannot be expected. We must therefore accept that the era of cheap
energy has gone.

What this means can be best illustrated by the Energy White Paper of 1967:
"Subject to the overriding considerations of adequacy and security of supplies,
the Government's basic objective can be summarised as cheap energy" (paragraph
79).

Now we may suspect that OPEC has merely anticipated the inexorable collision
between growing demand and shrinking supply due to the accelerating exhaustion
of an irreplaceable natural resource leading to eventual scarcity, but that
process would have been foreseeable and slow. It would have allowed gradual
adjustment. Cheap oil might have easily lasted until, say, nuclear fusion
released plentiful energy. It is the abruptness of the change which has been
especially damaging.

In another sense, however, this pre-occupation with energy saving is new. We have never before been faced with anything like the increase in international oil prices - more than fourfold - which we have experienced since October 1973. We have never before had to cope with the enormous balance of payments effects of such a rapid change; for the U.K. in 1974 it amounted to an extra £2,500m for 2 per cent less oil. We have never before experienced a situation in which a cheap, plentiful commodity of basic importance to our economy was converted overnight into an expensive one.

There is very little that can be done in the short term about substitution, but we must speed up preparation for the long term as quickly as is practicable. The acceptance of the urgency of energy conservation has been a striking feature of the last 18 months. If you study major pronouncements on energy before October 1973, (e.g. the White Paper on Fuel Policy in 1967), you will not find it mentioned, yet today it is one of the chief areas of concern to Governments in the field of energy policy, along with the development of indigenous resources and the pursuit of international cooperation on energy matters. There is no shortage of authoritative bodies - NEDC, CPRS, the Select Committee on Science and Technology, EEC, OECD, even NATO - willing to spend good time and money finding out what can be done to promote it. I shall try myself to see what the Department of Energy can do to lessen demand for energy.

I should perhaps begin with a word of reassurance. I am not about to hand out "Save It" badges or to demonstrate to you how to install 3-inch cylinder jackets. My concern today is to take you behind the scenes a little and to consider what things Governments can, and cannot, reasonably expect to do in this area and what issues they may encounter in the process, but first I must consider why energy saving has become so important.

In the decade up to 1973, policy-makers largely gave their attention to questions concerned with energy supply owing to the successful campaign by the great oil companies to discover new and even more plentiful resources; the troubled future of the coal industry, the place of nuclear power, the development of the North Sea, the investment programmes of the nationalised industries - these were the issues that concerned them. The new situation which I have described has introduced a new direction for their work: the replacement of imported and expensive energy by economically produced home substitutes. I

do not intend today to go into our strategy for energy production. No doubt
you have touched on the main areas in other sessions: the coal industry, our
policies for North Sea oil and gas development, our decisions on nuclear
reactor choice and so on. My central point is that in this changed situation,
energy conservation has a much greater role to play in:

- keeping down our import bill;
- easing the burden on national resources of the investment
  needed to provide indigenous energy;
- improving the capacity of the economy to cope with interruptions in energy supply; and
- spinning out our energy reserves against the ultimate decline
  in physical supply which many observers foresee for the next
  century.

What is clear is that the resulting need for energy conservation will be no
9-day wonder. The prospect is that energy will continue to be an expensive
resource for the indefinite future. Although it is only imported oil which
is such a burden on our balance of payments, the need for conservation extends
to other forms of energy. In the immediate future, we are unable to supply
all the energy we need from domestic resources. We shall be using imported
oil at the margin of our energy economy and all the savings which we can
achieve in other forms of energy can be matched, by substitution, by a reduction in oil imports. Moreover, the need for energy saving will still remain
even when we achieve self-sufficiency in terms of indigenous fuel and nuclear
power. For any savings we can achieve, this can be reflected in extra oil
exports and in the retention of our energy reserves for use later.

It is particularly important to see our North Sea oil reserves' need in this
context. I believe it would be the greatest mistake to think that these
reserves - which are, without doubt, a fortunate blessing - will solve all
our problems, restore our international manufacturing competitiveness, and
enable us to live forever on exporting less and consuming more imports. They
will not solve the deficiencies in our industrial investment. What they may
do is aid us powerfully in these important and difficult areas. But we shall
have every incentive to use them to best advantage and with the greatest
efficiency.

I have given you some idea of why energy conservation has come so much to the

forefront of attention. I should now like to consider, first, some questions
of approach to energy saving and, second, the scope for Government action.

SOME QUESTIONS OF APPROACH TO ENERGY SAVING

One question which is not, perhaps, asked often enough is: why concentrate so
much on conservation of energy rather than other resources? Energy in its
variety of forms as 'useful heat' is, after all, neither more nor less a saleable commodity like other saleable commodities. Why then should Government
be so concerned about its influence in this area?

I have already given one answer to these questions: the rapid rate of increase
in the price of oil which, even in these days of high inflation, has been
extreme. There are a number of other considerations, too, which tend to concentrate attention on the energy sector: the massive quantities of capital
which the energy industries need for investment purposes; the long lead times
before that investment takes effect and the great length of time for which
investment decisions last, once taken; the fact that the energy industries
draw on natural resources which will one day be exhausted; above all, the
integral role which energy use plays in every aspect of national life. Of
course, it would be wrong to treat energy as fundamentally different from all
other goods and services, and it is by no means the only resource which needs
to be used efficiently, but, whether Governments like it or not, they are
deeply involved in these basic questions and it would be very difficult now
to ignore the importance which energy saving is assuming.

What, then, should the Government be aiming at? The elimination of waste in
energy use? I ask this second question because 'waste' is a concept which
frequently crops up in discussing energy use, particularly as a stick with
which to beat the Government or the energy industries. Yet it is not an easy
word to define. Consider the everyday instance of a person using a heating
appliance to heat the fabric of a house and all the air and furniture in it,
merely to keep himself warm. How much of that is necessary and how much is
'wasted'? Again, take the instance of a person who wants to watch television

and who causes the CEGB system to increase its burn of coal in the Midlands
in order to supply the electricity. This involves a fuel consumption equivalent to, say, 500 watts and yet the energy used consists of the almost immea-
surably small amount represented by the energy of the photons falling on the
retina of that person's eye. Almost all the 500 watts of energy is dissipated
in the power station, cables, transmitter and receiver. This can be regarded
as 'waste' in the sense that the heat is not wanted or used yet it is a necessary part of the process of converting a piece of coal to meet the requirements of the consumer. Clearly the concept of 'waste' sounds deceptively
simple and is not always a clear guide for Government action.

Should we, then, be adopting 'energy accounting'? This is an approach which
studies the total energy inputs needed to produce commodities, including the
energy used in producing the raw materials and intermediate products which go
into them. We could embark on a major input - output analysis of the energy
flows in the economy as a whole, calculating how the energy inputs ultimately
contribute to broad classes of intermediate or final product. We could carry
out extensive process analyses, taking particular commodities and re-tracing
their production processes to find their energy inputs at each stage.

Energy accounting is certainly an interesting area, one where a lot of work
is going on at the moment inside and outside Government, but it is important
to recognise its limitations as well as its uses. Energy is in no way the
only important resource whose use needs to be optimised. It would be highly
misleading to think that energy has some kind of value of its own, different
from capital and labour for instance, which in effect, makes it the source of
all other values. The public sector must be concerned not just with energy
but with total resource costs; and these include, besides energy, all the
other factors of production such as other raw materials, land, labour and
social costs not included in market prices.

If we were to concentrate solely on one commodity we would be likely to come
to some very odd conclusions. Consider a power station, for example. In the
normal course of operations it will consume several times more energy than it
produces. Are we, then, to assume that we have made a mistake and it should
not have been built? The answer, of course, is that despite its high cost in
terms of energy and other resources, a power station's output of energy has a

higher resource value, in terms of consumer convenience, efficiency in use, cost of equipment needed to use it and so on, than its input energy. It can be helpful for us to have a picture of the total energy requirements of a process or policy if we are concerned with the effect of a change in energy prices on the relative costs of commodities or with selecting energy-intensive areas as targets for increasing efficiency in energy use. On its own, however, energy accounting does not tell us the different resource values of different fuels, nor does it provide a guide to how much we should spend in other resources to save a given amount of energy.

THE SCOPE FOR GOVERNMENT ACTION

The Government's approach to influencing energy use is conditioned by three important factors.

First, the test of any particular energy saving measure must be the relationship between the value of the savings made and the costs of achieving them. This means that we want to concentrate on those measures which are economic and pay for themselves in terms of the savings which they yield. Obviously, the more expensive energy becomes, the more savings can be made and the more measures pass the test. That is why pricing policy is at the heart of energy saving. On the whole we are still pricing energy as if we wanted to stimulate its use and not the reverse and this applies not only to the subsidies given to the energy industries as one of the main inflationary measures, but more generally in the structure of electricity and gas charges.

Second, energy use is very fragmented. The great majority of decisions about energy consumption are and can only be taken by individuals and businesses in every sector of society, and these are outside direct Government control. Leaving nationalised industries and defence questions on one side, the Government directly controls less than one per cent of U.K. energy consumption in its own buildings and establishments. This represents an important constraint on Government action. Nonetheless, across the remaining 99 per cent of consumption outside Government control, action to secure savings is still possible under one or more of three headings:

- the setting of appropriate economic signals, which includes
    pricing and taxation policy;
- the provision to consumers of information, advice and help;
- the imposition of mandatory constraints.

Third, the patterns of energy consumption in our economy are immensely complex
and energy enters into, or in some way affects, a tremendous range of functions
in our daily life. This, too, is a powerful consideration, for in many areas
where Government might act to save energy it can only do so by depriving the
community of some other benefit. Consider the simple example of floodlighting
of buildings. Many people ask why we permit this to contine when we want to
save smaller quantities of energy in their homes. The answer is that such
floodlighting consumes very little of our total energy consumption and in
return has a valuable contribution to make to our tourist trade and to the
security of some buildings.

Government ability to influence energy use is, therefore, rather limited.
Each of the main channels that I have mentioned - economic signals, information
and advice, mandatory constraints - has its limits. For instance, information
and advice are not likely to be effective unless the basic economic incentive
to save energy is present; many areas of energy use are not suitable for mandatory controls, very often for no other reason than that of enforcement.
Moreover, even within the limits of what is possible there are a number of
factors which slow down consumer response to the need to save energy. For
instance, although 'good housekeeping' measures can usually be adopted almost
at once, the application of existing technologies of energy saving may have to
be deferred until capital equipment falls due for replacement or until the
margin of cost savings justifies new capital expenditure.

Nonetheless, despite these limitations, there is a considerable amount that
Government can do - and is doing - to influence energy users by information
and advice, mandatory constraints and economic signals. The list of measures
which we have taken so far is a long one, though perhaps not sufficiently
drastic: a major publicity campaign has been launched, lower speed limits on
some roads, restrictions on space heating (except living accommodation),
restrictions on the use of electricity for daylight advertising, a doubling

of thermal insulation standards for new homes, an Energy Saving Loan Scheme, introduction of 100 per cent initial allowance for the insulation of industrial buildings, pamphlets, guidance for local authorities, consultations with many of the expert bodies in the field, I will not weary you with the complete list. Our objectives, with the help of the Advisory Council on Energy Conservation, is to create and build up a wide-ranging programme of measures which will speed the adjustment of consumer attitudes to the new fact of high-cost energy.

A central ingredient in this approach is, of course, the concept of 'economic energy pricing'. The Chancellor announced last November the Government's decision to phase out subsidies to the nationalised industries as fast as possible. We have started from a position in which the nationalised energy industries alone, without price increases, would have required subsidies amounting to around £1,200 million. The constraints on public expenditure are such that, quite apart from energy conservation, we can no longer afford these vast subsidies which we have been pouring into these industries through adherence to successive Governments' policies on price restraint. With the price increases we have implemented and the expected gas price increases in the autumn we are hoping for a subsidy figure of only about £30 million.

We fully recognise that this return to "economic" prices is a hard and painful affair. In the period since March 1974, the price of industrial coal has gone up by 133 per cent and domestic coal by 56 per cent, domestic electricity by 86 per cent and industrial electricity by 48 per cent. You will all know about the price of petrol. Even at a time of high inflation these increases are very heavy, but, even when we have achieved the near break-even position for our energy industries that we hope for by the end of this financial year, we cannot afford to be complacent. We will still have some way to go if our energy industries are to earn a proper return on capital which takes due account of the impact of inflation. Here, in passing, I would only ask you to note that there is a considerable variety of ways in which 'economic pricing' can be interpreted. Far too little attention has been paid to the problem of whether in a situation of shortage we ought not to invert the scale of charges, but here I trespass on a painful theme, a minefield for a junior Minister confronted by a determined bureaucracy.

CONCLUSION

The ability of Government to influence energy use is, in the last resort, a
question of persuasion. As in other fields we can bully, cajole and bribe,
but ultimately the success or failure of our efforts will depend on the actions
of individual consumers and their willingness to adjust. It would be wrong
to hope for too much from what we are trying to do. Energy saving cannot,
for instance, do more than make a useful contribution to reducing the cost of
oil imports and we shall face a considerable challenge in maintaining the
momentum of what we have done so far.

The forces which have brought energy conservation back into the limelight are,
nevertheless, likely to be with us for as far as we see in the future. There
is no doubt that Governments from now on will take a much keener interest in
their influence on energy use.

DISCUSSION

Question: Do you think the present fuel pricing policy should be changed so
            that natural gas would be reserved for premium use, its non-
            premium use being taken over by other fuels, to conserve supplies?

Answer: The pricing policy of the nationalised energy supply industries
            has been influenced largely by governments attempting to reduce
            inflation by restricting fuel price increases. That has not
            worked and now all nationalised industries, including energy sup-
            pliers, will have to charge economic prices for their goods and
            services. The price of natural gas will be determined by its cost
            though its price could, in principle, be raised artificially for
            the reason you suggest, but nothing so drastic is yet in mind.

Question: Would your department be able, if it saw fit, to divert the money
            spent on nuclear fission into the development of fossil fuel tech-
            nology, thus enabling us to bridge the coming energy gap with
            fossil fuel and solve the attendant pollution problem?

Answer: This was Professor Feld's point I think. I do not think that by
            diverting one's effort, one can guarantee a particular result.
            It is not only a technological problem; how do we persuade people
            to go underground to dig out the coal we will need?

Question: We have been told that, in government economic forecasts, a discount
            rate of 10% is used. Would a rate of, say, 3% be more appropriate
            to encourage conservation of resources? How does the present
            inflation rate of 25% or so affect discounting?

Answer: The 10% discount rate is a rate in real terms, i.e. the revenues
            and costs are calculated in terms of the base year with no allow-
            ance for inflation. To allow for inflation, the discount rate
            would have to be 30% or so to be 10% in real terms. The value of
            10%, which the nationalised energy suppliers use, is based on the
            profit large private industrial concerns expected from their
            investments. I agree that, when conservation of resources is con-
            cerned, a rate of less than 10% would be appropriate. The policy
            on depletion will come to the fore for North Sea oil soon; it
            seems sensible not to use it too quickly. There is a rush to use
            North Sea oil at the moment because we are amassing large debts
            and the oil companies can develop as they choose any fields they
            discover by the end of this year.

Question: Why duplicate a large established system?

Answer: It is considered politic that not all of such an important business should be in the private sector.

Question: What place do you think direct government expenditure has in energy conservation?

Answer: The government view is that fuel price should cover cost and its expenditure should be limited to demonstrating to consumers how money can be saved by saving energy.

Question: Will there be enough capital for the long-term solutions to the energy crisis e.g. wave and fusion power?

Answer: If there is no other way to obtain energy and potential users need that energy, they will pay the cost, capital cost included.

Question: It is said that a government spends its first two years discovering that it cannot fulfil its election promises, the third year attempting to govern, and the final two years devising a new set of election promises. In other words, it is no wonder that governments find it difficult to look ahead more than five years. At this School, we have heard that we should be looking decades ahead (for radioactive waste storage, thousands of years ahead), because of the very long lead times needed to introduce new technologies on a large scale. Can you suggest how governments can be persuaded to do this? You said that the government was not giving serious consideration to zero economic growth because people generally want growth to continue; can the government be persuaded to inform people of what could happen if the exponential growth that successive governments strive for continues? Anybody who dares to question the wisdom of continuous growth is labelled "environmentalist" or "doomwatcher" and assigned idiot status.

Answer: You are right in that our political system does not encourage long-term planning. Presumably, educated informed opinion will eventually force governments to do this and reconsider growth policies but this does not seem to be the majority opinion at present.

Question: Our economic structure appears to be governed by market forces in isolation from other considerations. Can economists be made to realise that this is not satisfactory?

Answer: It does seem that economics is not equipped to handle resource exhaustion. Malthus worried a long time ago about the population expanding faster than it could support itself. Science and technology have always provided solutions, so economists have become used to dealing only with the short term. We now need to consider the long term and to complement the work of economists with that of scientists and technologists i.e. those who can make reasonable forecasts of conditions in years hence, but I am not clear how to enmesh the two.

Question: On fuel pricing, do you favour the reversal of present tariff structure, which appears to encourage use?

Answer: Why not apply this argument to all resources? It implies that we know what the reserves and prices situation in 10 or 20 years' time will be and can price accordingly today.

Question: Professor Feld suggested we should limit our energy consumption to about 4kW/capita. Is your department studying the social and economic implications of zero growth?

Answer: Not at present. Most people want growth.

Question: With the establishment of the British National Oil Company, will the government become involved in the oil business right down to retail level?

Answer: Yes, but progressively. Initially, North Sea oil will be sold to the trade.

ENERGY SOURCES FOR THE FUTURE
AND THEIR EFFECTIVE UTILIZATION

B. T. Feld

Physics Department, Massachusetts Institute of Technology

The demands for energy, both in the developed and in the less developed regions of the world, have been increasing exponentially, with a doubling time on the order of ten years, for the past century. Nor does this demand show any signs of saturating in the foreseeable future.

However, there are a number of limiting factors, some already coming into play and others which may be expected to become increasingly important as we approach the end of this century. These include:

a) The exhaustion of available sources of fossil fuels -- oil already in sight and coal still some hundreds of years away.

b) The seriously detrimental effects of environmental despoilage and atmospheric pollution arising from the extraction and burning of fossil fuels, which would become intolerable, or prohibitively expensive to prevent, if there were to be a many-fold expansion in the energy production from fossil fuels.

c) The very limited number, energy potential and distribution of accessible hydro-sources for the direct production of hydroelectric energy. However, it is important to note, in this regard, that there remain substantial, untapped hydro-sources in the (underdeveloped) regions of the world where they are now most needed. Considering also that the environmental injunctions against the expansion of the use of fossil fuels are a very long way from applying to most of the developing world, the needs of this large (roughly two-thirds) part of mankind could be met, at least for the rest of this century, without evoking any new sources or any drastic alterations in the pattern of utilization of the old ones. In the long run, however, all mankind is in the same energy boat.

d) The possibly insuperable problems associated with the unlimited use of the energy of nuclear fission. These difficulties arise from the problems of disposal of vast and inevitable accumulations of radioactive wastes inherently present in operating nuclear fission power plants and the concomitant fuel-element processing and fabrication plants (note that a 1GWt reactor, in steady state operation, contains 1,000 megacuries of stored fission product radioactivity); from the dangers of hugely lethal dispersals of radioactive fission products and the highly toxic plutonium, as a consequence of rupture of reactor containment vessels from natural calamities, sabotage, military action, or accident; and from the possible diversion of plutonium, by irresponsible governmental or non-governmental groups, to the production of nuclear bombs. (See Table I for a rough estimate). This last problem, at least, would be greatly exacerbated by a widespread commitment to a plutonium breeder fission energy economy. On the other hand, the development of practical nuclear fusion energy would avoid these problems to a very large degree.

e) Last and, in the final analysis, the consideration that will eventually place an absolute upper limit on the world-wide conversion of stored (potential) energy via power into heat -- i.e. on the use of fossil, fission, fusion and, to a somewhat lesser extent, geothermal power -- and also on the density of local power production, is the effect on world-wide (and local) climate patterns arising from an additional heat

TABLE I.   IAEA NUCLEAR POWER PROJECTIONS AND APPROXIMATE
EQUIVALENT POTENTIAL BOMB PRODUCTION RATES
(assumed breeding ratio of 1/2)

	1980		2000	
	GWe	bombs/yr	GWe	bombs/yr
Less Developed Countries	25	1,000	750	30,000
Developed Countries	290	10,000	4,550	200,000

energy input approaching a few per cent of that which normally impinges on the earth from the sun.   We are already within a factor of one hundred or less, world-wide, of this limit, and much closer in local areas of dense industrialization.   Fortunately, sources that derive their energy directly from the solar radiation -- direct solar energy conversion, wind and tidal power -- are not, in a first approximation, subject to this limitation, although they could still produce serious local climatic changes by altering significantly the detailed balance between the absorption and re-emission of solar radiation.

If we accept the thesis that there is an upper limit to the total heat production permitted to mankind, and that this limit is already in sight, then the distribution of energy consumption becomes of critical importance.   As things now stand, the industrialized (developed) one-third of the earth's population consumes, per capita, almost ten times as much energy as the less-developed two-thirds.   Even within the industrialized third, the average American consumes over three times as much as the average citizen in another part of the developed world; and there is an even wider spread in the power consumption among the peoples of the developing world.   (See Table II for some illustrative data.)

TABLE II.   WORLD ENERGY CONSUMPTION, 1972 FIGURES

	average (KWt per person)	population (10^9)		total (1,000 GWt = 10^{13} KWh/yr)	
LDCs	0.2	2.4	(67%)	0.5	(9.5%)
DCs	4	1.2	(33%)	4.8	(91%)
worldwide	1.5	3.6		5.3	
Indonesia	0.1	0.1	(2.8%)	0.01	(0.2%)
United States	12	0.2	(5.5%)	2.4	(45%)

It is important to note that a continuing, uniform, 5 per cent annual growth of energy consumption worldwide, combined with the continuation of current population growth patterns, would result in a widening of the gap between the developed countries (DC)

and the less-developed ones (LDC) -- from the factor of 20 in 1972 to a projected
factor of 30 in the year 2000 (see Table III).   Furthermore, the achievement of such
uniform growth appears to require a massive increase in the use of fission energy
(Table IV).   However, most if not all of the nuclear energy would be required to
satisfy the growth demands of the DCs (compare Tables III and IV).

TABLE III.   ENERGY CONSUMPTION IN YEAR 2000
(assuming uniform growth)

	Total (1,000 GWt)	Average (KWt per capita)
LDCs	1.9	0.32
DCs	19	9.6
Worldwide	21	2.6

TABLE IV.   IAEA ENERGY USE PROJECTION

Fuel	1980 (1000 GWt)	2000 (1000 GWt)	Factor of Increase
solid (coal)	2.4	4.1	1.7
liquid (oil)	4.1	5.2	1.3
natural gas	2.0	3.3	1.7
hydro	0.6	1.6	2.7
nuclear	0.6	9.4	16
	9.7	23.6	2.4

If it were possible to bring all the world's population up to today's American level of
power consumption, this would already bring the world too close for comfort to the
limits of permitted thermal pollution, leaving very little to spare for new energy
requirements -- e.g. for the recycling of scarce and badly needed raw materials.
Even the more modest goal of a uniform worldwide energy consumption of 4KWt per
capita (zero growth for the DCs) would imply a vast increase for the LDCs (see Table
V), necessitating a 14 per cent per annum growth rate in the LDCs, and the consumption
by them of all the available energy projected by the IAEA (Table IV).

TABLE V.   ENERGY CONSUMPTION IN THE YEAR 2000
(4KWt per capita worldwide)

	Total (1,000 GWt)	Factor over 1972
LDCs	24	50
DCs	8	1.7
Worldwide	32	6

However, aside from the unavailability of usable energy sources for doing this on
such a short-time scale, the sheer magnitude of the economic and technological
requirements, especially in the developing countries, sets a time scale of the order
of a century at least for the accomplishment of this goal.   On such a time scale, the
development of new energy sources -- in particular, nuclear fusion and direct solar
energy conversion -- become both feasible and imperative.

In the meanwhile, on the intermediate time-scale (5-30 years), a number of parallel
steps need to be taken to avoid the periodic re-occurrence of energy "crises" both in
the developed and the developing worlds:
a)   The scarce oil sources should be, as far as possible, reserved for special and
unique uses (e.g., petro-chemicals) for future generations.   Immediate needs should,
as far as possible, be filled by maximum utilization of hydro-sources and coal. The
efficient and non-polluting utilization of coal, however, will require the expenditure
of appreciable research efforts and funds (for, e.g., practical development of in situ
coal gasification and liquefication, efficient waste "scrubbing" equipment, etc.).
b)   Nuclear fission energy should be regarded as, at best, an interim solution to
meet the energy needs in the industrialized world (for the reasons detailed in (d) above).
In particular, large further investment in the development of new technologies -- i.e.
the liquid metal fast breeder reactors -- should be avoided, since such developments
will tend to commit us -- both economically and psychologically -- to a fission breeder
economy, and to discourage or draw off the necessary intellectual and financial
investment in reasonable long-term solutions such as fusion and direct solar energy.
c)   A very large effort must be devoted to ensuring the much more efficient utilization
of available energy in the highly industrialized nations, and to curb the apparently
insatiable appetites for unlimited growth in energy consumption.   Although there is
an undeniable correlation between per capita energy consumption and individual well-
being (standard of living), there remains considerable leeway for the reduction of
individual energy consumption without appreciably affecting the living standard. Thus,
while the average American uses around three times more energy than the average
West European, it is difficult to argue that he lives three times as well, or that he is
three times as comfortable or happy.   On the other hand, the ten-fold gap in energy
consumption between the average Indian and the average European does reflect a
serious gap in the general well-being; which is not to imply that Indians need to try
to emulate the European life-style in their programmes for development.

My purpose in dwelling at some length on the near- and intermediate-range energy
problem has been two-fold:
a)   First, to make the point that long-range solutions will not be sufficient for the
avoidance of the very serious crises in energy, natural resources, ecology, food,
and rapid world-wide population growth (not to speak of the serious problems of mal-
distribution between the rich and the poor in almost all nations) that we are facing in
the next fifty years;
b)   and second, to emphasize that, in the long run, none of the currently available
energy sources is satisfactory for the needs of mankind over the coming centuries.
As far as now can be predicted, there are two promising energy sources for the future --
nuclear fusion and direct solar -- and these are remarkably complementary in their
properties, so that both will be needed for the assurance of a uniform standard of
well-being for future generations.

However, before considering the properties and prospects for these sources in somewhat greater detail, one additional technical point requires some elaboration. One of the major factors in the inefficient utilization of currently available energy arises from the difficulty and expense of the transport of the fuel or electricity from the source to the point of utilization. The large-scale transmission of electrical power over long distances is very expensive and inefficient, thus encouraging its production as close as possible to the site of its use in units matched to its consumption. On the other hand, efficiency of production generally increases with the size of the producing unit. (This is very likely to be true for nuclear fusion, although it is not quite so obvious in the case of solar energy). Oil, which is now the most easily transportable energy source (and therefore has been over-used for transportation and local heating purposes), is rapidly running out, aside from being prohibitively polluting. Hence, what is most acutely needed, both now and in the future, is a compact, cheap, efficient and readily available energy carrier. Hydrogen appears to be the answer to this need, either in liquid form (which has the disadvantage, however, of requiring efficient cryogenic facilities) or, more likely, in the form of suitable chemically active compounds of reasonable liquefication temperature, which can provide compact sources of high energy content; hydrogen is available in the oceans in essentially infinite quantities; and it is non-polluting, the main product of its combustion being water. The development of a cheap and efficient technology of hydrogen production and transport must therefore have the highest priority in programmes of research and development aimed at solving the energy crises of the present and future.

Coming finally to the energy sources for the future: as noted in (b) above, nuclear fusion appears to be one of the most promising. The raw material -- heavy hydrogen (deuterium) is, for all practical purposes, available in the oceans in essentially infinite quantity; the heat energy is necessarily produced at a high temperature, making for greater efficiency in the cycle of conversion to electricity, or production of hydrogen by dissociation of water, or in the desalinization of sea water; although by no means free from problems of associated radioactivity (see below), these are inherently much less formidable than in the case of fission energy, since fusion does not involve production of the multi-megacurie amounts of fission products, some of very long half-life, that are inescapably associated with nuclear fission.

However, although tantalizingly close to realization, the feasibility of use of controlled thermonuclear reactions for power production remains to be proved. Once demonstrated, there will still remain very formidable technical problems -- notably the provision of suitable materials for withstanding the high temperatures and radiation levels required in the practical operation of a fusion power plant. The main radioactive by-product (tritium, i.e. hydrogen of atomic weight three), while presenting problems of containment, can be re-used in the energy cycle, so that it presents no problems of disposal. On the other hand, thermonuclear reactions produce high fluxes of neutrons, which will be absorbed in the structural materials of the reactor and in the shielding materials required to prevent their escape into the surroundings (where they could produce dangerous levels of radioactivity). Thus, structural and containment material will become highly radioactive, and will be relatively inaccessible for long periods after shut-down of the reactor, unless these materials are very carefully chosen so that the half-lives of the nuclei produced by neutron reactions are relatively short. Finally, in common with other sources of stored energy, fusion power gives rise to thermal pollution and, consequently, the total amount of energy that could be produced in this way will be determined by the global climatological limitations discussed at the beginning of this paper. (There is a possible escape from this limitation, however: it is theoretically possible that methods could be developed for the direct conversion of fusion to electrical energy, thereby raising very considerably the thermal limits). Since fusion seems to favour very large installations, (although this may not turn out to be universally the case for all schemes of thermonuclear power under consideration -- e.g., the implosion of small deuterium pellets by intense laser beams) its use should be mainly contemplated for those situations that require very high power concentrations, e.g., refining, smelting, etc. of large quantities of ores into metals, desalinization of water.

Solar energy, on the other hand, owing to the nature of its arrival on the earth's surface, is most suitable for limited local uses spread over wide areas -- e.g., heating and air conditioning, food production. Utilized reasonably efficiently, over wide areas of the earth, the quantity available is sufficient to supply most of the ordinary energy needs, as well as food and other organic requirements (e.g., wood) of mankind for many generations to come (assuming containment of the population explosion). It is inherently non-polluting, although there could be problems of local disequilibrium for concentrated applications. The problems that need to be solved for its useful application involve the provision of cheap and efficient converters (solar energy cells) or inexpensive materials for its efficient absorption over relatively large areas, the technology for the conversion to accessible energy of low-temperature heat, and the provision of effective means of energy transport (cheap electric batteries, hydrogen). However, there do not appear to be any inherent technical limitations to their solution.

In summary:
a) The major near-term energy problem is the replacement of oil by coal (used non-pollutingly), by hydropower and possibly by other simple but indirect forms of solar energy -- wind, tidal -- together with the provision of alternative fuels for use in transportation.
b) On the intermediate time-scale, nuclear fission power will be available, and the use of hydrogen as an energy carrier should be made practical.
c) Geothermal energy is also a possibly promising energy source on the intermediate (and possibly longer) time scale, but its availability and feasibility require extensive further exploration and research.
d) However, over the next fifty years or so, the most pressing problems will be to curb the energy appetites of the industrialized nations, so as not to exceed natural ecological limits, while accelerating the availability of suitable energy sources to bridge the gap between the energy-disadvantaged portions of mankind and the large energy consumers.
e) In the long run, nuclear fusion and direct solar energy conversion can provide for all the reasonable energy needs of mankind into the indefinite future.

DISCUSSION

Question: How can Pu be kept safe from terrorist action?

Answer: I have thought about this problem for about 30 years now. If we
             accept that fission power is with us for the intermediate future,
             the most reasonable solution is to decide that Pu recycling is
             out of the question and that it should be left in the spent fuel
             elements. This makes the wastes only slightly more dangerous than
             they would have been otherwise. (Highly enriched uranium is still
             a problem but 10% enrichment is quite adequate for reactors and
             cannot be used for bombs). Any terrorist group without a compli-
             cated, remotely-operated separation plant would kill itself before
             it could use the Pu. Any other solution is temporary, tentative
             and insufficient but, if you insist on using Pu, you will have to
             adopt the Fort Knox approach and only two transits would be made:
             of spent fuel elements to the reprocessing plant and of refabri-
             cated fuel elements back to the reactor. The latter would be
             denatured by mixing with elements that do not inhibit the thermal
             process but increase the critical mass.

Question: We have heard that an expanding fission programme would involve
             a very large primary energy investment before a net energy gain
             would be achieved. Would this also be true of other large scale
             programmes e.g. geothermal and fusion?

Answer: This is true but the world will need a mix of low and high invest-
             ment energy sources basically to provide low- and high-grade heat
             and one has to decide on the balance. The sources are there but
             we cannot use energy at the rates people contemplate, e.g. fission
             energy yield increasing by a factor 100 over the next century,
             these are unattainable.

Question: Bearing in mind capital costs, how should an African country deve-
             lop an energy programme?

Answer: At the large scale, the capital costs/unit of power are relatively
             invariant with type. Solar power units in the 100 MW region have
             not been developed yet but solar power units are quite feasible
             at the intermediate technology level for less developed countries.
             An African country would be better to concentrate on its indigenous
             resources, e.g. hydro, solar, rather than mortgage its future to
             enriched uranium and fission technology with implied external
             control.

Question: Can anything be done about the smugness and complacency of the
             developed countries? They believe that continued economic growth,
             and growth in energy use, are essential to the survival of their
             societies. We have heard at the School that the energy supply
             industries are eager to cater for our energy demand. We have seen
             how   these escalating demands and diminishing fossil resources
             will lead to an energy gap within a generation or so which the
             nuclear industry will be delighted to bridge with a massive breeder
             programme. The proponents of growth ridicule their opponents yet
             the great dangers of growth and the breeder/plutonium solution
             have been made clear to us here.

Answer: That is the $64 question. There are two ways of avoiding the
             Malthusian catastrophe, i.e. exponential growth to the point where
             the system can no longer sustain it and collapse occurs. One way
             is to see the danger and build in controls to flatten the

exponential curve. The other, pessimistically, is that a small
Malthusian catastrophe may occur and shock people into changing
the system. The problem is that such a catastrophe is unlikely
to be small.

Question: Why do you think there has been no general condemnation by govern-
             ments of the recent decision by West Germany to sell a nuclear
             reprocessing plant to Brazil?

Answer: In my view, this decision is an unmitigated disaster. This is not
             only because it will give Brazil the ability to make its own nuclear
             weapons in the future – it had already announced its intention to
             do this and has not signed the Non-Proliferation Treaty. Brazil
             has announced its intention to produce nuclear devices for peace-
             ful explosions; whether such a device is for peace or war is a
             question of semantics only. The disaster is that this announcement
             coincided with the end of a conference to review the Treaty, at
             which conference it was decided that it was permissable to sell
             nuclear capability to countries that did not adhere to the Treaty
             provided that these countries would submit to uniform safeguards
             on those plants (nothing was said about future plants developed
             from the imported technology), and the West German sale immedia-
             tely put the final seal of sanction on this. There is now no
             effective Non-Proliferation Treaty.

EPILOGUE

I. M. Blair

Energy Technology Support Unit, Harwell, and Chairman of the School

It would not do justice to such a wide ranging and intensive a School as this to produce a summary in a few pages. I shall not even attempt to do so. I shall content myself by setting down just a few of the residual impressions left on me by the School following two weeks of continuous exposure. I should emphasise that these are personal impressions and do not necessarily reflect the views of the Energy Technology Support Unit or the Department of Energy.

During the School one noticed two, apparently incompatible, themes:-

(1)   The "Establishment" view, linked with continually advancing life style and increasing energy consumption, hell-bent on ever higher technology, particularly committed to an intensive nuclear programme.

(2)   The "Anti-Establishment" view, pressing for a low energy, low technology, life style, and bitterly opposed to nuclear power.

On the face of it, this is not a likely scenario for a profitable discussion but in the event it proved manifestly to be so. It became clear that they were but facets of an exceedingly complex problem, whose solution must emerge as a linear combination of them both. When intelligent people gather to discuss a problem, the diachromatism of polarised opinion merges into the continuous spectrum of consensus.

The Establishment/Anti-Establishment conflict appeared to be rooted in the mistaken impression that our affairs are being run by a ruling class, peopled by a race of infallible supermen, impervious to the wishes and opinions of we mere mortals. Intensive research which I have conducted over nearly four decades has failed to produce any evidence for the existence of such a class; when I have met certain persons charged with the responsibility of making decisions on our behalf, they have looked remarkably like us. They need, and seek, advice and opinion from anyone competent to give it to enable them to discharge their responsibility effectively.

Though this leads to a "weighted egalitarianism" which might cause offence to those with excessively pure ideals, it is clearly right to give more credence to the opinion of those who have devoted themselves to a study of the issue in question than to those whose expertise lies elsewhere.

Such was the consensus which appeared to me to arise from discussion at the
School. Though it might inspire relief it should also inspire responsibility.
Let each of us, within our sphere of competence, make sure that our views are
forcefully and rationally presented. Perhaps the imperfections in our society
are due not to "them" but to "us".

The debate on nuclear power became progressively more productive as mutual
understanding of differing views increased. Even the strongest advocates were
by no means unaware of the problems involved in this technology, nor were the
strongest opponents suggesting that these problems were so severe that all
reactors must be switched off forthwith. Thus no-one would seriously object
to one reactor, it is only when large numbers of them are envisaged that the
debate hots up. The problems associated with nuclear power, such as safety,
capital cost and energy intensiveness, were balanced against the consequences
of not having nuclear power. Thus a consensus view emerged, not for or against
nuclear power in an absolute sense, but rather that an optimum balance point
must be reached. The issues involved would thus dictate the scale and rate of
growth of a nuclear programme.

How much energy do we actually need? This was the crucial question around which
much of the debate on energy consumption and the sources we must develop to meet
it revolved. It involved issues which, though important, were difficult to
quantify, such as quality of life, personal expectations, etc. Intuitively one
feels that quality of life is not a monotonic function of rate of energy consumption, but rather must pass through some optimum value. One is reminded of
Professor Thring's much-quoted adage:-

> "Happiness is three tons of coal equivalent"

Much concern was expressed over the ever increasing demand in the developed
countries for energy and other manifestations of material wealth. It was questioned how much longer this could continue, and, even if it was desirable, that
it should. It was suggested that such a life style, based on selfishness and
greed, should be replaced by a simpler but more satisfying life style based on
a lower material standard in which all people could share equitably. This
touches upon moral and ethical issues which were not, unfortunately, included
in the formal presentations though they were discussed informally. What is
being asked for is nothing short of a change in human nature, and on a scale
comparable to a religious reformation the like of which the world has never

seen. Let the reader form his own view on the probability of this happening, and hence the wisdom or otherwise of using this as the basis on which to build an energy policy. Of course it might happen, but I have yet to meet a man who would bet ale on it.

So what happens next? We have indulged in much interesting talk and discussion. Let us hope that this was not merely an enjoyable intellectual exercise, for when all the assessments, paper studies, scenario building, etc., etc., have been done what is required at the end of the day to solve the problems is real practical hardware that works in the real world to exploit new energy sources, to make our existing technology safer, and to make more effective use of our conventional fuels. Who is going to produce it? The participants in this School represent a cross-section of our young energy technologists on the threshold of their careers. If they, and others who might read these pages, have been so inspired to determine to convert what they have learned here into practice, then the venture has been well worthwhile.

STUDENT CONTRIBUTIONS

B. R. Clayton

It is clear that many aspects of energy conversion are receiving detailed attention. It is also evident however that progress, nationally, in all fields is
haphazard and disorganised. A major reason for this, now happily diminishing,
is the competition between the nationalised energy suppliers. In the Universities and other establishments many research programmes rely on grants from
the Treasury distributed by Government Departments whose policies often appear
to make quite erratic changes in course. It is surely time to formulate and
operate a national policy on energy, if necessary by applying legal sanctions
to prevent wastefulness, in order to avoid the muddled actions which seem to
prevail at present. Various aspects of such a policy will inevitably take time
to complete so it is important that interference at a party political level
should be vigorously resisted.

Many of those engaged on energy research topics are associated with Universities, yet the role of Universities in tackling national and indeed global problems receives scant attention. The autonomy of British Universities is a
jealously (and quite properly) guarded privilege and this includes the freedom
to pursue individual lines of research. Nevertheless, there is a strong case
to be made for greater collaboration on energy problems, not only between different Universities, but between different departments within a given University
and between Universities and Industry. Furthermore, this collaboration must
be coordinated for maximum benefit; after all we expect optimum performance
from the separate components comprising an energy conversion device. Positive
encouragement in the form of finance, experimental facilities, secondment to
other research groups, and so on must be forthcoming. Under these conditions
regional groups of Universities, say, could embark on extensive investigations
sharing resources and expertise and eliminating much of the duplication which
is now commonplace. Such schemes are well established in the field of internal
combustion engines, for example, but for success the full and free exchange of
ideas and developments must come from the participants. Undoubtedly, this sort
of (naiive) attitude is unlikely to be accepted readily by many working in
Universities, especially in the present, somewhat tense, atmosphere.

The subject of education in the energy sector is not adequately discussed, yet
this surely warrants high priority as an immediate and future investment.
Merely asking the nation to 'Save It' just scratches the surface of our difficulties. A programme must be devised to cover the whole educational spectrum
and perhaps no requirement is more urgent at the moment than that satisfying
the needs of the main industrial consumers. The training and preparation of
'energy managers' can be done most effectively in Institutes of Higher Education where expertise in short 'post-experience' and one-year M.Sc. courses is
already well established in many subjects.

R. Critoph

I am worried about the apparent lack of co-ordination and co-operation between
the energy supply industries with respect to long term policy. I understand
that each energy supplier makes his own long term plans based on assumptions
relating to the plans of the other energy industries which he derives independently. Thus the Coal Board has an electricity supply and demand model, which
influences its projected coal supply and demand. Similarly the Central Electricity Generating Board has a coal model and so on. The nationalised concerns,
viz. the gas, coal, and electricity suppliers, submit their individual plans to
the Department of Energy. We must assume that inside the Department of Energy
these separate plans are brought together and that they are consistent. Presumably, if they were incompatible, the Department of Energy would inform the
industries, but it seems a very inefficient way of formulating a long term
strategy. Also, I object to the fact that the Department of Energy makes none
of the combined information public, say as a broad strategy for future energy
policy. An unknowledgeable observer might be misled into thinking that we had
no long-term energy policy.

I feel that there is a strong case for re-organising the top level of the three
industries, amalgamating their planning processes, and giving them a new brief.
One could imagine an organisation, "British Energy", which combined the National
Coal Board, C.E.G.B., and British Gas. Its brief might be to produce energy
for the consumer at least cost, whilst ensuring future continuity of supply.
I make no mention of the oil industry as it seems that its future is very much
in the policital melting pot at the moment. However, should the government
decide to proceed with nationalisation, then it would be logical to incorporate
it into an amalgamated energy industry.

It can be argued that competition between the energy industries is a useful
thing and, to some extent, I would agree. However, the share of the national
market allotted to each should be in the national interest rather than the
result of straight competition. We do not want competition to increase to the
extent it did a few years ago when, for example, housing estates were built
without a gas supply because artificially reduced electricity connection prices
were charged on this condition. This is clearly a counter-productive example
of competition.

DESIGN FOR AN ALTERNATIVE TECHNOLOGY

H. L. Liddell

"If an activity has been branded as uneconomic, its right to existence is not
merely questioned but energetically denied. Anything that is found to be an
impediment to economic growth is a shameful thing and, if people cling to it,
they are thought of as either saboteurs or fools. Call a thing immoral, ugly,
soul-destroying, a degradation of man, a peril to the peace of the world or to
the well-being of future generations; as long as you have not shown it to be
'uneconomic' you have not really questioned its right to exist, grow and
prosper".   E. F. Schumacher (1).

THE GROWTH OF GROWTH

Over the past ten years many people have questioned growth in general and the
nuclear scenario in particular; these people range from Commoner and Illich to
Ehrlich and Meadows (2). They follow a very distinguished line of predecessors,
including Bertrand Russell, Einstein and many Nobel prizewinners.

It is somewhat disconcerting, therefore, to discover that no sector of the
Energy Industry in Britain is seriously considering alternatives to growth in
any of their future plans, and that the amount of research being carried out
is, at best, peripheral to the main issues and, at worst, dilettante. It is
difficult to reconcile the calls for more funding for the nuclear fusion programme in order to "keep Britain in the forefront" with the view that, considering the funding in the U.S.A. for solar energy research (compare with, say,
nuclear research) "We will wait until others have developed the systems and
then join in" - especially when both those attitudes apply to the same group
of people. To call this cynical would be flattering; the feasibility of using
solar and solar-related energy in Britain is based on very different principles from those for any of the three main climatic regions of the U.S.A. Even
if the Americans are generous enough to present us with their findings, it is
possible that very little information will be directly applicable in Britain.

What is it that makes the establishment's myopic commitment to Nuclear Technology so disconcerting? Schumacher expresses his doubts in terms of being confronted with the proposition "that future scientists and technologists will be
able to devise safety rules and precautions of such perfection that the using,
transporting, processing and storing of radioactive materials in ever-increasing
quantities will be made entirely safe; also that it will be the task of politicians and social scientists to create a world society in which wars or civil
disturbances can never happen. It is a proposition to solve one problem simply
by shifting it to another sphere, the sphere of everyday human behaviour" (3).

Like so many of the assumptions being made by growth protagonists the proposition is technologically sophisticated but sociologically naive. A growing body
of informed opinion is convinced that sooner or later there will inevitably be
some form of man-made disaster which will make Aberfan, Flixborough and the
Windscale worries pale into insignificance. It will not take very much to fan
the flames of existing public prejudice against technology. The producers of
the BBC 2 'Horizon' series characterised the 10 years of the programme's exist-

ence as spanning from "Science as the great hope to Science as the great threat".
David Dickson attributes this distrust to a number of factors. "The many social
benefits which technology has helped to bring about are being increasingly
counterbalanced by the social problems associated with its use. These range
from the oppression and manipulation of the individual to the widespread destruction of the natural environment and the depletion of the world's finite
supply of natural resources. At the same time, man's technological skills have
so far failed to provide effective solutions to many of the world's major problems, in particular those of mass poverty, starvation and international conflict.
Technology is no longer seen as the omnipotent God that it was even ten years
ago" (4).

It is possible that all the eminent people questioning the growth syndrome are
just an out-of-touch élite, but there seem to be too many for that; they are
certainly no more remote from public opinion and scrutiny than, for example,
the nuclear technologists.

THE MONOPOLIES

Perhaps one of the most worrying aspects of the energy industry is the monopoly
held in the various sectors (and these sectors are not small), e.g. the scale
of the Central Electricity Generating Board, the British Gas Corporation, etc.
is beyond comprehension. "Gigathink" is no respecter of persons and its dominance is all-pervading. The CEGB spends much money to placate the "environmental lobby" so how can Friends of the Earth, the Council for the Preservation
of Rural England or the Conservation Society compete? Goliath has donned his
steel helmet.

A cynic might see even more worrying aspects of the current activities of the
energy monopolies. They engage in research into alternative energy technologies on a scale which is insignificant by comparison with their total research
budget, but which is very significant in terms of what is being spent nationally. Artists' conceptions in the "CEGB Research" journal of Solar Space Stations beaming microwave energy to grid do nothing to instil confidence that
these alternative technologies are being taken seriously by the monopolies.
Solar energy is already distributed; why centralise and re-distribute? One
might be forgiven for bringing to mind one method of pest control where a few
sterilized males are introduced and allowed to associate freely with the rest
of the pests; their forceful neutrality ultimately reduces the species to vir-
tual extinction. This analogy represents perhaps an over-cynical view, but
the only way to remove the mistrust of a monopoly is to remove the monopoly
itself.

DESIGNING ALTERNATIVES

The cry for some clear statement of a reasonable alternative inevitably follows
any criticism of the status quo and there is no intention here of shirking that
responsibility. Before setting out such a statement it is worth saying that
the alternative will probably seem quite unreasonable to a number of people;
however, if it is only seen as no less unreasonable than the establishment
(growth) scenario with all its implicit costs, there is the basis for a serious
dialogue and for some useful work to contribute to that dialogue.

Probably the simplest and clearest statement of the ideals and intentions of
Alternative (or 'soft') Technology is contained in 35 principles set out by
Robin Clarke (5) - a few of these are set out in Table 1 below.

TABLE 1

"Hard" Technology Society	"Soft" Technology Society
1. Ecologically unsound	Ecologically sound
2. Large energy input	Small energy input
3. Non-reversible use of materials and energy sources	Reversible materials and energy sources only
4. Mass production	Craft industry
5. High specialisation	Low specialisation
6. Technology liable to misuse	Safeguards against misuse
7. Growth-orientated economy	Steady state economy
8. Centralist	Decentralist

In his definitive book on Alternative Technology, Dickson summarises the criteria in the following terms:- "The tools and machines required to maintain
this alternative would necessarily embody a very different set of social and
cultural values from those we possess at present. These tools and machines,
together with the techniques by which they are used, form what is generally
meant by the term 'alternative technology'" (6). The central point is that the
technology and sociology are inextricably interdependent; in such a system
scientists and technologists cannot claim moral immunity from the results of
the application of their designs. 'Design', an holistic descipline, not
'Science', a series of specialist disciplines, makes the initiative.

1975 sees the end of Buckminster Fuller's "World Design Science Decade", and
it is significant that this programme was launched in Design rather than in
Science Schools. It is unfortunate that, at the end of such a period, Britain,
which has proportionally the same number of scientists per capita as the U.S.A.
and Europe, should have only 1/6th the number of designers and yet the record
of this relatively small band of people has been fairly impressive. The justification for the design-orientated approach has been expressed by Victor Papanek
as follows:- "At present industrial and environmental designers are the logical
foci in any design team. Their logical status as key synthesist in a design
situation is not because they are superior beings, better informed, or necessarily more creative, but rather because they assume their status as comprehensive synthesist by the default of the other disciplines. For in this age
in America, education in all the other areas is a matter of increasing vertical
specialisation. Only in industrial and environmental design is education horizontally cross-disciplinary" (7).

Those involved in future projections can only base their extrapolations on what
is already designed, but the designer is in the business of "breakthroughs"
and these, as with sudden changes in public opinion, can so often make a nonsense of such projections. Clear examples of this occurred at this School
where a number of commentators dismissed solar energy as being too capital-
intensive and too low-grade to be significant. These conclusions were based
on current technology which is in its infancy but which can be reasonably expected to have one or two breakthroughs (this is no more, and is probably less,
optimistic than are the protagonists of the almost "science fiction" nuclear
fusion programme). Preliminary feasibility studies show that as much as 50%

saving in energy use could be made in urban buildings from a combination of
solar-related sources and conservation techniques - within current Government
cost yardsticks (8). This involves a number of innovatory factors, particularly
in terms of cross-disciplinary integration, but it is all within "reasonable"
limits and challenges statements that only 20% saving could be achieved in this
situation.

                          JUST IN CASE ....

The Guardian Science correspondent reports as follows:- "Sir Alan Cottrell,
former Chief Scientific Advisor to the Government, has gone firmly on the record
to the effect that - in his view - the public will not accept the hazards associated with a large fast-breeder reactor programme" (9). This opinion is shared
by Professor Bernard Feld who believes there will be some major nuclear catastrophe in the next decade or two - at which point "there will be a sudden pull-
ing up in horror on the part of the leaders" (10). The Guardian continues:-
"Not before time, some will say, but it can be seen that the situation is intriguing since the fast-breeder is the only Great White Hope of the Cabinet Office
and the Ministry of Energy. In the meantime, over in the United States in its
latest assessment the U.S. Energy Research and Development Agency has lifted
solar energy research from near the bottom to joint top of the list with fast-
breeder systems. New estimates of the solar contribution to U.S. energy needs,
envisage a solar component of 720 gigawatts equivalent by 2020. That is roughly
four times the present total useful output of the British National electrical
system".

Until some substantial support is given to serious investigation of alternatives by independent groups (i.e. not related to the energy monopolies) the
discussion is going to be very one-sided. What is needed is investment now,
of both time and money, in a truly alternative scenario so that we are ready
to take the opportunity .... just in case .... or in any case!

                            REFERENCES

1) E. F. Schumacher - "Small is Beautiful", Blond and Briggs, 1973, p.37.
2) Commoner - "The Closing Circle", Jonathan Cape, 1972.
    Illich - "Energy and Equity", Penguin, 1973.
    Ehrlich - "The Population Bomb", Pan/Ballantine, 1970.
    Meadows - "The Limits to Growth", Potamae Associates, 1972.
3) E. F. Schumacher - "Small is Beautiful", Blond and Briggs, 1973, p.16.
4) D. Dickson - "Alternative Technology and the Politics of Technical Change",
       Fontana, 1974, p.9.
5) ibid. p.103 (Dickson).
6) ibid. p.96 (Dickson).
7) V. Papanek - "Design for the Real World", Paladin, 1971, p.177.
8) Hull School of Architecture - "Low Impact Community Housing", 1975 (unpub.).
9) Anthony Tucker - "Putting the Sun on the Hot Spot", Guardian, 22/9/75.
10) Prof. B. Feld - "Science for Peace", New Scientist, 24/7/75.

DESIGN WITH NATURE

H. L. Liddell

H. L. Liddell

INTRODUCTION

Over the course of the past week two points in particular have caused me some concern. The first is the prevalent adherence to the growth scenario, something seriously challenged by the Meadows Report and many others since (1). The second point concerns the lack of a total systems approach to problem solving.

In 1912 Woodrow Wilson said the following:-

"What I fear is a government of experts. What are we for if we are to be scientifically taken care of by a small number of gentlemen who are the only ones who understand the job? Because if we do not understand the job, then we are not free people" (2).

I do not propose to spend much time describing the dangers of the growth scenario as it has been done so often and so much better by a number of other people. Suffice to say that a continuing growth in energy demand is no more inevitable than the U.K. population reaching 76 million by the year 2000. This prediction of 2 years ago has been drastically reduced a number of times since and we now see this month that the number of deaths is expected to exceed the number of births for the first time since the Plague in the 17th century.

PARTIAL SOLUTIONS

Before dealing with the question of positive alternatives to growth I would like to proceed to my second theme by illustrating how a specialised approach to problem solving does not always solve the fundamental problem itself.

As an illustration I will use the example already mentioned by Mr. Peters at this School and being heavily promoted by the Electricity Boards called "Integrated Environmental Design" (IED). This is a system which employs thermal balance air-conditioning with a combination of heat recovery methods (including the reclamation of refrigeration equipment condenser heat and the extraction of exhaust air through the light fittings - which draws heat from the lamps, etc). The system is automatically self-balancing and, as a principle, heat addition and removal never coincide. This would appear to be just what the energy conservationist would want but this system has been developed for deep-plan offices, which can only exist by virtue of air-conditioning and high lighting levels (sometimes in excess of 2,000 lux, which compares with the minimum requirement for schools of 200 lux).

In the view of an increasing number of people, IED can be associated with the maxim: "Don't do well what you shouldn't be doing at all".

Studies by Milbank et al at the Building Research Establishment have shown that the cost of electricity for the lighting of air-conditioned offices is about 50% more than all the energy costs put together for a conventional centrally-

831

heated office (3). All that IED does is to make air-conditioning more efficient but it is using electricity as a heating source and, since a major part
of the heating comes from the lighting, this of course means that 75% of the
energy has already been lost - "locking the stable door after the horse has
bolted" as Mr. Peters put it.

Whilst there is basically only one main climate in the U.K. (excluding parts of
Scotland), there are at least 3 in the U.S.A. It was because of one of these -
the hot, humid climate in the Florida region - that air-conditioning technology
developed. This climate does not exist anywhere in the U.K. but, because one
area in the U.S.A. had air-conditioning, the rest wanted it for prestige reasons,
and because the Americans had it, the British wanted it. In the U.K. it has
become a status symbol based on false premises.

Milbank et al have shown that air-conditioning in the U.K. can only be justified for about 2 weeks in any one year and even that tends to be for excessively glazed buildings; more subtle designs render it virtually unnecessary
except for a few specialist activities. Not only does this solution not help
to conserve energy (the Electricity Boards' advertising notwithstanding), it
actually contributes to the problem by using more primary energy. This is
because the designers only looked at part of the problem - not the total context.

HOLISTIC SOLUTIONS

What are the positive alternatives? I would like to present some case studies
from my own profession.

As a designer of the built environment, I would find it difficult to avoid
being labelled an "environmentalist" - especially in the U.K. where practically
the whole landscape is man-made and therefore the internal and external environments are equally at the mercy of the design professions, viz. Planners, Landscape Architects, Architects, etc. This is an important distinction from the
image of the environmentalist which has been presented by a number of the speakers
at this School. Instead of being cast in the role of somewhat negative "pre-
servationists", my definition implies a positive, creative role.

It is also important to note that these professions, as in all other sectors of
society, are undergoing a change in attitude from unchecked pioneering to the
more considered approach of conservation. (This is distinct from "preservation". The first is an active process, the second is sterile).

The devastation caused by various types of pollution has created a new position of humility towards natural systems. This is developing an attitude for
designing with nature and not in spite of it.

What does "Designing with Nature" mean?

Amongst other things it means looking at traditional buildings which have adapted to their context without the benefit of the CEGB and others. The igloo is
an outstanding example. The use of air-conditioning for the hot, humid climate
of Florida has already been mentioned - but the Malaysians managed to solve the
same problem in their vernacular styles of building by encouraging the wind to
give them the required number of air changes through permeable walls (see Fig.1)

A hot <u>arid</u> climate requires a completely different approach. In these regions
they use a large thermal mass to absorb heat slowly in the day-time and give it
out slowly at night, thereby significantly reducing the diurnal variation in
temperature. This is important in such regions since the lack of clouds at
night means a very high radiative heat loss to the sky vault and temperatures
often fall well below freezing.

Much work on the use of thermal capacity in buildings has been done by Baer (4)
in the New Mexico Desert. The principal design element in his house is a south-
facing wall of water-filled tanks which absorb heat during the day and, insulated on the outside by a winched-up shutter (the inside of which helps reflect
the sun on to the tanks in the day), give off heat to the house during the
night, (see Fig.2).

The situation in the desert is of course more straightforward than, for example,
the U.K. but although the U.K. climate is unpredictable, it is not extreme and
we do receive a substantial amount of both direct and diffuse solar energy, as
already mentioned by Dr. Brinkworth at this School. It also receives considerable indirect solar energy in the form of winds, waves and of course photosynthesis. Plants do grow well in the U.K.

In the U.K., we have to be a bit more subtle but climate-related Architecture
<u>does work</u>. Apart from the various vernacular styles, there is at least one
very good example of a contemporary solar-heated building in the U.K. which
has been in operation for about 15 years. This is the much-maligned St. George's
School at Wallasey (see Fig.3) and I think it important here to make 3 points
to put the record straight. Work by Dr. Davis (5) of Liverpool University has
shown that:-

(a)   The annual energy costs of approximately £4.00 per pupil compare with an
      average of approximately £6.50 for conventional schools in the area. Of
      the £2.50 saves, £1.00 can be attributed to higher insulation standards
      and £1.50 (i.e. 23%) can be attributed to solar gain.
(b)   The complaints about body odour from the Gymnasium and the smell from the
      lavatories are justified, but are not essentially to do with the fact that
      the building is designed for Solar Energy - these parts of the building
      could have been separately ventilated with little or no effect on the
      building performance. It was a planning mistake which could still be
      solved relatively easily.
(c)   The building was built within the normal cost yardstick for schools.

The principle of the building is very simple - it has a double-glazed south-
facing wall and high mass (externally insulated) internal and external walls,
floors and roof - the fabric therefore maintains a good Mean Radiant Temperature which is then supplemented by heat from solar gain, body heat and lighting - there is also a relatively low ventilation rate.

Finally I would like to describe some work which we have been doing in Hull.
This is a case which emphasises the environmental designer's involvement with
both the internal and the external environment. Work by Patricia Pringle has
shown that there are a number of ways in which we could be more efficient in
the food industry (6). In the same way that it is easier to ask the population
to save energy, rather than asking the CEGB to increase its efficiency, it would
be easier to ask the population to waste less food, rather than exhort the
farmers to produce 2% more, as has been done recently. (We currently waste 20%
of our food as against 5% during the 2nd World War).

VERNACULAR FORM OF HOUSE
FOR A HOT-HUMID CLIMATE
FIG. 1

Close-thatch roof
for fast rain
run-off.

open
Roof space
increases cool
effect

Louvred (open slat)
walls cause
increased flow
of moisture-
laden air.

Stilts cater for
flooding, moisture
venomous fauna
& increased airflow

BAER'S CLIMATE-RELATED
HOUSE IN HOT-ARID ZONE
FIG 2   (after P. Steadman)

use of re-cycled &
local materials

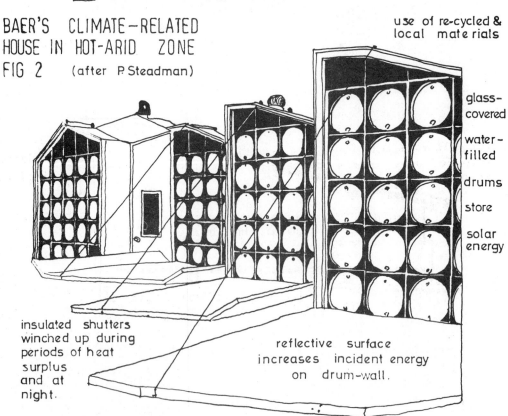

glass-
covered

water-
filled

drums

store

solar
energy

insulated shutters
winched up during
periods of heat
surplus
and at
night.

reflective surface
increases incident energy
on drum-wall.

St. GEORGE'S SOLAR
SCHOOL , WALLASEY
(after Banham) FIG. 3

shutters on internal skin of
south wall allow for
seasonal adjustment.

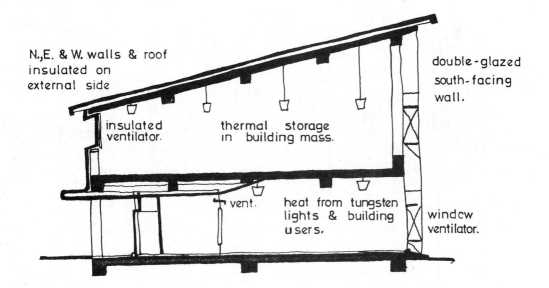

N.,E. & W. walls & roof
insulated on
external side

double-glazed
south-facing
wall.

insulated
ventilator.

thermal storage
in building mass.

vent.

heat from tungsten
lights & building
users.

window
ventilator.

It is also shown that only 14% of gardens are used for growing food compared
with a potential 66% - and a mere doubling to 28% would reduce the balance of
payments by £114m. But what was of most interest to us in Hull was the fact
that research has shown that, at a density of 12-15 houses to the acre, people
start growing vegetables instead of flowers, etc. This happens to be the density at which, according to a number of experts, average sized families could
be self-sufficient in vegetables. This also happens to be the average density
of Hull housing. With such a series of coincidences, the desire to carry out
a feasibility design at that density was more than we could withstand and, not
only did we find that with a judicious reduction in roads we could grow a surplus of vegetables or some wheat, we also discovered that this was about optimum density for 2-storey terrace housing development using integrated solar and
wind energy and high insulation (saving about 40-50% of energy costs). This
was a new development but further studies on the conversion of parts of the
city's existing building stock have shown promise (see Fig.4) (7).

This scenario has two very important implications, viz. the two points with
which I started this paper.

(1) This is a proposal for a community that would make very few energy demands
     and certainly very much less than the present norm.
(2) The designs proposed here are applied basic principles across a number of
     disciplines and integrated at the conceptual stage.

The important point in these case studies is that there is very little change,
in either the life-style or the city structure, implied in the proposals and
it would be possible to go for a much lower level of consumption without too
much of a "Culture Shock".

Growth is not inevitable.

There are alternative futures available.

REFERENCES

1) Meadows "The Limits to Growth", Potomae Associates 1972.

2) Ralph Lapp "The New Priesthood", New York 1965.

3) Milbank, Dowdall, Slate, BRE Current Paper 38/71.

4) P. Steadman "Energy, Environment and Building", Cambridge University Press 1975.

5) Davis "The Wallasey School", ISES Conference, February, 1975.

    Banham "Architecture of the Well-Tempered Environment", Architectural Press,

    1969.

6) P. Pringle, "Undercurrents", 9/74.

7) C. Watterson "Low Impact Community Housing", unpublished.

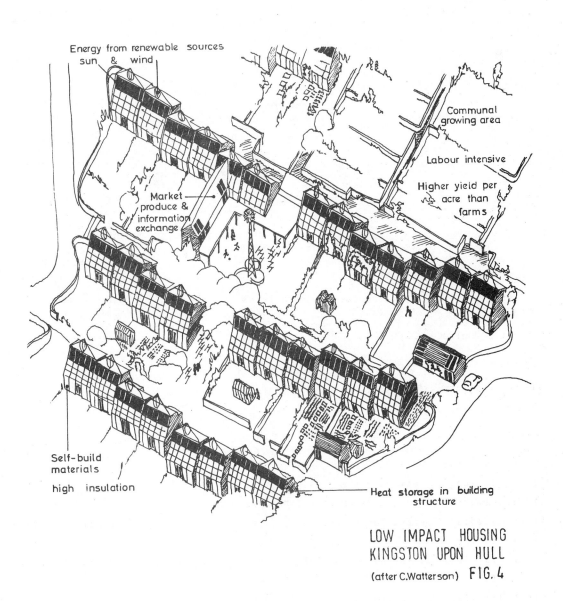

Energy from renewable sources
sun & wind

Communal
growing area

Labour intensive

Higher yield per
acre than
farms

Market
produce &
information
exchange

Self-build
materials

high insulation

Heat storage in building
structure

LOW IMPACT HOUSING
KINGSTON UPON HULL
(after C.Watterson) FIG. 4

ENERGY SCENARIOS:
EFFECTS OF UNCERTAINTY

G. D. Madeley

One of the aspects of energy conversion which has not been discussed fully is the effect on energy forecasts of uncertainty. What effects do changes in our assumptions of fossil fuel reserves make, or what happens if the demand for energy is altered?

Various people have produced scenarios of energy demand into the future. Their estimates vary considerably. Some examples are given below.

FORECASTS OF ENERGY DEMAND

Forecaster	Date of Forecast	Average Growth in Primary Energy Demand. % p.a.	Primary Energy Consumption in 1990. m.t.c.e.
Boley (1)	Sept. '73	1.4 - 2.6	430 - 530
Madeley	July '74	1.0 - 2.0	410 - 490
Bainbridge (2)	Feb. '75	1.0 - 2.4	395 - 510
Dept. of Energy (3)	May '75	1.9 - 2.4	464 - 505

For the purpose of this discussion I will take as the central case an assumed growth in primary energy consumption of 1.5% p.a.

The other aspect of an energy scenario that has to be considered is the method of supplying the energy to meet the demand. We have considerable indigenous energy resources, so it is interesting to see how much of the proposed demand they can meet. Figure 1 shows the combined production rates for indigenous fuels compared with the assumed demand growth. For the purposes of this graph, I have based the curves for oil and gas on the probable reserves given in the latest Department of Energy estimates (4), and a slightly more pesimistic estimate of coal production than the constant level of production proposed by the Government. For nuclear power I have assumed all existing ordered and proposed reactors will be operating from at least 1985 onwards with no further orders.

If one considers this graph, it falls naturally into three sectors. The first sector, up to about 1980, shows a situation of growing supplies of indigenous energy and is a period in which the difference between demand and indigenous supply can be met by importing oil. The second period, covering most of the 1980's shows an energy surplus. The basic options for dealing with this surplus are either to sell it, probably as oil or oil products, or to keep it to extend the life of the reserves. It is likely, in my opinion, that most of the surplus will be sold, but even if it is conserved for future use it is unlikely to

AVAILABILITY OF INDIGENOUS PRIMARY ENERGIES COMPARED WITH A DEMAND FORECAST — A POSSIBLE SCENARIO.

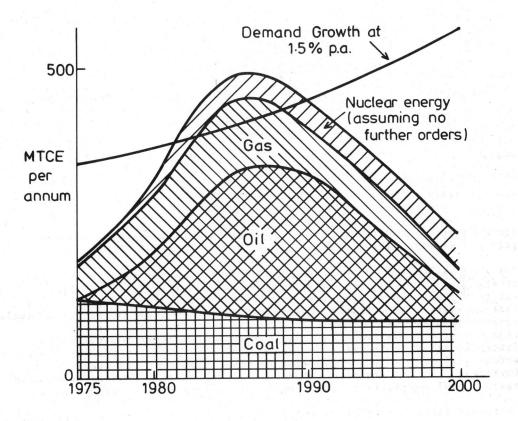

extend the period of our self-sufficiency for more than, say, 3 - 5 years. The
third period, starting around 1990, is a time of falling indigenous energy supplies and a rapidly growing energy deficit to be filled. I think it is readily
apparent from a consideration of this figure that the principal effect of altering the assumptions of the growth of energy demand or of the availability of
indigenous fossil fuels is to make slightly earlier or later the onset of the
deficit in energy supply.

Studies like this suggest that the most likely time for this deficit to start
is between the late 1980's and the turn of the century.

The obvious problem will be in filling the energy gap that will be opening up.
The four obvious solutions are:-

1) To import a fuel, probably oil but perhaps coal. However the world produc-
    tion of oil may by that time be declining (5) and the world demand for ener-
    gy will still be increasing. It may not be economic to import fossil fuels
    because of the high price induced by scarcity and high demand.
2) To use some renewable energy source, such as solar power. This has obvious
    attractions but the use of this type of energy source in the U.K. is in its
    infancy at present. It is most unlikely to have reached a stage of econo-
    mic and technical development to be able to provide more than a marginal
    impact by the turn of the century.
3) To increase the production of indigenous coal. However it is unlikely that
    the coal industry could expand to make more than a fairly small impact in
    the 1990's. It is even possible that unless the Government applies very
    massive support to the industry in the 1980's, the coal industry may have
    great difficulty in surviving this period (6).
4) To use nuclear power. At the present time this is the only source of ener-
    gy that is apparently capable of substantially filling the energy gap.
    Following the arguments of Leslie and Dale earlier in this School, this
    nuclear power will have to be obtained mainly from fast breeder reactors.

In conclusion then, using a range of assumptions, the two most important factors
in the energy scenarios are the size of the energy gap when it does happen, and
the fact that it is likely to occur within the next 15 - 25 years. On the evidence we have at present the only economic way of filling that gap is to use
fast breeder reactors.

Finally, I would like to stress that the views presented here are my own, and
do not necessarily represent the views of the British Gas Corporation.

REFERENCES
1) T. A. Boley. Energy Demand in the U.K. Through the 1980's.
    F.T. World Energy Conference, London, September, 1973.

2) G. R. Bainbridge. Selected British Energy Options.
    Energy Centre, University of Newcastle-upon-Tyne, March, 1973.
    Report produced for the Castle Morpeth Council.

3) M. Kenward. Energy File, New Scientist, 15th May, 1975.

4) Department of Energy. Development of the Oil and Gas Resources of the
    United Kingdom. H.M.S.O. April, 1975.

5) H. R. Warman. The Future Availability of Oil.
    Financial Times, World Energy Conference, London, September, 1973.

6) C. Robinson. The Energy Crisis and British Gas. Institute of Economic
    Affairs. Hobart Paper 59, 1974.

W. R. Stewart

A society in a state of stable zero economic growth will certainly have an economic system different from what we know today and, by implication, different political relationships. Our present political institutions have been assumed as a static factor by many speakers at the School, just as economic growth has been assumed. Energy is a form of wealth but when we have been assessing whether various technologies of energy conversion are economic we have ignored the fact that a large sector of the public does not agree that this produces an equitable distribution of wealth. When assessing the value of an energy technology, we should ask ourselves "value to whom?", and not forget we live in a class society.

The counter-culture of the extreme left has its own definition of "alternative technology", based on the view that the expropriation of technology has led to the growth of the State, the money system, the class system, the repression of women and minorities and the rise of permanent armies. The counter-culture therefore argues that an essential aspect of freedom from domination implies the adoption of soft technology with increased independence for the individual and emphasis on decentralisation. This would support the "pyramid of organisation" or federalist ideal for society proposed by those who see the State as an instrument first of domination and only secondly of organisation.

I believe it is essential to recognise that we are very rapidly approaching the end of humankind's industrial phase. The recognition of this, together with the planning involved, constitutes humanity's first major test as a species.

INDEX